Insecticide, Acaricide, Repellent and Antimicrobial Development

Insecticide, Acaricide, Repellent and Antimicrobial Development

Editor

Giovanni Benelli

MDPI • Basel • Beijing • Wuhan • Barcelona • Belgrade • Manchester • Tokyo • Cluj • Tianjin

Editor
Giovanni Benelli
University of Pisa
Italy

Editorial Office
MDPI
St. Alban-Anlage 66
4052 Basel, Switzerland

This is a reprint of articles from the Special Issue published online in the open access journal *Molecules* (ISSN 1420-3049) (available at: https://www.mdpi.com/journal/molecules/special_issues/Insecticide_Acaricide).

For citation purposes, cite each article independently as indicated on the article page online and as indicated below:

LastName, A.A.; LastName, B.B.; LastName, C.C. Article Title. *Journal Name* **Year**, *Volume Number*, Page Range.

ISBN 978-3-0365-3238-7 (Hbk)
ISBN 978-3-0365-3239-4 (PDF)

Contents

About the Editor

Giovanni Benelli serves as an Associate Professor of General and Applied Entomology at the Department of Agriculture, Food and Environment, University of Pisa, Italy. He teaches Agrarian Zoology, Biological Control, Trends and Challenges in the Management of Vineyard Pests, and Biotechnologies for Managing Animal Parasites. He obtained an International Ph.D. in Agrarian and Veterinary Sciences at University of Pisa and Sant'Anna School of Advanced Studies. Giovanni has worked in several international institutions, including University of Hawaii at Manoa (USA) and University of Jaén (Spain). Giovanni's research focuses on insect behaviour, biological control, chemical ecology (with special reference to sex pheromones and mating disruption), and insect-inspired robotics, covering agricultural pests as well as vectors of medical and veterinary importance. He has cooperated with a large number of researchers worldwide on various research projects (e.g., iGuess-MED PRIMA, STRADIOL). He is actively engaged in third-mission activities, through agricultural extension services focused on IPM and the biological control of olive and vineyard insect pests. Giovanni serves as the Editor in Chief/Associate Editor/Editorial Board Member for many top-ranked international journals in the field of general and applied entomology. He has been awarded with various research prizes from international and national organizations, including the Odile Bain Memorial Prize 2018 (Parasites and Vectors & Boehringer Animal Health) and the Antico Fattore Prize 2016 (Accademia dei Georgofili, Firenze). He was appointed as a Member of Accademia dei Georgofili in December 2021.

Preface to "Insecticide, Acaricide, Repellent and Antimicrobial Development"

Nowadays, arthropod pest and vector species are still managed though the use of synthetic insecticides and acaricides. However, in an Integrated Pest/Vector Management framework, substantial efforts are directed to design and validate environmentally sustainable and selective products with multiple modes of action, which make the develop of insecticide and acaricide resistance unlikely. A similar scenario applies to antimicrobials.

Further, bites from bloodsucker insects and mites can be avoided using repellents. In this scenario, discovering novel and effective products to repel mosquitoes, ticks, tabanids and kissing bugs, just to cite some hot-topic examples, is a challenge for public health. Natural products represent a huge source of highly effective active ingredients to be used for repellent purposes.

From this perspective, the present book, a reprint of the *Molecules* Special Issue "Insecticide, Acaricide, Repellent and Antimicrobial Development", is dedicated to the development of effective and eco-friendly insecticides, acaricides, repellents, and antimicrobials, including products of natural origin.

Giovanni Benelli
Editor

molecules MDPI

Editorial

Insecticide, Acaricide, Repellent and Antimicrobial Development

Giovanni Benelli

Department of Agriculture, Food and Environment, University of Pisa, via del Borghetto 80, 56124 Pisa, Italy; giovanni.benelli@unipi.it; Tel.: +39-0502216141

Citation: Benelli, G. Insecticide, Acaricide, Repellent and Antimicrobial Development. *Molecules* **2022**, *27*, 386. https://doi.org/10.3390/molecules27020386

Received: 29 December 2021
Accepted: 4 January 2022
Published: 8 January 2022

The quick spread of invasive arthropod species worldwide, sometimes boosted by global warming and urbanization [1–4], outlines again the need for effective and timely pest and vector management tools [5]. However, most of them rely on the use of synthetic insecticides and acaricides. This represents a major problem, since synthetic molecules often rely on a single mechanism of action, making resistance development quick and hard to deal with [6,7]. Similarly, fast resistance development to widely used antimicrobials has been detected in a wide number of microbial pathogens and parasites [8,9]. The massive, often inappropriate, employ of synthetic pesticides also leads to serious non-target effects on human health and the environment [10].

Further, bites from bloodsucker insects and mites can be avoided using repellents. In this scenario, discovering novel and effective products to repel mosquitoes, ticks and tabanids, just to cite some hot examples, is a challenge for public health [11–14]. Natural products represent a huge source of highly effective active ingredients to be used for repellent purposes (e.g., *Eucalyptus citriodora* and the related molecule *p*-menthane-3,8-diol) [15].

In this framework, the present Special Issue is dedicated to the development of effective and eco-friendly insecticides, acaricides, repellents and antimicrobials, including products of natural origin (e.g., plant extracts, essential oils, selected bacterial and fungal metabolites). Research efforts shedding light on the modes of action, behavioural modifications and non-target effects of the above-mentioned natural products have been welcomed. It has been recommended to the authors to include a positive control in the experiments [16], as well as detailed information on the chemical composition of the tested products [17]. Both original research and reviews have been included in the Special Issue.

Herein, contributions on the following topics have been included:

(a) Laboratory evaluation of the insecticidal, acaricidal and/or antimicrobial activity of plant essential oils [18,19].
(b) Isolation of pure constituents from plant extracts, and assessment of their insecticidal [20–22], acaricidal [23] and/or antimicrobial activities [24], including toxicological stability assays [25].
(c) Synthesis and characterization of novel semisynthetic insecticides, along with their in vitro evaluation on insect cells [26].
(d) Exploitation of invasive plant species as sources of effective insecticidal products [27].
(e) Evaluation of the impact of selected plant-borne compounds on the behaviour of key insect pests, with special reference to aphids [28].
(f) Development of botanical-based insecticidal formulations (including nanoformulations) characterized by an improved bioactivity and stability over time [29,30].

Finally, the Special Issue ends with two reviews. The first summarized current knowledge on the use of diatomaceous earths in crop protection, stored product, and urban pest control, presenting a number of challenges for future research [31]. The second one highlights current prospects and challenges about the use of plant-borne products as pesticides for agricultural purposes [32].

In conclusion, despite the relevant research efforts undertaken in this field for discovering new insecticides, acaricides and repellents of natural origin, the road to their large-scale use in the real world appears long and windy, complicated by costly and complex authorization requirements [33], and with limited commercialization outcomes [34]. In this scenario, I sincerely hope that the present Special Issue will be useful in inspiring future research and even extension efforts on the topic, particularly among young researchers.

Funding: This article received no external funding.

Acknowledgments: I am grateful to Ashley Chen and Carey Yuan for their assistance in preparing this Special Issue for *Molecules*.

Conflicts of Interest: The author declares no conflict of interest.

References

1. Wilke, A.B.B.; Benelli, G.; Beier, J.C. Beyond frontiers: On invasive alien mosquito species in America and Europe. *PLoS Negl. Trop. Dis.* **2020**, *14*, e0007864. [CrossRef]
2. Madder, M.; Thys, E.; Achi, L.; Touré, A.; De Deken, R. *Rhipicephalus* (*Boophilus*) *microplus*: A most successful invasive tick species in West-Africa. *Exp. Appl. Acarol.* **2011**, *53*, 139–145. [CrossRef]
3. Benelli, G.; Wilke, A.B.B.; Beier, J.C. *Aedes albopictus* (Asian tiger mosquito). *Trends Parasitol.* **2020**, *36*, 942–943. [CrossRef]
4. Li, R.; Xu, L.; Bjørnstad, O.N.; Liu, K.; Song, T.; Chen, A.; Xu, B.; Liu, Q.; Stenseth, N.C. Climate-driven variation in mosquito density predicts the spatiotemporal dynamics of dengue. *Proc. Natl. Acad. Sci. USA* **2019**, *116*, 3624–3629. [CrossRef] [PubMed]
5. Hemingway, J.; Ranson, H.; Magill, A.; Kolaczinski, J.; Fornadel, C.; Gimnig, J.; Coetzee, M.; Simard, F.; Roch, D.K.; Hinzoumbe, C.K.; et al. Averting a malaria disaster: Will insecticide resistance derail malaria control? *Lancet* **2016**, *387*, 1785–1788. [CrossRef]
6. Benelli, G.; André, A.B.B.; Bloomquist, J.R.; Desneux, N.; Beier, J.C. Overexposing mosquitoes to insecticides under global warming: A public health concern? *Sci. Total. Environ.* **2021**, *762*, 143069. [CrossRef] [PubMed]
7. Benelli, G. Managing mosquitoes and ticks in a rapidly changing world—Facts and trends. *Saudi J. Biol. Sci.* **2019**, *26*, 921–929. [CrossRef] [PubMed]
8. Ashley, E.A.; Dhorda, M.; Fairhurst, R.M.; Amaratunga, C.; Lim, P.; Suon, S.; Sreng, S.; Anderson, J.M.; Mao, S.; Sam, B.; et al. Spread of artemisinin resistance in Plasmodium falciparum malaria. *N. Engl. J. Med.* **2014**, *371*, 411–423. [CrossRef]
9. Naß, J.; Efferth, T. Development of artemisinin resistance in malaria therapy. *Pharmacol. Res.* **2019**, *146*, 104275. [CrossRef]
10. Desneux, N.; Decourtye, A.; Delpuech, J.M. The sublethal effects of pesticides on beneficial arthropods. *Annu. Rev. Entomol.* **2007**, *52*, 81–106. [CrossRef]
11. Bissinger, B.W.; Roe, M.R. Tick repellents: Past, present, and future. *Pest. Biochem. Physiol.* **2010**, *96*, 63–79. [CrossRef]
12. Lupi, E.; Hatz, C.; Schlagenhauf, P. The efficacy of repel- lents against *Aedes, Anopheles, Culex* and *Ixodes* spp.—A literature review. *Travel Med. Infect. Dis.* **2013**, *11*, 374–411. [CrossRef]
13. Benelli, G.; Pavela, R. Beyond mosquitoes-essential oil toxicity and repellency against bloodsucking insects. *Ind. Crops Prod.* **2018**, *117*, 382–392. [CrossRef]
14. Štefanidesová, K.; Škultéty, L'.; Sparagano, O.A.E.; Špitalská, E. The repellent efficacy of eleven essential oils against adult *Dermacentor reticulatus* ticks. *Ticks Tick Borne Dis.* **2017**, *8*, 780–786. [CrossRef] [PubMed]
15. Benelli, G.; Maggi, F.; Canale, A.; Mehlhorn, H. Lyme disease is on the rise—How about tick repellents? A global view. *Entomol. Gen.* **2019**, *39*, 61–72. [CrossRef]
16. Isman, M.B.; Grieneisen, M.L. Botanical insecticide research: Many publications, limited useful data. *Trends Plant Sci.* **2014**, *19*, 140–145. [CrossRef]
17. Pavela, R.; Maggi, F.; Iannarelli, R.; Benelli, G. Plant extracts for developing mosquito larvicides: From laboratory to the field, with insights on the modes of action. *Acta Trop.* **2019**, *193*, 236–271. [CrossRef] [PubMed]
18. Dai, D.N.; Chung, N.T.; Huong, L.T.; Hung, N.H.; Chau, D.T.M.; Yen, N.T.; Setzer, W.N. chemical compositions, mosquito larvicidal and antimicrobial activities of essential oils from five species of *cinnamomum* growing wild in north central vietnam. *Molecules* **2020**, *25*, 1303. [CrossRef]
19. Debbabi, H.; El Mokni, R.; Chaieb, I.; Nardoni, S.; Maggi, F.; Caprioli, G.; Hammami, S. Chemical composition, antifungal and insecticidal activities of the essential oils from tunisian *Clinopodium nepeta* subsp. *nepeta* and *Clinopodium nepeta* subsp. *glandulosum*. *Molecules* **2020**, *25*, 2137. [CrossRef]
20. Rahman-Soad, A.; Dávila-Lara, A.; Paetz, C.; Mithöfer, A. Plumbagin, a potent naphthoquinone from *Nepenthes* plants with growth inhibiting and larvicidal activities. *Molecules* **2021**, *26*, 825. [CrossRef] [PubMed]
21. Ahmed, M.; Qin, P.; Ji, M.; An, R.; Guo, H.; Shafi, J. Spinasterol, 22,23-Dihydrospinasterol and fernenol from *Citrullus Colocynthis* L. with aphicidal activity against cabbage aphid *Brevicoryne Brassicae* L. *Molecules* **2020**, *25*, 2184. [CrossRef]
22. Karthi, S.; Uthirarajan, K.; Manohar, V.; Venkatesan, M.; Chinnaperumal, K.; Vasantha-Srinivasan, P.; Krutmuang, P. Larvicidal enzyme inhibition and repellent activity of red mangrove *Rhizophora mucronata* (Lam.) leaf extracts and their biomolecules against three medically challenging arthropod vectors. *Molecules* **2020**, *25*, 3844. [CrossRef]

23. Hussain, A.; AlJabr, A.M. Potential synergy between spores of *Metarhizium anisopliae* and plant secondary metabolite, 1-Chlorooctadecane for effective natural acaricide development. *Molecules* **2020**, *25*, 1900. [CrossRef] [PubMed]
24. Hasan, M.; Zafar, A.; Shahzadi, I.; Luo, F.; Hassan, S.G.; Tariq, T.; Zehra, S.; Munawar, T.; Iqbal, F.; Shu, X. Fractionation of biomolecules in *Withania coagulans* extract for bioreductive nanoparticle synthesis, antifungal and biofilm activity. *Molecules* **2020**, *25*, 3478. [CrossRef] [PubMed]
25. Moura, E.D.S.; Faroni, L.R.D.; Heleno, F.F.; Rodrigues, A.A.Z. Toxicological stability of *Ocimum basilicum* essential oil and its major components in the control of *Sitophilus zeamais*. *Molecules* **2021**, *26*, 6483. [CrossRef] [PubMed]
26. Pereira, R.B.; Pinto, N.F.S.; Fernandes, M.J.G.; Vieira, T.F.; Rodrigues, A.R.O.; Pereira, D.M.; Sousa, S.F.; Castanheira, E.M.S.; Fortes, A.G.; Gonçalves, M.S.T. Amino alcohols from eugenol as potential semisynthetic insecticides: Chemical, biological, and computational insights. *Molecules* **2021**, *26*, 6616. [CrossRef] [PubMed]
27. Hoi, T.M.; Huong, L.T.; Chinh, H.V.; Hau, D.V.; Satyal, P.; Tai, T.A.; Dai, D.N.; Hung, N.H.; Hien, V.T.; Setzer, W.N. Essential oil compositions of three invasive *Conyza* species collected in vietnam and their larvicidal activities against *Aedes aegypti*, *Aedes albopictus*, and *Culex quinquefasciatus*. *Molecules* **2020**, *25*, 4576. [CrossRef]
28. Stec, K.; Kozłowska, J.; Wróblewska-Kurdyk, A.; Kordan, B.; Anioł, M.; Gabryś, B. Effect of naringenin and its derivatives on the probing behavior of *Myzus persicae* (Sulz.). *Molecules* **2020**, *25*, 3185. [CrossRef]
29. Kavallieratos, N.G.; Nika, E.P.; Skourti, A.; Ntalli, N.; Boukouvala, M.C.; Ntalaka, C.T.; Maggi, F.; Rakotosaona, R.; Cespi, M.; Perinelli, D.R.; et al. Developing a *Hazomalania voyronii* essential oil nanoemulsion for the eco-friendly management of *Tribolium confusum*, *Tribolium castaneum* and *Tenebrio molitor* larvae and adults on stored wheat. *Molecules* **2021**, *26*, 1812. [CrossRef]
30. López, S.; Tapia, A.; Zygadlo, J.; Stariolo, R.; Abraham, G.A.; Cortez Tornello, P.R. *Zuccagnia punctata* cav. essential oil into poly(ε-caprolactone) matrices as a sustainable and environmentally friendly strategy biorepellent against *Triatoma infestans* (Klug) (Hemiptera, Reduviidae). *Molecules* **2021**, *26*, 4056. [CrossRef]
31. Zeni, V.; Baliota, G.V.; Benelli, G.; Canale, A.; Athanassiou, C.G. Diatomaceous earth for arthropod pest control: Back to the future. *Molecules* **2021**, *26*, 7487. [CrossRef] [PubMed]
32. Souto, A.L.; Sylvestre, M.; Tölke, E.D.; Tavares, J.F.; Barbosa-Filho, J.M.; Cebrián-Torrejón, G. Plant-derived pesticides as an alternative to pest management and sustainable agricultural production: Prospects, applications and challenges. *Molecules.* **2021**, *26*, 4835. [CrossRef] [PubMed]
33. Pavela, R.; Benelli, G. Essential oils as ecofriendly biopesticides? Challenges and constraints. *Trends Plant Sci.* **2016**, *21*, 1000–1007. [CrossRef] [PubMed]
34. Isman, M.B. Botanical insecticides in the twenty-first century-fulfilling their promise? *Annu. Rev. Entomol.* **2020**, *65*, 233–249. [CrossRef]

molecules

MDPI

Article

Amino Alcohols from Eugenol as Potential Semisynthetic Insecticides: Chemical, Biological, and Computational Insights

Renato B. Pereira [1,†], Nuno F. S. Pinto [2,†], Maria José G. Fernandes [2], Tatiana F. Vieira [3,4], Ana Rita O. Rodrigues [5], David M. Pereira [1], Sérgio F. Sousa [3,4], Elisabete M. S. Castanheira [5], A. Gil Fortes [2] and M. Sameiro T. Gonçalves [2,*]

1 REQUIMTE/LAQV, Laboratory of Pharmacognosy, Department of Chemistry, Faculty of Pharmacy, University of Porto, R. Jorge Viterbo Ferreira, 228, 4050-313 Porto, Portugal; rjpereira@ff.up.pt (R.B.P.); dpereira@ff.up.pt (D.M.P.)

2 Centre of Chemistry, Department of Chemistry, University of Minho, Campus of Gualtar, 4710-057 Braga, Portugal; nuno_pinto1993@hotmail.com (N.F.S.P.); mjfernandes@quimica.uminho.pt (M.J.G.F.); gilf@quimica.uminho.pt (A.G.F.)

3 Associate Laboratory i4HB—Institute for Health and Bioeconomy, Faculty of Medicine, University of Porto, 4200-319 Porto, Portugal; tatianafvieira@gmail.com (T.F.V.); segiofsousa@med.up.pt (S.F.S.)

4 UCIBIO—Applied Molecular Biosciences Unit, BioSIM—Department of Biomedicine, Faculty of Medicine, University of Porto, 4200-319 Porto, Portugal

5 Centre of Physics of Minho and Porto Universities (CF-UM-UP), University of Minho, Campus of Gualtar, 4710-057 Braga, Portugal; ritarodrigues@fisica.uminho.pt (A.R.O.R.); ecoutinho@fisica.uminho.pt (E.M.S.C.)

* Correspondence: msameiro@quimica.uminho.pt; Tel.: +351-253-604-372

† These authors contributed equally to this work.

Citation: Pereira, R.B.; Pinto, N.F.S.; Fernandes, M.J.G.; Vieira, T.F.; Rodrigues, A.R.O.; Pereira, D.M.; Sousa, S.F.; Castanheira, E.M.S.; Fortes, A.G.; Gonçalves, M.S.T. Amino Alcohols from Eugenol as Potential Semisynthetic Insecticides: Chemical, Biological, and Computational Insights. *Molecules* **2021**, *26*, 6616. https://doi.org/10.3390/molecules26216616

Academic Editor: Giovanni Benelli

Received: 17 September 2021
Accepted: 26 October 2021
Published: 31 October 2021

Publisher's Note: MDPI stays neutral with regard to jurisdictional claims in published maps and institutional affiliations.

Abstract: A series of β-amino alcohols were prepared by the reaction of eugenol epoxide with aliphatic and aromatic amine nucleophiles. The synthesized compounds were fully characterized and evaluated as potential insecticides through the assessment of their biological activity against *Sf9* insect cells, compared with a commercial synthetic pesticide (chlorpyrifos, CHPY). Three derivatives bearing a terminal benzene ring, either substituted or unsubstituted, were identified as the most potent molecules, two of them displaying higher toxicity to insect cells than CHPY. In addition, the most promising molecules were able to increase the activity of serine proteases (caspases) pivotal to apoptosis and were more toxic to insect cells than human cells. Structure-based inverted virtual screening and molecular dynamics simulations demonstrate that these molecules likely target acetylcholinesterase and/or the insect odorant-binding proteins and are able to form stable complexes with these proteins. Encapsulation assays in liposomes of DMPG and DPPC/DMPG (1:1) were performed for the most active compound, and high encapsulation efficiencies were obtained. A thermosensitive formulation was achieved with the compound release being more efficient at higher temperatures.

Keywords: eugenol derivatives; amino alcohols; semisynthetic insecticides; biopesticides; bioinsecticides; phenylpropanoids; *Spodoptera frugiperda*

1. Introduction

The use of synthetic pesticides for decades to manage pest control in crops has resulted in an accumulation of various residues with adverse effects on many organisms and potential negative impact in human health. At the same time, crop destruction by pests, mainly by insects, is one of the main problems responsible for losses in agricultural production. Pesticides from natural sources are an effective alternative to synthetic pesticides and are becoming more important for pest management in agriculture and also public health. In this respect, plants offer a wide variety of secondary metabolites with efficacy against insects [1,2]. In recent years, essential oils (EOs) became an important natural source of pesticides. Many EOs present insecticidal, repellent, fumigant, and antifeedant activities against a wide variety of insects [3,4]. Essential oil components and their derivatives are considered to be an alternative way of insect control. In particular, phenylpropanoids, one of the main constituents of some EOs, have proved to present efficacy against insects [4]. Eugenol is a phenylpropanoid and a major constituent of clove essential oil with many applications in pharmaceutical, food, agricultural, and cosmetics industries [5], and it has been shown to be biologically active as antioxidant [5,6], antiviral [7] anti-inflammatory [8] and antimicrobial [9]. Enan [10] showed that eugenol mimicked octopamine in increasing intracellular calcium levels in cloned cells from the brain of *Periplaneta americana* and *Drosophila melanogaster*, and this was also found to be mediated via octopamine receptors. Structural changes of eugenol are known to be a useful strategy in order to improve biological activity and to obtain new analogues with reduced side effects [11].

Further, epoxides are important intermediates in pharmaceutical and agrochemical industries. The three-membered heterocyclic ring is strained and susceptible to attack by a range of nucleophiles, including nitrogen (e.g., ammonia, amines, azides), oxygen (e.g., water, alcohols, phenols, acids), and sulfur (thiol)-containing compounds, leading to bifunctional molecules of great industrial value. The β-amino alcohols are used in the synthesis of β-blockers, insecticidal agents, and oxazolines, as well as chiral ligands in asymmetric synthesis [12–17]. β-Amino alcohol functionality is found in many biologically active compounds, being an important pharmacophore [14,18], and *N*-substituted β-amino alcohols are important building blocks in the preparation of added-value chemicals [16,19]. Salbutamol and propranolol are on the World Health Organization List of Essential Medicines and represent the most important examples of therapeutic agents having this structural feature [20].

Following our research interests in plant-inspired alternatives to synthetic pesticides [21,22], and specifically the study in which some eugenol derivatives have shown potential as biopesticides [22], in the present work, eugenol was converted to the corresponding epoxide with *m*-chloroperoxybenzoic acid (*m*-CPBA) in dichloromethane (DCM) and further reacted with a series of amine nucleophiles to afford the corresponding β-amino alcohols. The β-amino alcohols were purified by column chromatography and fully characterized by ^{1}H and ^{13}C NMR spectroscopy and HRMS (high-resolution mass spectrometry). The obtained compounds were evaluated as potential insecticides through the assessment of their biological activity against the *Sf9* insect cell lines compared with chlorpyrifos, which is a commercial synthetic pesticide. In addition, computational studies were performed to identify the most likely protein targets responsible for the observed insecticide activity of these molecules through the application of structure-based inverted virtual screening protocol combined with molecular dynamics simulations and free energy calculations. Nanoencapsulation of the most promising compound considering its insecticidal activity was performed in liposomes of DMPG (dimyristoylphosphatidylglycerol) and DPPC/DMPG (dipalmitoylphosphatidylcholine/dimyristoylphosphatidylglycerol) (1:1), aiming at obtaining thermosensitive formulations. DMPG has a gel to liquid–crystalline phase transition temperature (T_m) at 23 °C, while for DPPC, it is 41 °C [23]. The increase in membrane fluidity upon phase transition is expected to promote an enhanced release of the encapsulated compounds, providing a triggered release by temperature above T_m of the formulation.

2. Results and Discussion

2.1. Synthesis

Eugenol **1** is easily obtained by hydrodistillation from clove, and is known for its various biological activities, as mentioned above, namely insecticidal. Following our recent interests in finding new biopesticides [22], the present work describes a strategy consisting of structural changes of eugenol in an attempt to obtain semisynthetic alternatives with improved insecticidal activity. Eugenol epoxide **2** was prepared from eugenol through reaction with *m*-CPBA in DCM using a known procedure [11,24] in 48% yield. Then, the epoxide was further reacted at 50 °C with a series of aliphatic and aromatic amine nucleophiles in ethanol/water as solvent [25], which is followed by column chromatography purification on silica gel using dichloromethane/methanol, mixtures of increasing polarity (**3b**, **3d–f**) as the eluent, or by evaporation of solvents under reduced pressure (**3a** and **3c**) to afford the corresponding β-amino alcohol derivatives **3a–f** as oil materials. Thus, the reaction of 2-methoxy-4-(oxiran-2-ylmethyl)phenol **2** with 2-methylpropan-2-amine, octan-2-amine, piperidine, aniline, 3-methoxyaniline, and 4-cyanoaniline gave 4-(3-(*tert*-butylamino)-2-hydroxypropyl)-2-methoxyphenol **3a**, 4-(2-hydroxy-3-(octan-2-ylamino)propyl)-2-methoxyphenol **3b**, (4-(2-hydroxy-3-(piperidin-1-yl)-propyl)-2-methoxyphenol **3c**, 4-(2-hydroxy-3-(methyl(phenyl)amino)propyl)- 2-methoxyphenol **3d**, 4-(2-hydroxy-3-((3-methoxyphenyl)amino)propyl)-2-methoxy- phenol **3e**, and 4-(2-hydroxy-3-(4-hydroxy-3-methoxyphenyl)propyl)amino)benzonitrile **3f**, respectively, in 9–97% yield, and they were fully characterized by ^{1}H and ^{13}C NMR spectroscopy and HRMS (Scheme 1).

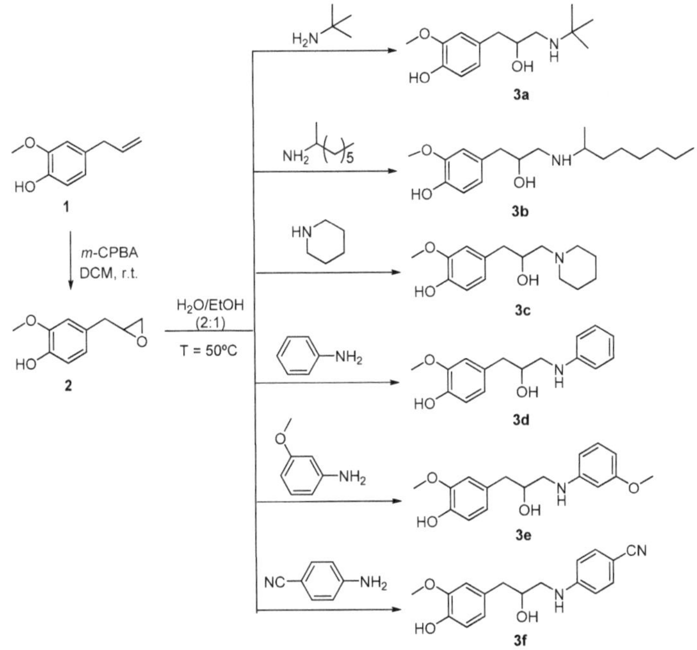

Scheme 1. Synthesis of eugenol amino alcohols **3a–f**.

The main ^{1}H-NMR features of compounds **3a–f** are the signals for protons of the CHOH, CH_2N, and OCH_3 groups. The CH_2N protons show up as two distinct signals as doublets of doublets or multiplets (δ 3.45–2.53 ppm); the CHOH proton displays as a multiplet in all compounds (δ 4.91–3.89 ppm); and the OCH_3 group shows up as a singlet (δ 3.89–3.81 ppm). The *tert*-butyl group in **3a** shows up as a singlet (δ 1.16 ppm), while in **3b** and **3c**, the *N*-alkyl chain corresponds to a series of multiplets (δ 2.88–0.88 ppm). The ^{13}C main features are the CH_2N (δ 41.24–41.10 ppm), CHOH carbon (δ 71.1–67.01 ppm), OCH_3 carbon (δ 55.93–55.72 ppm), and the additional OCH_3 in **3e** (δ 54.96 ppm). In addition, the *tert*-butyl carbons in **3a** shows up (δ 26.79 and 24.58 ppm), while in **3b**, the methyl terminal is highlighted (δ 13.98 ppm).

2.2. Toxicity Assessment in Insect Cells

All molecules obtained were evaluated for their impact in the viability of the *Sf*9 cells at 100 μg/mL (i.e., **1**—6.09×10^{-4} M; **2**—5.55×10^{-4} M; **3a**—3.95×10^{-4} M; **3b**—3.23×10^{-4} M; **3c**—3.77×10^{-4} M; **3d**—3.66×10^{-4} M; **3e**—3.30×10^{-4} M; **3f**— 3.35×10^{-4} M;

CHPY– 2.85 × 10^{-4} M) by the means of a resazurin-based method. For benchmarking purposes, the insecticide chlorpyrifos was used at the same concentration. As shown in Figure 1, the only molecule devoid of toxicity was **3c**, which incidentally was also the only one bearing a piperidine ring. A second group of molecules, which elicited residual toxicity (under 25% of viability loss), was **1**, **2**, **3a**, and **3b**. Eugenol **1** was the starting material, and the results show that the replacement of the terminal methylene group by the epoxide had no effect upon the biological activity of the molecule. Finally, the most potent molecules were **3d**, **3e**, and **3f**, which caused losses of 40%, 30%, and 50% viability in insect cells, respectively. These three molecules were also the only ones bearing a benzene ring next to the nitrogen atom. Considering the unsubstituted ring, **3d**, its methoxylation resulted in decreased potency, while the presence of the cyanide group increased it. In light of these results, we decided to advance our studies solely with the two most potent molecules, **3d** and **3f**, as they were more potent than the benchmark used, chlorpyrifos.

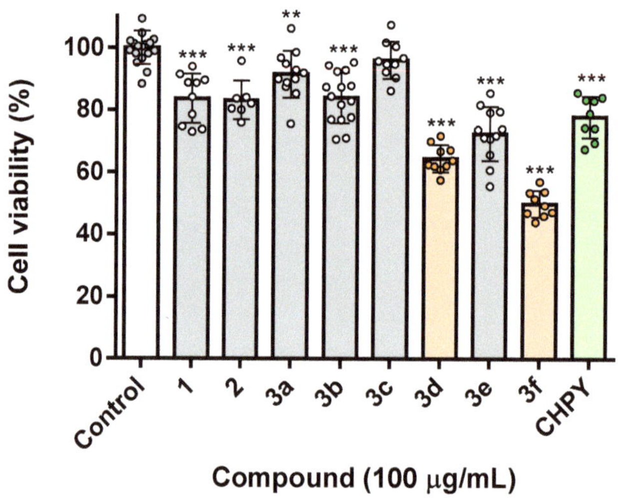

Figure 1. Viability of the *Sf9* cells after incubation with the presented molecules (100 μg/mL), medium (control), or the reference insecticide chlorpyrifos (CHPY, 100 μg/mL). Cells were incubated for 24 h, after which viability was evaluated. ** $p < 0.01$, *** $p < 0.001$.

2.3. *Amino Alcohols **3d** and **3f** Activate Caspase-Like Proteases in the Sf9 Cells*

After establishing the toxicity of the selected molecules toward insect cells, we investigated the mechanism of action behind this effect. In fact, the loss of viability could be a consequence of an array of different biological processes, from necrosis to cell cycle arrest and apoptosis, among others. Necrosis is a process of uncontrolled cell death that encompasses the destruction of cell membranes, with consequent leakage of cytoplasmic content to the surrounding tissues; for this reason, it is usually avoided in biological contexts [26]. To assess the potential unfolding of this event, we assessed the levels of leaked lactate dehydrogenase (LDH) in cells incubated with the selected molecules. Being a cytoplasmic enzyme, the finding of extracellular LDH is widely used as a marker of necrosis. As shown in Figure 2A, the incubation of cells with a lysis solution (LS) resulted in a three to four-fold increase in extracellular LDH. Conversely, the incubation of cells with **3d** and **3f** had no detectable impact in LDH levels in culture media. In light of this, we concluded that the impact of these molecules in the viability of the *Sf9* cells was not a consequence of an ongoing necrotic process. Next, we assessed if a process of organized cell death, such as apoptosis, could be taking place. Given the pivotal role of cysteine-aspartic proteases in most forms of apoptosis, we investigated the impact of **3d** and **3f** in the insect equivalent of mammal caspases, in this case DRACE, using a proluminescent substrate of this target.

As shown in Figure 2B, both **3d** and **3f** significantly increased the caspase-like activity in treated cells, the latter having a more pronounced effect. This result suggests that both **3d** and **3f** elicit their cytotoxic effect toward the *Sf9* cells by triggering an organized process of cell death with the involvement of cysteine-aspartic proteases.

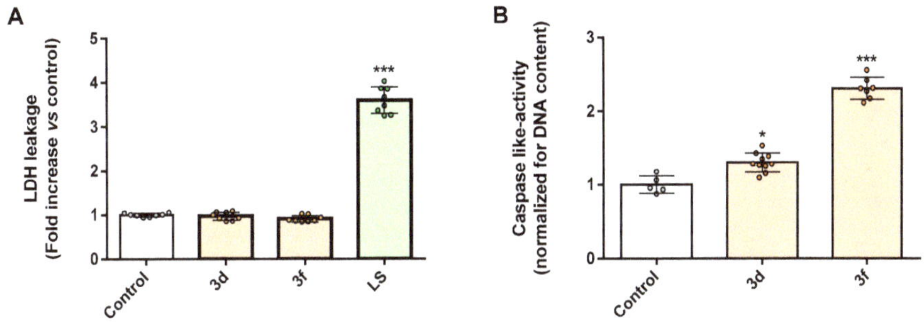

Figure 2. (**A**) LDH activity found in the culture media of the *Sf9* cells treated with compounds **3d** and **3f** (100 µg/mL) for 24 h. Lysis solution (LS) was used as positive control to generate a maximum LDH release. (**B**) Caspase-like activity of the *Sf9* cells after incubation with compounds **3d** and **3f** (100 µg/mL) for 24 h. Results are normalized for DNA content. * $p < 0.05$, *** $p < 0.001$.

2.4. Amino Alcohols 3d and 3f Are More Toxic to Insect Cells Than Human Cells

Up to this point, we had already identified two molecules that presented higher potency than the commercial insecticide chlorpyrifos and that were shown the be non-necrotic and pro-apoptotic. In addition to these traits reported herein, it is also important that prospective new insecticides present some degree of selectivity, specifically low toxicity to human cells. To this end, we assessed the impact of the **3d** and **3f** in 2D models of human cells. We chose human keratinocytes (HaCaT cell line), as one of the most relevant routes of human contact with pesticides is usually via the skin, where keratinocytes are the first population of living cells in the skin. As shown in Figure 3, both molecules elicited a weak loss of viability, around 20%. Importantly, both molecules were less toxic than the benchmark chlorpyrifos and, relevantly, they were less toxic to human cells than insect cells.

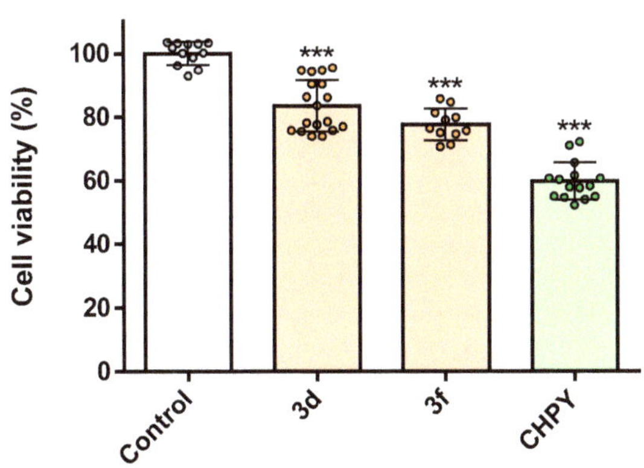

Figure 3. Viability of HaCaT cells exposed to compounds **3d** and **3f** (100 µg/mL), medium (control), or the reference insecticide chlorpyrifos (CHPY, 100 µg/mL). Cells were incubated for 24 h, after which viability was evaluated. *** $p < 0.001$.

These results are promising and pave the way for further developments in the field, as the chemical diversity obtained allows drawing some structure–activity relationships, as addressed above.

2.5. Inverted Virtual Screening Results

After identifying the most promising molecules and the biological processes involved in their cytotoxic effect, we were interested in shedding light on the possible molecular targets. To this end, an array of computational methods was used.

Table 1 presents the average scores obtained for compounds **3d** and **3f** for each potential target with each scoring function. Regarding the different scoring functions, it is important to mention that they are based on different metrics and scales. The score for all the GOLD scoring functions is dimensionless, and the higher the score, the better the binding affinity. The Vina scoring function, on the other hand, uses a metric that approximates that of binding free energies, so a more negative value means better affinity. The PDB structure with the best score was selected for each potential target, and they were ranked from the best target to worst, according to the predictions of the different docking programs/scoring functions.

Table 1. Average scores of the amino alcohol derivatives of eugenol obtained for all PDB structures with the five different scoring functions and overall ranking of the most likely protein targets for interaction.

Target	PDB	PLP	ASP	ChemScore	GoldScore	Vina	Overall Ranking
Ecdysone receptor	1R20 1R1K	68.82 74.97	36.26 35.04	30.28 35.86	57.54 56.38	−7.40 −8.60	5
Chitinase	3WL1 3WQV	75.87 73.47	50.28 45.81	30.745 31.21	63.06 59.7	−8.05 −8.15	4
β-*N*-acetyl-D-hexosaminidase OfHex1	3NSN 3OZP	78.01 72.91	57.04 51.61	36.30 31.11	70.90 67.35	−7.10 −8.05	3
N-Acetylglucosamine-1-phosphate uridyltransferase (GlmU)	2V0K 2VD4	65.88 58.64	30.02 28.58	25.93 25.97	55.305 46.5	−7.15 −6.20	13
Acetylcholinesterase (AChE)	1QON 4EY6 1DX4	89.26 79.52 85.66	58.71 51.20 51.32	40.87 39.12 37.48	71.41 62.63 67.86	−8.90 −8.45 −8.70	2
Prophenoloxidase (PPO)	3HHS	77.89	40.86	31.48	62.17	−6.85	6
p-Hydroxyphenylpyruvate dioxygenase	6ISD	70.33	37.53	31.03	55.00	−7.35	9
Voltage-gated sodium channel	6A95	68.72	31.27	28.95	57.86	−6.75	12
Octopamine receptor	4N7C	57.07	37.56	31.60	61.72	−5.55	11
Sterol carrier protein-2 (HaSCP-2)	4UEI	65.70	37.22	31.64	50.255	−7.40	10
Peptide deformylase	5CY8	76.49	33.49	29.60	65.65	−7.75	7
α-Esterase-7	5TYJ 5TYP	68.31 68.54	39.23 42.26	33.26 32.22	53.04 55.565	−6.95 −6.95	8
Odorant Binding Protein	5V13 2GTE 3N7H 3K1E	86.69 79.14 82.54 91.51	52.52 41.46 44.31 48.43	41.24 38.55 36.01 41.90	66.46 69.11 69.11 74.58	−6.90 −7.55 −6.90 −9.30	1

Globally, considering the results obtained with the different scoring functions, the odorant binding proteins class (OBP) and acetylcholinesterase (AChE) are the protein targets with the highest affinity toward compounds **3d** and **3f**. This tendency was quite clear with all the different scoring functions evaluated, which further strengthens our results.

2.6. Molecular Dynamics Simulations and Free Energy Calculations Results

To validate the inverted screening results, we evaluate the protein flexibility and characterize the molecular interactions formed, and molecular dynamics simulations were performed for the complexes formed with compound **3d** and compound **3f** and the two groups of targets predicted at the inverted VS stage: OBP and AChE. Structures with the best score from each group were selected (3K1E for OBP and 1QON for AChE). The stability of AChE: compound **3d**, AChE: compound **3f**, OBP: compound **3d**, and OBP: compound **3f** complexes was evaluated using RMSD calculations for the Cα atoms of each complex and for the ligands, Solvent-Accessible Surface Area (SASA) analysis, and hydrogen bonding analysis.

All systems and ligands presented relatively low RMSD values, as seen in Table 2 (and Figure S1), showing that the target–ligand complexes are well equilibrated and that the eugenol derivatives evaluated maintain their binding conformation predicted from the docking.

Table 2. Average protein and ligand RMSD values (Å), ligand RMSD (Å), average ligand SASA (Å), percentage of potential ligand SASA buried, and average number of ligand–target hydrogen bonds obtained from the MD simulations. ΔG binding energy determined using MM/GBSA and per-residue decomposition, which were calculated for the last 90 ns of the simulation.

		Average RMSD of the Complex (Å)	Average RMSD of the Ligand (Å)	Ligand SASA (Å^2)	Percentage of Potential Ligand SASA Buried (%)	Average Number H-bonds	ΔG_{bind} (kcal/mol)	Main Contributors
AChE	**3d**	4.6 ± 0.6	1.6 ± 0.4	59.6 ± 16.9	88	0.2 ± 0.1	−18.3 ± 0.1	Trp83 (−2.4 ± 0.8) Tyr370 (−1.3 ± 0.4) His480 (−1.3 ± 0.6)
	3f	3.1 ± 0.2	1.4 ± 0.2	36.8 ± 10.2	93	0.5 ± 0.6	−28.2 ± 0.2	Tyr370 (−2.4 ± 0.1) Tyr374 (−2.5 ± 0.8) Trp83 (−1.9 ± 0.4)
OBP	**3d**	2.2 ± 0.3	1.7 ± 0.3	27.2 ± 10.4	95	0.2 ± 0.5	−31.7 ± 0.2	Leu67 (−2.5± 0.5) Trp105 (−2.1± 0.4) Ala79 (−1.7± 0.5)
	3f	2.2 ± 0.2	2.1 ± 0.3	33.2 ± 8.5	94	0.9 ± 0.9	−41.6 ± 0.2	Met75 (−2.9 ± 0.4) Trp105 (−2.5 ± 0.4) Phe114 (−1.8 ± 0.8)

When analyzing the percentage of potential SASA area buried for compound **3d** and compound **3f** when complexed with AChE and OBP, it can be seen that the two molecules remain tightly bound to the two targets evaluated and well protected from the solvent with average buried areas over 90% (Table 2). A small decrease was noticed for compound **3d** bound to AChE, in relation to the initial configuration predicted from docking, with an average buried area oscillating between 80 and 90%. These results demonstrate that compounds **3d** and **3f** remain well bound to the two targets evaluated, even after 100 ns. In particular, the eugenol derivatives evaluated in complex with OBP remain very well protected from the solvent throughout time.

Hydrogen bonding analysis is important to understand the stability of the interactions between the targets and ligands throughout time. The results presented in Table 2 show that both ligands maintain a stable hydrogen bonding profile with the targets evaluated, maintaining between one and three hydrogen bonds with AChE and one and four hydrogen bonds with OBP. Globally, the profile observed shows that compounds **3d** and **3f** establish more hydrogen bonds with OBP and with AChE.

Table 2 summarizes the results discussed so far and presents the values for the Gibbs binding free energy calculated using MM/GBSA. The analysis of the residue contribution to the eugenol derivatives' binding free energy to the two protein targets evaluated highlights the interaction profile of compounds **3d** and **3f** against AChE and OBP, showing the most important amino acid residues involved in ligand stabilization.

AChE is a serine hydrolase, and it is a very common target for pesticides as it is an enzyme vital for the regulation of acetylcholine in several organisms, from insects to mammals. Since this is an enzyme transversal to many species, the use of anticholinesterase insecticides can cause serious health and environmental problems. In addition, there are reports of insect resistance due to mutation of the AChE gene [27].

For AChE, compound **3d** binding is stabilized mostly by residues Trp83 (-2.4 ± 0.8), Tyr370 (-1.3 ± 0.4), and His480 (-1.3 ± 0.6) through non-polar interactions. For compound **3f**, the residues contributing more toward AChE binding are Tyr370 (-2.4 ± 0.1), Tyr372 (-2.5 ± 0.8), and Trp83 (-1.9 ± 0.4), with non-polar interactions playing an important role and π–π stacking with Trp83. Figure 4 represents the average structure of the dominant cluster of the AChE-eugenol derivatives complexes obtained from the analysis of the MD trajectory, illustrating the binding pocket and main interactions formed.

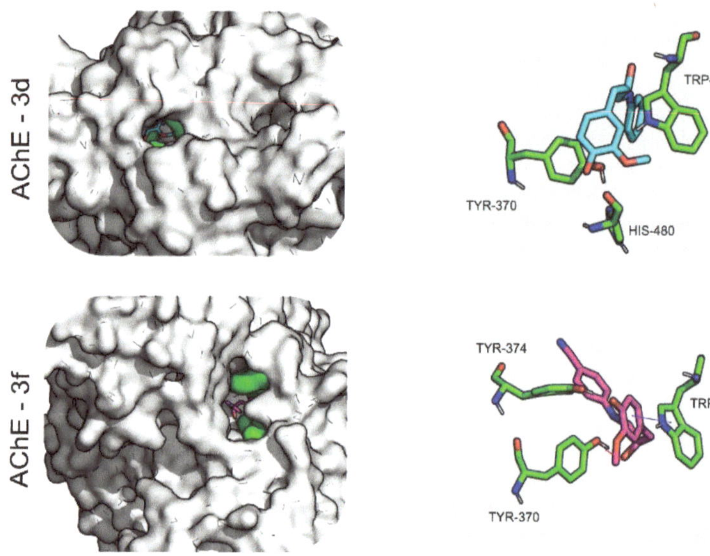

Figure 4. Compound **3d** (cyan licorice) and compound **3f** (pink licorice) interaction map with AChE. The most important residues for the interaction are highlighted in green. Blue arrow indicates π–π stacking with the ring of Trp-83. Red lines represent hydrogen bonding.

For the sake of warranting the potential off-target effect that could pose a toxicity risk, human AChE was also analyzed, the docking scores (Table S1) being inferior to the ones obtained for insect AChE, hence suggesting that the eugenol derivatives evaluated favor binding to the insect AChE considered over that of human AChE. When comparing the sequence of amino acids between insect and human AChE, there is only 53–54% sequence identity, even though their 3D structures are very similar. The active-site gorge in the insect enzyme is narrower, and the amino acid residues are different. Moreover, the residues in the opening of the gorge are also different [27], which might explain the difference in affinity of the eugenol derivatives.

The results show that the most stable complexes are OBP–compound **3d** and OBP–compound **3f**, with binding free energy values of -31.7 ± 0.2 and -41.6 ± 0.2 kcal/mol, respectively. This is consistent with the results presented so far and indicates that eugenol derivatives have indeed a high affinity toward OBP. In fact, there is a structure deposited in the PDB of a bee OBP14 from bound to eugenol [28].

OBP are a large and diverse family of insect proteins. They are involved in the transport of hydrophobic odorant and pheromone molecules toward the olfactory receptors. They are abundant in the insect family and different in structure but carry out similar roles.

In the *Drosophila melanogaster*, there are 52 different types of OBPs alone. Even though diverse in number and sequence, they present some common features. They are small, have six conserved cysteine residues joined by three disulfide bridges, and have six alpha-helical domains [29–32].

Compounds **3d** and **3f** have a higher molecular weight than eugenol (273.33 g/mol, 298.34 g/mol, respectively, versus 164.20 g/mol), but they are also lipophilic and if volatile, they can in fact be capable of binding OBP. The precise mechanism of action still needs to be further validated.

For OBP1, compound **3d** binding is stabilized mostly by residues Leu67 (−2.5 ± 0.5), Trp105 (−2.1 ± 0.4), and Ala79 (−1.7 ± 0.5), through van der Waals interactions. For compound **3f**, the residues contributing more toward OBP binding are Met75 (−2.9 ± 0.4) and Phe114 (−1.8 ± 0.8) through van der Waals interactions and Trp105 (−2.5 ± 0.4) through a hydrogen bond with the backbone. Figure 5 illustrates the average structure of the dominant cluster of the OBP1 binding pocket and main interactions formed between compound **3d**-OBP1 and compound **3f**-OBP1, respectively.

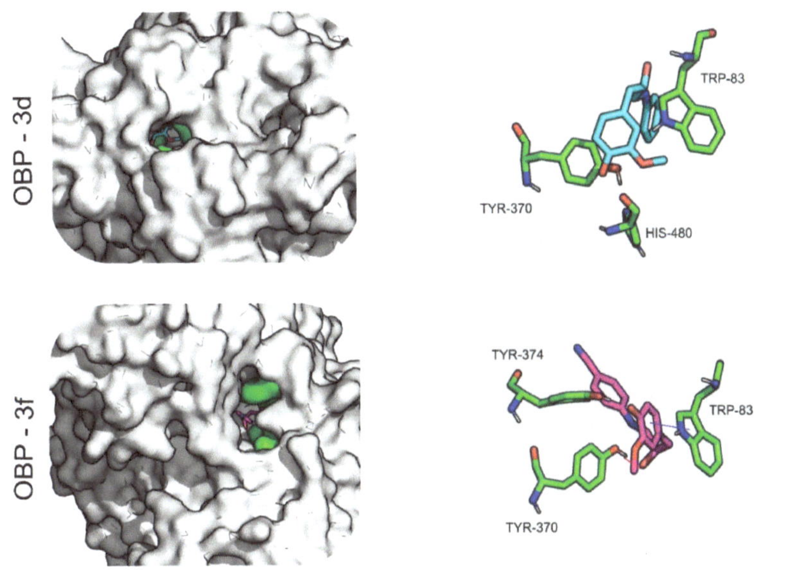

Figure 5. Compound **3d** (cyan licorice) and compound **3f** (pink licorice) interaction map with OBP1. The most important residues are highlighted in green.

2.7. Nanoencapsulation and Release Assays

The most active compound against the *Sf9* cells, compound **3f**, was encapsulated in liposomal systems of the phospholipids DMPG (100%) and DMPG/DPPC (1:1). The size (hydrodynamic diameter), polydispersity index, and zeta potential of the compound-loaded liposomes were determined by Dynamic and Electrophoretic Light Scattering (Table 3). These properties can affect the bulk properties, performance, processability, and stability of a nanoformulation. Particularly, the surface charge highly influences the stability of the liposomes, and zeta potential values more negative than −30 mV or more positive than +30 mV are considered optimal values for good stabilization of a nanodispersion [33]. In view of this fact, a negatively charged lipid, phosphatidylglycerol, was chosen for the liposomal formulation. Specifically, the phospholipid DMPG has a phase transition temperature (23 °C) near room temperature [23], allowing obtaining an enhanced release at summer temperatures (around or above 30 °C), where the lipid is in the fluid (liquid-crystalline) phase. However, the relatively short hydrocarbon chain of DMPG and the tendency to form leaky vesicles [34] can hamper a high encapsulation efficiency of compound **3f**. Therefore, the DPPC/DPPG (1:1) formulation was also tested.

Table 3. Hydrodynamic diameter (D_h), polydispersity index (PDI), and zeta (ζ) potential of DMPG (100%) and DMPG:DPPC (1:1) liposomes (SD from three independent measurements).

	$D_h \pm$ SD (nm)	PDI $\pm$ SD	ζ-Potential $\pm$ SD (mV)
DMPG (100%)	196 ± 10	0.245 ± 0.007	−43.8 ± 1.8
DMPG:DPPC (1:1)	223 ± 17	0.264 ± 0.014	−55.2 ± 2.4

The values in Table 3 show that both liposome formulations are small in size, with hydrodynamic diameters around 200 nm, presenting also a low polydispersity. A PDI value below 0.3 is considered to be acceptable, indicating a homogenous population of phospholipid vesicles [35]. The zeta potential values indicate a highly negative surface charge, anticipating a low aggregation (due to the electrostatic repulsion) and high colloidal stability.

The compound **3f** is a fluorescent molecule (Figure 6) in several solvents and in liposomes. This is a great advantage for the determination of the encapsulation efficiency and drug release, due to the high sensitivity (and selectivity) of fluorescence spectroscopy.

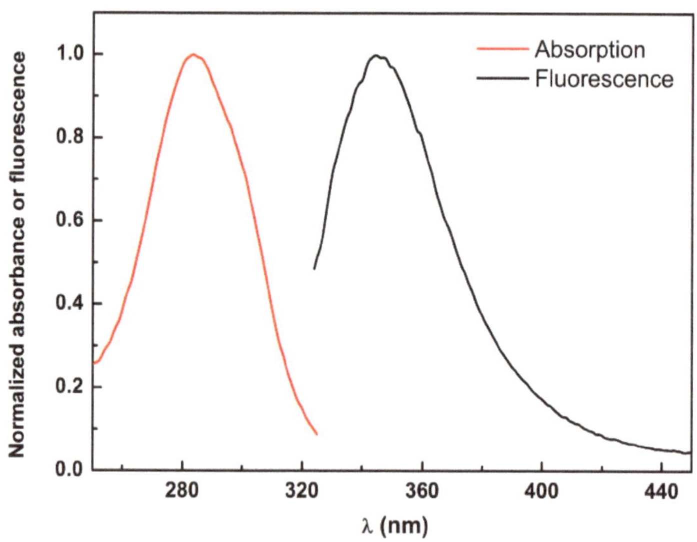

Figure 6. Normalized absorption and fluorescence emission (excitation at 290 nm) spectra of **3f** solution (2×10^{-5} M for absorption and 5×10^{-6} M for emission).

The encapsulation efficiency (EE%) of compound **3f** in liposomes was determined by fluorescence measurements (Table 4). In these assays, it was kept in mind that the compound is active against the *Sf9* cells at a concentration of 100 µg/mL (3.35×10^{-4} M) and, therefore, at least this concentration must be encapsulated. The encapsulation efficiencies are high, the system DPPC/DMPG being the most advantageous for **3f** encapsulation. Nevertheless, the EE% values show that both formulations are able to encapsulate **3f** at concentrations that may guarantee an insecticidal activity (if compound release is effective).

Table 4. Encapsulation efficiency (EE%) of compound **3f** in DMPG (100%) and DMPG/DPPC (50:50) liposomes and concentration of encapsulated compound.

	EE (%)	Encapsulated Concentration (M)
DMPG (100%)	65 ± 7	4.39×10^{-4} M
DMPG:DPPC (50:50)	92 ± 1	6.16×10^{-4} M

Compound release from both liposomal formulations was studied at 20 °C and 35 °C, to investigate the temperature dependence of the release profile (Figure 7). The experimental data were analyzed with the Weibull model (Table S2 and Figure S4 in Supporting Information) and the cumulative concentration released was compared in terms of liposome formulation and temperature.

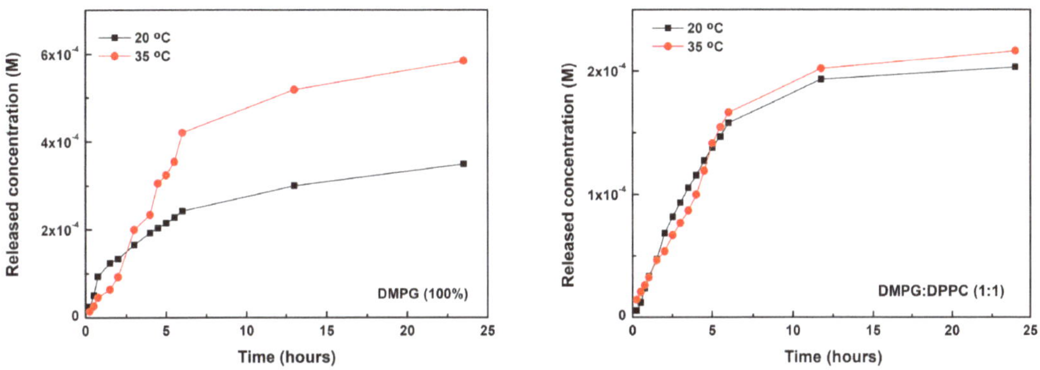

Figure 7. Cumulative release of compound **3f** at 20 °C (black squares) and 35 °C (red dots) from liposomes of DMPG (100%) (**left**) and DMPG:DPPC (**right**).

An enhanced release of **3f** was observed from liposomes of DMPG, reaching a cumulative release of 62% in 24 h at 35 °C, while, at 20 °C, a 36% release was attained. This is a result of the higher membrane fluidity at 35 °C (above transition temperature) for DMPG liposomes, which provide a thermosensitive formulation. Moreover, at both temperatures, the compound released is higher than 100 μg/mL in 24 h. The rigidity of DPPC at temperatures below its gel to liquid-crystalline phase transition (41 °C) justifies the much lower compound release from DPPC/DMPG liposomes, with 14% and 16% of released compound at 20 °C and 35 °C, respectively, not displaying a significant sensitivity to temperature.

The parameter b of the Weibull model can be related to the release mechanism [36]. If $b \leq 0.75$, the release is due to a Fickian diffusion, which is the case for DMPG liposomes at 20 °C (Table S2). If $b > 1$, a complex release mechanism takes place, with multiple mechanisms contributions, which is verified in the other cases.

3. Materials and Methods

3.1. Chemicals and Reagents

Dichloromethane, ethanol, methanol, ethyl acetate, light petroleum, and *m*-chloroperbenzoic acid were purchased from Fisher Scientific (Geel, Belgium). The anhydrous magnesium sulfate was PanReac Applichem (Barcelona, Spain) products. Chloroform-d was produced by Eurisotop (Cambridge, England). Thin-layer chromatography (TLC) analyses were carried out on 0.25 mm thick, precoated silica plates (Merck Fertigplatten Kieselgel 60F254, Germany), and spots were visualized under UVlight. Chromatography on silica gel was carried out on Merck Kieselgel (230–240 mesh).

3.2. Analytical Instruments

NMR spectra were obtained on a Bruker Avance III (Bruker Corporation, Billerica, MA, USA) at an operating frequency of 400 MHz for ^{1}H NMR and 100.6 MHz for ^{13}C NMR using the solvent peak as internal reference at 25 °C. All chemical shifts are given in ppm using δ Me_4Si = 0 ppm as reference, and J values are given in hertz. Assignments were made by comparison of chemical shifts, peak multiplicities, and J values and were supported by spin decoupling-double resonance and bidimensional heteronuclear correlation techniques. High-resolution mass spectrometry analyses were performed at the "CACTI—Unidade de Masas e Proteómica, at University of Santiago de Compostela", Spain.

3.3. Synthesis of 2-methoxy-4-(oxiran-2-ylmethyl)phenol **2**

A solution of eugenol **1** (0.500 g 3.0 mmol; 1 equiv) dissolved in dichloromethane (18 mL) was added dropwise to a suspension of 70% *m*-chloroperbenzoic acid (0.750 g; 4.3 mmol; 1 equiv) in dichloromethane (10 mL) at 0 °C. After stirring for 1 h, *m*-chloroperbenzoic acid was again added (1 equiv), and the reaction mixture was stirred for another 24 h at room temperature. A 10% aqueous solution of sodium sulfate (2 × 20 mL) was added, and the resulting solution was washed with 5% aqueous solution of sodium hydrogen carbonate (2 × 20 mL). The organic phase was dried with anhydrous magnesium sulfate, and the solvent was evaporated to afford the compound **2** as a dark yellow oil (0.239 g; 48%). Rf = 0.27 (DCM). ^{1}H-NMR δ_H ($CDCl_3$, 400 MHz): 6.87 (d, 1H, J = 8 Hz, H-6), 6.73–6.78 (m, 2H, H-3 and H-5), 5.54 (s, 1H, OH), 3.90 (s, 3H, OCH_3), 3.12–3.16 (m, 1H, CH oxirane), 2.79–2.82 (m, 3H, CH_2Ph and CH_2 oxirane), 2.55 (q, J = 2.8 Hz, 1H, CH_2 oxirane) ppm. ^{13}C-NMR δ_C ($CDCl_3$, 100.6 MHz): 146.46 (C-2), 144.39 (C-1), 129.03 (C-4), 121.64 (C-5), 114.32 (C-6), 111.54 (C-3), 55.90 (OCH_3), 52.67 (CH oxirane), 46.79 (CH_2 oxirane), 38.37 (CH_2Ph) ppm.

3.4. Synthesis of Amino Alcohols **3a–f**

3.4.1. Synthesis of 4-(3-(tert-butylamino)-2-hydroxypropyl)-2-methoxyphenol **3a**

To a suspension of 2-methoxy-4-(oxiran-2-ilmethyl)phenol **2** (0.163 g; 0.90 mmol; 1 equiv) in H_2O/EtOH 2:1 (2 mL) was added 2-methylpropan-2-amine (0.325 g; 4.44 mmol), and the resulting mixture was heated at 50 C for 5 h. The solvents and the amine were evaporated under reduced pressure to afford a compound **3a** as an orange oil (0.097 g; 0.38 mmol; 42%), Rf = 0.30 (MeOH/DCM 1:9). ^{1}H-NMR δ_H ($CDCl_3$, 400 MHz): 6.78 (d, J = 8.0 Hz, 1H, Ar-H), 6.74 (d, J = 1.6 Hz, 1H, Ar-H), 6.45 (dd, J = 8.4 Hz, 2.0 Hz 1H, Ar-H), 4.04–3.98 (m, 1H, $CH_2CH(OH)$), 3.81 (s, 3H, OCH_3), 3.45 (s, 1H, CH_2NH), 2.81 (dd, J = 12.0 Hz, 2.4 Hz, 1H, CH_2NH), 2.77–2.62 (m, 2H, $CH_2CH(OH)$), 1.16 (s, 9H, *t*-Bu) ppm. ^{13}C-NMR δ_C ($CDCl_3$, 100.6 MHz): 146.63 (Ar-C),144.34 (Ar-C), 121.83 (Ar-C), 114.4 (Ar-C), 112.13 (Ar-C), 69.37 ($CH_2CH(OH)$), 55.8 (OCH_3), 47.06 (CH_2), 41.25 (CH_2), 26.79 (3×CH_3), 24.58 ($C(CH_3)$) ppm. HRMS (ESI-TOF): calcd for $C_{14}H_{24}NO_3$ [M^+ + H]: 254.1751; found 254.1753.

3.4.2. Synthesis of 4-(2-hydroxy-3-(octan-2-ylamino)propyl)-2-methoxyphenol **3b**

To a suspension of 2-methoxy-4-(oxiran-2-ilmethyl)phenol **2** (0.163 g; 0.90 mmol; 1 equiv) in H_2O/EtOH 2:1 (2 mL) was added octan-2-amine (0.502 g; 3.89 mmol) and the resulting mixture was heated at 50 °C for 4 h. Then, water (2 mL) was added, and the resulting mixture extracted with EtOAc (2 mL), The organic phase was collected, dried with anhydrous $MgSO_4$, and the solvent evaporated to afford an oil (0.202 g), which was subjected to column chromatography using DCM/MeOH as eluent of increasing polarity to give the compound **3b** as a brown oil (0.165 g; 0.53 mmol; 59%). Rf = 0.45 (MeOH/DCM 10:90). 1H-NMR δ_H ($CDCl_3$, 400 MHz): 6.78 (d, J = 8.0 Hz, 1H, Ar-H), 6.72 (ls, 1H, Ar-H), 6.48 (d, J = 8.0 Hz, 1H, Ar-H), 3.92–3.89 (m, 1H, $CH_2CH(OH)$), 3.81 (s, 3H, OCH_3), 2.84–2.53 (m, 6H, 2×CH_2 and CH_2NH), 1.49 (m, 1H, $NHCHCH_3$), 1.24 (m, 10H, 5×CH_2), 1.07 (m, 3H, $NHCHCH_3$), 0.88 (m, 3H, $CH_2CH_2CH_3$) ppm. ^{13}C-NMR δ_C ($CDCl_3$, 100.6 MHz): 146.69 (Ar-Cq),144.36 (Ar-Cq), 129.59 (Ar-Cq), 121.76 (Ar-C), 114.55 (Ar-C), 111.99 (Ar-C), 70.25

(CH), 69.93 (CH), 55.72 (OCH3), 51.48 (CH_2), 41.32 (CH_2), 35.61 (CH_2), 31.68 (CH_2), 29.21 (CH_2), 25.77 (CH_2), 22.51 (CH_2), 19.17 (CH_3), 13.98 (CH_3) ppm.

3.4.3. Synthesis of 4-(2-hydroxy-3-(piperidin-1-yl)propyl)-2-methoxyphenol **3c**

To a suspension of 2-methoxy-4-(oxiran-2-ilmethyl)phenol **2** (0.1 g; 0.56 mmol; 1 equiv) in H_2O/EtOH 2:1 (2 mL) was added piperidine (0.047 g; 0.56 mmol), and the resulting mixture was heated at 50 °C for 5 h. The solvent was evaporated under reduced pressure to afford compound **3c** as a brown oil (0.142 g; 0.54 mmol; 97%). Rf = 0.35 (MeOH/DCM 10:90). ^{1}H-NMR δ_H ($CDCl_3$, 400 MHz): 6.83 (d, J = 8.4 Hz, 1 H, Ar-H), 6.77 (d, J = 2.0 Hz, 1H, Ar-H), 6.67 (dd, J = 8.0 Hz, 2.0 Hz, 1H, Ar-H), 4.18–4.12 (m, 1H, $CH_2CH(OH)$), 3.89 (s, 3H, OCH_3), 2.88–2.81 (m, 4H, CH_2 and CH_2NH), 2.65–2.52 (m, 4H, 2×CH_2), 1.82–1.70 (m, 4H, 2×CH_2) 1.58–1.47 (m, 4H, 2×CH_2) ppm. ^{13}C-NMR δ_C ($CDCl_3$, 100.6 MHz): 146.52 (Ar-Cq), 144.31 (Ar-Cq), 129.52 (Ar-Cq), 121.80 (Ar-C), 114.27 (Ar-C), 111.80 (Ar-C), 67.01 (CH), 63.93 (CH_2), 55.93 (OCH_3), 54.76 (CH_2), 41.30 (CH_2), 24.26 (CH_2), 23.05 (CH_2) ppm. HRMS (ESI-TOF): calcd for $C_{15}H_{24}NO_3$ [M $^+$ +H]: 266.1751, found 266.1752.

3.4.4. Synthesis of 4-(2-hydroxy-3-(phenylamino)propyl)-2-methoxyphenol **3d**

To a suspension of 2-methoxy-4-(oxiran-2-ilmethyl)phenol **2** (1 equiv) in H_2O/EtOH 2:1 (2 mL) was added aniline (0.4 mL; 3.9 equiv), and the resulting mixture was heated at 50 °C for 5.5 h. Then, water (2 mL) was added, and the resulting mixture was extracted with EtOAc (2 mL). The organic phase was collected, dried with anhydrous $MgSO_4$, and the solvent was evaporated to afford an oil (0.284 g), which was subjected to column chromatography using DCM/MeOH as eluent of increasing polarity to give compound **3d** as a yellow oil (0.095 g; 0.35 mmol; 36%). Rf = 0.7 (MeOH/DCM 5:95). ^{1}H-NMR δ_H ($CDCl_3$, 400 MHz): 7.19 (t, J = 7.2 Hz, 2 H, Ar-H), 6.88 (d, J = 8 Hz, 1H, Ar-H), 6.75 (m, 3H, Ar-H), 6.65 (d, J = 8 Hz, 2H, Ar-H), 4.09–4.02 (m, 1H, $CH_2CH(OH)$), 3.87 (s, 3H, OCH_3), 3.31 (dd, J = 12.4 Hz, 7.2 Hz, 1H, CH_2NH), 3.10 (dd, J = 12.4 and 8 Hz, 1H, CH_2NH), 2.83 (dd, J = 14 Hz and 5.2 Hz, 1H, $CH_2CH(OH)$), 2.75 (dd, J = 14 and 8 Hz, 1H, $CH_2CH(OH)$) ppm. ^{13}C-NMR δ_C ($CDCl_3$, 100.6 MHz): 147.9 (Ar-C), 146.6 (Ar-C), 144.4 (Ar-C), 129.4 (Ar-C), 129.3 (Ar-C), 122.0 (Ar-C), 118.1 (Ar-C), 114.5 (Ar-C), 113.5 (Ar-C), 111.8 (Ar-C), 71.1 (CH), 55.9 (OCH_3), 49.5 (CH_2), 41.2 (CH_2) ppm HRMS (ESI-TOF): calcd for $C_{16}H_{20}NO_3$ [M^+ +H]: 274.1438; found 274.1430.

3.4.5. Synthesis of 4-(2-hydroxy-3-((3-methoxyphenyl)amino)propyl)-2-methoxyphenol **3e**

To a suspension of 2-methoxy-4-(oxiran-2-ilmethyl)phenol **2** (0.162 g; 1 equiv) in H_2O/EtOH 2:1 (2 mL) was added 3-methoxyaniline (0.544 mg; 4.42 mmol), and the resulting mixture was heated at 50 °C for 4 h. Then, water (2 mL) was added, and the resulting mixture was extracted with DCM (2 mL). The organic phase was collected, dried with anhydrous $MgSO_4$, and the solvent was evaporated to afford an oil (0.579 g), which was subjected to column chromatography using light petroleum/EtOAc as an eluent of increasing polarity to give compound **3e** as a brown oil (0.098 g; 0.32 mmol; 36%). Rf = 0.35 (ethyl acetate/light petroleum 1:1). ^{1}H-NMR δ_H ($CDCl_3$, 400 MHz): 7.08 (t, J = 8.0 Hz, 1H, Ar-H), 6.86 (d, J = 8 Hz, 1H, Ar-H), 6.73–6.71 (m, 1H, Ar-H), 6.30 (dd, J = 8 Hz, 1.6 Hz, 1H, Ar-H), 6.24 (dd, J = 8 Hz, 2.4 Hz, 1H, Ar-H), 6.20–6.16 (m, 1H, Ar-H), 4.05–3.99 (m, 1H, $CH_2CH(OH)$), 3.84 (s, 3H, OCH_3), 3.76 (s, 3H, OCH_3), 3.27 (dd, J = 13.2 Hz, 3.6 Hz, 1H, CH_2NH), 3.05 (dd, J = 12.8 Hz, 7.6 Hz, 1H, CH_2NH), 2.83–2.70 (m, 2H, $CH_2CH(OH)$) ppm. ^{13}C-NMR δ_C ($CDCl_3$, 100.6 MHz): 160.69 (Ar-Cq), 149.42 (Ar-Cq), 146.57 (Ar-Cq), 144.27 (Ar-Cq), 129.99 (Ar-Cq), 129.91 (Ar-CH), 121.85 (Ar-CH), 114.48 (Ar-CH), 111.81 (Ar-CH), 106.35 (Ar-CH), 102.95 (Ar-CH), 102.88 (Ar-CH), 71.07 (CH), 55.76 (OCH_3), 54.96 (OCH_3), 49.32 (CH_2), 41.08 (CH_2) ppm. HRMS (ESI-TOF): calcd for $C_{14}H_{24}NO_3$ [M^+ + H]: 304.1543, found 304.1547.

3.4.6. Synthesis of 4-(2-hydroxy-3-(4-hydroxy-3methoxyphenyl)propyl)amino)benzonitrile **3f**

To a suspension of 2-methoxy-4-(oxiran-2-ilmethyl)phenol **2** (0.162 g; 0.90 mmol; 1 equiv) in H_2O/EtOH 2:1 (2 mL) was added 4-aminobenzonitrile (0.528 mg; 4.47 mmol), and the resulting mixture was heated at 50 °C for 37 h. Then, water (2 mL) was added, and the resulting mixture was extracted with EtOAc (2 mL), The organic phase was collected, dried with anhydrous MgSO4, and the solvent was evaporated to afford an oil (0.265 g), which was subjected to column chromatography using DCM/MeOH as an eluent of increasing polarity to give compound **3f** as a dark yellow oil (0.025 g; 0.08 mmol; 9%). Rf = 0.45 (MeOH/DCM 5:95). 1H-NMR δ_H ($CDCl_3$, 400 MHz): 7.39 (d, J = 8.8 Hz, 2H, Ar-H), 6.87 (d, J = 8.4 Hz, 1H, Ar-H), 6.72–6.69 (m, 2H, Ar-H), 6.58–6.50 (m, 2H, Ar-H), 4.07–4.01 (m, 1H, $CH_2CH(OH)$), 3.85 (s, 3H, OCH_3), 3.31 (dd, J = 12.8 Hz, 7.2 Hz, 1H, CH_2NHPhe), 3.12 (dd, J = 12.8 Hz and 7.4 Hz, 1H, CH_2NHPhe), 2.82 (dd, J = 13.6 Hz, 5.2 Hz, 1H, $CH_2CH(OH)$), 2.73 (dd, J = 13.6 Hz and 8 Hz, 1H, $CH_2CH(OH)$) ppm. ^{13}C-NMR δ_C ($CDCl_3$, 100.6 MHz): 151.20 (Ar-Cq), 146.68 (Ar-Cq), 144.53 (Ar-Cq), 136.33 (Ar-Cq), 133.65 (Ar-CH), 128.83 (Ar-Cq), 121.88 (Ar-CH), 121.33 (Ar-CH), 114.62 (Ar-CH), 112.58 (Ar-CH), 111.72 (Ar-CH), 98.91 (Ar-CH), 70.97 (Ar-CH), 55.87 (OCH_3), 48.10 (CH_2), 41.20 (CH_2) ppm.

3.5. *Cell Culture*

Insect cells (*Sf9*, *Spodoptera frugiperda*) cells were maintained as a suspension culture and cultivated in Grace's medium with 10% FBS and 1% penicillin/streptomycin, at 28 °C with agitation. Cells were used in experiments while in the exponential phase of growth. On the other hand, HaCaT (human keratinocytes) cells were culture in DMEM supplemented with 10% FBS and 1% penicillin/streptomycin at 37 °C in a humidified atmosphere of 5% CO_2.

3.6. *Viability Assessment*

For the assessment of viability, a resazurin-based method was used. The *Sf9* and HaCaT cells were plated at a density of 3.0×10^4 and 1.5×10^4 cells/well, respectively, incubated for 24 h, and then exposed to the molecules under study (at 100 μg/mL in Grace's medium) for 24 h. After this period, a commercial solution of resazurin was added (1:10), and the kinetic reaction of fluorescence increase was monitored at 560/590 nm. For HaCaT and the *Sf9* cells, 30 and 60 min of incubation were used, respectively.

3.7. *LDH Assay*

The Sf9 cells were cultured at the same density described above for the viability assessment. To assess the release of the stable cytosolic enzyme lactate dehydrogenase (LDH) into the media, 24 h after the incubation of cells with the molecules under study (at 100 μg/mL in Grace's medium), 50 μL of culture media were removed to a 96-well plate. The LDH released was determined using a CytoTox 96® assay kit (Promega; Madison, WI, USA) according to the manufacturer's instructions. A lysis solution (LS) was used as a positive control to generate a maximum LDH release (45 min). Absorbances were measured at 490 nm in a Multiskan GO plate reader (Thermo Fisher Scientific; Waltham, MA, USA), and results correspond to the fold increase of absorbance in treated vs. untreated cells of four independent experiments performed in duplicate.

3.8. *Caspase-Like Activity*

The *Sf9* cells were plated at the same density described for viability studies and exposed to the molecules under study for the designated time. Generally, the same method described before by some of us was used [37]; however, it has been adapted toward insect cells, as previously reported [21,22]. After the incubation period, caspase-3/7 substrate was added to wells, and cells were incubated for 20 min at 22 °C. The luminescent signal was measured in a microplate reader (Cytation™ 3, BioTek, Winooski, VT, USA), and three independent experiments were performed in duplicate. Then, to normalize the results, DNA quantification was performed in a triplicate pool using a Qubit™ 1X dsDNA HS Assay Kit according to a previously described procedure [38] and manufacturer's instructions.

3.9. Statistical Analysis

For biological assays, the Shapiro–Wilks normality test was performed in the data to ensure that it followed a normal distribution. Comparison between the means of controls and each experimental condition was performed using one-way ANOVA. Outliers were identified by the Grubbs' test. Data were expressed as the mean ± standard deviation (SD) of at least three independent experiments. GraphPad Prism 7.0 software was used, and values were considered statistically significant with a $p < 0.05$.

3.10. Molecular Docking and Inverted Virtual Screening Studies

To identify possible molecular targets of the amino alcohols derived from eugenol, an inverted virtual screening protocol was applied. A search on Scopus was performed for papers describing virtual screening (VS) studies involving targets and molecules with insecticidal activity using the keywords: "virtual screening" and "pesticides". The selection criteria were relevance of the target and year of publication. In the 18 studies found, 23 PDB structures were identified and downloaded, enabling the creation of a structural database of putative insecticide targets. These are listed in Table S3.

The 23 PBD structures were prepared for docking starting using the Autodock Vina plugin for Pymol [39]. Crystallographic waters were removed. Then, the crystallographic ligands were saved in separate files and used as reference for active site coordinates as well as for validation in the re-docking steps. In the absence of ligands, the active site coordinates were based on the most important residues described in the literature. Re-docking was used to evaluate the ability of the docking software to reproduce the geometry and orientation of the crystallographic pose, as well as the quality of the docking protocol, and to optimize the docking protocol.

The docking programs/scoring functions used were GOLD [40] (PLP, ASP, ChemScore, and GoldScore scoring functions), and AutoDock Vina [41]. With each docking program/scoring function, the protocol was optimized for each protein target, to minimize the RMSD in the docking predictions of the reference ligand in re-docking, by comparison with the crystallographic structure of the corresponding complex.

The optimized parameters for each program/scoring function were as follows: Vina-docking box position, docking box dimension, exhaustiveness; GOLD (PLP, ASP, ChemScore, GoldScore)-binding pocket center, docking region radius, search efficiency, number of runs. The final optimized conditions were used for the subsequent stages. Structures for the two eugenol amino alcohol derivatives with the highest insecticide activity were prepared for docking using Datawarrior [42] and OpenBabel [43] and were docked into each structure with the optimized protocol with all the five scoring functions. A ranked list of most likely targets was prepared based on the average scores obtained for each target with the different scoring functions.

3.11. Molecular Dynamics Simulations and Free Energy Calculations

Molecular dynamics simulations were performed using the Amber18 software (University of California, San Francisco, USA) for the two compounds identified from the experimental studies to have the highest insecticide activity (compounds **3d** and **3f**), which are bound to the two most promising targets identified from the inverted virtual screening study (odorant binding protein 1–3KIE and acetylcholinesterase-1QON). Since 1QON presented a gap in the structure, a homology model was generated using SWISS-MODEL [44]. A total of 1466 templates were found to match the original sequence, but only the top 50 were used to build the model (Figure S5 in Supplementary Information).

Models for the MD simulations were prepared starting from the pose predicted for these complexes in the docking experiments during the inverted virtual screening protocol with GOLD/PLP and treated with the Leap module of AMBER [45]. The protein targets were described with the ff14SB force field [46], while the eugenol derivatives were parameterized using ANTECHAMBER, with RESP HF/6-31G (d) charges calculated with Gaussian16 [47] and the General Amber Force Field (GAFF) [48]. The overall charge on the system was neutralized through the addition of counter-ions (Na^+ or Cl^-), and the systems were placed in TIP3P water boxes with a minimum distance of 12 Å between the protein surface and the side of the box.

In order to remove the clashes, the systems were submitted to four consecutive minimizations stages, which were followed by an equilibration and production. In the first four minimization stages, the procedure was applied to (1) water molecules (2500 steps); (2) hydrogens atoms (2500 steps); (3) side chains of all the amino acid residues (2500 steps); and (4) the full system (10,000 steps). After the complete minimization, the systems were equilibrated by a procedure, which was divided into two stages: in the first stage, NVT ensemble, the systems were gradually heated to 298 K using a Langevin thermostat at constant volume (50 ps); in the second stage, the density of the systems was further equilibrated at 298 K (subsequent 50 ps). Finally, the productions runs were performed during 100 ns. Production was executed with an NPT ensemble at constant temperature (298 K, Langevin thermostat) and pressure (1 bar, Berendsen barostat), with periodic boundary conditions. An integration time of 2.0 fs using the SHAKE algorithm was used to constrain all covalent bonds involving hydrogen atoms. The nonbonded interactions were cut off at 10 Å throughout the entire molecular simulation procedure. The final trajectories were analyzed in terms of RMSD to obtain confirmation that both systems were well equilibrated after the initial 10 ns. The last 90 ns of the simulation were considered for hydrogen bonding analysis, and cluster analysis of the conformations was generated. This overall procedure has been previously used with success in the treatment of several biomolecular systems [49–57].

The Molecular Mechanics-Generalized Born Surface Area (MM-GBSA) method [27] was applied to estimate the binding free energies of compounds **3d** and **3f** to the odorant binding protein 1 and to acetylcholinesterase, considering a salt concentration of 0.100 $mol.dm^{-3}$. In addition, the energy decomposition method was employed to estimate the contribution of all the amino acid residues for each of these binding free energies. From each MD trajectory, a total of 1400 conformations taken from the last 70 ns of simulation were considered for the MM-GBSA calculations.

3.12. Nanoencapsulation Studies

The most active compound against the *Sf9* cells, compound **3f**, was encapsulated in liposomes composed of the phospholipids 1,2-dimyristoyl-*sn*-glycero-3-phospho- (1′-*rac*-glycerol) (sodium salt) (DMPG) and 1,2-dipalmitoyl-*sn*-glycero-3-phosphocholine (DPPC), which are either composed of DMPG (100%) or DPPC/DMPG (1:1). The liposomes (2 mM total lipid concentration) were prepared by the ethanolic injection method [58] above the transition temperature of each lipid, as previously described [21]. The compound (at an initial concentration of 6.7×10^{-4} M) was encapsulated by co-injection with the ethanolic lipid solution. The size, polydispersity, and zeta potential of compound-loaded liposomes were measured in a Litesizer 500 Dynamic Light Scattering apparatus from Anton Paar (Anton Paar GmbH, Graz, Austria).

The encapsulation efficiency (EE%) was determined as previously reported [21] and calculated through the Equation (1):

$$EE\% = \frac{(C_{total\ compound} - C_{free\ compound})}{C_{total\ compound}} \times 100 \quad (1)$$

using a calibration curve of fluorescence intensity vs. concentration (C), taking advantage of the fluorescence emission of the compound.

The compound release was followed during 24 h at 20 °C and 35 °C, below and above the phase transition temperature of DMPG. The Weibull model (a distribution function) was used to study the transport mechanism involved in the compound release [36], being expressed in terms of the compound fraction accumulated (m) in solution at time t (Equation (2)):

$$m = 1 - \exp\left[\frac{-(t-T_i)^b}{a}\right] \quad (2)$$

where a is a scale parameter that defines the timescale of the process, T_i represents the latency time of the release process (often being zero), and b is a formal parameter that characterizes the type of curve ($b = 1$ is exponential; $b > 1$ is sigmoid, with ascendant curvature delimited by an inflection point; and $b < 1$ is parabolic, displaying high initial slope and a consistent exponential character).

4. Conclusions

A series of β-amino alcohols were prepared by reaction of eugenol epoxide with various aliphatic and aromatic amines. The obtained eugenol derivatives were subjected to biological activity evaluation in the *Sf9* cell line, in comparison with the corresponding precursors, in order to evaluate their application as potential natural based insecticides.

We identified that the three derivatives bearing a terminal benzene ring, either substituted or unsubstituted, were those showing higher potency, in some cases higher than the benchmark used. We further clarified that the molecules were eliciting their effect by triggering organized cell death, and they were selective for insect cells.

Inverted virtual screening studies with five independent methods suggest that these molecules display their insecticide activity most likely by targeting the insect acetylcholinesterase and/or the insect odorant binding proteins. Molecular dynamics simulations and free energy calculations confirm that these two molecules bind strongly to both targets forming very stable complexes with well-defined molecular interactions that are maintained through time.

Nanoencapsulation studies allow obtaining very reasonable encapsulation efficiencies and a controlled release. Liposomes of DMPG provide a temperature-sensitive compound release, which is more effective than the DPPC/DMPG (1:1) formulation.

Supplementary Materials: The following are available online, Figure S1: Protein and ligand RMSD (Å) of the AChE and OBP–ligand complexes, Figure S2: Percentage of the potential solvent accessible surface area of the ligands that is buried by the protein targets evaluated, Figure S3: Number of ligand–target hydrogen bonds formed during the simulations for compound **3d** and **3f** when complexed with AchE and OBP, Figure S4: Fitting of the release profiles to the Weibull model, Figure S5: Homology model built for 1QON, Table S1: Docking scores for compounds **3d** and **3f** in complex with human and insect AChE, Table S2: Parameters of the Weibull model for the release of compound **3f** from liposomes and corresponding coefficients of determination (R^2),Table S3: List of targets selected for the inverted virtual screening study.

Author Contributions: Conceptualization, M.S.T.G.; A.G.F.; D.M.P.; S.F.S. and E.M.S.C.; methodology, A.G.F.; M.S.T.G.; D.M.P.; S.F.S. and E.M.S.C.; validation, A.G.F.; M.S.T.G.; D.M.P.; S.F.S. and E.M.S.C.; investigation, N.F.S.P.; R.B.P.; T.F.V.; M.J.G.F. and A.R.O.R.; writing—original draft preparation, A.G.F.; M.S.T.G.; D.M.P.; S.F.S.; E.M.S.C. and R.B.P.; writing—review and editing, M.S.T.G.; A.G.F.; D.M.P.; S.F.S.; E.M.S.C. and R.B.P.; supervision, A.G.F.; M.S.T.G.; D.M.P.; S.F.S. and E.M.S.C.; project administration, M.S.T.G. All authors have read and agreed to the published version of the manuscript.

Funding: This research was funded by project PTDC/ASP-AGR/30154/2017 (POCI-01-0145-FEDER-030154) of the COMPETE 2020 program, co-financed by the FEDER and the European Union. The authors acknowledge also the Foundation for Science and Technology (FCT, Portugal) and FEDER-COMPETE-QREN-EU for financial support to the research centers CQ-UM (UID/QUI/00686/2020), CF-UM-UP (UIDB/04650/2020) and REQUIMTE (UIDB/50006/2020). The NMR spectrometer Bruker Avance III 400 is part of the National NMR Network and was purchased within the framework of the National Program for Scientific Re-equipment, contract REDE/1517/RMN/2005 with funds from POCI 2010 (FEDER) and FCT.

Institutional Review Board Statement: Not applicable.

Informed Consent Statement: Not applicable.

Data Availability Statement: Not applicable.

Conflicts of Interest: The authors declare no conflict of interest.

Sample Availability: Not applicable.

References

1. Pavela, R.; Maggi, F.; Iannarelli, R.; Benelli, G. Plant extracts for developing mosquito larvicides: From laboratory to the field, with insights on the modes of action. *Acta Trop.* **2019**, *193*, 236–271. [CrossRef]
2. Salman, M.; Abbas, R.Z.; Israr, M.; Abbas, A.; Mehmood, K.; Khan, M.K.; Sindhu, Z.D.; Hussaind, R.; Saleemie, M.K.; Shaha, S. Repellent and acaricidal activity of essential oils and their components against Rhipicephalus ticks in cattle. *Vet. Parasitol.* **2020**, *283*, 109178. [CrossRef] [PubMed]
3. Mossa, A.-T.H. Green pesticides: Essential oils as biopesticides in insect-pest management. *J. Environ. Sci. Technol.* **2016**, *9*, 354–378. [CrossRef]
4. Tripathi, A.K.; Upadhyay, S.; Bhuiyan, M.; Bhattacharya, P.R. A review on prospects of essential oils as biopesticide in insect-pest management. *J. Pharm. Phytother.* **2009**, *1*, 52–63.
5. Cortés-Rojas, D.F.; Fernandes de Souza, C.R.; Pereira Oliveira, W. Clove (Syzygium aromaticum): A precious spice. *Asian Pac. J. Trop. Biomed.* **2014**, *4*, 90–96. [CrossRef]
6. Evambe, G.; Canales, L.; Banik, B.K. Antimicrobial activity of eugenol derivatives. *Heterocycl. Lett.* **2011**, *1*, 2231–3087.
7. Sun, W.-J.; Lv, W.-J.; Li, L.-N. Eugenol confers resistance to tomato yellow leaf curl virus (TYLCV) by regulating the expression of SlPer1 in tomato plants. *New Biotechnol.* **2016**, *33*, 345–354. [CrossRef]
8. Fonseca, D.V.; Salgado, P.R.R.; Neto, H.C.A.; Golzio, A.M.F.O.; Filho, M.R.D.C.; Melo, C.G.F.; Leite, F.C.; Piuvezam, M.R.; Pordeus, L.C.M.; Filho, J.M.B. Ortho-eugenol exhibits anti-nociceptive and anti-inflammatory activities. *Int. Immunopharmacol.* **2016**, *38*, 402–408. [CrossRef]
9. Awasthi, P.K.; Dixit, S.C.; Dixit, N.; Sinha, A.K. Eugenol derivatives as future potential drugs. *Drugs* **2008**, *1*, 215–220.
10. Enan, E.E. Molecular and pharmacological analysis of an octopamine receptor from American cockroach and fruit fly in response to plant essential oils. *Arch. Insect Biochem. Physiol.* **2005**, *59*, 161–171. [CrossRef]
11. Silva, F.F.M.; Monte, F.J.Q.; Lemos, T.L.G.; Nascimento, P.G.G.; Costa, A.K.M.; Paiva, L.M.M. Eugenol derivatives: Synthesis, characterization, and evaluation of antibacterial and antioxidant activities. *Chem. Cent. J.* **2018**, *12*, 34. [CrossRef]
12. Corey, E.J.; Zhang, F.-Y. re- and si-Face-selective nitroaldol reactions catalyzed by a rigid chiral quaternary ammonium salt: A highly stereoselective synthesis of the HIV protease inhibitor amprenavir (Vertex 478). *Angew. Chem. Int. Ed.* **1999**, *38*, 1931–1934. [CrossRef]
13. Ager, D.J.; Prakash, I.; Schaad, D.R. 1,2-Amino alcohols and their heterocyclic derivatives as chiral auxiliaries in asymmetric synthesis. *Chem. Rev.* **1996**, *96*, 835–876. [CrossRef]
14. O'Brien, P. Sharpless asymmetric aminohydroxylation: Scope, limitations, and use in synthesis. *Angew. Chem. Int. Ed. Engl.* **1999**, *38*, 326–329. [CrossRef]
15. Li, G.; Chang, H.-T.; Sharpless, K.B. Catalytic asymmetric aminohydroxylation (AA) of olefins. *Angew. Chem. Int. Ed. Engl.* **1996**, *35*, 451–454. [CrossRef]
16. Manda, B.R.; Prasad, A.N.; Thatikonda, N.R.; Lacerda, V., Jr.; Barbosa, L.R.; Santos, H.; Romão, W.; Pavan, F.R.; Ribeiro, C.M.; dos Santos, E.A.; et al. Synthesis, antibacterial and antitubercular evaluation of cardanol and glycerol based β-amino alcohol derivatives. *J. Braz. Chem. Soc.* **2018**, *29*, 639–648. [CrossRef]
17. Shi, C.; Ren, C.; Zhang, E.; Jin, H.; Yu, X.; Wang, S. Synthesis of β-amino alcohols using the tandem reduction and ring-opening reaction of nitroarenes and epoxides. *Tetrahedron* **2016**, *72*, 3839–3843. [CrossRef]
18. Ullmann, N.; Caggiano, S.; Cutrera, R. Salbutamol and around. *Ital. J. Pediatr.* **2015**, *41*, A74. [CrossRef]
19. Ang, W.; Ye, W.; Sang, Z.; Liu, Y.; Yang, T.; Deng, Y.; Luo, Y.; Wei, Y. Discovery of novel bis-oxazolidinone compounds as potential potent and selective antitubercular agents. *Bioorg. Med. Chem. Lett.* **2014**, *24*, 1496–1501. [CrossRef]
20. Cioc, R.C.; van der Niet, D.J.H.; Janssen, E.; Dr Ruijter, E.; Prof Dr Orru, R.V.A. One-pot synthesis of N-substituted β-amino alcohols from aldehydes and isocyanides. *Chem. Eur. J.* **2015**, *21*, 7808–7813. [CrossRef]
21. Lopes, A.I.F.; Monteiro, M.; Araújo, A.R.L.; Rodrigues, A.R.O.; Castanheira, E.M.S.; Pereira, D.M.; Olim, P.; Fortes, A.G.; Gonçalves, M.S.T. Cytotoxic plant extracts towards insect cells: Bioactivity and nanoencapsulation studies for application as biopesticides. *Molecules* **2020**, *25*, 5855. [CrossRef]
22. Fernandes, M.J.G.; Pereira, R.B.; Pereira, D.M.; Fortes, A.G.; Castanheira, E.M.S.; Gonçalves, M.S.T. New eugenol derivatives with enhanced insecticidal activity. *Int. J. Mol. Sci.* **2020**, *21*, 9257. [CrossRef]
23. Silvius, J.R. Thermotropic phase transitions of pure lipids in model membranes and their modifications by membrane proteins. *Lipid-Protein Interact.* **1982**, *2*, 239–281.
24. Modjinou, T.; Versace, D.-L.; Abbad-Andaloussi, S.; Langlois, V.; Renard, E. Antibacterial and antioxidant photoinitiated epoxy co-networks of resorcinol and eugenol derivatives. *Mater. Today Commun.* **2017**, *12*, 19–28. [CrossRef]
25. Azizi, N.; Saidi, M.R. Highly chemoselective addition of amines to epoxides in water. *Org. Lett.* **2015**, *7*, 3649–3651. [CrossRef]

26. Fink, S.L.; Cookson, B.T. Apoptosis, pyroptosis, and necrosis: Mechanistic description of dead and dying eukaryotic cells. *Infect. Immun.* **2005**, *73*, 1907–1916. [CrossRef] [PubMed]
27. Lang, G.-J.; Yan, Z.K.; Zhang, C.-X. Can acetylcholinesterase serve as a target for developing more selective insecticides? *Curr. Drug Targets.* **2012**, *13*, 495–501. [CrossRef] [PubMed]
28. Spinelli, S.; Lagarde, A.; Iovinella, I.; Legrand, P.; Tegoni, M.; Pelosi, P.; Cambillau, C. Crystal structure of Apis mellifera OBP14, a C-minus odorant-binding protein, and its complexes with odorant molecules. *Insect Biochem. Mol. Bio.* **2012**, *42*, 41–50. [CrossRef]
29. González-González, A.; Palma-Millanao, R.; Yáñez, O.; Rojas, M.; Mutis, A.; Venthur, H.; Quiroz, A.; Ramírez, C.C. Virtual screening of plant volatile compounds reveals a high affinity of Hylamorpha elegans (Coleoptera: Scarabaeidae) odorant-binding proteins for sesquiterpenes from Its native host. *J. Insect Sci.* **2016**, *16*, 30. [CrossRef] [PubMed]
30. Sun, J.S.; Xiao, S.; Carlson, J.R. The diverse small proteins called odorant-binding proteins. *Open Biol.* **2018**, *8*, 180208. [CrossRef]
31. Pelosi, P.; Iovinella, I.; Zhu, J.; Wang, G.; Dani, F.R. Beyond chemoreception: Diverse tasks of soluble olfactory proteins in insects. *Biol. Rev.* **2018**, *93*, 184–200. [CrossRef]
32. Pelosi, P.; Mastrogiacomo, R.; Iovinella, I.; Tuccori, E.; Persaud, K.C. Structure and biotechnological applications of odorant-binding proteins. *Appl. Microbiol. Biotechnol.* **2014**, *98*, 61–70. [CrossRef] [PubMed]
33. Raval, N.; Maheshwari, R.; Kalyane, D.; Youngren-Ortiz, S.R.; Chougule, M.B.; Tekade, R.K. Importance of physicochemical characterization of nanoparticles in pharmaceutical product development. In *Advances in Pharmaceutical Product Development and Research: Basic Fundamentals of Drug Delivery*; Chapter 10; Academic Press: Cambridge, MA, USA, 2019; pp. 369–400.
34. Barroso, R.P.; Perez, K.R.; Cuccovia, I.M.; Lamy, M.T. Aqueous dispersions of DMPG in low salt contain leaky vesicles. *Chem. Phys. Lipids* **2012**, *165*, 169–177. [CrossRef] [PubMed]
35. Danaei, M.; Dehghankhold, M.; Ataei, S.; Davarani, F.H.; Javanmard, R.; Dokhani, A.; Khorasani, S.; Mozafari, M.R. Impact of particle size and polydispersity index on the clinical applications of lipidic nanocarrier systems. *Pharmaceutics* **2018**, *10*, 57. [CrossRef]
36. Papadopoulou, V.; Kosmidis, K.; Vlachou, M.; Macheras, P. On the use of the Weibull function for the discernment of drug release mechanisms. *Int. J. Pharm.* **2006**, *309*, 44–50. [CrossRef] [PubMed]
37. Chouiter, M.I.; Boulebd, H.; Pereira, D.M.; Valentão, P.; Andrade, P.B.; Belfaitah, A.; Silva, A.M. New chalcone-type compounds and 2-pyrazoline derivatives: Synthesis and caspase-dependent anticancer activity. *Future Med. Chem.* **2020**, *12*, 493–509. [CrossRef]
38. Amorim, C.; Veloso, S.R.S.; Castanheira, E.M.S.; Hilliou, L.; Pereira, R.B.; Pereira, D.M.; Martins, J.Á.; Jervis, P.J.; Ferreira, P.M.T. Bolaamphiphilic bis-dehydropeptide hydrogels as potential drug release systems. *Gels* **2021**, *7*, 52. [CrossRef]
39. Seeliger, D.; de Groot, B.L. Ligand docking and binding site analysis with PyMOL and Autodock/Vina. *J. Comput. Aided Mol. Des.* **2010**, *24*, 417–422. [CrossRef]
40. Jones, G.; Willett, P.; Glen, R.C.; Leach, A.R.; Taylor, R. Development and validation of a genetic algorithm for flexible docking. *J. Mol. Biol.* **1997**, *267*, 727–748. [CrossRef]
41. Trott, O.; Olson, A.J. AutoDock Vina: Improving the speed and accuracy of docking with a new scoring function, efficient optimization, and multithreading. *J. Comput. Chem.* **2009**, *31*, 455–461. [CrossRef]
42. Sander, T.; Freyss, J.; von Korff, M.; Rufener, C. DataWarrior: An open-source program for chemistry aware data visualization and analysis. *J. Chem. Inf. Model.* **2015**, *55*, 460–473. [CrossRef]
43. O'Boyle, N.M.; Banck, M.; James, C.A.; Morley, C.; Vandermeersch, T.; Hutchison, G.R. Open Babel: An open chemical toolbox. *J. Cheminform.* **2011**, *3*, 33. [CrossRef]
44. Waterhouse, A.; Bertoni, M.; Bienert, S.; Studer, G.; Tauriello, G.; Gumienny, R.; Heer, F.T.; de Beer, T.A.P.; Rempfer, C.; Bordoli, L.; et al. SWISS-MODEL: Homology modelling of protein structures and complexes. *Nucleic Acids Res.* **2018**, *46*, W296–W303. [CrossRef]
45. Wang, J.; Wolf, R.M.; Caldwell, J.W.; Kollman, P.A.; Case, D.A. Development and testing of a general amber force field. *J. Comp. Chem.* **2004**, *25*, 1157–1174. [CrossRef]
46. Maier, J.A.; Martinez, C.; Kasavajhala, K.; Wickstrom, L.; Hauser, K.E.; Simmerling, C. ff14SB: Improving the accuracy of protein side chain and backbone parameters from ff99SB. *J. Chem. Theory Comput.* **2015**, *11*, 3696–3713. [CrossRef]
47. Frisch, M.J.; Trucks, G.W.; Schlegel, H.B.; Scuseria, G.E.; Robb, A.; Cheeseman, J.R.; Scalmani, G.; Barone, V.; Mennucci, B.; Petersson, G.A.; et al. *Gaussian 09, RevisionD.02*; Gaussian Inc.: Wallingford, CT, USA, 2016.
48. Wang, J.; Wang, W.; Kollman, P.A.; Case, D.A. Automatic atom type and bond type perception in molecular mechanical calculations. *J. Mol. Graph. Mode.* **2006**, *25*, 247–260. [CrossRef] [PubMed]
49. Lapaillerie, D.; Charlier, C.; Fernandes, H.S.; Sousa, S.F.; Lesbats, P.; Weigel, P.; Favereaux, A.; Guyonnet-Duperat, V.; Parissi, V. In silico, in vitro and in cellulo models for monitoring SARS-CoV-2 Spike/Human ACE2 complex, viral entry and cell fusion. *Viruses* **2021**, *3*, 365. [CrossRef]
50. Martins, F.G.; Melo, A.; Sousa, S.F. Identification of new potential inhibitors of quorum sensing through a specialized multi-level computational approach. *Molecules* **2021**, *26*, 2600. [CrossRef]
51. Moorthy, N.S.H.N.; Sousa, S.F.; Ramos, M.J.; Fernandes, P.A. Molecular dynamic simulations and structure-based pharmacophore development for farnesyltransferase inhibitors discovery. *J. Enzym. Inhib. Med. Chem.* **2016**, *31*, 1428–1442. [CrossRef] [PubMed]
52. Sousa, S.F.; Cerqueira, N.M.F.S.A.; Brás, N.; Fernandes, P.A.; Ramos, M.J. Enzymatic "tricks": Carboxylate shift and sulfur shift. *Int. J. Quantum Chem.* **2014**, *114*, 1253–1256. [CrossRef]

53. Sousa, S.F.; Coimbra, J.T.; Paramos, D.; Pinto, R.; Guimarães, R.S.; Teixeira, V.; Fernandes, P.A.; Ramos, M.J. Molecular dynamics analysis of a series of 22 potential farnesyltransferase substrates containing a CaaX-motif. *J. Mol. Model.* **2013**, *19*, 673–688. [CrossRef] [PubMed]
54. Marcial, B.L.; Sousa, S.F.; Barbosa, I.L.; Dos Santos, H.F.; Ramos, M.J. Chemically modified tetracyclines as inhibitors of MMP-2 matrix metalloproteinase: A molecular and structural study. *J. Phys. Chem. B* **2012**, *116*, 13644–13654. [CrossRef] [PubMed]
55. Sousa, S.F.; Fernandes, P.A.; Ramos, M.J. Molecular dynamics simulations on the critical states of the farnesyltransferase enzyme. *Bioorg. Med. Chem.* **2009**, *17*, 3369–3378. [CrossRef] [PubMed]
56. Sousa, S.F.; Fernandes, P.A.; Ramos, M.J. Enzyme flexibility and the catalytic mechanism of farnesyltransferase: Targeting the relation. *J. Phys. Chem. B* **2008**, *112*, 8681–8691. [CrossRef]
57. Wang, E.; Sun, H.; Wang, J.; Wang, Z.; Liu, H.; Zhang, J.Z.H.; Tingjun, H.T. End-point binding free energy calculation with MM/PBSA and MM/GBSA: Strategies and applications in drug design. *Chem. Rev.* **2019**, *119*, 9478–9508. [CrossRef] [PubMed]
58. Jaafar-Maalej, C.; Diab, R.; Andrieu, V.; Elaissari, A.; Fessi, H. Ethanol injection method for hydrophilic and lipophilic drug-loaded liposome preparation. *J. Liposome Res.* **2010**, *20*, 228–243. [CrossRef]

molecules MDPI

Article

Toxicological Stability of *Ocimum basilicum* Essential Oil and Its Major Components in the Control of *Sitophilus zeamais*

Eridiane da Silva Moura [1], Lêda Rita D'Antonino Faroni [1,*], Fernanda Fernandes Heleno [1,2] and Alessandra Aparecida Zinato Rodrigues [1,3]

1 Department of Agricultural Engineering, Universidade Federal de Viçosa, Viçosa 36570-900, MG, Brazil; annne.moura@gmail.com (E.d.S.M.); fernandafhy@yahoo.com.br (F.F.H.); alessandra.rodrigues@ufv.br (A.A.Z.R.)

2 Serviço Autônomo de Água e Esgoto, Senador Firmino 36540-000, MG, Brazil

3 Department of Chemistry, Universidade Federal de Viçosa, Viçosa 36570-900, MG, Brazil

* Correspondence: lfaroni@gmail.com; Tel.: +55-31-3612-4022; Fax: +55-31-3899-2732

Abstract: Essential oils (EOs) are widely recognized as efficient and safe alternatives for controlling pest insects in foods. However, there is a lack of studies evaluating the toxicological stability of botanical insecticides in stored grains in order to establish criteria of use and ensure your efficiency. The objective of this work was to evaluate the toxicological stability of basil essential oil (*O. basilicum*) and its linalool and estragole components for *Sitophilus zeamais* (Motschulsky) adults in corn grains by fumigation. The identification of the chemical compounds of the essential oil was performed with a gas chromatograph coupled to a mass selective detector. Mortality of insects was assessed after 24 h exposure. After storage for six (EO) and two months (linalool and estragole) under different conditions of temperature (5, 20, and 35 °C) and light (with and without exposure to light), its toxicological stability was evaluated. Studies revealed that the essential oil of *O. basilicum* and its main components exhibited insecticidal potential against adults of *S. zeamais*. For greater toxicological stability, suitable storage conditions for them include absence of light and temperatures equal to or less than 20 °C.

Keywords: storage; monoterpenes; bioinsecticide; insect pest; toxicity

Citation: Moura, E.d.S.; Faroni, L.R.D.; Heleno, F.F.; Rodrigues, A.A.Z. Toxicological Stability of *Ocimum basilicum* Essential Oil and Its Major Components in the Control of *Sitophilus zeamais*. *Molecules* **2021**, *26*, 6483. https://doi.org/10.3390/molecules26216483

Academic Editor: Giovanni Benelli

Received: 29 September 2021
Accepted: 25 October 2021
Published: 27 October 2021

1. Introduction

Essential oils (EOs) are classified as secondary metabolites produced by various parts of the plant such as seeds, stems, leaves, and flowers. They are mixtures of volatile, natural substances, characterized by strong odor and, in most cases, have lipophilic constitution [1]. As they are composed of volatile terpenoids such as monoterpenes (C_{10}) and sesquiterpenes (C_{15}) and phenylpropenes (derived from the phenyl group junction (aromatic ring) and a three-carbon side chain (propyl group) [2], which usually originate from various biosynthesis pathways [3], there are a wide variety of possible applications of essential oils [4]. Among the current applications of EOs is their use as an alternative to synthetic insecticides, as EOs have great biocidal potential, presenting insect toxicity [5].

EOs and their compounds are believed to have a higher barrier to pest resistance and lower risk to human health and environmental contamination compared to conventional insecticides [6]. Among the essential oils with insecticidal activity is the essential oil of *Ocimum basilicum*, aromatic and medicinal plant of the Lamiaceae family [7], composed mainly of linalool and estragole [8,9].

The toxicity of *O. basilcium* essential oil has already been proven for *Acanthoscelides obtectus* (Coleoptera: Chrysomelidae) [10], *Rhyzopertha dominica* (Coleoptera: Bostrichidae) [11], *Sitophilus zeamais* (Coleoptera: Curculionidae) [12], *Tribolium castaneum* (Coleoptera: Tenebrionidae) [13], *Zabrotes subfasciatus* (Coleoptera: Chrysomelidae) [14], *Anopheles funestus* (Diptera: Culicidae) [15] and *Sitophilus Oryzae* (Coleoptera: Curculionidae) [16]. Because this essential oil is mainly composed of linalool and estragole, its toxicity to stored grain pest insects can be explained by the action of mixed inhibition of the enzyme acetylcholinesterase (AChE) caused by such compounds, especially when they are applied by fumigation [17,18].

Linalool, the most well-known monoterpene of *O. basilicum* essential oil, is present in the essential oil of various medicinal plants, mainly of the Lamiaceae family [19]. It is insect repellent [20], inhibits the reproduction of *Acanthoscelides obtectus* (Say) (Coleoptera: Chrysomelidae) [21] and exhibits larvicidal activity against *Culex quinquefasciatus* and *Aedes stephensi* larvae (Diptera: Culicidae) [22,23].

Estragole, a volatile monoterpenoid ether found in numerous plants [24]. This component and its biotransformation products have toxic potential because they are genotoxic, mutagenic or carcinogenic [25]. Nevertheless, it was considered safe (GRAS—Generally Recognized As Safe) by FEMA (Flavor and Extract Manufacturer's Association, 2008) as it does not pose a risk to human health in small quantities (0.6 mg kg^{-1} day^{-1}). Estragole inhibits the growth of *Aedes aegipyti* (Diptera: Culicidae) larvae and has antiparasitic and antihelmintic actions [26]. This compound has reported potential insecticide for *Oryzaephilus surinamensis* (Coleoptera: Silvanidae), *Lasioderma serricorne* (Coleoptera: Anobiidae), *Liposcelis bostrychophila* (Psocoptera: Liposcelididae) and *Tribolium castaneum* (Coleoptera: Tenebrionidae) [27–31].

Given that *O. basilicum* EO and its major components linalool and estragole are toxic to stored grain pest insects and knowing that once deprived of the protective compartmentalization of the plant, essential oil constituents are especially prone to oxidative damage, chemical transformations, or polymerization through enzymatic or chemically triggered reactions by external factors such as temperature and light [32]. the objective of this work was to determine the toxicological stability of *O. basilicum* EO and its linalool and estragole components on fumigation *Sitophilus zeamais* in corn grains, after storage as a function of temperature and luminosity for a period of six (EO) and two months (linalool and estragol).

2. Material and Methods

2.1. Insect Colony

The insects were raised on maize grains with water content of 12.1% (wet basis) under constant conditions of temperature (25 ± 2 °C), relative humidity (70 ± 5%) and scotophase 24 h. For the creation were used 3 L glass vials, closed with perforated plastic lid and internally coated with organza to allow gas exchange.

2.2. Essential Oil

The essential oil used in the research was acquired through the company Mundo dos Óleos (Brasília, DF, Brazil). 100% pure and natural oil extracted from *O. basilicum* leaves by steam distillation, obtained from selected raw material, to preserve the main properties of each extracted element, as well as enhance its flavor, color, and aroma characteristics. All the essential oil used in the research was acquired on the same date, thus belonging to the same manufacturing batch, in order to avoid interference in the research due to compositional variability.

2.3. Essential Oil Analysis

The analysis of the chemical composition of the essential oil was performed at the Department of Chemistry of the Federal University of Viçosa in Viçosa, Minas Gerais, Brazil. *O. basilicum* essential oil was analyzed by mass spectrometry coupled gas chromatography (GC-MS) on a QP2010 model equipment (Shimadzu, Tokyo, Japan) under the following conditions: fused silica capillary column (30 m in length) and 0.25 mm internal diameter) with RTX®-5MS stationary phase (0.25 μm film thickness) and helium as a carrier gas with a flow rate of 1.0 mL/min. Injector temperature of 220 °C, the initial column temperature was 60 °C, with programming to increase by 2 °C until reaching a temperature of 200 °C, and 5 °C until reaching a maximum temperature of 250 °C. Mass spectra were obtained by electron impact at 70 eV, with 29 to 400 (m/z) scan. 1 μL of the prepared oil solution was injected at a concentration of 10 mg mL^{-1} with a split ratio of 1:20. The main constituents were identified and quantified by their retention index (IR) relative to the hydrocarbon standard (C_7–C_{30}) (99%, Supelco, Bellefonte, PA, USA) and confirmed by comparing the mass spectrum of the compounds with the NIST 14 spectrotheque.

2.4. Exposure to Temperature and Light Radiation

For the evaluation of the effect of temperature on the stability of the essential oil and its major compounds, clear glass containers wrapped in foil and properly sealed, 20 mL of *O. basilicum* essential oil and 1 mL of each compound were under different temperature conditions. The flasks were divided into three lots and packaged for six months for EO and two months for linalool and estragole, in the following environments: refrigerator at 5.0 ± 1 °C (low temperature); in an incubator chamber (model 347, CD, Fanem, São Paulo, SP, Brazil) at a temperature of 20 ± 2 °C (average temperature) and an incubator chamber at a temperature of 35 ± 2 °C (high temperature).

For the evaluation of light stability, clear glass vials containing the essential oil and its linalool and estragole compounds were kept in a B.O.D. (model 347 CD, Fanem, São Paulo, SP, Brazil) at a temperature of 20 ± 2 °C and subjected to light from cold white lamps (100 W each) (Philips, São Paulo, SP, Brazil) for six months for EO and two months for linalool and estragole.

2.5. Toxicological Stability

The fumigation bioassays were performed in 0.8 L (8 cm diameter × 15 cm high) glass vials with 50 non-sexed *S. zeamais* adults, in four replications. The concentrations of *O. basilicum* essential oil stored under different conditions ranged from 8 to 40 μL L^{-1} of air. Working solutions of the essential oil were prepared with toluene solvent (Sigma-Aldrich, 99.9%, Baden-Württemberg, Germany) and applied with a microsyringe (Hamilton, Reno, NV, USA) on 4 mm diameter paper filter discs. 4 cm placed in Petri dishes (6.5 cm diameter). Petri dishes were covered with organza type tissue and placed at the base of the flasks. Pure solvent (toluene) was used as a control. The vials were sealed with a screw-on metal cap and sealed with parafilm (PM996, American, NV, USA) after insect distribution to prevent oil vapor leakage during the exposure period. The flasks were kept in an incubator chamber at a temperature of 27 ± 2 °C for 24 h. After this period, dead and living insects were counted. Corrected mortality was calculated by Abbott's formula [33].

Pure linalool and estragole were purchased from Sigma-Aldrich (Burlington, MA, USA). Toxicity assays were performed at concentrations ranging from 8 to 40 μL L^{-1}. Each filter paper disc (4.4 cm) was treated with 25 μL of toluene diluted linalool and estragole solution and placed in a Petri dish (6.5 cm in diameter), covered with organza and inserted into the base of glass pots with a capacity of 0.8 L. A total of 50 non-sexed adults were placed by pot to expose the insects to the fumigant activity of the compounds for 24 h. Each treatment consisted of four repetitions. As a control 25 μL of pure toluene was used.

2.6. Statistical Analysis

Toxicity data were subjected to probit analysis using SAS software (SAS Institute, Cary, NC, USA), generating concentration-mortality curves. Mortality data were submitted to ANOVA and Tukey test with Statistica 8 software (StatSoft Inc., Tulsa, OK, USA).

3. Results and Discussion

3.1. Essential Oil Composition

The relative chemical composition of the essential oil compounds of *O. basilicum* leaves were performed by GC-MS. The major constituents were identified by their retention index (RI) relative to a homologous series of n-alkanes and confirmed by comparing the mass spectrum of the compounds with the NIST 14 spectrotheque. Chromatographic analysis showed that estragole ($H_2C{=}CHCH_2C_6H_4OCH_3$) and linalool ((CH_3) $2C{=}CHCH_2CH_2C$

(CH_3)(OH)CH=CH_2) were the major components of *O. basilicum* essential oil (Figure 1), representing 85% and 12% of the identified compounds, respectively.

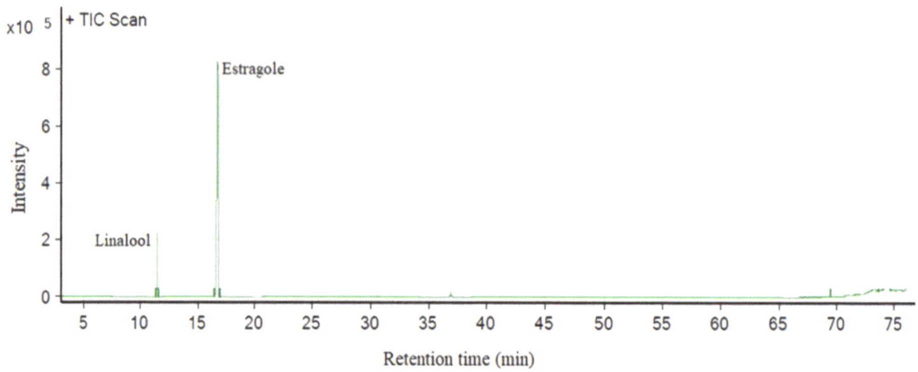

Figure 1. Chromatogram of *Ocimum basilicum* essential oil (10 mg mL^{-1} in toluene).

These compounds are responsible for most of the composition of this essential oil [8,9]. Generally, linalool and estragole are the major components of *O. basilicum* EO, but factors such as soil type, altitude, temperature, insolation period, cultivation, drying conditions and storage influence its composition [31], explaining variations in the amount of the compounds.

3.2. *Toxicological Stability of Essential Oil*

The Probit model was adequate for the concentration-mortality data of all fumigation treatments, based on the low χ^2 value and the high p value obtained from the *O. basilicum* essential oil curves stored under different conditions and their components. linalool and estragole over *S. zeamais*. For the untreated essential oil, the values of χ^2 = 0.14 and p = 0.98 were obtained. Lethal concentrations to cause 50 and 95% insect mortality (LC_{50} and LC_{95}) were 25.4 μL L^{-1} and 178.4 μL L^{-1} of air, respectively (Table 1). The slope of the curve was (1.94 ± 0.37), which indicates genetic homogeneity among individuals of the *S. zeamais* population.

Table 1. Lethal concentrations of *Ocimum basilicum* essential oil stored under different conditions and their major components for fumigation *Sitophilus zeamais*.

Components	LC_{50} (FI 95%) (μL L^{-1} of Air)	LC_{95} (FI 95%) (μL L^{-1} of Air)	Inclination (±MSE [1])	χ^2 (*df*)	*p*
Essential oil	25.4 (23.1–28.6)	178.4 (161.2–196.2)	1.94 ± 0.37	0.14 (7)	0.98
EO at 5 °C	25.7 (23.7–29.1)	172.3 (154.3–187.2)	1.99 ± 0.37	0.09 (7)	0.99
EO at 20 °C	25.8 (24.1–30.6)	151.3 (139.4–192.3)	2.13 ± 0.38	0.85 (7)	0.83

LC = Lethal Concentration (μL L^{-1} of air); FI = Fiducial Interval; MSE [1] = Mean square error; χ^2 = Chi square; p = Probability; df = degrees of freedom; EO = Essential oil.

O. basilicum EO was lethal to *S. zeamais* by fumigation [34], but the effectiveness of essential oils depends on factors such as dose or concentration, insect species, application surface, penetration pathway, method of application and composition of oil, season, ecological conditions, method and extraction time, plant part and storage conditions [35,36]. Both the untreated EO and the EO stored under different temperatures (5 and 20 °C) and exposed to light for a period of six months had lower LC_{50} when applied by fumigation.

The LC_{50} value of untreated *O. basilicum* essential oil when applied by fumigation (25.4 μL L^{-1} air) was lower than that of *Minthostachys verticillata* (28.2 μL L^{-1} air) and *Eucalyptus globulus* essential oil (335.7 μL L^{-1} of air) [37] and higher than *Melaleuca al-*

ternifolia essential oil (7.7 μL L^{-1} of air) for *S. zeamais* [38]. The fumigant activity of *O. basilicum* EO on *S. zeamais* can be explained by the fact that monoterpenoids inhibit the acetylcholinestrase (AChE) nerve conduction enzyme [37]. In addition, studies have shown that essential oils can significantly inhibit the activity of two detoxifying enzymes in *S. zeamais*, glutathione S-transferase (GST) and carboxylesterase (CarE), as well as negatively regulating differentially expressed genes (DEGs) in response to fumigation [37].

O. basilicum EO caused higher mortality of *S. zeamais* when compared to the negative control, showing that it has higher fumigant activity (Figure 2). The *O. basilicum* EO stored at 5 to 20 °C and without storage, differed statistically from each other in only three concentrations (8; 16 and 40 μL L^{-1} of air), indicating that temperatures up to 20 °C do not interfere significantly on the toxicological stability of EO for *S. zeamais* adults when stored for six months. When comparing the EO without storage and EO stored at 35 °C and in light exposure, there was a statistical difference in all concentrations (Figure 2). The EO stored at 35 °C was more stable than the EO stored in light exposure, causing higher mortality of *S. zeamais* adults, which shows that light exposure decreases the toxicity of *O. basilicum* EO on *S. zeamais*.

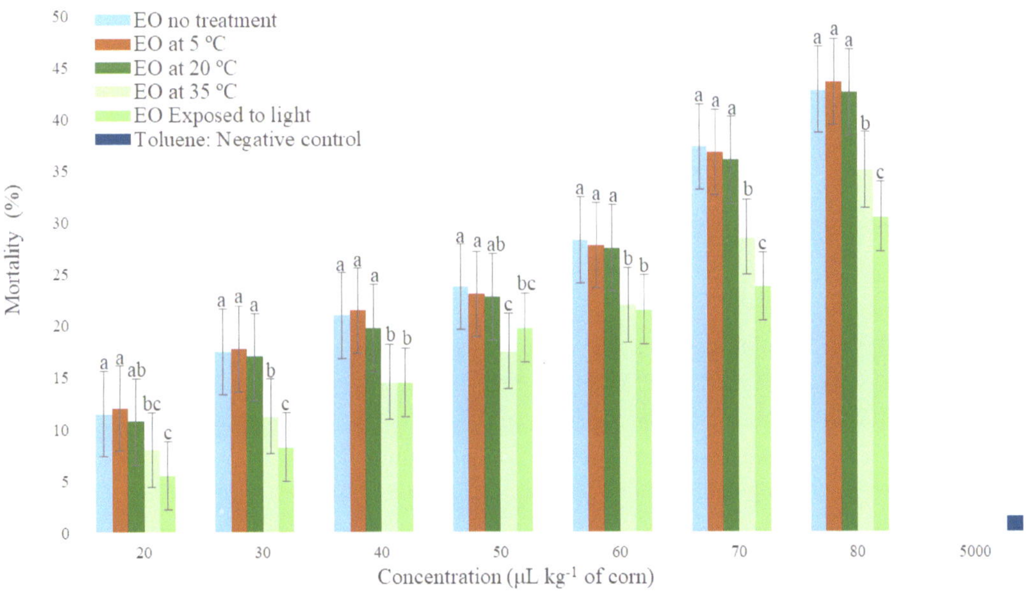

Figure 2. Toxicological stability by fumigation of *Ocimum basilicum* essential oil stored under different conditions and their major components for *Sitophilus zeamais*. Means followed by the same letter in the column do not differ at 5% probability by Tukey test.

When comparing the untreated EO and the EO stored at 35 °C in the fumigation applications, there was a statistical difference in all concentrations (Figures 2 and 3). This indicates that storage at high temperatures for a period of six months affects its toxicological stability on *S. zeamais* adults. The temperature plays a crucial role in the degradation process of essential oils, which directly affects their stability. This decisively influences the stability of the essential oil in several respects [38]. Generally, chemical reactions accelerate with increasing heat due to temperature dependence of the reaction rate, as expressed by the Arrhenius equation [39]. Based on this, Van't Hoff's law states that a temperature increase of 10 °C doubles chemical reaction rates, a ratio that can be consulted to predict stability at different temperatures [40].

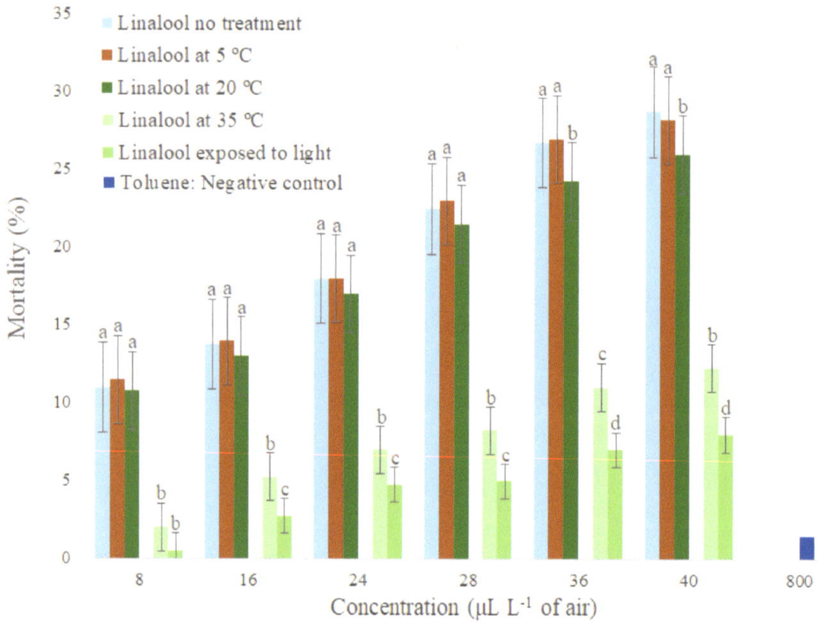

Figure 3. Toxicological stability by fumigation of linalool stored under different conditions for *Sitophilus zeamais*. Means followed by the same letter in the column do not differ at 5% probability by Tukey test.

Increasing temperature advances the self-oxidation and decomposition processes of hydroperoxides, as heat can contribute to free radical formation [41]. Essential oils vary in their susceptibility to self-oxidation at different storage temperatures. In general, monitoring of volatile plant extracts and essential oil composition demonstrates that stability decreases with prolonged storage time, as well as a temperature rise from 0 to 28 °C [42], 4 to 25 °C [43] and 23 to 38 °C [44].

There was a statistical difference between the untreated EO and the EO stored in light exposure at all concentrations (Figures 2 and 3). This indicates that six-month storage in light exposure affects its toxicological stability in *S. zeamais* adults. This is possibly due to the presence of ultraviolet light (UV) and visible light (Vis) being responsible for accelerating the self-oxidation processes in the essential oils, triggering what results in free radical formation [44]. Auto-oxidation involves a succession of chemical reactions that alter the initial composition of the oil, leading to the production of low molecular weight compounds and oxidized polymers, as well as the destruction of important fatty acids and the formation of other compounds, compromising their stability [42].

Comparison of EO stored at 35 °C and in light exposure shows that the toxicological stability of *O. basilicum* EO over *S. zeamais* was most affected by storage in light exposure. The light is much more important than temperature in the oxidation of essential oils [42], although the effect of light on oil oxidation is lessened with increasing temperature [45]. The effect of sunlight for 2 h caused degradation of the quality of ginger oil, while it remained stable when stored in the dark for the same period of time [46].

Processing and storage of oils in exposure to light can lead to the generation of a wide range of undesirable compounds, some of which are harmful to health because of their high toxicity, thereby altering their stability [47]. Among the components of essential oils, monoterpenes have been shown to degrade rapidly under the influence of visible light [48]. The same study also showed that there were transformation reactions in marjoram oil during storage under visible light, which led to the formation of several unidentified elements and smaller components.

3.3. Toxicological Stability of Linalool and Estragole

The Probit model was adequate for concentration-mortality data, based on the low χ^2 values and the high p values obtained on the linalool and estragole curve stored under different conditions over *S. zeamais*. For untreated linalool, the values of $\chi^2 = 5.57$ and $p = 0.34$ were obtained. Lethal concentrations to cause 50% and 95% insect mortality (LC_{50} and LC_{95}) were 34.6 µL L^{-1} and 330.3 µL L^{-1} of air, respectively (Table 2). The slope of the curve was (1.67 ± 0.22), which indicates genetic homogeneity among individuals of the *S. zeamais* population.

Table 2. Lethal concentrations of linalool stored under different conditions for fumigation *Sitophilus zeamais*.

Components	LC_{50} (FI 95%) (µL L^{-1} of Air)	LC_{95} (FI 95%) (µL L^{-1} of Air)	Inclination (±MSE [1])	χ^2 (*df*)	*p*
Linalool	34.6 (31.6–41.2)	330.3 (269.8–463.6)	1.67 ± 0.22	5.57 (5)	0.34
Linalool at 5 °C	33.9 (30.7–42.1)	339.3 (301.3–498.9)	1.64 ± 0.22	6.07 (5)	0.29
Linalool at 20 °C	39.4 (33.5–48.3)	382.7 (365.2–529.8)	1.51 ± 0.22	6.91 (5)	0.22
Linalool at 35 °C	132.4 (110.3–198.7)	495.6 (427.5–612.6)	1.39 ± 0.19	1.56 (5)	0.98
Linalool exposed to light	178.9 (152.4–215.6)	534.2 (518.3–725.5)	1.53 ± 0.23	1.84 (5)	0.96

LC = Lethal Concentration (µL L^{-1} of air); FI = Fiducial Interval; MSE [1] = Mean square error; χ^2 = Chi square; p = Probability; df = degrees of freedom.

Linalool caused higher mortality of *S. zeamais* adults when compared to the negative control, showing that it has higher fumigant activity (Figure 3). Studies have shown that essential oil components are able to inhibit cellular respiration enzymes, nervous system enzymes such as acetylcholinesterase (AChE), and detoxification system enzymes such as P450 and esterase [49], which weakens the insecticide metabolism in insects. Linalol acts together with other compounds in the cholinergic system of insects, promoting the rapid breakdown of the nervous system [50]. Linalool stored at 5 to 20 °C and without storage differed statistically from each other in only two concentrations (36 and 40 µL L^{-1} of air), indicating that temperatures up to 20 °C do not significantly affect its toxicological stability for adults of *S. zeamais* during storage for two months. When comparing non-stored linalool and linalool stored at 35 °C and in light exposure, there was a statistical difference in all concentrations except one (8 µL L^{-1} of air) (Figure 3). Linalool stored at 35 °C was more stable than linalool stored in light exposure, causing higher mortality of *S. zeamais* adults, which shows that light exposure is more detrimental to linalool stability than temperature increase.

For untreated estragole, the values of $\chi^2 = 5.48$ and $p = 0.35$ were obtained. Lethal concentrations to cause 50% and 95% insect mortality (LC_{50} and LC_{95}) were 38.13 µL L^{-1} and 314.01 µL L^{-1} of air, respectively (Table 3). The slope of the curve was (1.79 ± 0.22), which indicates genetic homogeneity among individuals of the *S. zeamais* population.

Table 3. Lethal concentrations of estragole stored under different conditions for *Sitophilus zeamais* by fumigationis.

Components	LC_{50} (FI 95%) (µL L^{-1} of Air)	LC_{95} (FI 95%) (µL L^{-1} of Air)	Inclination (±MSE [1])	χ^2 (*df*)	*p*
Estragole	38.1 (35.6–45.7)	314.0 (298.3–465.6)	1.79 ± 0.22	5.48 (5)	0.35
Estragole at 5 °C	37.2 (34.3–43.8)	305.4 (287.1–419.4)	1.79 ± 0.22	6.08 (5)	0.29
Estragole at 20 °C	41.0 (38.9–49.2)	335.0 (303.5–490.2)	1.80 ± 0.23	4.38 (5)	0.49
Estragole at 35 °C	56.5 (47.5–63.4)	395.2 (347.6–511.5)	1.94 ± 0.25	3.44 (5)	0.63
Estragole exposed to light	53.6 (45.9–61.3)	362.1 (323.7–501.7)	1.98 ± 0.17	6.96 (5)	0.43

LC = Lethal Concentration (µL L^{-1} of air); FI = Fiducial Interval; MSE [1] = Mean square error; χ^2 = Chi square; p = Probability; df = degrees of freedom.

Estragole caused higher mortality of *S. zeamais* adults when compared to the negative control, showing that it has higher fumigant activity (Figure 4). Estragole stored at 5 to 20 °C and without storage differed statistically from only one concentration (32 µL L^{-1} of air), indicating that temperatures up to 20 °C do not significantly affect its toxicological stability for adults of *S. zeamais* during storage for two months. Comparing estragole without storage and linalool stored at 35 °C and in light exposure, there was a statistical difference in all concentrations, indicating that the increase of storage temperature decreases the toxicity of estragol for adults of *S. zeamais* (Figure 4). Estragole stored at 35 °C differed statistically from estragole stored in light exposure by only one concentration (16 μL L^{-1} of air), which shows that both treatments decrease the toxicity of estragole for *S. zeamais* adults to the same extent.

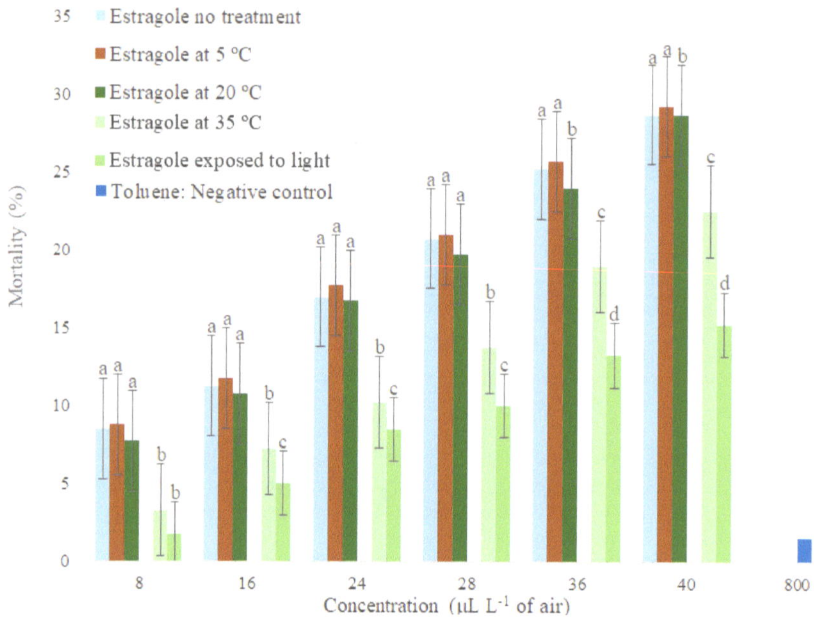

Figure 4. Toxicological stability by estragole fumigation stored under different conditions for *Sitophilus zeamais*. Means followed by the same letter in the column do not differ at 5% probability by Tukey test.

Linalool LC_{50} and LC_{95} increased from 34.56 to 132.4 μL L^{-1} of air and from 330.3 to 495.6 μL L^{-1} of air respectively when stored at 35 °C for two months. The same occurred with estragol LC_{50} and LC_{95} which increased from 38.1 to 56.5 μL L^{-1} of air and from 314.0 to 395.2 μL L^{-1} of air, respectively. This shows that storage at 35 °C decreases the toxicity of this compound for *S. zeamais* adults. This can be explained by the fact that temperature directly influences the stability of volatile compounds [41]. Generally, chemical reactions accelerate with increasing heat due to temperature dependence of the reaction rate, as expressed by the Arrhenius equation [42]. Terpenoids, especially terpenes and aldehydes, are known to be susceptible to rearrangement processes at elevated temperatures. Terpenic conversion reactions by heating have been reported for both isolated compounds [51,52] and for essential oils [53]. Mircene, for example, suffered degradation as it was exposed to higher storage temperatures for 120 days. The initial percentage of mircene fell from 17.38% (6 °C) to 3.99% at 37.5 °C [54].

In this work, the comparison between linalool stored at 35 °C and stored at 20 °C in light exposure shows that it was more toxicologically stable on *S. zeamais* when exposed to 35 °C, indicating that light exposure is more detrimental to the stability of linalool than increase in temperature. This is because some monoterpenes are more thermodynamically stable while others demonstrate rapid degradation under the influence of visible light [50]. Linalol, for example, is 5.9 kJ mol $^{-1}$ more stable than geraniol [55].

In contrast to linalool, the toxicological stability of estragole on *S. zeamais* was more affected by increased storage temperature (35 °C) than by exposure to light. This can occur due to the evaporation process of low boiling compounds, mainly hydrocarbons and sesquiterpenes [56].

There was no significant difference between linalol and estragole stored at 5 and 20 °C, which shows that they can be stored in this temperature range for two months without decreasing their toxicological stability on *S. zeamais*. Three temperatures (4 °C in a cold room, −20 °C in a freezer and 25 °C at room temperature) were used to assess the stability of *Thymus daenensis* essential oil for three months [57]. The results indicated that at room temperature, the amounts of thymol and carvacrol increased considerably by 26.6% and 23% after 3 months, respectively. The increase in thymol and carvacrol by storage at room temperature represents an increase in oil quality index. In addition, oil compositions exhibited the smallest changes and maintained primary quality when stored at low temperatures, particularly at 20 °C [57].

4. Conclusions

The essential oil of *O. basilicum* and its linalool and estragole components components exhibited insecticidal potential against *S. zeamais* adults in corn grains by fumigation. Increasing temperature (35 °C) and exposure to light during storage negatively affects the stability of *O. basilicum* EO, reducing its toxicity against *S. zeamais*. Aiming at the higher toxicity of *O. basilicum* EO to *S. zeamais*, the storage conditions suitable for it are at temperatures of maximum 20 °C and without exposure to light.

Author Contributions: E.d.S.M. performed the experiments, collected the data and wrote the manuscript. L.R.D.F. conceived the idea, supervised the work and obtained funding for the research. F.F.H. analyzed the data and participated in the writing of the manuscript. A.A.Z.R. participated in the CG-MS analysis of the essential oil end edited the final version of the manuscript. All authors have read and agreed to the published version of the manuscript.

Funding: This research was funded by Fundação de Amparo à Pesquisa do Estado de Minas Gerais (FAPEMIG), Conselho Nacional de Desenvolvimento Científico e Tecnológico (CNPq), and Coordenação de Aperfeiçoamento de Pessoal de Nível Superior—Brazil (CAPES)—Finance Code 001. The APC was funded by "Graduate Program in Agricultural Engineering" of Universidade Federal de Viçosa (UFV)—PPGEA-UFV.

Acknowledgments: The authors are grateful to the Department of Chemistry and Department of Agricultural Engineering of the Federal University of Viçosa (Brazil) for support and research facilities. The authors would like to thank the Brazilian Agencies: Fundação de Amparo à Pesquisa do Estado de Minas Gerais (FAPEMIG), Conselho Nacional de Desenvolvimento Científico e Tecnológico (CNPq). Also, this study was financed in part by the Coordenação de Aperfeiçoamento de Pessoal de Nível Superior–Brazil (CAPES)–Finance Code 001.

Conflicts of Interest: The authors declare that they have no conflict of interest.

Sample Availability: Samples of the compounds are not available from the authors.

References

1. Bakkali, F.; Averbeck, S.; Averbeck, D.; Idaomar, M. Biological effects of essential oils—A review. *Food Chem. Toxicol.* **2008**, *46*, 446–475. [CrossRef]
2. Ahn, W.C.; Kalish, S.A.; Gelman, D.L.; Medin, C.; Luhmann, S.; Atran, J.D.; Coley, P.; Shafto, P. Why essences are essential in the psychology of concepts: Commentary on Strevens. *Cognition* **2001**, *15*, 237–262. [CrossRef]
3. Franz, C.M.; Novak, J. Sources of essential oils. In *Handbook of Essential Oils. Science, Technology, and Applications*; Başer, K.H., Buchbauer, G., Eds.; CRC Press: Boca Raton, FL, USA, 2010; pp. 39–81.
4. Dornic, N.; Ficheux, A.S.; Roudot, A.C. Qualitative and quantitative composition of essential oils: A literature based database on contact allergens used for safety assessment. *Regul. Toxicol. Pharm.* **2016**, *80*, 226–232. [CrossRef]
5. Oliveira, J.V.; França, S.M.; Barbosa, D.R.S.; Dutra, K.A.; Araújo, A.M.N.; Navarro, D.M.A.F. Fumigation and repellency of essential oils against *Callosobruchus maculatus* (Coleoptera: Chrysomelidae: Bruchinae) in cowpea. *Pesqui. Agropecu. Bras.* **2017**, *52*, 10–17. [CrossRef]
6. Correa, Y.D.C.G.; Faroni, L.R.A.; Haddi, K.; Oliveira, E.E.; Pereira, E.J.G. Locomotory and physiological responses induced by clove and cinnamon essential oils in the maize weevil *Sitophilus zeamais*. *Pestic. Biochem. Phys.* **2015**, *125*, 31–37. [CrossRef]

7. Costa, C.M.G.R.; Santos, M.S.; Barros, H.M.M.; Agra, P.F.M.; Farias, M.A.A. Efeito inibitório do óleo essencial de manjericão sobre o crescimento in vitro de Erwinia carotovora. *Rev. Tecnol. Ciên. Agropec.* **2009**, *3*, 35–38.
8. Nagaia, A.; Duarte, L.M.L.; Santos, D.Y.A.C. Influence of Viral Infection on Essential Oil Composition of *Ocimum basilicum* (Lamiaceae). *Nat. Prod. Commun.* **2011**, *6*, 1189–1192. [CrossRef]
9. Padalia, R.C.; Verma, R.S.; Chauhan, A.; Goswami, P.; Singh, V.R.; Verma, S.K.; Darokar, M.P.; Kurmi, A.; Singh, M.; Saikia, D.; et al. Essential Oil Composition and Antimicrobial Activity of Methyl cinnamate-Linalool Chemovariant of *Ocimum basilicum* L. from India. *Rec. Nat. Prod.* **2017**, *11*, 193–204.
10. Mazzonetto, F.; Vendramim, J.D. Efeito de Pós de Origem Vegetal sobre *Acanthoscelides obtectus* (Say) (Coleoptera: Bruchidae) em Feijão Armazenado. *Neotrop. Entomol.* **2003**, *32*, 145–149. [CrossRef]
11. López, M.D.; Jordán, M.J.; Pascual-Villalobos, M.J. Toxic compounds in essential oils of coriander, caraway and basil active against stored rice pests. *J. Stored Prod. Res.* **2008**, *44*, 273–278. [CrossRef]
12. Asawalam, E.F.; Emosairue, S.O.; Hassanali, A. Essential oil of *Ocimum grattissimum* (Labiatae) as *Sitophilus zeamais* (Coleoptera: Curculionidae) protectant. *Afr. J. Biotechnol.* **2008**, *7*, 3771–3776.
13. Mishra, B.B.; Tripathi, S.P.; Tripathi, C.P.M. Repellent effect of leaves essential oils from *Eucalyptus globulus* (Mirtaceae) and *Ocimum basilicum* (Lamiaceae) against two major stored grain insect pests of Coleopterons. *Nat. Sci.* **2012**, *10*, 50–54.
14. França, S.M.; Oliveira, J.V.; Esteves Filho, A.B.; Oliveira, C.M. Toxicity and repellency of essential oils to *Zabrotes subfasciatus* (Boheman) (Coleoptera, Chrysomelidae, Bruchinae) in *Phaseolus vulgaris* L. *Acta Amaz.* **2012**, *42*, 381–386. [CrossRef]
15. Belong, P.; Ntonga, P.A.; Bakwo Fils, E.M.; Dadji, G.A.F.; Tamesse, J.L. Chemical composition and residue activities of *Ocimum canum* Sims and *Ocimum basilicum* L essential oils on adult female *Anopheles funestus ss. J Anim. Plant Sci.* **2013**, *19*, 2854–2863.
16. Popovic, Z.; Kostic, M.; Popovic, S.; Skorcic, S. Bioactivities of Essential Oils from Basil and Sage to *Sitophilus oryzae* L. *Biotechnol Biotec. Eq.* **2015**, *20*, 36–40. [CrossRef]
17. Almeida, F.A.C.; Almeida, S.A.; Santos, N.R.; Gomes, J.P.; Araújo, M.E.R. Efeito de extratos alcoólicos de plantas sobre o caruncho do feijão vigna (*Callosobruchus maculatus*). *Rev. Bras Eng. Agrícola Ambient.* **2005**, *9*, 585–590. [CrossRef]
18. Lopez, M.D.; Pascual-Villalobos, M.J. Mode of inhibition of acetylcholinesterase by monoterpenoids and implications for pest control. *Ind. Crops Prod.* **2010**, *31*, 284–288. [CrossRef]
19. Mughal, M.H. Linalool: A mechanistic treatise. *J. Nut. Food Res. Technol.* **2019**, *2*, 1–5. [CrossRef]
20. Chapman, R.F.; Bernays, E.A.; Simpson, S.J. Attraction and repulsion of the aphid *Cavariella aegopodii* by plant odors. *J. Chem. Ecol.* **1981**, *7*, 881. [CrossRef] [PubMed]
21. Regnault-Roger, C.; Hamraoui, A. Fumigant toxic activity and reproductive inhibition induced by monoterpenes on *Acanthoscelides obtectus* (Say) (Coleoptera), a bruchid of kidney bean (*Phaseolus vulgaris* L.). *J. Stored Prod. Res.* **1995**, *31*, 291–299. [CrossRef]
22. Cheng, S.S.; Liu, J.Y.; Tsai, K.H.; Chen, W.J.; Chang, S.T. Chemical composition and mosquito larvicidal activity of essential oils from leaves of different *Cinnamomum osmophloeum* provenances. *J. Agric. Food Chem.* **2004**, *52*, 4395–4400. [CrossRef] [PubMed]
23. Tiwary, M.; Naik, S.N.; Tewaryb, D.K.; Mittalc, P.K.; Yadavc, S. Chemical composition and larvicidal activities of the essential oil of *Zanthoxylum armatum* DC (Rutaceae) against three mosquito vectors. *Vector-Borne Zoonot.* **2007**, *44*, 198–204.
24. Smith, R.L.; Adams, T.B.; Doull, J.; Feron, V.J.; Goodman, J.I.; Marnett, L.J.; Portoghese, P.S.; Waddell, W.J.; Wagner, B.M.; Rogers, A.E.; et al. Safety assessment of allylalkoxybenzene derivatives used as flavouring substances, methyl eugenol and estragole. *Food Chem. Toxicol.* **2002**, *40*, 851–870. [CrossRef]
25. De Vicenzi, M.; Silano, M.; MaialettI, F.; Scazzocchio, B. Constituents of aromatic plants: II. Estragole. *Fitoterapia* **2000**, *71*, 725–729. [CrossRef]
26. Morais, S.M.; Catunda-Junior, F.E.A.C.; Da Silva, A.R.A.; Neto, J.S.M.; Rondina, D.; Leal-Cardoso, J.H. Antioxidant activity of essential oils from Northeastern Brazilian Croton species. *Quím. Nova* **2006**, *29*, 907–910. [CrossRef]
27. Lopez, M.D.; Pascual-Villalobos, M.J. Are monoterpenoids and phenylpropanoids efficient inhibitors of acetylcholinesterase from stored product insect strains? *Flavour Fragr. J.* **2015**, *30*, 108–112. [CrossRef]
28. Wang, C.F. Components and insecticidal activity against the maize weevils of *Zanthoxylum schinifolium* fruits and leaves. *Molecules* **2011**, *16*, 3077–3088. [CrossRef] [PubMed]
29. Mbata, G.N.; Pascual-Villalobos, M.J.; Payton, M.E. Comparative mortality of diapausing and nondiapausing larvae of *Plodia interpunctella* (Lepidoptera: Pyralidae) exposed to monoterpenoids and low pressure. *J. Econ. Entomol.* **2012**, *105*, 679–685. [CrossRef]
30. Zhao, N.N. Evaluation of the toxicity of the essential oils of some common Chinese spices against *Liposcelis bostrychophila. Food Control* **2012**, *26*, 486–490. [CrossRef]
31. Kim, S.I.; Lee, D.W. Toxicity of basil and orange essential oils and their components against two coleopteran stored products insect pests. *J. Asia Pac. Entomol.* **2014**, *17*, 13–17. [CrossRef]
32. Treibs, W. Gewinnung atherischer Ole durch Destillation, Extraktion und Pressung. A. Einleitung. In *Die atherischen Ole Band I*, 4th ed.; Gildemeister, E., Hoffmann, F., Eds.; Akademie-Verlag: Berlin, Germany, 1956; pp. 307–309.
33. Abbott, W.S. A method of computing the effectiveness of on insecticide. *J. Econ. Entomol.* **1925**, *18*, 265–267. [CrossRef]
34. Souza, A.P.; Vendramim, J.D. Atividade inseticida de extratos aquosos de meliáceas sobre *Bemisia tabaci* (Genn) biótipo B. *Neotrop. Entomol.* **2001**, *30*, 133–137. [CrossRef]
35. Don-Pedro, K.N. Investigation of single and joint fumigant insecticidal action of citruspeel oil components. *Pestic. Sci.* **1996**, *46*, 79–84. [CrossRef]

36. Lee, S.E.; Lee, B.H.; Choi, W.S.; Park, B.S.; Kim, J.G.; CampbelL, B.C. Fumigant toxicity of volatile natural products from Korean spices and medicinal plants towards the rice weevil, *Sitophilus oryzae* (L). *Pest Manag. Sci.* **2001**, *57*, 548–553. [CrossRef]
37. Arena, J.S.; Peschiutta, M.L.; Calvimonte, H.; Zygadlo, J.A. Fumigant and repellent activities of different essential oils alone and combined against the maize weevil (*Sitophilus zeamais* motschulsky). *MOJ Biorg. Org. Chem.* **2017**, *1*, 249–253.
38. Liao, M.; Xiao, J.; Zhou, L.; Liu, Y.; Wu, X.; Hua, R.; Wang, G.; Cao, H. Insecticidal Activity of *Melaleuca alternifolia* Essential Oil and RNA-Seq Analysis of *Sitophilus zeamais* Transcriptome in Response to Oil Fumigation. *PLoS ONE* **2016**, *9*, e0167748. [CrossRef]
39. Abdelgaleil, S.A.M.; Mohamed, M.I.E.; Badaway, M.E.I.; El-Arami, S.A.A. Fumigant and contact toxicities monoterpenes to *Sitophilus oryzae* (L.) and *Tribolium castaneum* (Herbst) and their inhibitory effects on acetylcholinesterase activity. *J. Chem. Ecol.* **2009**, *35*, 518–525. [CrossRef] [PubMed]
40. Díaz-Maroto, M.C.; Pardo, E.A.; Castillo-Muñoz, N.; Díaz-Maroto, I.J.; Pérez-Coello, M.S. Effect of storage conditions on volatile composition of dried rosemary (*Rosmarinus officinalis* L.) leaves. *Flavour Fragr. J.* **2009**, *24*, 245–250. [CrossRef]
41. Atkins, P.W. *Kurzlehrbuch Physikalische Chemie*, 3rd ed.; Wiley-VCH: Weinheim, Germany, 2002.
42. Choe, E.; Min, D.B. Mechanisms and factors for edible oil oxidation. *Compr. Rev. Food Sci. Food Saf.* **2006**, *5*, 169–186. [CrossRef]
43. Glasl, H. Uber die Haltbarkeit von Terpenoiden in Extrakten und Losungen mit unterschiedlichem Alkoholgehalt. *Arch. Pharm.* **1975**, *308*, 88–93. [CrossRef]
44. Gopalakrishnan, N. Studies on the storage quality of CO_2-extracted cardamom and clove bud oils. *J. Agric. Food Chem.* **1994**, *42*, 796–798. [CrossRef]
45. Braun, M.; Franz, G. Quality criteria of bitter fennel oil in the German Pharmacopoeia. *Pharm. Pharmacol. Lett.* **1999**, *9*, 48–51.
46. Turek, C.; Stintzing, F.C. Impact of different storage conditions on the quality of selected essential oils. *Int. Food Res. J.* **2012**, *46*, 341–353. [CrossRef]
47. Velasco, J.; Dobarganes, C. Oxidative stability of virgin olive oil. *Eur. J. Lipid Sci. Technol.* **2002**, *104*, 661–676. [CrossRef]
48. Brandares, M.F.T.; Dar Juan, B.B.; Vulelban, A.M.; Anzaldo, F.E. Stability studies of essential oils from some Philippine plants I. *Zingiber Officinale Rosc. Philipp. J. Sci.* **1987**, *116*, 1–11.
49. Walallawita, W.K.U.S.; Bopitiya, D.; Sivakanthan, S.; Jayawardana, N.W.I.A.; Madhujith, T. Comparison of oxidative stability of sesame (*Sesamum indicum*), soybean (*Glycine max*) and mahua (mee) (*Madhuca longifolia*) oils against photo-oxidation and autoxidation. *Procedia Food Sci.* **2016**, *6*, 204–207. [CrossRef]
50. Misharina, T.A.; Polshkov, A.N.; Ruchkina, E.L.; Medvedeva, I.B. Changes in the composition of the essential oil of marjoram during storage. *Appl. Biochem. Microbiol.* **2003**, *39*, 311–316. [CrossRef]
51. Fazolin, M.; Estrela, J.L.V.; Monteiro, A.F.M.; Silva, I.M.; Gomes, L.P.; Silva, M.S.F. Synergistic potential of dillapiole-rich essential oil with synthetic pyrethroid insecticides against fall armyworm. *Ciênc. Rural* **2016**, *46*, 382–388. [CrossRef]
52. Marques, M.G.; Landim, T.N.; Silva, S.M.; Cunha, J.P.A.R.; Sampaio, M.V. Toxicity of *Ocimum basilicum* essential oil combined with thiamethoxam to cotton aphid. *Rev. Bras. Cienc. Agrar.* **2019**, *14*, e565. [CrossRef]
53. Mcgraw, G.W.; Hemingway, R.W.; Ingram, L.L., Jr.; Canady, C.S.; Mcgraw, W.B. Thermal degradation of terpenes: Camphene, Δ3-carene, limonene, and α-terpinene. *Environ. Sci. Technol.* **1999**, *33*, 4029–4033.
54. Ternes, W. *Naturwissenschaftliche Grundlagen der Lebensmittelzubereitung*, 3rd ed.; Behr's Verlag: Hamburg, Germany, 2008.
55. Turek, C.; Stintzing, F.C. Evaluation of selected quality parameters to monitor essential oil alteration during storage. *J. Food Sci.* **2011**, *76*, 1365–1375. [CrossRef] [PubMed]
56. Guimarães, L.G.L.; Cardoso, M.G.; Zacaroni, L.M.; Lima, R.K. Influência da luz e da temperatura sobre a oxidação do óleo essencial de Capim-Limão (*Cymbopogon citratus* (D.C.) STAPF). *Quim. Nova* **2008**, *31*, 1476–1480. [CrossRef]
57. Brodkorb, D.; Gottschall, M.; Marmulla, R.; Luddeke, F.; Harder, J. Linalool Dehydratase-Isomerase, a Bifunctional Enzyme in the Anaerobic Degradation of Monoterpenes. *J. Biol. Chem.* **2010**, *285*, 30436–30442. [CrossRef] [PubMed]

molecules

Article

Zuccagnia punctata Cav. Essential Oil into Poly(ε-caprolactone) Matrices as a Sustainable and Environmentally Friendly Strategy Biorepellent against *Triatoma infestans* (Klug) (Hemiptera, Reduviidae)

Sandra López [1], Alejandro Tapia [1,*], Julio Zygadlo [2], Raúl Stariolo [3], Gustavo A. Abraham [4] and Pablo R. Cortez Tornello [4,*]

1 Instituto de Biotecnología-Instituto de CienciasBásicas, Universidad Nacional de San Juan, Av. Libertador General San Martín 1109 (O), San Juan 5400, Argentina; slopez@unsj.edu.ar
2 Instituto Multidisciplinario de Biología Vegetal, Cátedra de Química Orgánica, Facultad de Ciencias Exactas, Físicas y Naturales, Universidad Nacional de Córdoba, Córdoba 5016, Argentina; jzygadlo@unc.edu.ar
3 Coordinación Nacional de Control de Vectores, Córdoba 5000, Argentina; stariolo@yahoo.com.ar
4 Instituto de Investigaciones en Ciencia y Tecnología de Materiales, INTEMA (UNMdP-CONICET), Mar del Plata 7600, Argentina; gabraham@fi.mdp.edu.ar
* Correspondence: atapia@unsj.edu.ar (A.T.); pablocortez@fi.mdp.edu.ar (P.R.C.T.); Tel.: +54-264-4211700 (A.T.); +54-223-6260600 (P.R.C.T.)

Citation: López, S.; Tapia, A.; Zygadlo, J.; Stariolo, R.; Abraham, G.A.; Cortez Tornello, P.R. *Zuccagnia punctata* Cav. Essential Oil into Poly(ε-caprolactone) Matrices as a Sustainable and Environmentally Friendly Strategy Biorepellent against *Triatoma infestans* (Klug) (Hemiptera, Reduviidae). *Molecules* **2021**, *26*, 4056. https://doi.org/10.3390/molecules26134056

Academic Editor: Giovanni Benelli

Received: 19 May 2021
Accepted: 28 June 2021
Published: 2 July 2021

Publisher's Note: MDPI stays neutral with regard to jurisdictional claims in published maps and institutional affiliations.

Abstract: The main strategies against *Triatoma infestans* (primary vector responsible for the Chagas disease transmission) are the elimination or reduction of its abundance in homes through the application of insecticides or repellents with residual power, and environmental management through the improvement of housing. The use of plant-derived compounds as a source of therapeutic agents (i.e., essential oils from aromatic plants and their components) is a valuable alternative to conventional insecticides and repellents. Essential oil-based insect repellents are environmentally friendly and provide reliable personal protection against the bites of mosquitoes and other blood-sucking insects. This study investigates, for the first time to our knowledge, the potential repellent activity of *Zuccagnia punctata* essential oil (ZEO) and poly(ε-caprolactone) matrices loaded with ZEO (ZEOP) prepared by solvent casting. The analysis of its essential oil from aerial parts by GC–FID and GC-MS, MS allowed the identification of 25 constituents representing 99.5% of the composition. The main components of the oil were identified as (−)-5,6-dehydrocamphor (62.4%), alpha-pinene (9.1%), thuja-2, 4 (10)-diene (4.6%) and dihydroeugenol (4.5%). ZEOP matrices were homogeneous and opaque, with thickness of 800 ± 140 μm and encapsulation efficiency values above 98%. ZEO and ZEOP at the lowest dose (0.5% wt./wt., 96 h) showed a repellency of 33 and 73% respectively, while at the highest dose (1% wt./wt., 96 h) exhibited a repellent activity of 40 and 66 %, respectively. On the other hand, until 72 h, ZEO showed a strong repellent activity against *T. infestans* (88% repellency average; Class V) to both concentrations, compared with positive control N-N diethyl-3-methylbenzamide (DEET). The essential oils from the Andean flora have shown an excellent repellent activity, highlighting the repellent activity of *Zuccagnia punctata*. The effectiveness of ZEO was extended by its incorporation in polymeric systems and could have a potential home or peridomiciliary use, which might help prevent, or at least reduce, Chagas' disease transmission.

Keywords: Chagas' disease transmission; triatomines; peridomiciliary use; Argentina

1. Introduction

Chagas disease is an extraordinarily complex zoonosis that is present throughout the territory of South America, Central America, and Mexico, and continues to represent a serious threat to the health of the countries of the region. This illness affects 6 to 7 million people around the world [1]. The main strategies used to interrupt vector transmission of

Trypanosoma cruzi by *Triatoma infestans* (primary responsible for the Chagas disease transmission) are the elimination or reduction of its abundance in homes through the application of insecticides with residual power, and environmental management the improvement of housing [1].

Pyrethroid insecticides have been used for over 20 years to control the vectors of Chagas disease in Argentina and other Latin American countries [2]. From them, deltamethrin has been intensively used for the chemical control of *T. infestans* "vinchucas" and in general has displayed an effect as a highly effective triatomicide [3]. In 2002, the health authorities for the control of vectors in Argentina reported failures in the chemical control of *T. infestans* associated with different levels of resistance to pyrethroids [4–6]. This fact could be caused by the rapid degradation of the active compound, as shown in several studies [7].

To overcome these problems, active agents have been incorporated in polymeric systems allowing their protection and sustained release. Polymer-based systems have shown a longer lasting effect than traditional suspension concentrate formulations, both under experimental and field conditions [8]. Nanofibrous mats containing citriodiol as biorepellent against *Aedes aegypti* mosquitoes and its incorporation into a layered fabric were recently studied [9]. Monolithic and core-enriched nanofibrous mats with repellent activity were successfully obtained, and core-enriched mats displayed a 100% of repellency for 34 days. Moreover, compounds of botanical origin such as the essential oils from aromatic plants and their components provide an alternative to conventional insecticides and repellents. Essential oil-based insect repellents are environmentally friendly and provide dependable personal protection against the bites of mosquitoes and other blood-sucking insects [10]. On the other hand, the essential oils have the advantage of being agents of low toxicity in mammals, little residual life in the environment, and fewer requirements imposed by the legal framework, because they enjoy social acceptance due to the widespread use of aromatic species [11].

The essential oils from the Andean flora growing in the province of San Juan located in the center-west of the Argentine have shown an excellent repellent activity repellents to *Triatoma infestans* (Klug) (Hemiptera, Reduviidae), the vector of Chagas disease, since they constitute a rich source of bioactive compounds that are biodegradable into nontoxic products [12–14], effect that could be enhanced by incorporating them into polymeric systems, which are considered a suitable strategy for time and distribution-controlled repellent delivery [13]. The use of repellents against *T. infestans* vectors might help prevent, or at least reduce, Chagas' disease transmission. The resinous species including the genus *Larrea* in Argentina (*Larrea ameghinoi, L. cuneifolia, L. divaricata,* and *L. nitida*), vernacular name "jarillas" and *Zuccagnia punctata,* commonly called "jarilla macho" are used extensively in traditional medicine in Argentina Andean communities for the treatment of injuries and bruises, and a good disinfectant of wounds, repellent of insects, for roof construction in rural areas and as a vegetable fuel for cooking food. [15,16].

On the other hand, poly(ε-caprolactone) (PCL) is a well-known aliphatic biocompatible polyester with a glass transition temperature at −60 °C and melting temperature between 59–64 °C. Its semicrystalline structure and hydrophobic character allow PCL to exhibit a long degradation time under humidity or physiological conditions of around 2 years. This is an attractive property for long-term applications in bioactive agent delivery [17–20]. There are reports from Peres et al. in which they propose the encapsulation of essential oil of fruit and leaves of *Xylopia aromatica* in PCL nanoparticles [21]. The nanoencapsulation of these bioactive compounds promotes their protection from environmental degradation and prolongs their biological activity. De Ávila et al. reported the preparation of PCL microparticles with encapsulated citronella oil through an emulsion technique followed by solvent evaporation [22]. Akolade et al. reported the microencapsulation of eucalyptol in poly(ethylene glycol) and PCL using particles from gas-saturated solutions [23]. Unalan et al. reported the fabrication and characterization of various concentrations of peppermint essential oil (PEP) loaded on PCL electrospun fiber mats for wound healing applications, where PEP was intended to impart antibacterial activity to the fibers [24].

This study investigated for the first time the potential repellent activity of *Zuccagnia punctata* essential oil and its incorporation in PCL matrices for increasing the duration of the repellent activity.

2. Results and Discussion

2.1. Essential Oil Composition, Yield, and Spectroscopy Characterization

The essential oil yield was 0.25% (v/wt.); δ^{25}: 0.96 g/mL. Regarding the chemical profile, a total of 25 compounds amounting 99.5% of the oil were identified according [25–27]. The main constituents are showed in Table 1 and include (−)-5,6-dehydrocamphor (62.4%), alpha-pinene (9.1%), thuja-2,4(10)-diene (4.6%), terpinen-4-ol (4.4%), verbenone (3.1%) and dihydroeugenol (4.5%). The monoterpenes represented the main portion of the oil accounting for 89.2% with a high percentage of oxygenated monoterpenes (69.9%). Hydrocarbon sesquiterpenes accounted for 4.7 %. Among them, the most abundant was epi-beta-santalene (2.1%). Epi-alpha-cadinol was the oxygenated sesquiterpenoid detected (0.7%).

Table 1. Chemical composition of the *Zuccagnia punctata* essential oil.

Peak	Component	RI	Area (%)	Identification Method
1	Alpha-thujene	928	t	1
2	Alpha-pinene	936	9.1	1, 2
3	Alpha-fenchene	950	0.3	1
4	Camphene	951	t	1, 2
5	Thuja-2,4(10)-diene	956	4.6	1
6	Myrcene	990	t	1, 2
7	Alpha-terpinene	1016	0.6	1
8	p-cymene	1025	t	1, 2
9	Limonene	1030	0.8	1, 2
10	Gamma-terpinene	1059	1.6	1
11	p-cymenene	1091	2.3	1
12	(−)-5,6-dehydrocamphor	1097	62.4	1
13	Terpinen-4-ol	1178	4.4	1, 2
14	Verbenone	1206	3.1	1
15	(E)-Cinnamyl alcohol	1304	0.4	1
16	Piperitenone	1343	t	1, 2
17	Eugenol (dihydro)	1369	4.5	1
18	(E)- caryophyllene	1419	0.8	1
19	Epi-beta-santalene	1447	2.1	1
20	Delta-amorphene	1512	0.6	1
21	(Z)-gamma-bisabolene	1515	0.1	1
22	Beta-curcumene	1516	0.2	1
23	Delta-cadinene	1523	0.4	1
24	(E)-gamma-bisabolene	1531	0.5	1
25	Epi-alpha-cadinol	1640	0.7	1
	Monoterpene hydrocarbons		19.3	
	Oxygenated monoterpenes		69.9	
	Phenylpropanoids		4.9	
	Sesquiterpenes hydrocarbons		4.7	
	Oxygenated sesquiterpenes		0.7	
Total			99.5	

Constituents listed in order of increasing retention indices (RI). Unidentified components less than 0.1% are not reported. Temperature-programmed RI referred to n-alkanes, determined on a HP-5MS capillary column. Percentage values less than 0.1% are denoted as t (traces). Method of identification of minor constituents: 1corresponds to comparison of

GC-MS data and RI with those of the volatile oil ADAMS, Wiley and NBS computer mass libraries, 2corresponds to comparison of GC-MS data and RI with those of authentic samples.

The chemical composition as well as the antifungal activity of the *Zuccagnia punctata* essential oil collected in the province of San Juan have been previously reported [27], standing out the presence of (−)-5,6-dehydrocamphor (56.5%), linalool (14.5%) and cis-linalool oxide THF (3.4%). The chemical composition of the essential oil reported here shows also that the main component is (−)-5,6-dehydrocamphor (62.4%), with some differences in minor components. The chemical composition is genetically determined (intrinsic factors) and on the other hand, environmental conditions (extrinsic factors) may be responsible for significant variations in the chemical composition of plants [28]. Essential oils can qualitatively and quantitatively change their chemical composition due to climatic factors, the composition of the soil, the plant organ, age, seasonality, and the phase of the circadian cycle [29–31].

2.2. Morphological Characterization

The Figure 1 shows a disc of 10 mm cut from the ZEOP matrix and the SEM image of ZEOP 1% sample. ZEOP matrices were homogeneous and opaque with thickness of 800 ± 140 μm, as measured with a low force caliper. The SEM micrograph exhibited a characteristic morphology of PCL matrices prepared by solvent casting. The surface porosity could favor the evaporation of essential oils from the polymer matrix regions with dispersed oil.

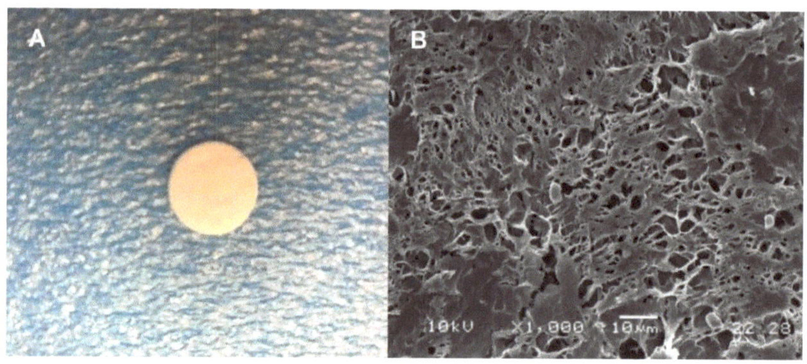

Figure 1. Morphological characterization, (**A**) Optical image of ZEOP 1% disc 10 mm, (**B**) SEM micrograph of ZEOP 1% (1000×).

2.3. Thermal Properties and Crystallinity

The Figure 2 shows the DSC thermograms. PCL pellets exhibited a characteristic thermogram with a melting temperature of 65.6 °C and crystallinity of 59.3%. PCL matrices showed a decrease in the melting point (Tm = 62 °C) and slight increase in crystallinity (63.8%) which can be attributed to the matrix formation during solvent casting. The incorporation of ZEO led to a decrease in Tm values with the increase in the oil content (ZEOP 0.5%, 61.4 °C and ZEOP 1%, 59.0 °C). This phenomenon agrees with the decrease in the crystallinity degree (ZEOP 0.5%, 61%, and ZEOP 1%, 59.6%), and it could be ascribed to the incorporation of oil, which makes difficult the crystallization process during the solvent evaporation. The ZEO thermogram did not show thermal events in the explored temperature by subheadings. It should provide a concise and precise description of the experimental results, their interpretation, as well as the experimental conclusions that can be drawn.

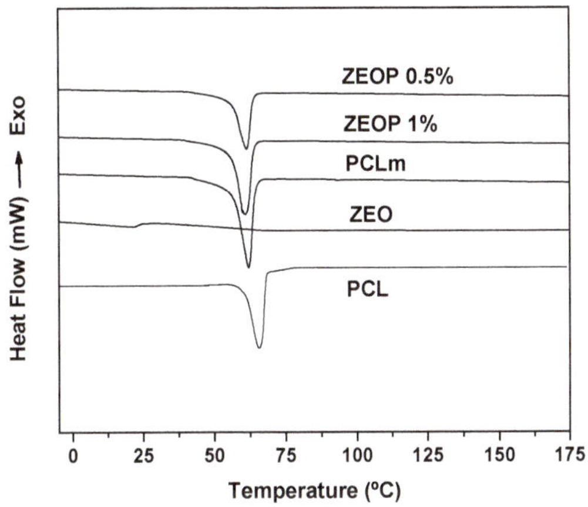

Figure 2. Differential scanning calorimetry thermograms corresponding to PCL, PCL matrix (PCLm), ZEO, ZEOP 0.5 % and ZEOP 1% formulations.

TGA curves of ZEO are shown in Figure 3. The results indicated that the thermal behavior of *Zuccagnia punctata* is simple and present only one thermal process. A continuous weight loss starting at 62.6 °C and continued until 358.2 °C. These values indicate that ZEO decomposition with temperature begins above room temperature, and therefore it is stable at the temperature of use of ZEOP matrices.

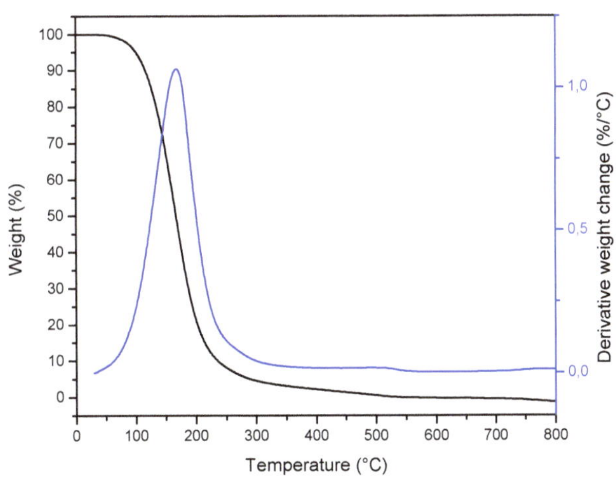

Figure 3. Thermal stability of ZEO: thermogravimetric and first-derivative of TGA curves.

2.4. ZEO Loading Capacity and Encapsulation Efficiency

Table 2 shows the ZEO content in ZEOP matrices per mass unit of sample (Mc), loading capacity (LC), and encapsulation efficiency (EE). LC values of ZEOP 0.5% (0.48%), and ZEOP 1% (0.97%) matrices showed that ZEO content were slightly lower than the amount present in the polymeric solution. On the other hand, EE values showed encapsulation of ZEO above 98%. These values were consistent with those reported by Peres et al. [21] for PCL nanoparticles containing the essential oils of *Xylopia* (95%), and higher than the obtained values for PCL microparticles loaded eucalyptol (77 %) [23] and PCL fibers with peppermint oil (37 %) [24]. These results indicate that PCL is a suitable polymer for ZEO encapsulation.

Table 2. ZEO content incorporated in ZEOP matrices per mass unit of sample (Mc), loading capacity (LC), and encapsulation efficiency (EE).

Sample	Mc (±s.d.) (mg/g)	LC (±s.d.) (%)	EE (±s.d.) (%)
ZEOP 0.5%	4.87 ± 0.30	0.48 ± 0.08	98.45 ± 0.03
ZEOP 1%	9.75 ± 0.20	0.97 ± 0.10	98.51 ± 0.02

2.5. *Repellent Activity against Triatoma Infestans Nymphs*

The results of the assay of repellency of the ZEO and ZEOP are shown in Tables 3 and 4.

Table 3. Repellent activity of ZEO, and ZEOP against *T. infestans* nymphs fifth instars (mean ± SD, n = 5) at 0.5% (wt./wt.).

	Repellency (%) at 0.5 % (wt./wt.)			
	Treatments			
Time (h)	ZEO	ZEOP	Control [3)]	DEET [4)]
1	97.0 ± 2.0	33.0 ± 11.1	−12.0 ± 5.4	100.0 ± 0.0
24	92.0 ± 10.4	60.0± 24.0	−20.0 ± 32.0	100.0 ± 0.0
72	76.0 ± 5.3	60.0 ± 24.0	−100.0 ± 0.0	100.0 ± 0.0
Average repellency [1)]	88.3 ± 5.0 ^{a}	51.0 ± 8.1 ^{a}	−44.0 ± 17.8 ^{b}	100.0 ± 0.0 ^{a}
Class [2)]	V	III	-	V
96	33.0 ± 23.1 ^{a}	73.0 ± 12.1 ^{a}	−100.0 ± 0.0 ^{b}	100.0 ± 0.0 ^{a}
Class [2)]	II	IV	-	V

[1)] Average value of repellency in the three times. [2)] Repellence class according to scale: Class 0 (0.01 to 0.01%), Class I (0.1 to 20%), Class II (20.1 to 40%), Class III (40.1 to 60%), Class IV (60.1 to 80%), and Class V (80.1 to 100). ^{a,b} indicate significant difference at the 0.05 level according to Tukey test. [3)] Blank control acetone. [4)] Positive control DEET at 0.5% (wt./v).

Table 4. Repellent activity of ZEO, and ZEOP against *T. infestans* nymphs fifth instars, the vector of Chagas disease (mean ± SD, n = 5) at 1.0% (wt./wt.).

	Repellency (%) at 1 % (wt./wt.)			
	Treatments			
Time (h)	ZEO	ZEOP	Control [3)]	DEET [4)]
1	100.0 ± 0.0	46.6 ± 11.1	−12.0 ± 34.4	100.0 ± 0.0
24	93.0 ± 11.5	60.0 ± 6.0	−28.0 ± 32.0	100.0 ± 0.0
72	73.3 ± 23.1	66.0 ± 23.0	−100.0 ± 0.0	100.0 ± 0.0
Average repellency [1)]	88.8 ± 5.0 ^{a}	55.5 ± 7.0 ^{a}	−46.7 ± 17.8 ^{b}	100.0 ± 0.0 ^{c}
Class [2)]	V	III	-	V
96	40.0 ± 13.1	66.6 ± 6.7 ^{a}	−100.0 ± 0.0 ^{b}	100.0 ± 0.0
Class [2)]	II	IV		V

[1)] Average value of repellency in the three times. [2)] Repellence class according to scale: Class 0 (0.01 to 0.01%), Class I (0.1 to 20%), Class II (20.1 to 40%), Class III (40.1 to 60%), Class IV (60.1 to 80%), and Class V (80.1 to 100%). ^{a,b,c} indicate significant difference at the 0.05 level according to Tukey test. [3)] Blank control acetone. [4)] Positive control DEET at 0.5% (wt./v).

The ZEO showed excellent repellent properties on *T. infestans* between 1 and 72 h, for the two concentrations of the oil tested (Tables 3 and 4). The percentage of repellence did not change significantly with time (no effect within subjects, $p > 0.05$), and no significant relationship was observed between time points and oil treatment ($p > 0.05$). The essential oil was Class V, which is the one with the highest repellency according to the methodology used; the mean values obtained were 88.3 and 88.8%, for concentrations of 0.5 and 1% (wt./wt.), respectively. For both concentrations, the repellent activity decays between 72 and 96 h until a Class II repellent activity. This short-term action regarding the duration of the effect may be what limits the use of insect repellent products based on essential oils, according to previous reports, it may be related to the rapid volatilization and short time of action [32]. On the other hand, significant differences in average percentage of repellence were observed between the ZEO treatment and blank control (effects between subjects, $p < 0.05$).

In a previous report, the chemical composition, anti-insect, and antimicrobial activity of *Baccharis darwinii* essential oil from Argentina, Patagonia were reported. The major components with recognized anti-insect and antimicrobial activity were identified, including limonene (47.1%), thymol (8.1%) and, 4-terpinelol (6.4%). The in vitro evaluation of the anti-insect properties showed promising insecticidal activity against *Ceratitis capitata* (LD50 19.9–31.0 g/fly for males and females respectively at 72 h) and repellent activity against *T. infestans* (average repellence 92%, Class V) [33]. The potential of the Andean medicinal flora of Argentina as a source of essential oils with repellent activity has been reported in the last decade, together with the chemical profile of volatile compounds [12–14].

Regarding ZEOP, the repellent activity showed a low activity during the first 24 h (Class III) and it was growing until 96 h (Class IV) to both concentrations assayed (Tables 3 and 4). On the other hand, significant differences in average percentage of repellence were observed between the ZEOP treatment and blank control (effects between subjects, $p < 0.05$).

Nanoproducts developed using natural products have been highlighted as ecologically and economically sustainable alternatives for effective control of crop pest and other vectors of human incidence such as mosquitoes and triatomines. The strong activity of limonene and β-pinene against *Tribolium castaneum* has been informed; however, the high volatility and hydrophobicity hinder the use of these monoterpenes as a large-scale pest control agent [34].Recently, has been reported the nano-emulsification of monoterpenes and essential oil allowed their incorporation into an aqueous matrix without losing its repellent activities [34], which gives support to the results reported here.The controlled release systems for repellents comprise polymer micro/nanocapsules, micro/solid lipid nanoparticles, nanoemulsions/microemulsions, liposomes/niosomes, nanostructured hydrogels and cyclodextrins [35]. There are many formulations based on micro and nanocapsules containing DEET and essential oils to increase repellent action time duration and decrease permeation and consequently, systemic toxicity [36]. Limonene essential oil successfully encapsulated in microcapsules of chitosan showed a slow and prolonged liberation profile by volatilization [36]. The ZEOP has also shown a slow and prolonged release during 96 h.

The oil from *Z. punctata*, one of the endemic resinous species in Argentina that is extensively used in the traditional medicine of Argentina and Andean people for various purposes, has shown significant potential as a biorepellent against the vector of Chagas disease. Repellent activity is prolonged significantly if the oil is supported in a polymeric system.

3. Materials and Methods

3.1. Chemicals

All solvents used were of analytical grade. Chloroform was purchased from Fisher (Walham, MA, USA); acetone and methanol (MeOH) grade UHPLC from J.T. Baker (Phillipsburg, NJ, USA) and dichloromethane (DCM) from Aldrich Chemical Co. (St. Louis, MO, USA). Poly(ε-caprolactone) (PCL, Mw 80000 g/mol) and *N,N*-diethyl-3-methylbenzamide (DEET) were purchased from Aldrich Chemical Co. (USA). Ultra-pure water (<5 μg/L) was obtained from a purification system Arium 61316-RO plus Arium 611 UV (Sartorius, Göttingen, Germany).

3.2. Plant Material

The aerial parts of *Zuccagnia punctata* Cav. (Fabaceae, Caesalpinoideae) were collected in January 2018, on Iglesia district, province of San Juan (Argentina) at an altitude of 1800 m above sea level. The species has been previously identified by Dr Gloria Barboza, IMBIV (Instituto Multidisciplinario de Biología Vegetal, Facultad de Ciencias Exactas, Físicas y Naturales, Universidad Nacional de Córdoba, Argentina). A voucher specimen has been previously deposited at the herbarium of the Botanic Museum of Córdoba (CORD 1125).

3.3. Essential Oil Extraction and Chemical Analysis

Fresh aerial parts (500 g) were subjected to hydrodistillation for 2 h using a Clevenger type apparatus. The yields were averaged over two experiments and calculated according to dry weight of plant material. Essential oils (ZEO) were stored at −1 °C in airtight micro-tubes prior to chemical analysis. Qualitative data were determined by GC–FID and GC-MS. Gas chromatography-mass spectrometry analyses were carried out on a Hewlett-Packard 5890 II gas chromatograph coupled to a Hewlett-Packard 5989 B mass spectrometer, using a methyl silicone HP-5MS (crosslinked 5% PH ME Siloxane) capillary column (30 m × 0.25 mm), film thickness 0.25 µm. Samples were analyzed using the following GC-MS conditions: oven temperature program: 50–250 °C at 3 °C/min, carrier gas: helium, 1.5 mL/min; injection temperature: 250 °C, FID detector temperature: 300 °C; split mode ratio of 1:60. Additional parameters in the mass spectrometer unit: ion source temperature of 250 °C; ionizing voltage of 70 eV; scan range from *m/z* 35 to *m/z* 300. The identification of components was performed with the use of the volatile oil ADAMS library together with retention indices of reference compounds and built-in Wiley and NBS peak matching library search systems. Quantitative percentage composition was determined from the GC peak areas without correction factors [25–27].

3.4. Preparation of *Zuccagnia punctata* Essential Oil Loaded Polymeric Systems

PCL solutions of 10 wt./v % were prepared by dissolving PCL pellets in a 5 mL of DCM:MeOH solvent mixture (50:50 by volume) under magnetic stirring. For the preparation of polymeric matrices of PCL containing *Zuccagnia punctata* essential oil (ZEOP), 0.5 and 1 % wt./wt. of ZEO with respect to PCL were added to the solution. The selected solvent mixture allowed the complete dissolution of ZEO and PCL.

ZEOP were prepared by solution casting onto a Petri dish (4.6 mm in diameter) and dried in a fume hood at room temperature for 24 h. Samples were subsequently vacuum dried to remove residual solvent. Disc samples of 10 mm were cut and stored at room temperature under vacuum until use.

3.5. PCL, ZEO, and ZEOP Matrices Characterization

The morphology of matrices was examined by scanning electron microscopy (SEM, JEOL JSM6460 LV, Peabody, MA, USA) operated at 15 kV. Samples were sputter-coated with gold during 15 min in a chamber evacuated to 500 mTorr (Sputter coater, Desk II, Denton Vacuum, Moorestown, NJ, USA). Thermal properties of PCL pellets, ZEO, ZEOP and PCL matrices were determined by differential scanning calorimetry (DSC, TA instrument, Model Q-2000, New Castle, DE, USA). Scans were carried out at a heating rate of 10 °C/min. Glass transition temperature was taken as the onset of the transition. The degree of crystallinity of PCL (Xc) was calculated as:

$$\text{Xc (\%)} = (\Delta\text{Hm experimental}/\Delta\text{Hmtheoretical}) \times 100 \tag{1}$$

where the theoretical melting heat (ΔHm) for pure high molecular weight PCL was taken as 148.05 J/g [37]. Thermogravimetric Analysis (TGA) was conducted to study the thermal stability of ZEO. TGA data were obtained using a thermogravimetric analyzer (TA instrument, Model Q-500, New Castle, DE, USA). A sample of 5–10 mg was accurately weighed in an aluminum pan and the measurement was conducted at heating rate of 10 °C/min under nitrogen purging.

Zuccagnia punctata essential oil content was determined by ultraviolet-visible spectroscopy using an Agilent 8453 spectrometer (Santa Clara, CA, USA) equipped with a diode array system. A predetermined amount of sample was dissolved in DCM:MeOH (1:1 by volume), and quantification was carried out observing the absorption band at λ = 280 nm. At least three measurements were performed.

The loading capacity (LC) was calculated from the ratio between the ZEO mass in the sample (mZ) and the polymer mass (mP) in ZEOP matrix.

$$LC\ (\%) = (mZ)/(mP) \times 100 \quad (2)$$

The encapsulation efficiency (EE) was calculated as:

$$EE\ (\%) = (mZf/mPCLf)/(mZi/mPCLi) \times 100 \quad (3)$$

where mZfis the mass of ZEO encapsulated, mZi the initial mass of ZEO, and mPCLf and mPCLi correspond to the final and initial mass of PCL, respectively.

3.6. Repellent Activity against *Triatoma infestans* Nymphs Fifth Instars

The bioassays were carried out according to [13,14]. *Triatoma infestans* nymphs fifth instar were provided by Servicio Nacional de Chagas (Córdoba, Argentina) and were used one day after receipt.

Filter paper discs (9 cm in diameter) divided by halves were used. One half was treated with 0.5 mL of acetone solutions of the essential oils (0.5% and 1% wt./wt.) while the other half remained untreated. As control, circular white filter papers divided in two halves, one treated with 0.5 mL of acetone and the other untreated, were used. After solvent evaporation, filter paper discs were placed covering the floor of a Petri dish. Five starved nymphs of *T. infestans* (fifth instar) were released in the center of each Petri dish and maintained under controlled conditions of temperature 24 ± 2 °C, 50 ± 5% RH and photoperiod of 16 h L/8h D. Experiments were performed by quintuplicate. The same procedure was carried out with the polymer discs containing the essential oil at the same concentrations (0.5% and 1% wt./wt.). Insect distribution was recorded at 1, 24, 72, and 96 h of treatment. *N,N*-diethyl-3-methylbenzamide (DEET) was used as positive control at 0.5% (wt./v) and acetone as blank control.

Data were transformed into repellency percentage (RP %) as:

$$RP\ \% = (Nc - 50) \times 2 \quad (4)$$

Nc corresponds to the percentage of nymphs in the blank half.

Positive values show repellence while negative values show attraction. Mean values were categorized according to the following scale: Class 0 (>0.01 to <0.1), I (0.1 to 20), II (20.1 to 40); III (40.1 to 60); IV (60.1 to 80), V (80.1 to 100) according to Talukder et al. [38]. Data were analyzed by repeated measures ANOVA to determine the overall significance of the repellence means between the time points and the effect of oil treatment as a factor between subjects. Data were analyzed with the statistical software SPSS 15.0 (SPSS Inc.).

4. Conclusions

Essential oil from *Zuccagnia punctata* Cav. (Caesalpinieae) growing in the province of San Juan, located in the center-west of the Argentine, may be a potential alternative repellent to *T. infestans* (Klug) (Hemiptera, Reduviidae), the vector of Chagas disease. This oil biorepellent constitutes a rich source of sustainable, bioactive, and biodegradable compounds, especially a high content of oxygenated monoterpenes, such as (−)-5,6-dehydrocamphor.

Polymeric matrices of PCL loaded with different amounts of *Zuccagnia punctata* were prepared and characterized. Essential oil content on polymeric matrices showed encapsulation efficiencies higher than 98%, and it is thermally stable. The essential oils from the Andean flora have shown an excellent repellent activity, highlighting the repellent activity of the essential oil of the medicinal species *Zuccagnia punctata*. The effectiveness of ZEO was extended by its incorporation in polymeric systems and could have a potential home or peridomiciliary use, which might help prevent, or at least reduce, Chagas' disease transmission.

Author Contributions: S.L., A.T., J.Z., G.A.A. and P.R.C.T. conceived and designed the experiments; S.L., A.T., R.S., obtained *Z. punctata* essential oil, physical-chemical parameters, and performed the repellence assays. J.Z. analyzed the date of GC-MS, G.A.A. and P.R.C.T. obtained and analyzed the poly(ε-caprolactone) matrices loaded with *Z. punctata* essential oil. All authors have read and agreed to the published version of the manuscript.

Funding: G.A.A. and P.R.C.T. received financial support from the National Scientific and Technical Research Council of Argentina (grant PIP 0153), and the Argentinean Agency for Scientific and Technological Promotion (grant PICT 2018/02334). J.Z. received financial support from SECyT-Universidad Nacional de Córdoba. This research was funded by CICITCA-Universidad Nacional de San Juan (S.L. and A.T.), Argentina.

Institutional Review Board Statement: Not applicable.

Informed Consent Statement: Not applicable.

Data Availability Statement: Not applicable.

Acknowledgments: G.A.A.; J.Z. and P.R.C.T. are researchers from CONICET, Argentina.

Conflicts of Interest: The authors declare no conflict of interest.

Sample Availability: Samples of the *Zuccagnia punctata* essential oil are available from the authors.

References

1. WHO. La Enfermedad de Chagas (Tripanosomiasis Americana). 2020. Available online: http://www.who.int/mediacentre/factsheets/fs340/es/ (accessed on 20 January 2020).
2. Zerba, E.N. Susceptibility and resistance to insecticides of Chagas disease vectors. *Medicina* **1999**, *59*, 41–46.
3. Rojas de Arias, A.; Lehane, M.J.; Schofield, C.J.; Maldonado, M. Pyrethroid insecticide evaluation on different house structures in a Chagas disease endemic area of the Paraguayan Chaco. *Mem. Inst. Oswaldo Cruz* **2004**, *99*, 657–662. [CrossRef]
4. Vassena, C.V.; Cueto, G.M.; Audino, P.G.; Alzogaray, R.; Picollo, M.I.; Zerba, E.N. Prevalence and Levels of Permethrin Resistance in Pediculus humanus capitis De Geer (Anoplura: Pediculidae) from Buenos Aires, Argentina. *J. Med. Emtomol.* **2003**, *40*, 447–450. [CrossRef]
5. Picollo, M.I.; Vassena, C.V.; Santo Orihuela, P.; Barrios, S.; Zerba, E.N. High resistance to pyrethroid insecticides associated with ineffective field treatments in *Triatoma infestans* (Hemiptera, Reduvidae) from the north of Argentina. *J. Med. Entomol.* **2005**, *42*, 637–642. [CrossRef] [PubMed]
6. Santo Orihuela, P.L.; Vassena, C.V.; Zerba, E.N.; Picollo, M.I. Relative contribution of monooxygenase and esterase to pyrethroid resistance in *Triatoma infestans* (Hemiptera: Reduviidae) from Argentina and Bolivia. *J. Med. Entomol.* **2008**, *45*, 298–306. [CrossRef]
7. Cycon, M.; Piotrowska-Seget, Z. Pyrethroid-Degrading Microorganisms and Their Potential for the Bioremediation of Contaminated Soils: A Review. *Front. Microbiol.* **2016**, *7*, 1463. [CrossRef]
8. Chiriac, A.P.; Rusu, A.G.; Nita, L.E.; Chiriac, V.M.; Neamtu, I.; Sandu, A. Polymeric Carriers Designed for Encapsulation of Essential Oils with Biological Activity. *Pharmaceutics* **2021**, *13*, 631. [CrossRef] [PubMed]
9. Muñoz, V.; Buffa, F.; Molinari, F.; Hermida, L.G.; García, J.J.; Abraham. G.A. Electrospun ethyl cellulose-based mats with insect-repellent activity. *Mater. Lett.* **2019**, *253*, 289–292. [CrossRef]
10. Chattopadhyay, P.; Dhiman, S.; Borah, S.; Rabha, B. Essential oil based polymeric patch development and evaluating its repellent activity against mosquitoes. *Acta Trop.* **2015**, *147*, 45–53. [CrossRef]
11. Toloza, A.C.; Zygadlo, J.; Cueto, G.; Biurrun, F.; Zerba, E.; Picollo, M.I. Fumigant and Repellent Properties of Essential Oils and Component Compounds Against Permethrin-Resistant *Pediculus humanus capitis* (Anoplura: Pediculidae). *J. Med. Entomol.* **2006**, *43*, 889–895. [CrossRef]
12. Lima, B.; López, S.; Luna, L.; Agüero, M.B.; Aragón, L.; Tapia, A.; Zacchino, S.; López, M.L.; Zygadlo, J.; Feresin, G.E. Essential oils of medicinal plants from the Central Andes of Argentina: Chemical composition, and antifungal, antibacterial, and insect-repellent activities. *Chem. Biodivers.* **2011**, *8*, 924–936. [CrossRef] [PubMed]
13. López, S.; Lima, B.; Aragón, L.; Ariza-Espinar, L.; Tapia, A.; Zacchino, S.; Zygadlo, J.; Feresin, G.E.; López, M.L. Essential Oil of *Azorella cryptantha* collected in two different locations from San Juan Province, Argentina: Chemical variability and anti-insect and antimicrobial activities. *Chem. Biodivers.* **2012**, *9*, 1452–1464. [CrossRef]
14. López, S.; Lima, B.; Agüero, M.B.; López, M.L.; Hadad, M.; Zygadlo, J.; Caballero, D.; Stariolo, R.; Suero, E.; Feresin, G.E.; et al. Chemical composition, antibacterial and repellent activities of *Azorella trifurcata*, *Senecio pogonias*, and *Senecio oreophyton* essential oils. *Arab. J. Chem.* **2018**, *11*, 181–187. [CrossRef]
15. Barradas, T.N.; Senna, J.P.; Júnior, E.R.; Mansur, C.R.E. Polymer-based drug delivery systems applied to insects repellents devices: A review. *Curr. Drug Deliv.* **2016**, *13*, 221–235. [CrossRef]

16. Gómez, J.; Simirgiotis, M.; Manrique, S.; Lima, B.; Bórquez, J.; Feresin, G.E.; Tapia, A. UHPLC-HESI-OT-MS-MS biomolecules profiling, antioxidant and antibacterial activity of the "orange-yellow resin" from *Zuccagnia punctata* Cav. *Antioxidants* **2020**, *9*, 123. [CrossRef] [PubMed]
17. Paneva, D.; Manolova, N.; Argirova, M.; Rashkov, I. Antibacterial electrospun poly(ε-caprolactone)/ascorbylpalmitate nanofibrous materials. *Int. J. Pharm.* **2011**, *416*, 346–355. [CrossRef]
18. Seremeta, K.A.; Höcht, C.; Taira, C.; Tornello, P.R.C.; Abraham, G.A.; Sosnik, A. Didanosine-loaded poly(epsilon-caprolactone) microparticles by a coaxial electrohydrodynamic atomization (CEHDA) technique. *J. Mater. Chem. B* **2015**, *3*, 102–111. [CrossRef]
19. Tornello, P.R.C.; Feresin, G.E.; Tapia, A.; Veiga, I.G.; Moraes, A.M.; Abraham, G.A.; Cuadrado, T.R. Dispersion and release of embelin from electrospun biodegradable, polymeric, membranes. *Polym. J.* **2012**, *44*, 1105–1111. [CrossRef]
20. Tornello, P.R.C.; Feresin, G.E.; Tapia, A.; Cuadrado, T.R.; Abraham, G.A. Multilayered electrospun nanofibrous scaffolds for tailored controlled release of embelin. *Soft Mater.* **2018**, *16*, 51–61. [CrossRef]
21. Peres, M.C.; de Souza Costa, G.C.; dos Reis, L.E.L.; da Silva, L.D.; Peixoto, M.F.; Alves, C.C.F.; Forim, M.R.; Quintela, E.D.; Araújo, W.L.; de Melo Cazal, C. In natura and nanoencapsulated essential oils from *Xylopia aromatica* reduce oviposition of *Bemisia tabaci* in *Phaseolus vulgaris*. *J. Pest Sci.* **2020**, *93*, 807–821. [CrossRef]
22. De Ávila, D.S.C.; Holz, J.P.; Carone, C.L.P.; Morisso, F.D.; Ligabue, R.A.; Pighinelli, L. Microcapsules PCL with essential oil citronella. *Adv. Tissue Eng. Regen.* **2017**, *2*, 159–162.
23. Akolade, J.O.; Balogun, M.; Swanepoel, A.; Ibrahim, R.B.; Yusuf, A.A.; Labuschagne, P. Microencapsulation of eucalyptol in polyethylene glycol and polycaprolactone using particles from gas-saturated solutions. *RSC Adv.* **2019**, *9*, 34039. [CrossRef]
24. Unalan, I.; Slavik, B.; Buettner, A.; Goldmann, W.H.; Frank, G.; Boccaccini, A.R. Physical and antibacterial properties of Peppermint essential oil loaded poly(e-caprolactone) (PCL) electrospun fiber matts for wound healing. *Front. Bioeng. Biotechnol.* **2019**, *7*, 346. [CrossRef]
25. Adams, R.P. *Identification of Essential Oil Components by Gas Chromatography/Mass Spectroscopy*, 4th ed.; Adams, R.P., Ed.; Allured Publishing Corporation: Carol Stream, IL, USA, 2007; ISBN 1932633219.
26. Stein, S.; Mirokhin, D.; Tchekhovskoi, D.; Mallard, G.; Mikaia, A.; Zaikin, V.; Sparkmanm, D. *The NIST Mass Spectral Search Program for the Nist/Epa/Nih Mass Spectra Library*; Standard Reference Data Program of the National Institute of Standards and Technology; National Institute of Standards and Technology: Gaithersburg, MD, USA, 2011.
27. Alvarez, S.L.; Cortadi, A.; Juarez, M.A.; Petenatti, E.; Tomi, F.; Casanova, J.; Van Baren, C.M.; Zacchino, S.; Vila, R. (−)-5, 6-Dehydrocamphor from the antifungal essential oil of *Zuccagnia punctata*. *Phytochem. Lett.* **2012**, *5*, 194–199. [CrossRef]
28. Alves, S.F.; Borges, L.L.; de Paula, J.A.M.; Vieira, R.F.; Ferri, P.H.; do Couto, R.O.; de Paula, J.R.; Bara, M.T.F. Chemical variability of the essential oils from fruits of Pterodon emarginatus in the Brazilian Cerrado. *Rev. Bras. Farmacogn.* **2013**, *23*, 224–229. [CrossRef]
29. Angioni, A.; Barra, A.; Coroneo, V.; Dessi, S.; Cabras, P. Chemical composition, seasonal variability, and antifungal activity of Lavandula stoechas L. ssp. stoechas essential oils from stem/leaves and flowers. *J. Agric. Food Chem.* **2003**, *54*, 4364–4370. [CrossRef]
30. Masotti, V.; Juteau, F.; Bessie're, J.M.; Viano, J. Seasonal and phenological variations of the essential oil from the narrow endemic species Artemisia molinieri and its biological activities. *J. Agric. Food Chem.* **2003**, *51*, 7115–7121. [CrossRef]
31. Bakkali, F.; Averbeck, S.; Averbeck, D.; Waomar, M. Biological effects of essential oils—A review. *Food Chem. Toxicol.* **2008**, *46*, 446–475. [CrossRef] [PubMed]
32. Paumgartten, F.J.R.; Delgado, I.F. Mosquito repellents, effectiveness in preventing diseases and safety during pregnancy. *Vigil. Sanit. Debate* **2016**, *4*, 97–104.
33. Kurdelas, R.R.; López, S.; Lima, B.; Feresin, G.E.; Zygadlo, J.; Zacchino, S.; López, M.L.; Tapia, A.; Freile, M.L. Chemical composition, anti-insect and antimicrobial activity of the *Baccharis darwinii* essential oil from Argentine Patagonia. *Ind. Crops Prod.* **2012**, *40*, 261–267. [CrossRef]
34. Lima, L.A.; Ferreira-Sá, P.S.; Garcia, M.D.N., Jr.; Pereira, V.L.P.; Carvalho, J.C.T.; Rocha, L.; Fernandes, C.P.; Souto, R.N.P.; Araújo, R.S.; Botas, G.; et al. Nano-emulsions of the essential oil of *Baccharis reticularia* and its constituents as eco-friendly repellents against *Tribolium castaneum*. *Ind. Crops Prod.* **2021**, *162*, 113282. [CrossRef]
35. Tavares, M.; da Silva, M.R.M.; de Siqueira, L.B.O.; Rodrigues, R.A.S.; Bodjolle-d'Almeidab, L.; dos Santos, P.E.; Ricci-Júnior, E. Trends in insectrepellent formulations: A review. *Int. J. Pharm.* **2018**, *539*, 190–209. [CrossRef] [PubMed]
36. Souza, J.M.; Caldas, A.L.; Tohidi, S.D.; Molina, J.; Souto, A.P.; Fangueiro, R.; Zille, A. Properties and controlled release of chitosan microencapsulated limonene oil. *Braz. J. Pharmacogn.* **2014**, *24*, 691–698. [CrossRef]
37. Van Krevelen, W.; teNijenhuis, K. *Properties of Polymers*, 4th ed.; Elsevier: Amsterdam, The Netherlands, 2009; p. 121.
38. Talukder, F.A.; Howse, P.E. Laboratory evaluation of toxic and repellent properties of the pithraj tree, *Aphanamixis polystachya* Wall & Parker, against *Sitophilus oryzae* (L.). *Int. J. Pest Manag.* **1994**, *40*, 274–279.

molecules

MDPI

Article

Developing a *Hazomalania voyronii* Essential Oil Nanoemulsion for the Eco-Friendly Management of *Tribolium confusum, Tribolium castaneum* and *Tenebrio molitor* Larvae and Adults on Stored Wheat

Nickolas G. Kavallieratos [1,*], Erifili P. Nika [1], Anna Skourti [1], Nikoletta Ntalli [2], Maria C. Boukouvala [1], Catherine T. Ntalaka [1], Filippo Maggi [3], Rianasoambolanoro Rakotosaona [4,5], Marco Cespi [3], Diego Romano Perinelli [3], Angelo Canale [6], Giulia Bonacucina [3] and Giovanni Benelli [6]

1 Laboratory of Agricultural Zoology and Entomology, Department of Crop Science, Agricultural University of Athens, 75 Iera Odos Str., 11855 Athens, Attica, Greece; erifilinika@aua.gr (E.P.N.); annaskourti@aua.gr (A.S.); mbouk@aua.gr (M.C.B.); p1171915@aua.gr (C.T.N.)
2 Laboratory of Efficacy Assessment of Pesticides, Scientific Directorate of Pesticides' Assessment and Phytopharmacy, Benaki Phytopathological Institute, 8 Stefanou Delta Str., 14561 Kifissia, Attica, Greece; nntali@agro.auth.gr
3 School of Pharmacy, University of Camerino, 62032 Camerino, Italy; filippo.maggi@unicam.it (F.M.); marco.cespi@unicam.it (M.C.); diego.perinelli@unicam.it (D.R.P.); giulia.bonacucina@unicam.it (G.B.)
4 Centre National d'Application de Recherches Pharmaceutiques, Ambodivoanjo Ambohijatovo, Rue RP Rahajarizafy Analamahitsy, BP 702, 101 Antananarivo, Madagascar; rravalison@gmail.com
5 Ecole Supérieure Polytechnique d'Antananarivo, University of Antananarivo, BP 1500, 101 Antananarivo, Madagascar
6 Department of Agriculture, Food and Environment, University of Pisa, Via del Borghetto 80, 56124 Pisa, Italy; angelo.canale@unipi.it (A.C.); giovanni.benelli@unipi.it (G.B.)
* Correspondence: nick_kaval@aua.gr; Tel.: +30-2105294569

Citation: Kavallieratos, N.G.; Nika, E.P.; Skourti, A.; Ntalli, N.; Boukouvala, M.C.; Ntalaka, C.T.; Maggi, F.; Rakotosaona, R.; Cespi, M.; Perinelli, D.R.; et al. Developing a *Hazomalania voyronii* Essential Oil Nanoemulsion for the Eco-Friendly Management of *Tribolium confusum, Tribolium castaneum* and *Tenebrio molitor* Larvae and Adults on Stored Wheat. *Molecules* **2021**, *26*, 1812. https://doi.org/10.3390/molecules26061812

Academic Editor: Vincenzo De Feo

Received: 28 February 2021
Accepted: 18 March 2021
Published: 23 March 2021

Abstract: Most insecticides commonly used in storage facilities are synthetic, an issue that generates concerns about food safety and public health. Therefore, the development of eco-friendly pest management tools is urgently needed. In the present study, a 6% (*w/w*) *Hazomalania voyronii* essential oil-based nanoemulsion (HvNE) was developed and evaluated for managing *Tribolium confusum*, *T. castaneum*, and *Tenebrio molitor*, as an eco-friendly wheat protectant. Larval and adult mortality was evaluated after 4, 8, and 16 h, and 1, 2, 3, 4, 5, 6, and 7 days, testing two HvNE concentrations (500 ppm and 1000 ppm). *T. confusum* and *T. castaneum* adults and *T. molitor* larvae were tolerant to both concentrations of the HvNE, reaching 13.0%, 18.7%, and 10.3% mortality, respectively, at 1000 ppm after 7 days of exposure. However, testing HvNE at 1000 ppm, the mortality of *T. confusum* and *T. castaneum* larvae and *T. molitor* adults 7 days post-exposure reached 92.1%, 97.4%, and 100.0%, respectively. Overall, the HvNE can be considered as an effective adulticide or larvicide, depending on the target species. Our results highlight the potential of *H. voyronii* essential oil for developing green nanoinsecticides to be used in real-world conditions against key stored-product pests.

Keywords: botanical-based insecticide; cereals; green grain protectant; essential oil nanoformulation; stored-product beetles; Tenebrionidae

1. Introduction

The confused flour beetle, *Tribolium confusum* (Jacquelin du Val) (Coleoptera: Tenebrionidae), is a cosmopolitan secondary stored-product pest of high economic importance [1]. It has been reported to infest 119 different commodities [2]. The contamination of infested stored products by body fragments and toxins (e.g., methyl-1,4-benzoquinone, ethyl-1,4-benzoquinone, methoxybenzoquinone) may have a negative impact on consumers [3,4]. Due to its ability to infest processed commodities, it is usually found in mills, bakeries, pet shops, and storage units [2]. Recently, Kavallieratos et al. [5] reported that the larval and pupal development period is a complex phenomenon depending on the geographical origin of *T. confusum* and the type of infested commodity.

The red flour beetle, *Tribolium castaneum* (Herbst) (Coleoptera: Tenebrionidae), is a key secondary pest of stored products that is globally distributed [1]. It has been reported to infest 246 different commodities, an issue that makes it one of the most polyphagous stored-product insect pests [2]. It can be routinely found in mills, storage units, pet stores, and retail stores [2]. As in the case of *T. confusum*, *T. castaneum* produces quinones that cause skin irritation [6]. Skourti et al. [7,8] recently outlined that different levels of temperature and different types of commodities, chiefly alter the immature developmental period.

The yellow mealworm beetle, *Tenebrio molitor* L. (Coleoptera: Tenebrionidae), is one of the biggest stored-product insects. It is categorized as a secondary pest and a scavenger [1]. It infests fewer commodities (i.e., at least 46) compared to the other two species [2]. *Tenebrio molitor* can be easily found in flour mills, storage units, and food shops [2]. Apart from being a pest of stored products, it is reared as pet food for birds, fish, and reptiles [1,9,10], and–more recently–widely considered as a promising food for humans, while it degrades polystyrene and plastic waste [11]. *Tenebrio molitor* contaminates the commodities with quinones, but it is not as severe as the contamination by the species belonging to the *Tribolium* genus [10]. Anyway, it can cause allergic reactions in humans [6]. It can complete the biological cycle in only 30 days, while as an adult, it can survive up to two years [1,3].

It is well known that tenebrionids can develop resistance to several insecticides [12–14]. Therefore, new insecticidal formulations are necessary to be developed [15–18]. Among natural products with efficacy against stored-product insects, essential oils (EOs) and their main constituents have been revealed to be promising [18–20]. However, in real-world conditions, EOs need to be encapsulated in micro- and nanoformulations to enhance their persistence and physio-chemical stability while maintaining their biological properties [21,22]. Nanoemulsions (NEs) can be considered one of the most promising ways for the encapsulation and formulation of EOs. Nanoemulsions are kinetically stable oil droplets in water systems with a surfactant-to-oil ratio (SOR) ranging from 1 to 2 [23], and a droplet diameter <100 nm. Thanks to the reduced size of the droplets of the internal phase and the consequent increment of the surface area, NEs allow a better interaction of the encapsulated compounds into the target site [23] by overcoming the EO's poor physicochemical stability and solubility issues [24]. This strategy can boost the applicability of EOs as natural insecticides [25–29].

Therefore, EO-based NEs are alternative solutions for pest management for a wide spectrum of insects of public health importance, and crop pests as well. For example, Ghosh et al. [30] treated 3rd instar larvae of *Aedes aegypti* L. (Diptera: Culicidae) with different NE concentrations from *Ocimum basilicum* L. (Lamiaceae) EO. Larval mortality was completely suppressed even after 15 min of exposure at 200 mL of 10-fold diluted NE. Duarte et al. [31] tested the larvicidal efficacy of 5% (*w*/*w*) NE of *Rosmarinus officinalis* L. (Lamiaceae) EO against the 4th instar larvae of *Ae. aegypti*. Mortality levels reached 80% after 24 h and 90% after 48 h at 250 ppm. Recently, Benelli et al. [27] proposed a 6% (*w*/*w*) NE of *Carlina acaulis* L. (Compositae) root EO against the European grapevine moth, *Lobesia botrana* (Denis and Schiffermüller) (Lepidoptera: Tortricidae), reaching 50% and 90% mortality of 1st instar larvae with 9.04 and 17.70 μL/mL, respectively. In a further recent study, Pavela et al. [32] evaluated NEs based on *C. acaulis* EO and found that less than 1200 μL/L caused 90% mortality to 3rd instar larvae of *Culex quinquefasciatus* Say (Diptera: Culicidae).

However, there is little published research on the utilization of NEs based on EOs as grain protectants [16,33]. For instance, Hashem et al. [19] used an NE of *Pimpinella anisum* L. (Apiaceae) EO for the management *T. castaneum* adults, over a wide spectrum of NE concentrations, on cracked wheat kernels. Nevertheless, there are no data on the efficacy of *Hazomalania voyronii* (Jum.) Capuron (Hernandiaceae) EO-based NE against *T. confusum*, *T. castaneum*, and *T. molitor*. However, the bioactivity of the pure EO of *H. voyronii* has been recently studied [20], showing that the raw *H. voyronii* EO exerted a rather limited toxicity as a grain protectant against important stored-product beetles. Even on selected species and instars (e.g., *Trogoderma granarium* Everts (Coleoptera: Dermestidae) adults), mortality rates reached about 79% after 7 days of exposure at 1000 ppm. In this framework, one may hypothesize that a way to boost the efficacy of this EO may be to develop highly stable EO-based nanoformulations [23]. As a model EO prototype, here we used the one obtained from *H. voyronii*, a traditional Malagasy plant (e.g., it is used to heal wounds, the drinkable bark decoction of stems is used for the treatment of malaria) with documented insecticidal efficacy [34]. Perilla aldehyde, the major compound of the *H. voyronii* EO, is used as a flavouring component to baked foods, sweets, meat products, dressing for salads, sauces, salted vegetables, and beverages [35]. Furthermore, perilla aldehyde is a "generally recognized as safe" (GRAS) substance [36].

To validate the hypothesis formulated above, the objective of the present study was the development of a 6% (*w*/*w*) *H. voyronii* EO-based NE for the effective and eco-friendly management of larvae and adults of three major stored-product beetles (i.e., *T. confusum*, *T. castaneum*, and *T. molitor*). To assess the applied potential to protect stored grains, the effectiveness of this NE as a grain protectant was investigated in small environments mimicking real wheat storage conditions.

2. Results

2.1. Development and Characterization of H. voyronii EO-Based NE

After a preliminary screening, the quantitative composition of the *H. voyronii* NE was selected as follows: 6% (*w*/*w*) of the EO phase was emulsified in the aqueous medium containing 4% (*w*/*w*) of surfactant (Polysorbate 80). A high-energy method (i.e., high pressure homogenization) was employed at the pressure of 130 MPa to obtain *H. voyronii* EO-based NE characterized by oil droplets with a size in the nanometric range. From DLS analysis, in fact, the sample showed a monomodal size distribution with a mean diameter (Z-average) of 53.54 ± 0.20 nm and a polydispersity index of 0.340 ± 0.013 after preparation (Figure 1). The absence of oil droplets with a diameter above 1 µm confirms the formation of a true NE. Indeed, the sample appeared homogenous upon observation by optical microscope.

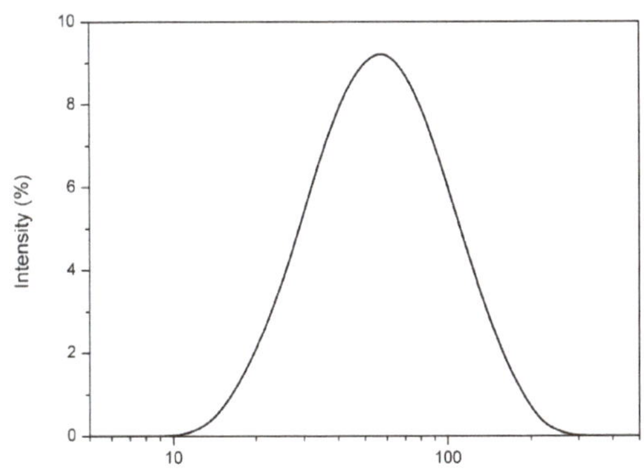

Figure 1. Size distribution (d.nm) of the prepared *Hazomalania voyronii*-based nanoemulsion (6% *w*/*w*) as obtained from dynamic light scattering.

2.2. Insecticidal Efficacy

When the insecticidal efficacy of the 6% (*w*/*w*) *H. voyronii* NE was evaluated, between exposure intervals, all main effects were significant, while the associate interaction was not significant (Table 1). Within exposure intervals, the main effect as well as the interaction exposure x insect species-stage were significant, while the interactions exposure x NE concentration and exposure x NE concentration x insect species-stage were not significant (Table 1).

Table 1. Evaluation of the insecticidal activity of a 6% (*w/w*) *Hazomalania voyronii* essential oil nanoemulsion: MANOVA parameters about the main effects and associated interactions leading to the observed mortality rates on *Tribolium castaneum*, *Tribolium confusum*, and *Tenebrio molitor* adults and larvae, between and within exposure intervals (error *df* = 96).

Between Exposure Intervals	*df*	*F*	*p*
Intercept	1	892.5	<0.01
Nanoemulsion concentration	1	8.4	<0.01
Insect species-stage	5	67.0	<0.01
Nanoemulsion concentration x insect species-stage	5	0.3	0.89
Within exposure intervals	*df*	*F*	*P*
Exposure	9	85.0	<0.01
Exposure x nanoemulsion concentration	9	0.9	0.51
Exposure x insect species-stage	45	6.1	<0.01
Exposure x nanoemulsion concentration x insect species-stage	45	1.1	0.40

Concerning *T. castaneum* adults, the mortality caused by *H. voyronii* EO-based NE was 0.0% until the 1st day post-treatment, then reached 2.2% after 2 days of exposure for both tested NE concentrations (Table 2). The mortality remained low and did not exceed 12.5% and 18.7% at 500 ppm and 1000 ppm, respectively, after 7 days of exposure. Although the mortality of *T. castaneum* larvae did not exceed 10% at 500 ppm and 12.6% at 1000 ppm 16 h post-exposure, it reached 84.1% at 500 ppm and 97.4% at 1000 ppm, after 7 days of exposure.

Regarding *T. confusum* adults, the mortality remained at low levels (Table 3). After 4 days of exposure at 500 ppm, the mortality was 3.3%, and after 7 days it reached 10.3%. The mortality at 1000 ppm was not significantly higher than the one at 500 ppm, after 2 and 7 days of exposure reaching 3.3% and 13.0%, respectively. As far as *T. confusum* larval mortality is concerned, 5 days of exposure to the two NE concentrations led to significantly higher mortality at 1000 ppm than 500 ppm (i.e., 45.0% and 26.4%, respectively). The *H. voyronii* NE killed 59.3% of larvae at 500 ppm, and 92.1% of larvae at 1000 ppm, 7 days post-exposure.

Mortality of *T. molitor* adults was <90% after the 7th day of exposure (94.8%) at 500 ppm, while at 1000 ppm after the 6th day of exposure it was 93.5% (Table 4). Complete mortality of this life stage was achieved testing 1000 ppm of *H. voyronii* NE after 7 days post-exposure. No mortality of *T. molitor* larvae was noted testing 500 ppm and 1000 ppm, 2 days and 1 day post-exposure, respectively. At the end of the experimental period, the overall larval mortality did not exceed 5.8% at 500 ppm, and 10.3% at 1000 ppm.

Table 2. Mean (%) mortality ± SE of *Tribolium castaneum* adults and larvae after 4 h, 8 h, 16 h, 1 day, 2 days, 3 days, 4 days, 5 days, 6 days, and 7 days of exposure to wheat treated with a 6% (*w*/*w*) *Hazomalania voyronii* essential oil nanoemulsion at two concentrations. Within each column, asterisks indicate significant differences, and in all cases $df = 16$; two-tailed t-test at $p = 0.05$. Within each row, means followed by the same uppercase letter are not significantly different, $df = 9, 89$; Tukey's HSD test at $p = 0.05$. Where no letters or no asterisks exist, no significant differences were recorded. Where dashes exist, no statistical analysis was performed.

Exposure	4 h	8 h	16 h	1 Day	2 Days	3 Days	4 Days	5 Days	6 Days	7 Days	F	p
	Adults											
500 ppm	0.0 ± 0.0^{B}	0.0 ± 0.0^{B}	0.0 ± 0.0^{B}	0.0 ± 0.0^{B}	2.2 ± 1.5^{AB}	3.3 ± 2.4^{AB}	3.3 ± 2.4^{AB}	6.7 ± 2.9^{AB}	9.1 ± 4.0^{AB}	12.5 ± 4.7^{A}	3.1	<0.01
1000 ppm	0.0 ± 0.0^{B}	0.0 ± 0.0^{B}	0.0 ± 0.0^{B}	0.0 ± 0.0^{B}	2.2 ± 1.5^{AB}	7.8 ± 3.2^{AB}	11.4 ± 4.0^{AB}	13.2 ± 3.8^{A}	16.5 ± 4.8^{A}	18.7 ± 5.4^{A}	6.1	<0.01
t	-	-	-	-	0	−1.0	−1.6	−1.1	−1.0	−0.6		
p	-	-	-	-	1.00	0.31	0.13	0.29	0.32	0.53		
	Larvae											
500 ppm	2.2 ± 1.5^{D}	8.9 ± 2.6^{CD}	10.0 ± 2.9^{C}	14.4 ± 2.9^{BC}	30.1 ± 4.6^{AB}	41.6 ± 5.2^{AB}	54.4 ± 5.0^{A}	58.2 ± 5.7^{A}	60.7 ± 6.8^{A}	84.1 ± 4.1^{A}	23.3	<0.01
1000 ppm	4.4 ± 1.8^{C}	10.0 ± 2.4^{C}	12.6 ± 3.3^{BC}	15.1 ± 4.2^{BC}	30.3 ± 4.9^{AB}	42.7 ± 5.4^{A}	54.9 ± 4.5^{A}	70.1 ± 3.5^{A}	$87.1 \pm 3.6^{A*}$	$97.4 \pm 1.7^{A*}$	19.7	<0.01
t	−1.0	−0.5	−0.5	0.4	<0.1	−0.2	−0.1	−1.9	−3.3	−2.9		
p	0.35	0.66	0.61	0.72	1.00	0.87	0.89	0.08	<0.01	0.01		

Table 3. Mean (%) mortality ± SE of *Tribolium confusum* adults and larvae after 4 h, 8 h, 16 h, 1 day, 2 days, 3 days, 4 days, 5 days, 6 days, and 7 days of exposure to wheat treated with a 6% (*w*/*w*) *Hazomalania voyronii* essential oil nanoemulsion at two concentrations. Within each column, asterisks indicate significant differences, in all cases $df = 16$; two-tailed t-test at $p = 0.05$. Within each row, means followed by the same uppercase letter are not significantly different, $df = 9, 89$; Tukey's HSD test at $p = 0.05$. Where no letters or no asterisks exist, no significant differences were recorded. Where dashes exist, no statistical analysis was performed.

Exposure	4 h	8 h	16 h	1 Day	2 Days	3 Days	4 Days	5 Days	6 Days	7 Days	F	p
	Adults											
500 ppm	0.0 ± 0.0^{C}	0.0 ± 0.0^{C}	0.0 ± 0.0^{C}	0.0 ± 0.0^{C}	0.0 ± 0.0^{C}	0.0 ± 0.0^{C}	3.3 ± 1.7^{BC}	7.9 ± 2.8^{AB}	10.3 ± 2.4^{A}	10.3 ± 2.4^{A}	11.3	<0.01
1000 ppm	0.0 ± 0.0^{B}	0.0 ± 0.0^{B}	1.1 ± 1.1^{B}	2.2 ± 2.2^{B}	3.3 ± 2.4^{AB}	4.4 ± 2.4^{AB}	5.6 ± 2.4^{AB}	9.3 ± 2.0^{A}	10.8 ± 2.5^{A}	13.0 ± 3.3^{A}	5.9	<0.01
t	-	-	−1	−1	−1.5	−2.0	−0.6	−0.8	−0.1	−0.3		
p	-	-	0.33	0.33	0.15	0.07	0.58	0.44	0.95	0.8		
	Larvae											
500 ppm	0.0 ± 0.0^{D}	0.0 ± 0.0^{D}	1.1 ± 1.1^{D}	4.7 ± 2.6^{CD}	6.9 ± 3.0^{CD}	13.0 ± 4.3^{BC}	20.7 ± 3.0^{AB}	26.4 ± 1.7^{A}	41.9 ± 2.3^{A}	59.3 ± 2.9^{A}	31.7	<0.01
1000 ppm	0.0 ± 0.0^{F}	1.1 ± 1.1^{F}	4.4 ± 2.4^{EF}	13.3 ± 4.7^{DE}	$21.1 \pm 6.1^{CD*}$	28.6 ± 6.2^{BCD}	36.0 ± 6.2^{ABC}	$45.0 \pm 6.7^{ABC*}$	$67.7 \pm 2.3^{AB*}$	$92.1 \pm 2.0^{A*}$	29.2	<0.01
t	-	−1	−1.2	−1.6	−2.3	−1.7	−1.9	−2.2	−7.6	−8.7		
p	-	0.33	0.26	0.14	0.03	0.12	0.07	0.05	<0.01	<0.01		

Table 4. Mean (%) mortality ± SE of *Tenebrio molitor* adults and larvae after 4 h, 8 h, 16 h, 1 day, 2 days, 3 days, 4 days, 5 days, 6 days, and 7 days of exposure to wheat treated with a 6% (*w*/*w*) *Hazomalania voyronii* essential oil nanoemulsion at two concentrations. Within each column, asterisks indicate significant differences, in all cases $df = 16$; two-tailed *t*-test at $P = 0.05$. Within each row, means followed by the same uppercase letter are not significantly different, $df = 9, 89$; Tukey's HSD test at $p = 0.05$. Where no letters or no asterisks exist, no significant differences were recorded. Where dashes exist, no statistical analysis was performed.

Exposure	4 h	8 h	16 h	1 Day	2 Days	3 Days	4 Days	5 Days	6 Days	7 Days	*F*	*p*
	Adults											
500 ppm	0.0 ± 0.0 E	0.0 ± 0.0 E	6.8 ± 2.4 D	15.8 ± 4.4 CD	25.9 ± 4.7 BC	43.1 ± 9.3 AB	55.3 ± 6.1 AB	70.8 ± 6.0 A	80.1 ± 4.9 A	94.8 ± 2.8 A	59.6	<0.01
1000 ppm	0.0 ± 0.0 E	0.0 ± 0.0 E	9.0 ± 2.0 D	21.7 ± 5.2 CD	41.3 ± 8.0 BC	63.3 ± 9.1 AB	72.5 ± 8.2 AB	84.3 ± 5.9 AB	93.5 ± 3.8 AB*	100.0 ± 0.0 A	49.1	<0.01
t	-	-	−0.9	−0.4	−0.4	−1.4	−1.3	−1.4	−2.1	−1.9		
p	-	-	0.38	0.68	0.68	0.17	0.21	0.18	0.05	0.08		
	Larvae											
500 ppm	0.0 ± 0.0B	0.0 ± 0.0 B	0.0 ± 0.0 B	0.0 ± 0.0 B	0.0 ± 0.0 B	1.1 ± 1.1 AB	2.2 ± 1.5 AB	3.3 ± 1.7 AB	5.6 ± 2.4 A	5.8 ± 2.5 A	3.0	<0.01
1000 ppm	0.0 ± 0.0 C	0.0 ± 0.0 C	0.0 ± 0.0 C	0.0 ± 0.0 C	4.4 ± 2.4 B	4.4 ± 2.4 B	5.6 ± 2.4 AB	5.6 ± 2.4 AB	5.6 ± 2.4 AB	10.3 ± 3.8 A	3.0	<0.01
t	-	-	-	-	−2.0	−1.2	−1.1	−0.6	0	−0.7		
p	-	-	-	-	0.07	0.26	0.31	0.58	1.00	0.52		

3. Discussion

Our results indicate that the NE-based on *H. voyronii* EO (6% *w/w*) is effective against *T. castaneum*, *T. confusum*, and *T. molitor*. This nanosystem led to high mortality levels on different life stages of the tested species, even at 500 ppm. Concerning *T. molitor* adults, the mortality reached 100% at 1000 ppm, and 94.8% at 500 ppm, after 7 days of exposure. The mortality of larvae was very low, even at the highest tested concentration (1000 ppm), reaching 10.3% after 7 days of exposure. Earlier research has revealed that the adult stage is the most susceptible life stage of *T. molitor*. For example, Kavallieratos et al. [37] found that when 0.504 ppm deltamethrin, 5 ppm pirimiphos-methyl, 1000 ppm silicoSec (which is a diatomaceous earth (DE)), and 1 ppm spinosad were applied on stored wheat as grain protectants, high adult mortality rates were achieved (92.2%, 100%, 100%, and 94,4%, respectively), while they caused moderate larval mortality (10.0%, 71.1%, 43.3%, and 28.9%, respectively). Apart from the wheat, barley and maize have also been tested on the same formulations as grain protectants, and the results followed the same pattern, with larvae being more tolerant than adults. In addition, testing the efficacy of 5 ppm pirimiphos-methyl on barley in a wide range of temperature and relative humidity levels, a greater mortality of *T. molitor* adults was found, if compared to the larvae [38].

In an earlier study, the raw EO of *H. voyronii* showed relevant efficacy on *T. granarium* adults, but not on the larvae [20]. The raw EO caused 78.9% adult mortality at 1000 ppm, but only 15.6% at 500 ppm after 7 days of exposure. Larvae were more tolerant and the mortality rates at 500 ppm and 1000 ppm after 7 days of exposure did not exceed 4.4% and 15.6%, respectively. Although the raw *H. voyronii* EO does not allow an adequate control of *T. molitor* and *T. granarium* larvae, the suppression of the adult stage of both pests is a very important finding, as adults are the vehicle of reproduction [20].

In contrast, the results of the present study outlined that the adults of *T. castaneum* and *T. confusum* remained practically unaffected by being exposed to the *H. voyronii* EO-based NE, given that the highest adult mortality was 18.7% for *T. castaneum* at 1000 ppm and 13.0% at 1000 ppm for *T. confusum*, while the larvae had a mortality of 97.4% for *T. castaneum* at 1000 ppm and 92.1% at 1000 ppm for *T. confusum* after 7 days of exposure. Previous research on *T. castaneum* has reported that the adults are tolerant to several insecticides. Fumigation studies on the species of four EOs extracted from the plants *Lantana camara* L. (Verbenaceae), *Cymbopogon nardus* (L.) Rendle (Poaceae), *Cinnamomum zeylanicum* Blume (Lauraceae), and *Trachyspermum ammi* (L.) Sprague (Apiaceae) showed that for all the tested EOs and exposure times, *T. castaneum* adults needed relatively high EO volumes (e.g., 14.56, 37.52, 4.40, and 14.86 μL, respectively, for each EO after 72 h of exposure) than the larvae (e.g., 5.00, 4.13, 2.48, and 2.73 μL, respectively, for each EO after 72 h of exposure) [39]. When the d-strain of *T. castaneum* was treated with 1 mg/L and the r-strain with 2 mg/L phosphine (PH_3), after 4 h of treatment, adults have been found more tolerant than the larvae. For both d- and r-strains of *T. castaneum*, after 24 h of exposure to <10 mg/L carbonyl sulphide (COS), the larvae were less tolerant than adults [40]. Earlier, Arthur [41] reported that larvae were less tolerant than the adults. Regarding contact toxicity, Deb and Kumar [42] reported that the *Artemisia annua* L. (Asteraceae) EO had greater larvicidal than adulticidal efficacy against *T. castaneum*. Mujeeb and Shakoori [43] suggested the larval stage as the preferable life stage to apply pirimiphos-methyl, outlining that this is the most susceptible stage. Adults of *T. confusum* were also more tolerant than the larvae. The natural insecticide silicoSec attained 100% larval mortality after 7 days of exposure, but when tested on adults it did not exceed 85% after 14 days of exposure [44]. Spinetoram and spinosad [45] had high larval mortality reaching 98.9% in a mixture of the two insecticides after 14 days on treated wheat kernels, while the adult mortality reached 67.8% for the same mixture after 14 days on treated wheat kernels. Similarly, studying the effectiveness of eight pyrrole derivatives, a better larvicide than adulticide action has been detected [46–48].

EO constituents have a wide spectrum of effectiveness on adults and larvae that could be partially explained by their different mode of actions [49]. For instance, perilla aldehyde, the main component of *H. voyronii* EO, showed insecticidal and inhibitory effects on the enzyme acetylcholinesterase (AChE) in *Drosophila suzukii* (Matsumura) (Diptera: Drosophilidae) [50], and it has been argued that its insecticidal activity could be related to the presence of exocyclic and endocyclic double bonds in the chemical structure [51]. Furthermore, 1,8-cineole and limonene, the other two main constituents of the *H. voyronii* EO, showed insecticidal efficacy against a wide spectrum of insects, including mosquitoes (e.g., *Culex pipiens* L. (Diptera: Culicidae), *Cx. quinquefasciatus*, *Ae. aegypti*, *Ae. albopictus* (Skuse) (Diptera: Culicidae)), houseflies (*Musca domestica* L. (Diptera: Muscidae)), and stored-product beetles (*Sitophilus granarius* (L.) (Coleoptera: Curculinonidae)) [52–55].

The significant toxicity of the EO-based NE developed here against *T. castaneum* and *T. confusum* larvae is important, since targeting the adults is crucial to achieve a major reduction of the overall population. Storage units can host several species in different developmental stages existing simultaneously [56–59]. Therefore, insecticides based on natural products such as EO-based NEs, which can manage a broad spectrum of stored-product insects, are highly desirable [16,33]. Our results showed that the *H. voyronii* EO-based NE is toxic against three tenebrionid species. Similarly, Hashem et al. [19] tested four different concentrations of a *P. anisum* EO-based NE against *T. castaneum* adults, showing that 7.5% and 10% v/v killed 51.2% and 74.3% of the exposed individuals, in comparison to the control after 9 days of exposure, respectively; 12 days post-exposure, the overall mortality rates reached 54.7% at 7.5% v/v and 81.3% at 10% v/v.

Our findings indicate that among the different species tested here, all belonging to the Tenebrionidae family, there is a wide variability of the performance of *H. voyronii* EO-based NE as a grain protectant. However, the level of efficacy depends on the life stage of the target species. A crucial factor for elevated effectiveness of insecticides is the timing of their application [7,60]. Therefore, the knowledge of the species and the life stage that infests grain commodities could maximize the management strategy if the *H. voyronii* EO-based NE is selected as a component of a management strategy against *T. castaneum*, *T. confusum*, and *T. molitor*.

4. Materials and Methods

4.1. Essential Oil

The dry bark of *H. voyronii* was obtained from trees growing in Kirindy Forest, Madagascar (coordinates: S 20°28′15.002″; E 44°17′56.06″; 62 m a.s.l.) in February, 2018. Then, it was subjected to hydrodistillation for 3 h using a Clevenger-type apparatus. The chemical composition was recently reported in our recent study by Benelli et al. [34]. The major components were oxygenated monoterpenes, namely perilla aldehyde (43%), 1,8-cineole (33.2%), and limonene (13%).

4.2. Insects

Tribolium confusum and *T. castaneum* were cultured on wheat flour and 5% brewer's yeast, and *T. molitor* on a combination of oat bran and potato slices for additional moisture [61], at 30 °C, 65% relative humidity, and continuous darkness [37,62,63]. The founding colonies were acquired from storage facilities in Greece. The two *Tribolium* species were maintained in the Laboratory of Agricultural Zoology and Entomology, Agricultural University of Athens, since 2003, and *T. molitor* since 2014. For the tests, unsexed adults <2 weeks old, and larvae that were 3rd–4th instar old (for *T. confusum* and *T. castaneum*) and 10–14 mm long (for *T. molitor*) were used [37,62].

4.3. Commodity

Hard wheat, *Triticum durum* Desf. (var. Claudio) (Poales: Poaceae) clean and free of infestations and pesticides was used in the experiments. Prior to the bioassays, moisture content of the wheat kernels was 12.4%, calibrated by a moisture meter (mini GAC plus, Dickey-John Europe S.A.S., Colombes, France).

4.4. Development and Characterization of *H. voyronii* EO-Based Nanoemulsion

Hazomalania voyronii EO NE was obtained through a high-energy method, by using a high-pressure homogenizer according to the procedure reported by Cappellani et al. [64]. A 6% (*w*/*w*) of EO was added dropwise to 4% (*w*/*w*) Polysorbate 80 (Sigma-Aldrich) aqueous solution under high-speed stirring (Ultraturrax T25 basic, IKAfi Werke GmbH & Co.KG, Staufen, Germany) for 5 min at 9500 rpm. The obtained emulsion was then homogenized with a French Pressure Cell Press (American Instrument Company, Aminco, MY, USA) for four cycles at a pressure of 130 MPa. Visual inspection of the formulation was performed by using a polarizing optical microscope (MT9000, Meiji Techno Co. Ltd., Chikumazawa, Miyoshi machi, Iruma-gun Saitama, Japan) equipped with a 3-megapixel CMOS camera (Invenio 3S, DeltaPix, Smørum, Denmark) to assess NEs formation. Dynamic light scattering (DLS) analyses were then carried out to determine lipophilic internal phase droplets size using a Zetasizer Nano S (Malvern Instruments, Worcestershire, UK) equipped with backscattered light detector working at 173°. A sample of 1 mL was analysed at 25 °C, following a temperature equilibration time of 180 s. The analyses were performed in triplicate.

4.5. Insecticidal Assays

The NE based on *H. voyronii* EO (6% *w*/*w*) was tested at the concentrations of 500 µL/kg wheat (=500 ppm) and 1000 µL/kg wheat (=1000 ppm). The test concentrations were selected based on preliminary tests on the three tenebrionid species. Test solutions were prepared in water at the final volume of 750 µL, and sprays were performed on plates, where 0.25 kg wheat lots were laid out [20]. Additional lots of 0.25 kg wheat treated with (i) water and (ii) carrier control (4% *w*/*w* surfactant dispersed in water) served as controls. The spraying of the wheat lots was conducted by an AG-4 airbrush (Mecafer S.A., Valence, France) on different trays. Controls were sprayed using different AG-4 airbrushes. Then treated lots were transferred to 1-L glass containers and were shaken for 10 min to equally distribute the NE on the total quantity of wheat. The same procedure was followed for controls. From each treated lot or controls, three samples of 10 g each were obtained with different scoops. A Precisa XB3200D compact balance (Alpha Analytical Instruments, Gerakas, Greece) was used to weigh the samples on a filter paper that was unique for each sample. Then, samples were placed in glass vials (7.5 cm × 12.5 cm diameter and height). The caps of the vials had a 1.5 cm diameter circular opening in the middle, covered with gauze, to sufficiently aerate the content of the vials. The upper inner necks of the vials were covered by polytetrafluoroethylene (60 wt % dispersion in water) (Sigma-Aldrich Chemie GmbH, Taufkirchen, Germany), to assure that the beetles would remain in the vials. Thereafter, 10 larvae or adults of each tenebrionid species were separately transferred into the vials. The containers remained in incubators set to 30 °C and 65% RH for the whole experimental period. Mortality rates' evaluation was accomplished by an Olympus stereomicroscope (SZX9, Bacacos S.A., Athens, Greece) at 57x total magnification after 4, 8, and 16 h and 1, 2, 3, 4, 5, 6, and 7 days of exposure, by gently nudging each individual insect using a fine brush (Cotman 111 No 000, Winsor and Newton, London, UK) to detect any movement. For each concentration of the NE and control, different brushes were used. The above procedure was repeated three times with new insects, wheat, and vials.

4.6. Data Analysis

Control mortality on wheat treated with water was low (<5%) for all insect species and life stages tested, therefore no correction was necessary. In contrast, the control mortality on wheat treated with the carrier control 4% *w*/*w*surfactant dispersed in water, was >5% for all tested species and life stages, ranging up to 17.8%. Therefore, mortality values were corrected by the Abbott formula (i.e., (1 − insect population in treated unit after treatment/insect population in control unit after treatment) × 100) [65]. Before conducting analysis, the mortality data were log(x + 1) transformed to normalize variance [66,67]. Statistical analyses were conducted by following the repeated-measures model [68]. Exposure interval was the repeated factor, and mortality was the response variable. The main effects were the concentration and insect species/developmental stage. The associated interactions of the main effects were considered in the analysis. All analyses were conducted using JMP v.14 software [69]. Means were separated using the Tukey–Kramer (HSD) test at 0.05 of significance [70]. The two-tailed *t*-test at *n*-2 *df* and 0.05 significance [71] was used to compare the two tested concentrations of *H. voyronii* EO-based NE at each species or life stage.

5. Conclusions

To eco-friendly manage several important insect pests of stored products, a combination of the *H. voyronii* EO-based NE with other natural insecticides (e.g., DEs) could potentially provide an enhanced level of protection for stored durable commodities against multi species infections. For example, the use of DE or natural zeolite as grain protectants caused 100% mortality on *T. castaneum* adults after 14 or 21 days of exposure, respectively [72]. A combination of *H. voyronii* EO-based NE with DE or natural zeolite could enhance the likelihood of the management of *T. castaneum*, *T. confusum*, and *T. molitor*, regardless of their being adults or larvae, an issue that merits further investigation. Furthermore, Athanassiou et al. [73] found that the mortality of *S. oryzae* was significantly higher when silica gel was combined with the EO of *Juniperus oxycedrus* L. ssp. *oxycedrus* (Pinales: Cupressaceae) compared to the silica gel alone. In this scenario, additional experimental efforts are necessary to shed light on the impact of *H. voyronii* EO-based NE as a multi-species killing agent.

Author Contributions: Conceptualization, N.G.K., F.M., and G.B. (Giovanni Benelli); methodology, N.G.K., N.N., F.M., G.B. (Giulia Bonacucina) and G.B. (Giovanni Benelli); software N.G.K., E.P.N., and G.B. (Giulia Bonacucina); validation, N.G.K., E.P.N., N.N., F.M., G.B. (Giulia Bonacucina), and G.B. (Giovanni Benelli); formal analysis, N.G.K., E.P.N., and G.B. (Giulia Bonacucina); investigation, N.G.K., E.P.N., A.S., N.N., M.C.B., C.T.N., F.M., R.R., M.C., D.R.P., A.C., G.B. (Giulia Bonacucina), and G.B. (Giovanni Benelli); resources, N.G.K. and G.B. (Giovanni Benelli); data curation, N.G.K., E.P.N., A.S., and G.B. (Giulia Bonacucina); writing—original draft preparation, N.G.K., E.P.N., F.M., G.B. (Giulia Bonacucina), and G.B. (Giovanni Benelli); writing—review and editing, N.G.K., E.P.N., A.S., N.N., M.C.B., C.T.N., F.M., R.R., M.C., D.R.P., A.C., G.B. (Giulia Bonacucina), and G.B. (Giovanni Benelli); visualization, N.G.K., E.P.N., F.M., G.B. (Giulia Bonacucina), and G.B. (Giovanni Benelli); supervision N.G.K., F.M., and G.B. (Giovanni Benelli); funding acquisition, N.G.K. and G.B. (Giovanni Benelli). All authors have read and agreed to the published version of the manuscript.

Funding: This research was partially funded by the 34.0401 Project (Special Account for Research Funds of the Agricultural University of Athens).

Institutional Review Board Statement: Not applicable.

Informed Consent Statement: Not applicable.

Data Availability Statement: All data are within this manuscript.

Conflicts of Interest: The authors declare no competing interests.

Sample Availability: Not available.

References

1. Rees, D. *Insects of Stored Products*; Manson Publishing: London, UK, 2004.
2. Hagstrum, D.W.; Subramanyam, B. *Stored-Product Insect Resource*; AACC International: St. Paul, MN, USA, 2009.
3. Robinson, W.H. *Urban Insects and Arachnids*; Cambridge University Press: Cambridge, UK, 2005.
4. Yezerki, A.; Gilmor, T.P.; Stevens, L. Genetic analysis of benzoquinone production in *Tribolium confusum*. *J. Chem. Ecol.* **2004**, *30*, 1035–1044. [CrossRef]
5. Kavallieratos, N.G.; Andrić, G.; Golić, M.P.; Nika, E.P.; Skourti, A.; Kljajić, P.; Papanikolaou, N.E. Biological features and population growth of two Southeastern European *Tribolium confusum* Jacquelin du Val (Coleoptera: Tenebrionidae) strains. *Insects* **2020**, *11*, 218. [CrossRef]
6. Krinsky, W.L. Beetles (Coleoptera). In *Medical and Veterinary Entomology*; Mullen, G.R., Durden, L.A., Eds.; Elsevier Academic Press Inc.: London, UK, 2019; pp. 129–143.

7. Skourti, A.; Kavallieratos, N.G.; Papanikolaou, N.E. Laboratory evaluation of development and survival of *Tribolium castaneum* (Herbst) (Coleoptera: Tenebrionidae) under constant temperatures. *J. Stored Prod. Res.* **2019**, *83*, 305–310. [CrossRef]
8. Skourti, A.; Kavallieratos, N.G.; Papanikolaou, N.E. Suitability of semolina, cracked wheat and cracked maize as feeding commodities for *Tribolium castaneum* (Herbst; Coleoptera: Tenebrionidae). *Insects* **2020**, *11*, 99. [CrossRef] [PubMed]
9. Hill, D.S. *Pests of Storage Foodstuffs and their Control*; Kluwer Academic Publishers: New York, NY, USA, 2003.
10. DeFoliart, G.R. Food, Insects as. In *Encyclopedia of Insects*; Resh, V.H., Cardé, R.T., Eds.; Elsevier Academic Press Inc.: London, UK, 2009; pp. 376–381.
11. Vigneron, A.; Jehan, C.; Rigaud, T.; Moret, Y. Immune defenses of a beneficial pest: The mealworm beetle *Tenebrio molitor*. *Front. Physiol.* **2019**, *10*, 138. [CrossRef] [PubMed]
12. Arthur, F.H. Grain Protectants: Current status and prospects for the future. *J. Stored Prod. Res.* **1996**, *4*, 293–302. [CrossRef]
13. Khaliq, A.; Ullah, M.I.; Afzal, M.; Ali, A.; Sajjad, A.; Ahmad, A.; Khalid, S. Management of *Tribolium castaneum* using synergism between conventional fumigant and plant essential oils. *Int. J. Trop. Insect Sc.* **2020**, *40*, 781–788. [CrossRef]
14. Cui, K.; Zhang, L.; He, L.; Zhang, Z.; Zhang, T.; Mu, W.; Lin, J.; Liu, F. Toxicological effects of the fungal volatile compound 1-octen-3-ol against the red flour beetle, *Tribolium castaneum* (Herbst). *Ecotox. Environ. Safe.* **2021**, *208*, 111597. [CrossRef]
15. Giunti, G.; Palermo, D.; Laudani, F.; Algeri, G.M.; Campolo, O.; Palmeri, V. Repellence and acute toxicity of a nano-emulsion of sweet orange essential oil toward two major stored grain insect pests. *Ind. Crops Prod.* **2019**, *142*, 111869. [CrossRef]
16. Giunti, G.; Campolo, O.; Laudani, F.; Zappalà, L.; Palmeri, V. Bioactivity of essential oil-based nano-biopesticides toward *Rhyzopertha dominica* (Coleoptera: Bostrichidae). *Ind. Crops Prod.* **2021**, *162*, 113257. [CrossRef]
17. Petrović, M.; Popović, A.; Kojić, D.; Šućur, J.; Bursić, V.; Aćimović, M.; Malenčić, Đ.; Stojanović, T.; Vuković, G. Assessment of toxicity and biochemical response of *Tenebrio molitor* and *Tribolium confusum* exposed to *Carum carvi* essential oil. *Entomol. Gen.* **2019**, *38*, 333–348. [CrossRef]
18. Kavallieratos, N.G.; Boukouvala, M.C.; Ntalli, N.; Kontodimas, D.C.; Cappellacci, L.; Petrelli, R.; Ricciutelli, M.; Benelli, G.; Maggi, F. Efficacy of the furanosesquiterpene isofuranodiene against the stored-product insects *Prostephanus truncatus* (Coleoptera: Bostrychidae) and *Trogoderma granarium* (Coleoptera: Dermestidae). *J. Stored Prod. Res.* **2020**, *86*, 101553. [CrossRef]
19. Hashem, A.S.; Awadalla, S.S.; Zayed, G.M.; Maggi, F.; Benelli, G. *Pimpinella anisum* essential oil nanoemulsions against *Tribolium castaneum*—insecticidal activity and mode of action. *Environ. Sci. Pollut. Res.* **2018**, *25*, 18802–18812. [CrossRef] [PubMed]
20. Kavallieratos, N.G.; Boukouvala, M.C.; Ntalli, N.; Skourti, A.; Karagianni, E.S.; Nika, E.P.; Kontodimas, D.C.; Cappellacci, L.; Petrelli, R.; Cianfaglione, K.; et al. Effectiveness of eight essential oils against two key stored-product beetles, *Prostephanus truncatus* (Horn) and *Trogoderma granarium* Everts. *Food Chem. Toxicol.* **2020**, *139*, 111255. [CrossRef]
21. Athanassiou, C.G.; Kavallieratos, N.G.; Benelli, G.; Losic, D.; Rani, P.U.; Desneux, N. Nanoparticles for pest control: Current status and future perspectives. *J. Pest Sci.* **2018**, *91*, 1–15. [CrossRef]
22. Benelli, G. On a magical mystery tour of green insecticide research: Current issues and challenges. *Molecules* **2020**, *25*, 5014. [CrossRef]
23. Pavoni, L.; Pavela, R.; Cespi, M.; Bonacucina, G.; Maggi, F.; Zeni, V.; Canale, A.; Lucchi, A.; Bruschi, F.; Benelli, G. Green micro- and nanoemulsions for managing parasites, vectors and pests. *Nanomaterials* **2019**, *9*, 1285. [CrossRef]
24. Turek, C.; Stintzing, F.C. Stability of essential oils: A review. *Compr. Rev. Food Sci. Food Saf.* **2013**, *12*, 40–53. [CrossRef]
25. Pavela, R.; Pavoni, L.; Bonacucina, G.; Cespi, M.; Kavallieratos, N.G.; Cappellacci, L.; Petrelli, R.; Maggi, F.; Benelli, G. Rationale for developing novel mosquito larvicides based on isofuranodiene microemulsions. *J. Pest Sci.* **2019**, *92*, 909–921. [CrossRef]
26. Pavela, R.; Benelli, G.; Pavoni, L.; Bonacucina, G.; Cespi, M.; Cianfaglione, K.; Bajalan, I.; Morshedloo, M.R.; Lupidi, G.; Romano, D.; et al. Microemulsions for delivery of Apiaceae essential oils—Towards highly effective and eco-friendly mosquito larvicides? *Ind. Crops Prod.* **2019**, *129*, 631–640. [CrossRef]
27. Benelli, G.; Pavoni, L.; Zeni, V.; Ricciardi, R.; Cosci, F.; Cacopardo, G.; Gendusa, S.; Spinozzi, E.; Petrelli, R.; Cappellacci, L.; et al. Developing a highly stable *Carlina acaulis* essential oil nanoemulsion for managing *Lobesia botrana*. *Nanomaterials* **2020**, *10*, 1867. [CrossRef]
28. Heydari, M.; Amirjani, A.; Bagheri, M.; Sharifian, I.; Sabahi, Q. Eco-friendly pesticide based on peppermint oil nanoemulsion: Preparation, physicochemical properties, and its aphicidal activity against cotton aphid. *Environ. Sci. Pollut. Res.* **2020**, *27*, 6667–6679. [CrossRef] [PubMed]
29. Rossi, P.; Cappelli, A.; Marinelli, O.; Valzano, M.; Pavoni, L.; Bonacucina, G.; Petrelli, R.; Pompei, P.; Mazzara, E.; Ricci, I.; et al. Mosquitocidal and anti-inflammatory properties of the essential oils obtained from monoecious, male, and female inflorescences of hemp (*Cannabis sativa* L.) and their encapsulation in nanoemulsions. *Molecules* **2020**, *25*, 3451. [CrossRef] [PubMed]
30. Ghosh, V.; Mukherjee, A.; Chandrasekaran, N. Formulation and characterization of plant essential oil based nanoemulsion: Evaluation of its larvicidal activity against *Aedes aegypti*. *Asian J. Chem.* **2013**, *25*, S321–S323.
31. Duarte, J.L.; Amado, J.R.R.; Oliveira, A.E.M.F.M.; Cruz, R.A.S.; Ferreira, A.M.; Souto, R.N.P.; Falcão, D.Q.; Carvalho, J.C.T.; Fernandes, C.P. Evaluation of larvicidal activity of a nanoemulsion of *Rosmarinus officinalis* essential oil. *Rev. Bras. Farmacogn.* **2015**, *25*, 189–192. [CrossRef]
32. Pavela, R.; Pavoni, L.; Bonacucina, G.; Cespi, M.; Cappellacci, L.; Petrelli, R.; Spinozzi, E.; Aguzzi, C.; Zeppa, L.; Ubaldi, M.; et al. Encapsulation of *Carlina acaulis* essential oil and carlina oxide to develop long-lasting mosquito larvicides: Microemulsions versus nanoemulsions. *J. Pest Sci.* **2021**. [CrossRef]

33. Lima, L.A.; Ferreira-Sá, P.S.; Garcia, M.D., Jr.; Pereira, V.L.P.; Carvalho, J.C.T.; Rocha, L.; Fernandes, C.P.; Souto, R.N.P.; Araújo, R.S.; Botas, G.; et al. Nano-emulsions of the essential oil of *Baccharis reticularia* and its constituents as eco-friendly repellents against *Tribolium castaneum*. *Ind. Crops Prod.* **2021**, *162*, 113282. [CrossRef]
34. Benelli, G.; Pavela, R.; Rakotosaona, R.; Nzekoue, F.K.; Canale, A.; Nicoletti, M.; Maggi, F. Insecticidal and mosquito repellent efficacy of the essential oils from stem bark and wood of *Hazomalania voyronii*. *J. Ethnopharmacol.* **2020**, *248*, 112333. [CrossRef]
35. Hobbs, C.A.; Taylor, S.V.; Beevers, C.; Lloyd, M.; Bowen, R.; Lillford, L.; Maronpot, R.; Hayashi, S. Genotoxicity assessment of the flavouring agent, perillaldehyde. *Food Chem. Toxicol.* **2016**, *97*, 232–242. [CrossRef]
36. Oser, B.L.; Ford, R.A. Recent progress in the consideration of flavoring ingredients under the Food Additives Amendment. 11. GRAS substances. *Food Technol.* **1978**, *32*, 60–70.
37. Kavallieratos, N.G.; Michail, E.J.; Boukouvala, M.C.; Nika, E.P.; Skourti, A. Efficacy of pirimiphos-methyl, deltamethrin, spinosad and silicoSec against adults and larvae of *Tenebrio molitor* L. on wheat, barley and maize. *J. Stored Prod. Res.* **2019**, *83*, 161–167. [CrossRef]
38. Kavallieratos, N.G.; Papanikolaou, N.E.; Kazani, A.N.; Boukouvala, M.C.; Malesios, C. Using multilevel models to explore the impact of abiotic and biotic conditions on the efficacy of pirimiphos-methyl against *Tenebrio molitor* L. *Environ. Sci. Pollut. Res.* **2020**. [CrossRef]
39. Kathirvelu, C.; Maline, A.S.; Raja, B.A.G.; Kavitha, T. Effect of fumigation on *Rhyzopertha dominica* F. and *Tribolium castaneum* H. in stored products using essential oils. *Ann. Agri Bio Res.* **2020**, *25*, 258–262.
40. Lee, H.K.; Jeong, G.; Kim, H.K.; Kim, B.S.; Yang, J.O.; Koo, H.N.; Kim, G.H. Fumigation activity against phosphine-resistant *Tribolium castaneum* (Coleoptera: Tenebrionidae) using carbonyl sulfide. *Insects* **2020**, *11*, 750. [CrossRef] [PubMed]
41. Arthur, F.H. Residual efficacy of aerosols to control *Tribolium castaneum* and *Tribolium confusum*.. In Proceedings of the 10th International Working Conference on Stored Product Protection, Estoril, Portugal, 27 June–2 July 2010; Carvalho, M.O., Fields, P.G., Adler, C.S., Arthur, F.H., Athanassiou, C.G., Campbell, J.F., Fleurat-Lessard, F., Flinn, P.W., Hodges, R.J., Isikber, A.A., et al., Eds.; Julius Kühn Institut, Bundesforschungsinstitut für Kulturpflanzen: Quedlinburg, Germany, 2010; Volume 425, pp. 789–792.
42. Deb, M.; Kumar, D. Bioactivity and efficacy of essential oils extracted from *Artemisia annua* against *Tribolium castaneum* (Herbst. 1797) (Coleoptera: Tenebrionidae): An eco-friendly approach. *Ecotox. Environ. Saf.* **2020**, *189*, 109988. [CrossRef]
43. Mujeeb, K.A.; Shakoori, A.R. Effect of an organophosphate, pirimiphos-methyl, on esterases of different developmental stages of stored grain pest red flour beetle, *Tribolium castaneum* (Herbst.)—spectrophotometric analysis. *Pak. J. Zool.* **2012**, *44*, 301–312.
44. Vayias, B.J.; Athanassiou, C.G. Factors affecting the insecticidal efficacy of the diatomaceous earth formulation silicoSec against adults and larvae of the confused flour beetle, *Tribolium confusum* DuVal (Coleoptera: Tenebrionidae). *Crop Prot.* **2004**, *23*, 565–573. [CrossRef]
45. Athanassiou, C.G.; Kavallieratos, N.G. Evaluation of spinetoram and spinosad for control of *Prostephanus truncatus*, *Rhyzopertha dominica*, *Sitophilus oryzae*, and *Tribolium confusum* on stored grains under laboratory tests. *J. Pest Sci.* **2014**, *87*, 469–483. [CrossRef]
46. Boukouvala, M.C.; Kavallieratos, N.G.; Athanassiou, C.G.; Hadjiarapoglou, L.P. Biological activity of two new pyrrole derivatives against stored-product species: Influence of temperature and relative humidity. *Bull. Entomol. Res.* **2016**, *106*, 446–456. [CrossRef]
47. Boukouvala, M.C.; Kavallieratos, N.G.; Athanassiou, C.G.; Hadjiarapoglou, L.P. Insecticidal effect of two novel pyrrole derivatives against two major stored product insect species. *Crop Prot.* **2016**, *84*, 1–7. [CrossRef]
48. Boukouvala, M.C.; Kavallieratos, N.G.; Athanassiou, C.G.; Benelli, G.; Hadjiarapoglou, L.P. Insecticidal efficacy of six new pyrrole derivatives against four stored-product pests. *Environ. Sci. Pollut. Res.* **2019**, *26*, 29845–29856. [CrossRef] [PubMed]
49. Jankowska, M.; Rogalska, J.; Wyszkowska, J.; Stankiewicz, M. Molecular targets for components of essential oils in the insect nervous system—A review. *Molecules* **2018**, *23*, 34. [CrossRef] [PubMed]
50. Park, C.G.; Jang, M.; Yoon, K.A.; Kim, J. Insecticidal and acetylcholinesterase inhibitory activities of Lamiaceae plant essential oils and their major components against *Drosophila suzukii* (Diptera: Drosophilidae). *Ind. Crops Prod.* **2016**, *89*, 507–513. [CrossRef]
51. You, C.X.; Wang, Y.; Zhang, W.J.; Yang, K.; Wu, Y.; Geng, Z.F.; Chen, H.P.; Jiang, H.Y.; Du, S.S.; Deng, Z.W.; et al. Chemical constituents and biological activities of the purple *Perilla* essential oil against *Lasioderma serricorne*. *Ind. Crops Prod.* **2014**, *61*, 331–337. [CrossRef]
52. Traboulsi, A.F.; El-Haj, S.; Tueni, M.; Taoubi, K.; Nader, N.A.; Mrad, A. Repellency and toxicity of aromatic plant extracts against the mosquito *Culex pipiens molestus* (Diptera: Culicidae). *Pest Manag. Sci.* **2005**, *61*, 597–604. [CrossRef] [PubMed]
53. Kordali, S.; Aslan, I.; Calmasur, O.; Cakir, A. Toxicity of essential oils isolated from three *Artemisia species* and some of their major components to granary weevil *Sitophilus granarius* (L.) (Coleoptera: Curculinonidae). *Ind. Crops Prod.* **2006**, *23*, 162–170. [CrossRef]
54. Kumar, P.; Mishra, S.; Malik, A.; Satya, S. Housefly (*Musca domestica* L.) control potential of *Cymbopogon citratus* Stapf. (Poales: Poaceae) essential oil and monoterpenes (citral and 1, 8-cineole). *Parasitol. Res.* **2013**, *112*, 69–76. [CrossRef]
55. Maggi, F.; Benelli, G. Essential oils from aromatic and medicinal plants as effective weapons against mosquito vectors of public health importance. In *Mosquito-borne Diseases*; Springer: Cham, Switzerland, 2018; pp. 69–129.
56. Nansen, C.; Subramanyam, B.; Roesli, R. Characterizing spatial distribution of trap captures of beetles in retail pet stores using SADIE® software. *J. Stored Prod. Res.* **2004**, *40*, 471–483. [CrossRef]
57. Athanassiou, C.G.; Kavallieratos, N.G.; Sciarretta, A.; Palyvos, N.E.; Trematerra, P. Spatial associations of insects and mites in stored wheat. *J. Econ. Entomol.* **2011**, *104*, 104–1752. [CrossRef]

58. Arthur, F.H.; Campbell, J.F.; Toews, M.D. Distribution, abundance, and seasonal patterns of stored product beetles in a commercial food storage facility. *J. Stored Prod. Res.* **2014**, *56*, 21–32. [CrossRef]
59. Kavallieratos, N.G.; Athanassiou, C.G.; Guedes, R.N.C.; Drempela, J.D.; Boukouvala, M.C. Invader competition with local competitors: Displacement or coexistence among the invasive khapra beetle, *Trogoderma granarium* Everts (Coleoptera: Dermestidae), and two other major stored-grain beetles? *Front. Plant Sci.* **2017**, *8*, 1837. [CrossRef] [PubMed]
60. Nika, E.P.; Kavallieratos, N.G.; Papanikolaou, N.E. Linear and non-linear models to explain influence of temperature on life history traits of *Oryzaephilus surinamensis* (L.). *Entomol. Gen.* **2021**. [CrossRef]
61. De Vosjoli, P. *The Lizard Keeper's Handbook*; Advanced Vivarium Systems: Irvine, CA, USA, 2007.
62. Kavallieratos, N.G.; Athanassiou, G.G.; Boukouvala, M.C. Insecticidal effect of chlorantraniliprole against major stored-product insect pests in different grain commodities under laboratory tests. *Pest Manag. Sci.* **2013**, *69*, 1141–1154. [CrossRef] [PubMed]
63. Boukouvala, M.C.; Romano, D.; Kavallieratos, N.G.; Athanassiou, C.G.; Stefanini, C.; Canale, A.; Benelli, G. Asymmetric courtship boosts male mating success in the red flour beetle, *Tribolium castaneum* (Herbst) (Coleoptera: Tenebrionidae). *J. Stored Prod. Res.* **2019**, *81*, 1–6. [CrossRef]
64. Cappellani, M.R.; Perinelli, D.R.; Pescosolido, L.; Schoubben, A.; Cespi, M.; Cossi, R.; Blasi, P. Injectable nanoemulsions prepared by high pressure homogenization: Processing, sterilization, and size evolution. *Appl. Nanosci.* **2018**, *8*, 1483–1491. [CrossRef]
65. Abbott, W.S. A Method of Computing the Effectiveness of an Insecticide. *J. Econ. Entomol.* **1925**, *18*, 265–266. [CrossRef]
66. Zar, J.H. *Biostatistical analysis*; Pearson: Upper Saddle River, NJ, USA, 2010.
67. Scheff, D.S.; Arthur, F.H. Fecundity of *Tribolium castaneum* and *Tribolium confusum* adults after exposure to deltamethrin packaging. *J. Pest Sci.* **2018**, *91*, 717–725. [CrossRef]
68. Sall, J.; Lehman, A.; Creighton, L. *JMP Start Statistics. A Guide to Statistics and Data Analysis Using JMP and JMP IN Software*; Duxbury Press: Belmont, CA, USA, 2001.
69. SAS Institute Inc. *Using JMP 14*; SAS Institute Inc.: Cary, NC, USA, 2018.
70. Sokal, R.R.; Rohlf, F.J. *Biometry*; Freeman & Company: New York, NY, USA, 1995.
71. Snedecor, G.W.; Cochran, W.G. *Statistical Methods*; Iowa State University Press: Ames, IA, USA, 1980.
72. Andrić, G.; Marković, M.M.; Adamović, M.; Daković, A.; Golić, M.P.; Kljajić, P.J. Insecticidal potential of natural zeolite and diatomaceous earth formulations against rice weevil (Coleoptera: Curculionidae) and red flour beetle (Coleoptera: Tenebrionidae). *J. Econ. Entomol.* **2012**, *105*, 670–678. [CrossRef]
73. Athanassiou, C.G.; Kavallieratos, N.G.; Evergetis, E.; Katsoula, A.M.; Haroutounian, S. Insecticidal efficacy of silica gel with *Juniperus oxycedrus* ssp. oxycedrus (Pinales: Cupressaceae) essential oil against Sitophilus oryzae (Coleoptera: Curculionidae) and Tribolium confusum (Coleoptera: Tenebrionidae). *J. Econ. Entomol.* **2013**, *106*, 1902–1910. [CrossRef]

molecules

MDPI

Article

Plumbagin, a Potent Naphthoquinone from *Nepenthes* Plants with Growth Inhibiting and Larvicidal Activities

Asifur Rahman-Soad [1], Alberto Dávila-Lara [1], Christian Paetz [2] and Axel Mithöfer [1,*]

[1] Research Group Plant Defense Physiology, Max Planck Institute for Chemical Ecology, 07745 Jena, Germany; msoad@ice.mpg.de (A.R.-S.); adavila-lara@ice.mpg.de (A.D.-L.)
[2] Research Group Biosynthesis/NMR, Max Planck Institute for Chemical Ecology, 07745 Jena, Germany; cpaetz@ice.mpg.de
* Correspondence: amithoefer@ice.mpg.de

Abstract: Some plant species are less susceptible to herbivore infestation than others. The reason for this is often unknown in detail but is very likely due to an efficient composition of secondary plant metabolites. Strikingly, carnivorous plants of the genus *Nepenthes* show extremely less herbivory both in the field and in green house. In order to identify the basis for the efficient defense against herbivorous insects in *Nepenthes,* we performed bioassays using larvae of the generalist lepidopteran herbivore, *Spodoptera littoralis*. Larvae fed with different tissues from *Nepenthes x ventrata* grew significantly less when feeding on a diet containing leaf tissue compared with pitcher-trap tissue. As dominating metabolite in *Nepenthes* tissues, we identified a naphthoquinone, plumbagin (5-hydroxy-2-methyl-1,4-naphthoquinone). When plumbagin was added at different concentrations to the diet of *S. littoralis* larvae, an EC_{50} value for larval growth inhibition was determined with 226.5 $\mu g\ g^{-1}$ diet. To further determine the concentration causing higher larval mortality, sweet potato leaf discs were covered with increasing plumbagin concentrations in no-choice-assays; a higher mortality of the larvae was found beyond 60 µg plumbagin per leaf, corresponding to 750 $\mu g\ g^{-1}$. Plant-derived insecticides have long been proposed as alternatives for pest management; plumbagin and derivatives might be such promising environmentally friendly candidates.

Keywords: naphthoquinones; plumbagin; *Spodoptera littoralis*; insect growth inhibition; carnivorous plants; *Nepenthes*

Citation: Rahman-Soad, A.; Dávila-Lara, A.; Paetz, C.; Mithöfer, A. Plumbagin, a Potent Naphthoquinone from *Nepenthes* Plants with Growth Inhibiting and Larvicidal Activities. *Molecules* **2021**, *26*, 825. https://doi.org/10.3390/molecules26040825

Academic Editor: Giovanni Benelli
Received: 30 December 2020
Accepted: 1 February 2021
Published: 5 February 2021

1. Introduction

Nepenthes is a tropical plant genus occurring mainly in Southeast Asia. Plants of this genus are carnivorous. They attract, catch, and digest insect prey in order to get additional nutrients, primarily, nitrogen and phosphate [1,2]. Therefore, *Nepenthes* species developed a pitfall trap (Figure 1), called pitcher, where insect prey falls inside due to a slippery surface and drown in a digestive fluid [1,2]. As in many other carnivorous plants, also the genus *Nepenthes* harbors a large chemical diversity; currently, several secondary metabolites are isolated for pharmaceutical, biotechnological, and ethnobotanical use [3,4]. Especially, *Nepenthes* species are well known in traditional medicine. Multiple reports are in the literature describing curative effects of *Nepenthes* extracts on diseases, e.g., on hypertension, cough, fever, urinary system infections [5], malaria [6,7], pain, asthma [7], *Staphylococcus* infection [8], celiac disease [9], and oral cancer cells [10].

However, up to now, most of the chemical analysis in *Nepenthes* has been done for the digestive pitcher fluid. Here, metabolites with antimicrobial properties have been found, e.g., naphthoquinones (NQ; droserone, 5–*O*–methyl droserone in *N. khasiana* [11]; plumbagin, 7-methyl-juglone in *N. ventricosa* [12]). Thus, it is hypothesized that such compounds mediate protection against microbes and preserve prey during digestion [11–14]. NQ derivatives are also described for tissues of various *Nepenthes* species including the pitchers [12,15–17]. In particular, plumbagin is of broad pharmaceutical interest because it is a candidate that may be used in therapies against various cancers or chronic diseases [18–21]. In addition to NQ, carotenoids, flavonoids, sterols, and triterpenes are described for *Nepenthes* leaves [1,22,23]. Recently, an untargeted metabolomics approach was performed in *N. x ventrata* comparing secondary metabolites of leaves and pitcher tissue before and after prey catches [24]. In that study, about 2000 compounds (MS/MS events) were detected in the two tissues showing enormous metabolome diversity, which was even higher in leaves. Strikingly, the tissue specificity of chemical compounds could significantly discriminate pitchers from leaves. Besides many yet unknown compounds, the common constituents were phenolics, flavonoids, and NQ [24]. These data suggest that the metabolite composition of the tissues can point to their function. In addition, the metabolite composition may represent mechanisms that promote the evolution of plant carnivory as well as enable the plants to cope with environmental challenges [14].

Figure 1. *Nepenthes x ventrata*. A natural hybrid of *N. ventricosa* and *N. alata*. Copyright © A. Rahman-Soad.

(A)biotic challenges include the attack of herbivorous insects. Interestingly, there are only a very few observations and studies published concerning the attack of insects on tissues of carnivorous pitcher plants. Recently, lepidopteran herbivory was described for some species of the new world pitcher plant *Sarracenia* [25,26]. There is only one investigation showing that *N. bicalcarata* plants are attacked by an insect, the weevil *Alcidodes spec.* [27]. Another study shows that in *N. gracilis* red pitchers experience less herbivory than green ones [28]. To the best of our knowledge, no other studies have been published yet that focus on herbivore damage in *Nepenthes*. Obviously, the carnivorous syndrome obtained much more attention. However, herbivory on *Nepenthes* tissue is obviously rare.

The reason for this is not known but it is unlikely that all herbivores are caught and digested. Instead, *Nepenthes* very likely has an efficient setting of defensive chemistry, which is not unusual in many plants [29]. In order to address this hypothesis and gain more insight in the ecological relevance of *Nepenthes* metabolites, we performed bioassays to study the effect of tissue of *N. x ventrata,* a robust natural hybrid of *N. alata* and *N. ventricosa,* on the feeding behavior and larval development of the generalist insect herbivore *Spodoptera littoralis.*

2. Results and Discussion

2.1. Effect of *Nepenthes x ventrata* Tissue on Insect Larvae Growth

The observation that *Nepenthes* plants are rarely infested by insect herbivores forced us to study this phenomenon. Therefore, freshly harvested tissues from *N. x ventrata* leaves and pitchers were added to an artificial diet and fed to larvae of the generalist herbivore *Spodoptera littoralis.* As can be seen in Figure 2A, starting at day 4 to 5, the presence of leaf but not pitcher tissue significantly affected the performance of the larvae, which gained less weight. At this point, it might be worth to mention that recently in *N. x ventrata* [24] and before in *N. khasiana* [15], the concentration of a NQ, very likely plumbagin, was determined to be significantly higher in leaves compared with pitchers, which may explain the result found in Figure 2A. We also could support these results by comparing plumbagin content in pitcher vs. leaf; by quantitative NMR analysis, we found a 5.2-fold higher plumbagin concentration in leaf compared with pitcher tissue (650 and 125 μg g^{-1} FW, respectively). Although significant, the growth inhibition effect was not very pronounced. Thus, the feeding experiment was repeated with dried leaf tissue in order to add more plant material to the diet, knowing that the water content of *N. x ventrata* tissue is about 90% [24]. Here, the effect of the plant tissue was more distinct (Figure 2B). Both quantities of leaf tissue, 10% and 15% (*w*/*w*), showed clear impairment on the growth and weight of the feeding *S. littoralis* larvae already at day 2. Starting from day 3 on, there was also a significant difference between the larvae feeding on either 10% or 15% of *Nepenthes* tissue that was included in the diet (Figure 2B).

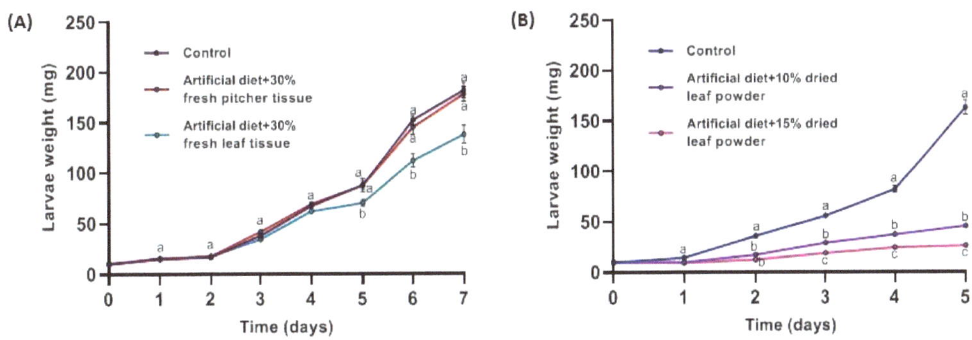

Figure 2. Performance of *Spodoptera littoralis* larvae feeding on artificial diet containing (**A**) fresh leaf powder of *Nepenthes x ventrata* leaf and pitcher (30% (*w*/*w*)) or (**B**) dried *N. x ventrata* leaf powder (10 and 15% (*w*/*w*)) Larvae were weighed every day for 7 days. Mean (± SE) labelled with different letters indicate significant difference ($p < 0.05$); two-way ANOVA, Šidák's multiple comparisons test; $n = 15$.

2.2. Plumbagin in *Nepenthes x ventrata* Tissue

In many carnivorous plants belonging to the order Nepenthales [14], a *sensu stricto* sister group to Caryophyllales [30] and including the plant families Droseraceae and Nepenthaceae, the presence of NQ has been described [31]. This includes species such as *Aldrovanda vesiculosus, Dionaea muscipula, Drosophyllum lusitanicum,* as well as the genera *Drosera* and *Nepenthes* [31]. Among their secondary compounds, in particular, plumbagin is slightly volatile; thus, its presence in plant tissue is often indicated by spontaneous sublimation, thereby staining the tissue surface or plastic material used for storage. We observed this effect with both leaf and pitcher tissue (Figure 3) stored in plastic vials. In order to proof its identity, a part of the compound was removed from the wall of the plastic vial by extraction with dichloromethane.

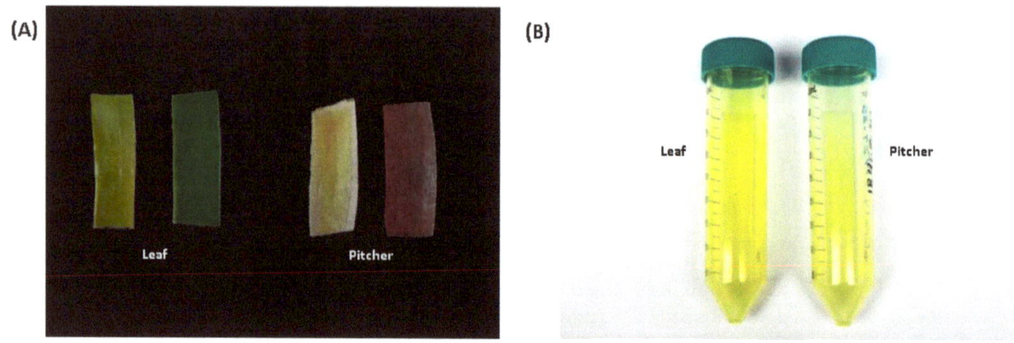

Figure 3. (**A**) Tissues of dry *Nepenthes x ventrata* leaf and pitcher stored for 6 months in a plastic tube. Sublimed compounds cover the dry material with a yellowish color (left) in comparison with freshly cut tissue (right). (**B**) Plastic tubes that stored the different tissue types for 6 months. New tubes do not show any color.

After evaporation of the solvent, the residue was used for NMR analysis. In parallel, leaf extracts from *N. x ventrata* were analyzed by ^{1}H-NMR as well (Figure 4). When compared with a reference, it could be confirmed that the sublimed volatile compound was indeed plumbagin, and this compound could also be proven in leaf material (Figure 4).

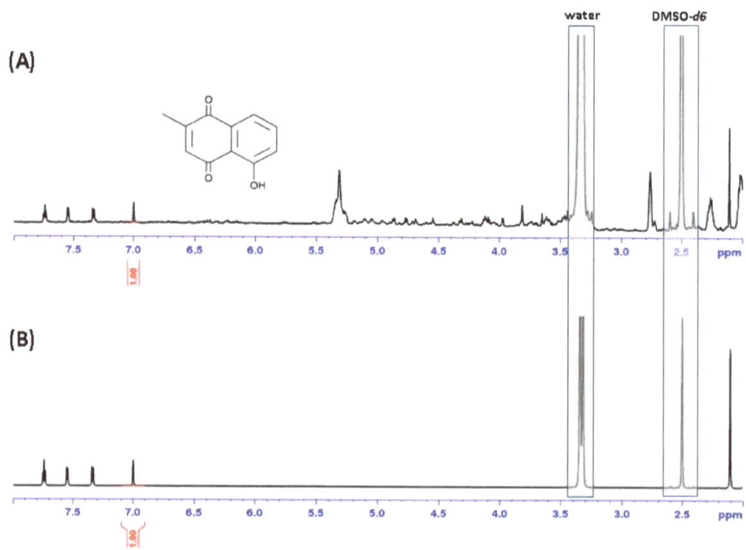

Figure 4. *Cont.*

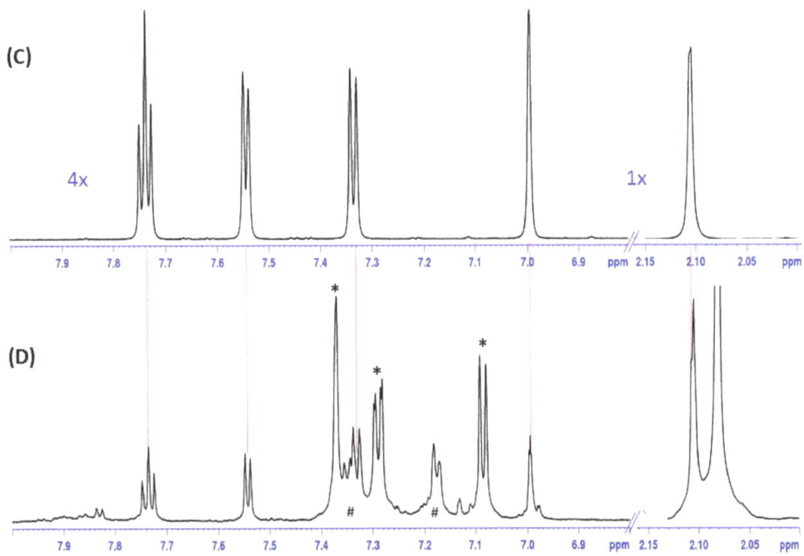

Figure 4. ^{1}H-NMR spectra in DMSO-d_6. (**A**) Plumbagin (see insert) extracted from *Nepenthes x ventrata* leaves and (**B**) a plumbagin reference. (**C**) Details of ^{1}H NMR spectra of a plumbagin reference and (**D**) the volatile exudate emitted by *N. x ventrata* pitcher material. Asterisks (*) indicate the presence of 4-tert-butylcatechol, a polymerization inhibitor probably extracted from the plastic material, and hashes (#) account for an unidentified impurity. The intensity of the aromatic range in (**C**) was increased as indicated by the factor.

These results raised the question of the function of plumbagin and other NQ in carnivorous plants and in *Nepenthes*. In general, NQ are highly bioactive compounds. Besides pharmacological properties against malaria, various cancers, inflammation, and much more [6,19,32–34], they have allelopathic effects as shown for the walnut trees (*Juglans* spp.) releasing the phytotoxin juglone (5-hydroxy-1,4-naphthalenedione) [35,36]. Many defense-related properties are associated with NQ, among them are activities against numerous microbes including human- and phytopathogenic parasites, bacteria, and fungi [31–33]. That means, the NQ might protect the plants from pathogen infection. In addition, for *N. khasiana*, it could be shown that droserone and its derivative 5–*O*–droserone provided antimicrobial protection in the pitcher fluid of [11,37]. Buch and coworkers identified plumbagin and 7-methyl-juglone in the pitcher fluid of *N. ventricosa* [12]. These results suggest a role for NQ in the pitcher fluid in order to control the microbiome in the digestive fluid, together with, e.g., pathogenesis-related proteins such as PR-1 [13,37].

2.3. Growth-Inhibiting and Larvicidal Activities of Plumbagin

Besides the hypothesis that NQ are involved in defense against microbial infection, there are several studies showing that these compounds can also affect insects [31–33,38–43]. We, therefore, performed feeding experiments with plumbagin-supplemented artificial diet and measured the weight of *S. littoralis* larvae every day. Knowing that the amount of plumbagin in *Nepenthes* leaves is about 0.05% of fresh weight [15], we covered a concentration range between 100 and 900 µg g^{-1}, representing 0.01–0.09% fresh mass, respectively. As shown in Figure 5, with increasing plumbagin concentrations, the larvae gained less weight. Based on these data the EC_{50} value was calculated indicating the plumbagin concentration necessary for 50% growth inhibition (weight gain), which was determined as 226.5 µg g^{-1} diet. For some lepidopteran species such as *Spodoptera litura*, *Achaea janata*, and *Trichoplusia ni*, it already has been shown that plumbagin affects the feeding behavior [38–41]. However, in those experiments, the focus of the analysis was on the level of feeding-avoidance rather than on the larval growth.

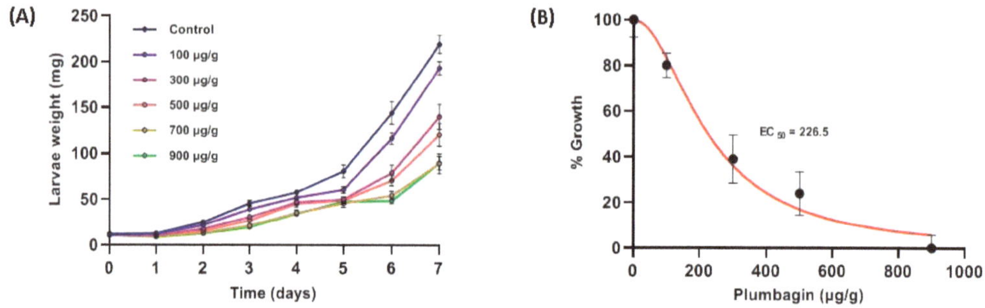

Figure 5. (**A**) Performance of *Spodoptera littoralis* larvae feeding on artificial diet containing various concentrations of plumbagin. Larvae were weighed every day for 7 days. Mean (±SE), $n = 15$. (**B**) Determination of EC_{50} value based on the data obtained in (**A**). EC_{50} was calculated with 226.5 and 1.2 µmol g^{-1} diet, respectively.

In contrast to most other bioassays that analyzed the antifeeding activity of plumbagin, here, the compound of interest was included in the food, not painted on leaves of various plant species. Nevertheless, in order to determine the mortality rate of larvae feeding on plumbagin, we also carried out an experiment using the approach with plumbagin-painted leaves. Therefore, a sweet potato cultivar (Tainong 66) that is known to be susceptible to herbivores and does not induce strong defense response upon attack was selected [44]. In first experiments, we observed that *S. littoralis* larvae even preferred cannibalism than feeding on those leaves. As a consequence, only individualized larvae were used. Up to a plumbagin concentration of 60 μg^{-1} leaf (13.3 µg cm^{-2}, 750 µg g^{-1} leaf) no larvicidal effect was determined for the period analyzed (Figure 6A). With 90 μg^{-1} leaf (20 µg cm^{-2}; 1.125 mg g^{-1} leaf) dead larvae could be found at the end of day 4 and the survival rate drop to 50% at the end of day 5. At 120 μg^{-1} leaf (26.7 µg cm^{-2}; 1.5 mg g^{-1} leaf), dead larvae were detected at day 3 and until the end of day 7, all larvae have died (Figure 6A). For *T. ni* feeding on plumbagin-covered cabbage leaves, an antifeeding effect was also determined in the low microgram per square centimeter range [41]. It also can be seen that the larvae avoided feeding on the leaves covered with high concentrations of plumbagin (Figure 6B,C). With respect to the results shown in Figure 5, it seems that larval growth is heavily affected at higher plumbagin concentrations of around 700 µg plumbagin g^{-1} diet. However, the larvae were affected in growth but still survived at all concentrations tested (up to 900 µg g^{-1}). The plumbagin concentrations used in the no-choice assay also showed no mortality up to 750 µg g^{-1} leaf tissue. Only at the used concentration of 1.125 µg g^{-1} leaf, we found the first larvae dying. This suggests that there might be a threshold of about 1 mg g^{-1} food before the *S. littoralis* larvae begin to die. The experiment is somehow comparable with a recent study by Hu and colleagues [42]. They investigated the mortality of *Pieris rapae* and *Helicoverpa armigera* feeding on cabbage leaves dipped into solutions with different concentrations of plumbagin and juglone, respectively. For plumbagin, IC_{50} values of 11 µg mL^{-1} (*P. rapae*) and 30 µg mL^{-1} (*H. armigera*) were calculated [42]. However, these data are hard to rank as it is not known how much of the compounds of interest was finally on or in the leaf disc. Nevertheless, for all the latter assays, it is difficult to discriminate whether the larvae really die either because of the ingested compounds or of hunger as they consequently avoid feeding. Other studies used topical assays where the compound was added directly onto the insect's (e.g., *S. litura*, *A. janata*, and *Musa domestica*) body to investigate the toxicity of compounds [38,43]. This approach is worth to carry out but not qualified for studies on activities of compounds that are incorporated during herbivory.

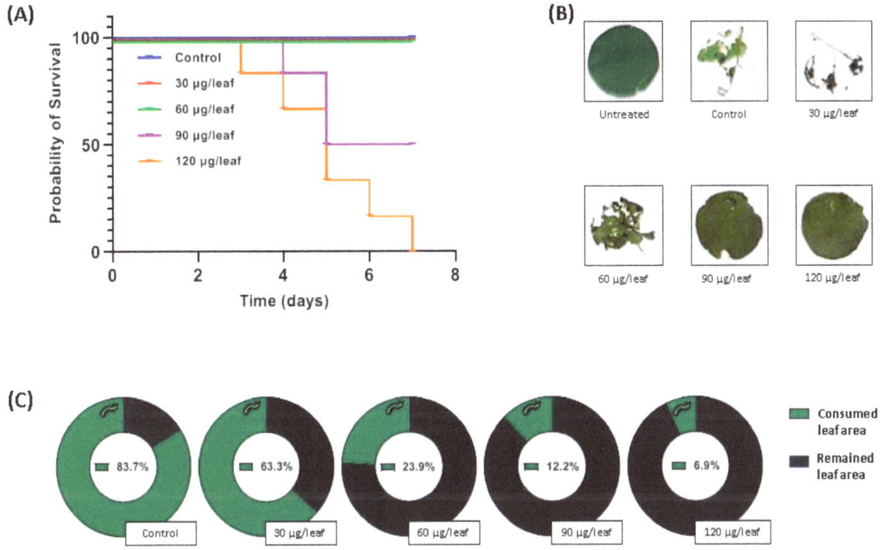

Figure 6. (**A**) Survival rate of *Spodoptera littoralis* larvae feeding on *Ipomoea batatas* (sweet potato) leaf discs painted with various concentrations of plumbagin ($n = 6$). (**B**) Representative leaf discs at the end of the feeding period of day three. Leaf disks were renewed every day. (**C**) Leaf areas consumed by *S. littoralis* larvae (indicated in green) at day 3 depending on the applied plumbagin concentration.

However, the mode of action of NQ is not completely known. In general, NQ are redox-active compounds that can generate oxidative stress [33]; moreover, there are hints for specific inhibition of enzymes and, hence, processes involved in insect development mainly the molting process in insects, e.g., the enzymes phenoloxidase [30], chitin synthetase [45], or ecdysone 20-monooxygenase [46]. The interaction with molting hormone pools is discussed as well [47]. Another study showed that in *Anopheles stephensi,* the level of certain enzymes such as esterases and SOD was decreased significantly in the presence of plumbagin, which also was active as repellent against *A. stephensi* at a concentration of 100 µg mL^{-1}. Further histological investigations showed that muscles, midgut, and hindgut were the most affected tissues [48]. However, most studies suggest that, most likely, the insecticidal activity of plumbagin is based on the inhibition of ecdysis. This also includes a certain specificity against insects compared with neurotoxic insecticidal compounds.

Botanical or plant-derived insecticides have long been touted as environmentally friendly alternatives to synthetic insecticides for pest and disease management [3]; NQ combine the advantage of both low toxicity, compared with conventional pesticides, and restricted environmental contamination and, thus, might be promising candidates for an ecological agriculture.

3. Materials and Methods

3.1. Insects and Plants

Spodoptera littoralis Boisd. (Lepidoptera: Noctuidae) were hatched from eggs kindly provided by Syngenta Crop Protection (Stein, Switzerland) and reared on artificial diet (500 g hackled beans, 9 g ascorbic acid, 9 g 4-ethylbenzoic acid, 9 g vitamin E Mazola oil mixture (7.1%), 4 mL formaldehyde, 1.2 L water, 1 g-sitosterol, 1 g leucine, 10 g AIN-76 vitamin mixture, and 200 mL (7.5%) agar-water solution) at 23–25 °C with a 14 h photoperiod. Sweet potato (*Ipomoea batatas* Lam. cv Tainong 66) scions were grown as described [40] under a 16/8 h light/dark regime at 28/25 °C, respectively, and 70% relative humidity. *Nepenthes x ventrata* (*N. alata x N. ventricosa* hybrid) plants were grown at 21–23 °C, 50–60% relative humidity, and a 16/8 h light/dark photoperiod. Pitcher and the associated leaf tissues were harvested at the time when the pitchers were just opened, directly frozen in liquid nitrogen and ground with mortar and pestle. Material was used directly (fresh) or freeze-dried before use.

3.2. *Feeding Assays*

For feeding assays, second to third instar larvae of *S. littoralis* were used. Ground fresh or dried plant material (leaves and pitcher) from *N. x ventrata* was added to the artificial diet with the indicated quantities (*w*/*w*). Plumbagin (5-hydroxy-2-methyl-1,4-naphthoquinone, $C_{11}H_8O_3$; Fischer Scientific, Schwerte, Germany) was dissolved in acetone and added to the diet. Controls were prepared in the same way without plumbagin. At all the time, it was made sure that acetone was evaporated. For these feeding assays, 15 independent repeats were done. No-choice leaf disks feeding assays according to [34] were further performed on sweet potato. Therefore, leaf discs of 24 mm in diameter were punched out with a cork borer put directly on wet filter paper in a petri dish (5.5 cm diameter). Plumbagin was solved as described before and diluted to the required concentration with 2.5% (*w*/*v*) PEG 2000 (Sigma-Aldrich, Taufkirchen, Germany). That solution was added onto the surface of the discs at the concentrations indicated. For the no-choice assays, 6 independent repeats were performed.

Every day fresh diet or leaf discs were provided. All assays were performed with individual larvae to avoid cannibalism. Larvae were reared for the indicated periods on the particular diets and weighed at the given time.

3.3. *Isolation of Plumbagin from Nepenthes x ventrata Leaves*

Freshly harvested *N. x ventrata* leaves (7.3 g) were immediately frozen in liquid N_2 and freeze-dried. Dried tissue was ground and extracted with 100 mL dichloromethane (DCM) for 15 min by stirring in Erlenmeyer flasks. After precipitation for 20 min, the clear supernatant (50 mL) was collected and another 50 mL DCM was added to the remaining material for re-extraction, which was repeated six times. Collected supernatants were filtered, combined, and DCM was removed using a rotary evaporator. The dried extract (9.3 mg) was dissolved in 2 mL DCM transferred into a HPLC vial and dried again under N_2 stream. For the whole procedure, only glassware was used. The NQ in the extract was identified by means of NMR spectroscopy by comparing spectral data with those of an authentic standard (plumbagin).

N. x ventrata leaf material was kept in 50 mL polypropylene tubes at room temperature over 6 months during which the NQ sublimed (Figure 4), leaving a yellowish stained plastic material. Absorbed compounds were extracted from closed tubes with DCM (10 mL) for 3 days at room temperature. The extract was transferred into a glass vial and evaporated using N_2 gas. The residue was reconstituted with DMSO-d_6 and subjected to NMR analysis.

Identity of the sublimed and extracted plumbagin was confirmed by ^{1}H-NMR spectroscopy. NMR spectra were measured on a Bruker Avance III HD spectrometer (Bruker BioSpin GmbH, Rheinstetten, Germany) equipped with a cryoplatform and a TCI 1.7 mm Micro-CryoProbe. Spectra were referenced to the residual solvent signal for DMSO-d_6 at δH 2.50. Spectrometer control and data processing was accomplished using Bruker TopSpin 3.6.1, and standard pulse programs as implemented in Bruker TopSpin 3.6.1 were used.

For a quantitative comparison of ^{1}H NMR spectra of extracts of *N. x ventrata* leaf and pitcher tissue, the spectral intensity was adjusted to equal solvent signal areas. The areas of signals accounting for plumbagin (range: δH 8.00–7.00) were determined and used for calculation based on the respective areas of a plumbagin standard. For preparation of the experiment, 729 mg (FW) of each tissue was ground in liquid N_2 and extracted with 20 mL of dichloromethane in closed vessels at room temperature with shaking. Extracts were filtered through Chromabond PTS phase separation cartridges (Macherey-Nagel, Düren, Germany) and the flow-through was evaporated with N_2 gas at room temperature within 30 min. Afterwards, the residue was reconstituted with 1.2 mL DMSO-*d6* and subjected to ^{1}H-NMR spectroscopy.

3.4. Statistical Analysis

Statistical calculations were performed using GraphPad Prism version 9.0.0 in all cases. Details are indicated in the particular figure legends. For EC_{50} analysis, the total response was normalized to run between 0% and 100% using control data. For growth experiments, larvae were picked randomly from a large population and all experiments were conducted out under highly standardized conditions to avoid investigator-included bias.

4. Conclusions

Naphthoquinones are known metabolites in several plant species. Among these are various carnivorous plants including the pitcher plant *Nepenthes*. Plumbagin is a prominent NQ in *Nepenthes x ventrata* and it was detected by ^{1}H-NMR in tissues in different concentrations (100 and 650 $\mu g\ g^{-1}$ fresh weight in pitcher and leaf, respectively). Plumbagin has known antimicrobial activities and is of pharmaceutical interest. Now, in different feeding assays with *Spodoptera littoralis* larvae the anti-feeding, growth-inhibiting and larvicidal activity of plumbagin or plumbagin-containing tissues was demonstrated at naturally occurring concentrations. Plumbagin as well as other NQ might become alternative compounds as natural insecticides in agriculture.

Author Contributions: A.R.-S., A.D.-L., and A.M. conceived the study and experiments. A.R.-S., A.D.-L., and C.P. performed the experiments and analyzed data. A.R.-S., A.D.-L., C.P., and A.M. discussed the data and wrote the manuscript. All authors have read and agreed to the published version of the manuscript.

Funding: A.D.-L. was supported by a PhD fellowship from the DAAD (German Academic Exchange Service).

Institutional Review Board Statement: Not applicable.

Informed Consent Statement: Not applicable.

Data Availability Statement: All data generated or analyzed during this study are included in the main text.

Acknowledgments: We thank the greenhouse team of the MPI for cultivating the plants, Syngenta for providing *Spodoptera littoralis* and Andrea Lehr for rearing larvae.

Conflicts of Interest: The authors declare no conflict of interest.

References

1. Juniper, B.E.; Robins, R.J.; Joel, D.M. *The Carnivorous Plants*; Academic Press: London, UK, 1989.
2. Thorogood, C.J.; Bauer, U.; Hiscock, S.J. Convergent and divergent evolution in carnivorous pitcher plant traps. *New Phytol.* **2018**, *217*, 1035–1041. [CrossRef]
3. Miguel, S.; Hehn, A.; Bourgaud, F. *Nepenthes*: State of the art of an inspiring plant for biotechnologists. *J. Biotechnol.* **2018**, *265*, 109–115. [CrossRef] [PubMed]
4. Legendre, G.; Darnowski, D.W. Biotechnology with carnivorous plants. In *Carnivorous Plants: Physiology, Ecology, and Evolution*; Ellison, A.M., Adamec, L., Eds.; Oxford University Press: Oxford, UK, 2018; pp. 270–282.
5. Chi, V.V. *Dictionary of Vietnamese Medicinal Plants*; Publishing House Medicine: Hanoi, Vietnam, 2012; Volume 2.
6. Likhitwitayawuid, K.; Kaewamatawong, R.; Ruangrungsi, N.; Krungkrai, J. Antimalarial naphthoquinones from *Nepenthes thorelii*. *Planta Med.* **1998**, *64*, 237–241. [CrossRef] [PubMed]
7. D'Amoto, P. *The Savage Garden*; Ten Speed Press: Berkeley, CA, USA, 1998.
8. Wiard, C.; Morgana, S.; Khalifah, S.; Mahan, M.; Ismael, S.; Buckle, M.; Narayana, A.K.; Sulaiman, M. Antimicrobial screening of plants used for traditional medicine in the state of Perak, Peninsula Malaysia. *Fitoterapia* **2004**, *75*, 68–73.
9. Rey, M.; Yang, M.; Lee, L.; Zhang, Y.; Sheff, J.G.; Sensen, C.W.; Mrazek, H.; Halada, P.; Man, P.; McCarville, J.L.; et al. Addressing proteolytic efficiency in enzymatic degradation therapy for celiac disease. *Sci. Rep.* **2016**, *6*, 30980. [CrossRef] [PubMed]
10. Tang, J.-Y.; Peng, S.-Y.; Cheng, Y.-B.; Wang, C.-L.; Farooqi, A.A.; Yu, T.-J.; Hou, M.-F.; Wang, S.-C.; Yem, C.-H.; Chan, L.-P.; et al. Ethyl acetat extract of *Nepenthes adrianii x clipeata* induces antiproliferation, apoptosis, and DNA damage against oral cancer cells through oxidative stress. *Environ. Toxcol.* **2019**, *34*, 891–901. [CrossRef]
11. Eilenberg, H.; Pnini-Cohen, S.; Rahamim, Y.; Sionov, E.; Segal, E.; Carmeli, S.; Zilberstein, A. Induced production of antifungal naphthoquinones in the pitchers of the carnivorous plant *Nepenthes khasiana*. *J. Exp. Bot.* **2010**, *61*, 911–922. [CrossRef]

12. Buch, F.; Rott, M.; Rottloff, S.; Paetz, C.; Hilke, I.; Raessler, M.; Mithöfer, A. 2013. Secreted pitfall-trap fluid of carnivorous *Nepenthes* plants is unsuitable for microbial growth. *Ann. Bot.* **2013**, *111*, 375–383. [CrossRef]
13. Mithöfer, A. Carnivorous pitcher plants: Insights in an old topic. *Phytochemistry* **2011**, *72*, 1678–1682. [CrossRef]
14. Hatcher, C.R.; Ryves, D.B.; Millett, J. The function of secondary metabolites in plant carnivory. *Ann. Bot.* **2020**, *125*, 399–411. [CrossRef]
15. Raj, G.; Kurup, R.; Hussain, A.A.; Baby, S. Distribution of naphthoquinones, plumbagin, droserone, and 5–*O*–methyl droserone in chitin-induced and uninduced *Nepenthes khasiana*: Molecular events in prey capture. *J. Exp. Bot.* **2011**, *62*, 5429–5436. [CrossRef]
16. Rischer, H.; Hamm, A.; Bringmann, G. Nepenthes insignis uses a C2-portion of the carbon skeleton of L-alanine acquired via its carnivorous organs, to build up the allelochemical plumbagin. *Phytochemistry* **2002**, *59*, 603–609. [CrossRef]
17. Schlauer, J.; Nerz, J.; Rischer, H. Carnivorous plant chemistry. *Acta Bot. Gall.* **2005**, *152*, 187–195. [CrossRef]
18. Liu, Y.; Cai, Y.; He, C.; Chen, M.; Li, H. Anticancer properties and pharmaceutical applications of plumbagin: A review. *Am. J. Chin. Med.* **2017**, *45*, 423–441. [CrossRef] [PubMed]
19. Tripathi, S.K.; Panda, M.; Biswal, B.K. Emerging role of plumbagin: Cytotoxic potential and pharmaceutical relevance towards cancer therapy. *Food Chem. Toxicol.* **2019**, *215*, 566–583. [CrossRef]
20. Yin, Z.H.; Zhang, J.J.; Chen, L.; Gio, Q.F.; Yang, B.C.; Zhang, W.; Kang, W.Y. Anticancer effects and mechanisms of action of plumbagin: Review of research advances. *Biomed Res. Int.* **2020**. [CrossRef] [PubMed]
21. Panichayupakaranant, P.; Ahmad, M.I. Plumbagin and its role in chronic diseases. *Adv. Exp. Med. Biol.* **2016**, *929*, 229–246.
22. Aung, H.; Chia, L.; Goh, N.K.; Chia, T.F.; Ahmad, A.A.; Pare, P.W.; Mabry, T.J. Phenolic constituents from the leaves of the carnivorous *Nepenthes gracilis*. *Fitoterapia* **2002**, *73*, 445–447. [CrossRef]
23. Wan, A.S.; Aexel, R.T.; Ramsey, R.B.; Nicholas, H.J. Sterols and triterpenes of the pitcher plant. *Phytochemistry* **1972**, *11*, 456–461. [CrossRef]
24. Dávila-Lara, A.; Rodríguez-López, C.E.; O'Connor, S.E.; Mithöfer, A. Metabolomics analysis reveals tissue-specific metabolite compositions in leaf blade and traps of carnivorous *Nepenthes* plants. *Int. J. Mol. Sci.* **2020**, *21*, 4376. [CrossRef]
25. Carmickle, R.N.; Horner, J.D. Impact of the specialist herbivore *Exyra semicrocea* on the carnivorous plant *Sarracenia alata*: A field experiment testing the effects of tissue loss and diminished prey capture on plant growth. *Plant Ecol.* **2019**, *220*, 553–561. [CrossRef]
26. Lamb, T.; Kalies, E.L. An overview of lepidopteran herbivory on North American pitcher plants (*Sarracenia*), with a novel observation of feeding on *Sarracenia flava*. *J. Lepid. Soc.* **2020**, *74*, 193–197.
27. Merbach, M.A.; Zizka, G.; Fiala, B.; Merbach, D.; Booth, W.E.; Maschwitz, U. Why a carnivorous plant cooperates with an ant–selective defense against pitcher-nutritional mutualism in a pitcher plant destroying weevils in the myrmecophytic pitcher plant *Nepenthes bicalcarata* Hook. F. *Ecotropica* **2007**, *13*, 45–56.
28. Gilbert, K.J.; Nitta, J.H.; Talavera, G.; Pierce, N.E. Keeping an eye on coloration: Ecological correlates of the evolution of pitcher traits in the genus *Nepenthes* (Caryophyllales). *Biol. J. Linn. Soc.* **2018**, *123*, 321–337. [CrossRef]
29. Mithöfer, A.; Boland, W. Plant defense against herbivores: Chemical aspects. *Annu. Rev. Plant Biol.* **2012**, *63*, 431–450. [CrossRef] [PubMed]
30. Fleischmann, A.; Schlauer, J.; Smith, S.A.; Givnish, T.J. Evolution of carnivory in angiosperms. In *Carnivorous Plants: Physiology, Ecology, and Evolution*; Ellison, A.M., Adamec, L., Eds.; Oxford University Press: Oxford, UK, 2018; pp. 22–42.
31. Devi, S.P.; Kumaria, S.; Rao, S.R.; Tandon, P. Carnivorous plants as a source of potent bioactive compounds: Naphthoquinones. *Tropical Plant Biol.* **2016**, *9*, 267–279. [CrossRef]
32. Bubla, P.; Adam, V.; Havel, L.; Kizek, R. Noteworthy secondary metabolites naphthoquinones—their occurrence, pharmacological properties and analysis. *Curr. Pharm. Anal.* **2009**, *5*, 47–68. [CrossRef]
33. Widhalm, J.R.; Rhodes, D. Biosynthesis and molecular actions of specialized 1,4-naphthoquinone natural products produced by horticultural plants. *Hortic. Res.* **2016**, *3*, 16046. [CrossRef]
34. Tripathi, K.S.; Rengasamy, K.R.R.; Biswal, B.K. Plumbagin engenders apoptosis in lung cancer celles via caspase-9 activation and targeting mitochondrial-mediated ROD induction. *Arch. Pharm. Res.* **2020**, *43*, 242–256. [CrossRef]
35. Soderquist, C.J. Juglone and allelopathy. *J. Chem. Educ.* **1973**, *50*, 782–783. [CrossRef]
36. Willis, R.J. Juglans spp., juglone and allelopathy. *Allelopath. J.* **2000**, *7*, 1–55.
37. Buch, F.; Pauchet, Y.; Rott, M.; Mithöfer, A. Characterization and heterologous expression of a novel PR-1 protein from traps of the carnivorous plant *Nepenthes mirabilis*. *Phytochemistry* **2014**, *100*, 43–50. [CrossRef] [PubMed]
38. Sreelatha, T.; Hymavathi, A.; Babu, K.S.; Murthy, J.M.; Pathipati, U.R.; Rao, J.M. Synthesis and insect antifeedant activity of plumbagin derivatives with the amino acid moiety. *Agric. Food Chem.* **2009**, *57*, 6090–6094. [CrossRef] [PubMed]
39. Tokunaga, T.; Takada, N.; Ueda, M. Mechanism of antifeedant activity of plumbagin, a compound concerning the chemical defense in carnivorous plant. *Tetrahedron Lett.* **2004**, *45*, 7115–7119. [CrossRef]
40. Tokunaga, T.; Dohmura, A.; Takada, N.; Ueda, M. Cytotoxic antifeedant from *Dionaea muscipula* Ellis: A defensive mechanism of carnivorous plants against predators. *Bull. Chem. Soc. Jpn.* **2004**, *77*, 537–541. [CrossRef]
41. Akhtar, Y.; Isman, M.B.; Niehaus, L.A.; Lee, C.-H.; Lee, H.-S. Antifeedant and toxic effects of naturally occurring and synthetic quinones to the cabbage looper, *Trichoplusia ni*. *Crop Prot.* **2012**, *31*, 8–14. [CrossRef]
42. Hu, W.; Du, W.; Bai, S.; Lv, S.; Chen, G. Phenoloxidase, an effective bioactivity target for botanical insecticide screening from green walnut husks. *Nat. Prod. Res.* **2018**, *32*, 2848–2851. [CrossRef]

43. Pavela, R. Efficacy of naphthoquinones as insecticides against the house fly, *Musca domestica* L. *Ind. Crops Prod.* **2013**, *43*, 745–750. [CrossRef]
44. Meents, A.K.; Chen, S.-P.; Reichelt, M.; Lu, H.-H.; Bartram, S.; Yeh, K.-W.; Mithöfer, A. Volatile DMNT systemically induces jasmonate-independent direct anti-herbivore defense in leaves of sweet potato (*Ipomoea batatas*) plants. *Sci. Rep.* **2019**, *9*, 17431. [CrossRef]
45. Kubo, I.; Uchida, M.; Klocke, J.A. An insect ecdysis inhibitor from the African medical plant *Plumbago capensis* (Plumbaginaceae); a naturally occurring chitin synthetase inhibitor. *Agric. Biol. Chem.* **1983**, *47*, 911–913.
46. Mitchell, M.J.; Smith, S.L. Effects of the chitin synthetase inhibitor plumbagin and its 2-dimethyl derivative juglone on insect ecdysone 20-monooxigenase activity. *Experientia* **1988**, *44*, 990–991. [CrossRef]
47. Gujar, G.T. Interaction of plumbagin with hormones in the cotton stainer, *Dysdercus koenigii* Fabricius. *Proc. Natl. Acad. Sci. India B* **1993**, *9*, 477–482.
48. Pradeepa, V.; Senthil-Nathan, S.; Sathish-Narayanan, S.; Selin-Rani, S.; Vasantha-Srinivasan, P.; Thanigaivel, A.; Ponsankar, A.; Edwin, E.S.; Sakthi-Bagavathy, M.; Kalaivani, K.; et al. Potential mode of action of a novel plumbagin as a mosquito repellent against the malarial vector *Anopheles stephensi*, (Culicidae: Diptera). *Pestic. Biochem. Physiol.* **2016**, *134*, 84–93. [CrossRef] [PubMed]

molecules

MDPI

Article

Essential Oil Compositions of Three Invasive *Conyza* Species Collected in Vietnam and Their Larvicidal Activities against *Aedes aegypti*, *Aedes albopictus*, and *Culex quinquefasciatus*

Tran Minh Hoi [1], Le Thi Huong [2], Hoang Van Chinh [3], Dang Viet Hau [4], Prabodh Satyal [5], Thieu Anh Tai [6], Do Ngoc Dai [7,8], Nguyen Huy Hung [6,9,*], Vu Thi Hien [10] and William N Setzer [5,11,*]

1 Department of Plant Resources, Institute of Ecology and Biological Resources, Vietnam Academy of Science and Technology, Hanoi 100000, Vietnam; tranhoiiebr@gmail.com

2 School of Natural Science Education, Vinh University, 182 Le Duan, Vinh City 43000, Vietnam; lehuong223@gmail.com

3 Faculty of Natural Sciences, Hong Duc University, 365 Quang Trung, Thanh Hoa 440000, Vietnam; hoangvanchinh@hdu.edu.vn

4 Center for Research and Technology Transfer, Vietnam Academy of Science and Technology, Hanoi 100000, Vietnam; hauhoahock20@gmail.com

5 Aromatic Plant Research Center, 230 N 1200 E, Suite 102, Lehi, UT 84043, USA; psatyal@aromaticplant.org

6 Department of Pharmacy, Duy Tan University, 03 Quang Trung, Da Nang 550000, Vietnam; anhtai0808qn@gmail.com

7 Graduate University of Science and Technology, Vietnam Academy of Science and Technology, 18-Hoang Quoc Viet, Cau Giay, Hanoi, 100000 Vietnam; daidn23@gmail.com

8 Faculty of Agriculture, Forestry and Fishery, Nghe An College of Economics, 51-Ly Tu Trong, Vinh City 460000, Vietnam

9 Center for Advanced Chemistry, Institute of Research and Development, Duy Tan University, 03 Quang Trung, Da Nang 550000, Vietnam

10 Faculty of Hydrometerology, Ho Chi Minh City University of Natural Resources and Environment, Ho Chi Minh City 70000, Vietnam; hiensphoa@gmail.com

11 Department of Chemistry, University of Alabama in Huntsville, Huntsville, AL 35899, USA

* Correspondence: nguyenhuyhung@duytan.edu.vn (N.H.H.); wsetzer@chemistry.uah.edu (W.N.S.)

Received: 20 September 2020; Accepted: 6 October 2020; Published: 7 October 2020

Abstract: Mosquito-borne infectious diseases are a persistent problem in tropical regions of the world, including Southeast Asia. Vector control has relied principally on synthetic insecticides, but these have detrimental environmental effects and there is an increasing demand for plant-based agents to control insect pests. Invasive weedy plant species may be able to serve as readily available sources of essential oils, some of which may be useful as larvicidal agents for control of mosquito populations. We hypothesize that members of the genus *Conyza* (Asteraceae) may produce essential oils that may have mosquito larvicidal properties. The essential oils from the aerial parts of *Conyza bonariensis*, *C. canadensis*, and *C. sumatrensis* were obtained by hydrodistillation, analyzed by gas chromatography–mass spectrometry, and screened for mosquito larvicidal activity against *Aedes aegypti*, *Ae. albopictus* and *Culex quinquefasciatus*. The essential oils of *C. canadensis* and *C. sumatrensis*, both rich in limonene (41.5% and 25.5%, respectively), showed notable larvicidal activities against *Ae. aegypti* (24-h LC_{50} = 9.80 and 21.7 μg/mL, respectively) and *Ae. albopictus* (24-h LC_{50} = 18.0 and 19.1 μg/mL, respectively). These two *Conyza* species may, therefore, serve as sources for alternative, environmentally-benign larvicidal control agents.

Keywords: *Erigeron*; *Conyza bonariensis*; *Conyza canadensis*; *Conyza sumatrensis*; mosquito; vector control

1. Introduction

Mosquito-borne infectious diseases have been a continuous health problem in Southeast Asia, including Vietnam. Dengue fever and dengue hemorrhagic fever are particularly problematic and chikungunya fever is an emerging threat in the country [1,2]. *Aedes aegypti* (L.) (Diptera: Culicidae), the yellow fever mosquito, is a recognized vector of dengue fever virus, chikungunya fever virus, Zika virus, and yellow fever virus [3]. *Aedes albopictus* (Skuse) (Diptera: Culicidae), the Asian tiger mosquito, is a key vector of several pathogenic viruses, including yellow fever virus [4], dengue fever virus [5], chikungunya virus [6], and possibly Zika virus [7]. *Culex quinquefasciatus* Say (Diptera: Culicidae), the southern house mosquito, is a vector of lymphatic filariasis [8] as well as several arboviruses such as West Nile virus and St. Louis encephalitis virus [9] and possibly Zika virus [10].

Several members of the genus *Conyza* Less. (Asteraceae) have been introduced throughout the tropics and subtropics where they have become invasive weeds [11–13]. *Conyza bonariensis* (L.) Cronquist (syn. *Erigeron bonariensis* L.), flaxleaf fleabane, probably originated in South America [14], but has been introduced throughout Asia, Africa, Mexico and the southern United States, Europe, and Oceania [13,15]. *Conyza canadensis* (L.) Cronquist (syn. *Erigeron canadensis* L.), Canada fleabane, is native to North America, but is also now naturalized throughout Europe, Asia, and Oceania [13]. *Conyza sumatrensis* (Retz.) E. Walker (syn. *Erigeron sumatrensis* Retz.) is probably native to South America, but this species has also been naturalized in tropical and subtropical regions [16].

Non-native invasive plant species are generally detrimental to the local environments where they have been introduced. They can outcompete native plant species and reduce biodiversity [17], they can alter ecosystem functions [18], and can have substantial economic impacts [19]. Control methods for invasive plants have generally included application of herbicides, physical cutting, or burning [20]. However, harvesting invasive species for beneficial uses as a method for control of invasive species may provide economic incentives to offset eradication costs [21]. For example, *Melaleuca quinquinervia* trees in south Florida have been cut and chipped for landscape mulch and boiler fuel [22]; it has been suggested that mechanical harvesting of invasive cattail (*Typha* spp.), common reed (*Phragmites australis*), and reed canary grass (*Phalaris arundinacea*) from coastal wetlands of Lake Ontario can be used as an agricultural nutrient source or as a biofuel [23]. The leaf essential oil of *Solidago canadensis*, an invasive plant in Europe, has been evaluated as a potential insecticide and demonstrated moderate larvicidal activity against *Cx. quinquefasciatus* [24].

The use of synthetic pesticides for mosquito control has had detrimental effects on the environment [25,26]. They tend to be persistent, toxic to non-target organisms, and insecticide resistance has been steadily increasing in mosquito species [27]. Essential oils have been suggested as viable, environmentally benign, and renewable alternatives to synthetic pesticides [28–32]. We have recently studied several introduced invasive plant species in Vietnam for potential use as mosquito vector control agents [33–35], and as part of our ongoing efforts in identifying readily-available essential oils for mosquito control, we have examined three *Conyza* species for larvicidal activity against *Aedes aegypti*, *Aedes albopictus*, and *Culex quinquefasciatus*, with the aim of identifying new mosquito-control essential oils and the components responsible for the activity.

2. Results and Discussion

2.1. Essential Oil Compositions

The essential oils from the aerial parts of *C. bonariensis*, *C. canadensis*, and *C. sumatrensis* were obtained by hydrodistillation in 1.10%, 1.37%, and 1.21% yield. The chemical compositions of the *Conyza* essential oils, determined using gas chromatography–mass spectrometry, are summarized in Table 1. *Conyza bonariensis* essential oil was dominated by sesquiterpenoids, especially *allo*-aromadendrene (41.2%), β-caryophyllene (13.3%), and caryophyllene oxide (12.2%). Concentrations of monoterpenoids (1.8%) and diterpenoids (trace) were relatively small. The essential oils of *C. canadensis* and *C. sumatrensis*, on the other hand, were rich in limonene (41.5% and 25.5%, respectively). The aerial parts essential oil of *C. sumatrensis* also had a large concentration of (Z)-lachnophyllum ester (20.7%). There is wide variation in the essential oil compositions of *Conyza* species, both between species and within the same species (see Table 2). This is not surprising given the very different geographical locations of the collection sites for these samples.

Table 1. Chemical compositions of the aerial parts essential oils of *Conyza bonariensis*, *Conyza canadensis*, and *Conyza sumatrensis* collected in Vietnam.

RI_{calc} [a]	RI_{db} [b]	Compound	Relative Content %		
			C. bonariensis	*C. canadensis*	*C. sumatrensis*
931	932	α-Pinene	0.5	0.5	0.2
948	950	Camphene	tr [c]	—	—
967	972	(3Z)-Octen-2-ol	—	—	tr
971	972	Sabinene	tr	0.1	0.1
976	978	β-Pinene	0.8	8.8	3.0
982	984	6-Methylhept-5-en-2-one	—	—	tr
987	989	Myrcene	tr	1.2	1.0
1023	1025	*p*-Cymene	tr	0.3	0.1
1028	1030	Limonene	0.2	41.5	25.5
1030	1031	β-Phellandrene	—	tr	—
1034	1034	(Z)-β-Ocimene	—	—	tr
1044	1045	(E)-β-Ocimene	—	tr	1.9
1049	1051	2,3,6-Trimethylhepta-1,5-diene	—	tr	—
1056	1057	γ-Terpinene	—	tr	—
1088	1091	*p*-Cymenene	—	0.1	—
1090	1091	Rosefuran	—	—	0.1
1093	1097	α-Pinene oxide	—	—	0.2
1097	1098	Perillene	—	0.1	—
1098	1101	Linalool	0.2	—	—
1101	1101	6-Methyl-3,5-heptadien-2-one	—	—	0.1
1103	1104	Nonanal	tr	—	—
1112	1113	4,8-Dimethylnona-1,3,7-triene	—	—	0.2
1118	1119	*endo*-Fenchol	tr	—	—
1120	1121	*trans-p*-Mentha-2,8-dien-1-ol	—	0.9	0.2
1124	1131	Cyclooctanone	—	0.8	—
1129	1130	4-Acetyl-1-methylcyclohexene	—	0.1	—
1131	1132	*cis*-Limonene oxide	—	0.6	0.2
1134	1137	*cis-p*-Mentha-2,8-dien-1-ol	—	1.2	0.3
1135	1137	*trans*-Limonene oxide	—	0.6	—
1137	1137	Nopinone	—	0.4	—
1137	1139	(E)-Myroxide	—	—	0.1
1139	1141	*trans*-Pinocarveol	tr	1.6	0.1
1150	1152	Citronellal	—	0.1	—
1160	1164	Pinocarvone	—	0.8	tr
1170	1170	Borneol	tr	—	—
1177	1179	2-Isopropenyl-5-methylhex-4-enal	—	0.3	—
1182	1184	*p*-Methylacetophenone	—	0.3	—
1185	1185	Cryptone	—	0.4	—
1185	1187	*trans-p*-Mentha-1(7),8-dien-2-ol	—	0.2	—
1189	1190	Methyl salicylate	tr	—	—
1193	1195	α-Terpineol	0.1	—	0.1
1193	1196	Myrtenal	—	1.4	—
1194	1195	Myrtenol	—	1.2	—
1196	1197	Methyl chavicol (=Estragol)	—	0.2	—
1198	1201	*cis*-Piperitol	—	0.8	0.1
1206	1207	Oct-3E-enyl acetate	—	—	0.1
1217	1218	*trans*-Carveol	—	3.8	0.2

Table 1. *Cont.*

RI_{calc} [a]	RI_{db} [b]	Compound	Relative Content %		
			C. bonariensis	*C. canadensis*	*C. sumatrensis*
1227	1228	*cis-p*-Mentha-1(7),8-dien-2-ol	—	0.1	—
1230	1232	*cis*-Carveol	—	1.1	0.1
1242	1242	Carvone	—	3.8	0.2
1247	1249	Linalyl acetate	tr	—	—
1266	1270	*iso*-Piperitenone	—	0.6	—
1273	1277	Perilla aldehyde	—	0.5	—
1287	1287	Limonene dioxide	—	0.7	—
1296	1299	Perilla alcohol	—	0.4	—
1303	—	Unidentified [d]	—	1.1	—
1316	1324	Limonene hydroperoxide	—	1.1	—
1343	1346	Limonene-1,2-diol	—	2.6	—
1344	1349	7-*epi*-Silphiperfol-5-ene	—	—	0.3
1345	1349	α-Cubebene	0.2	—	—
1355	1340	*p*-Mentha-6,8-diene-2-hydroperoxide	—	1.2	—
1367	1371	α-Ylangene	tr	—	—
1374	1375	α-Copaene	4.5	—	0.1
1376	1380	Daucene	—	—	0.4
1377	1374	Isoledene	—	—	0.3
1379	1382	Modheph-2-ene	—	—	0.4
1381	1382	β-Bourbonene	tr	—	—
1385	1387	β-Cubebene	0.4	—	0.1
1386	1385	α-Isocomene	—	—	0.1
1387	1390	β-Elemene	0.3	—	0.4
1392	1394	Sativene	—	—	0.1
1398	1405	(*Z*)-Caryophyllene	0.2	—	—
1404	1406	α-Gurjunene	0.1	—	—
1408	1411	β-Isocomene	—	—	0.1
1418	1417	(*E*)-Caryophyllene	13.3	—	5.5
1427	1430	β-Copaene	0.2	—	0.2
1430	1433	*trans*-α-Bergamotene	—	—	1.1
1432	1440	6,9-Guaiadiene	—	—	0.2
1433	1436	α-Guaiene	1.8	—	—
1436	1438	Aromadendrene	0.2	—	0.1
1445	1449	(*E*)-Lachnophyllum acid	—	—	0.2
1451	1452	(*E*)-β-Farnesene	—	—	6.7
1453	1454	α-Humulene	5.4	0.3	0.7
1457	1463	*cis*-Cadina-1(6),4-diene	—	—	0.4
1460	1458	*allo*-Aromadendrene	41.2	—	—
1469	—	Unidentified [e]	—	—	1.3
1472	1472	*trans*-Cadina-1(6),4-diene	0.5	—	0.2
1476	1479	α-Amorphene	0.1	—	—
1478	1483	Germacrene D	0.3	—	2.1
1481	1483	*trans*-β-Bergamotene	—	—	0.2
1486	1489	β-Selinene	0.5	—	—
1488	1491	Viridiflorene	0.2	—	—
1492	1497	Bicyclogermacrene	—	—	0.3
1493	1497	α-Selinene	0.3	—	—
1495	1497	α-Muurolene	0.4	—	0.1
1498	1505	α-Bulnesene	1.8	—	—
1501	1505	(*E,E*)-α-Farnesene	—	—	0.1
1504	1514	(*Z*)-Lachnophyllum acid	—	0.2	0.8
1507	1510	(*E*)-Lachnophyllum ester	—	—	0.4
1510	1512	γ-Cadinene	0.4	—	0.1
1515	1515	(*Z*)-Lachnophyllum ester	—	5.5	20.7
1515	1518	δ-Cadinene	0.6	—	—
1518	1519	*trans*-Calamenene	0.3	—	—
1521	1523	β-Sesquiphellandrene	—	—	0.3
1531	1532	Tridec-11-yn-1-ol	—	—	0.3
1533	1538	α-Cadinene	0.1	—	—
1538	1541	α-Calacorene	0.1	—	—

Table 1. *Cont.*

RI_{calc} [a]	RI_{db} [b]	Compound	Relative Content %		
			C. bonariensis	*C. canadensis*	*C. sumatrensis*
1556	1557	Germacrene B	—	—	0.1
1558	1560	(*E*)-Nerolidol	—	0.2	1.8
1559	1564	β-Calacorene	0.1	—	—
1565	1566	1,5-Epoxysalvial-4(14)-ene	—	—	0.2
1566	1568	Dendrolasin	—	—	0.1
1567	1567	Palustrol	0.1	—	—
1574	1576	Spathulenol	1.3	—	5.2
1580	1577	Caryophyllene oxide	12.2	1.1	5.8
1582	1590	Globulol	0.4	—	0.5
1589	1593	Salvial-4(14)-en-1-one	—	0.1	0.2
1590	1594	Viridiflorol	0.8	—	0.3
1593	1599	Cubeban-11-ol	0.2	—	—
1599	1601	Carotol	—	—	1.1
1601	1605	Ledol	0.6	—	—
1606	1611	Humulene epoxide II	2.2	2.9	0.4
1624	1628	1-*epi*-Cubenol	0.2	—	—
1629	1629	*iso*-Spathulenol	—	—	0.6
1633	1635	Caryophylla-4(12),8(13)-dien-5β-ol	0.2	—	—
1635	1632	Muurola-4,10(14)-dien-1β-ol	—	—	0.7
1638	1643	τ-Cadinol	0.2	—	0.4
1640	1644	τ-Muurolol	0.1	—	0.3
1643	1643	α-Muurolol	0.2	—	—
1643	1644	*allo*-Aromadendrene epoxide	—	0.3	—
1652	1655	α-Cadinol	0.6	0.3	0.4
1655	1655	Eudesma-4(15),7-dien-1α-ol	—	—	0.1
1661	1664	*cis*-Calamenen-10-ol	0.1	—	—
1666	1666	14-Hydroxy-9-*epi*-(*E*)-caryophyllene	0.1	—	—
1669	1677	Cadalene	0.1	—	—
1686	1685	Eudesma-4(15),7-dien-1β-ol	—	0.4	0.1
1698	1704	*cis*-Thujopsenol	0.1	—	—
1717	—	Unidentified [f]	—	1.0	—
1738	1740	8α,11-Elemodiol	0.1	—	—
1751	1748	Khusimol	1.5	—	—
1790	1792	14-Hydroxy-δ-cadinene	—	—	0.2
1800	—	Unidentified [g]	1.1	—	—
1833	1836	Neophytadiene	—	—	0.2
1857	1860	Platambin	0.1	0.5	0.1
1882	1884	Corymbolone	0.2	—	—
2103	2102	Phytol	tr	—	0.1
		Monoterpene hydrocarbons	1.5	52.7	31.8
		Oxygenated monoterpenoids	0.3	26.4	1.9
		Sesquiterpene hydrocarbons	73.7	0.3	20.7
		Oxygenated sesquiterpenoids	21.3	5.7	18.5
		Diterpenoids	trace	—	0.4
		Others	trace	7.2	22.9
		Total Identified	96.8	92.3	96.1

[a] RI_{calc} = Retention Index calculated with respect to a homologous series of n-alkanes on a ZB-5 column. [b] RI_{db} = Retention Index from the databases [36–39]. [c] tr = trace (< 0.05%). [d] MS(EI): 150(3%), 135(51%), 121(29%), 119(38%), 109(42%), 107(66%), 93(97%), 91(89%), 81(50%), 79(100%), 69(82%), 67(37%), 55(65%), 53(40%), 43(75%), 41(85%). [e] MS(EI): 204(25%), 189(3%), 161(100%), 147(9%), 133(28%), 120(48%), 119(25%), 105(51%), 91(47%), 69(20%), 57(19%), 55(21%), 41(20%). [f] MS(EI): 175(3%), 135(11%), 111(48%), 93(20%), 83(19%), 67(19%), 55(26%), 43(100%), 41(20%). [g] MS(EI): 218(29%), 203(28%), 189(100%), 175(46%), 147(34%), 133(61%), 119(38%), 105(70%), 91(90%), 79(42%), 67(43%), 55(34%), 41(52%).

Table 2. Major components of *Conyza bonariensis*, *Conyza canadensis*, and *Conyza sumatrensis* essential oils from different geographical locations.

***Conyza* Species (Collection Site)**	**Major Components (>5%)**	**Ref.**
C. bonariensis aerial parts EO (Chapada dos Guimarães, Mato Grosso, Brazil)	limonene (6.9%), (*E*)-caryophyllene (14.4%), (*E*)-β-farnesene (23.3%), germacrene D (15.3%), bicyclogermacrene (8.3%), spathulenol (7.6%)	[40]
C. bonariensis aerial parts EO (Melgaço, Pará, Brazil)	limonene (22.9%), (*E*)-caryophyllene (13.3%), *trans*-α-bergamotene (5.3%), (*E*)-β-farnesene (20.1%), bicyclogermacrene (6.6%), spathulenol (6.3%)	[40]
C. bonariensis aerial parts EO (Peixe-Boi, Pará, Brazil)	(*E*)-caryophyllene (13.3%), *trans*-α-bergamotene (8.1%), (*E*)-β-farnesene (30.9%)	[40]
C. bonariensis aerial parts EO (alta Floresta, Mato Grosso, Brazil)	limonene (12.6%), (*E*)-caryophyllene (13.0%), (*E*)-β-farnesene (19.1%), germacrene D (13.2%), bicyclogermacrene (6.3%), spathulenol (5.7%)	[40]
C. bonariensis aerial parts EO (Macapá, Amapá, Brazil)	limonene (58.4%), (*E*)-β-farnesene (7.0%)	[40]
C. bonariensis aerial parts EO (Rio de Janeiro, Brazil)	limonene (45.0%), (*E*)-β-ocimene (13.0%), (*E*)-β-farnesene (6.6%), germacrene D (6.4%)	[41]
C. bonariensis leaf EO (Minas Gerais State, Brazil)	limonene (29.6%), *trans*-α-bergamotene (10.3%), matricaria methyl ester (8.3%), β-copaen-4α-ol (7.4%)	[42]
C. bonariensis aerial parts EO (Athens, Greece)	limonene (8.3%), (*E*)-β-ocimene (11.5%), (*E*)-β-farnesene (8.1%), (*Z*)-lachnophyllum ester (21.2%), matricaria ester (17.5%)	[43]
C. bonariensis aerial parts EO (Southwestern Misiones Province, Argentina)	limonene (13.5%), (*E*)-β-ocimene (13.3%), *p*-mentha-1,3,8-triene (5.2%), germacrene D (14.6%), bicyclogermacrene (6.6%)	[44]
C. bonariensis leaf EO (Monastir, Tunisia)	limonene (5.8%), terpinolene (5.3%), (*E*)-β-farnesene (7.5%), matricaria ester (17.8%), caryophyllene oxide (7.8%)	[45]
C. bonariensis aerial parts EO (Cagliari, Sardinia, Italy)	limonene (5.1%), carvacrol (9.8%), α-curcumene (10.2%), spathulenol (18.6%), caryophyllene oxide (18.7%), neophytadiene (6.1%)	[46]
C. bonariensis leaf EO (Mérida State, Venezuela)	limonene (5.1%), (*Z*)-β-ocimene (5.1%), (*E*)-β-ocimene (20.7%), (*E*)-β-farnesene (37.8%), α-farnesene (5.6%), β-sesquiphellandrene (9.8%)	[47]
C. bonariensis leaf EO (Kabianga, Kericho, Kenya)	β-pinene (5.4%), limonene (8.3%), 2,6,7,7a-tetrahydro-1,5-dimethyl-1*H*-indene-3-carboxaldehyde (49.1%) [a]	[48]
C. bonariensis aerial parts EO (Parana State, Brazil)	limonene (66.3%), 2-heptyl acetate (6.9%)	[49]
C. bonariensis aerial parts EO	(*E*)-caryophyllene (13.3%), α-humulene (5.4%), *allo*-aromadendrene (41.2%), caryophyllene oxide (12.2%)	this work
C. canadensis aerial parts EO (Plovdiv, Bulgaria)	limonene (77.7–89.4%)	[50]
C. canadensis aerial parts EO (Łódź, Poland)	limonene (76.3%)	[51]
C. canadensis aerial parts EO (Alps, France)	limonene (83.2%)	[51]

Table 2. *Cont.*

Conyza Species (Collection Site)	Major Components (>5%)	Ref.
C. canadensis aerial parts EO (Rome, Italy)	limonene (70.3%), (*E*)-β-ocimene (5.5%)	[51]
C. canadensis aerial parts EO (Seville, Spain)	limonene (51.4%), (*E*)-β-ocimene (13.4%), *trans*-α-bergamotene (11.9%)	[51]
C. canadensis aerial parts EO (Belgium)	limonene (68.0%), (*E*)-β-ocimene (5.1%), *trans*-α-bergamotene (5.4%), germacrene D (7.3%) (*Z*,*Z*)-matricaria ester (6.1%)	[51]
C. canadensis aerial parts EO (Plovdiv, Bulgaria)	limonene (87.9%)	[51]
C. canadensis aerial parts EO (Vilnius, Lithuania)	limonene (77.7%), *trans*-α-bergamotene (5.5%)	[51]
C. canadensis aerial parts EO (Israel)	limonene (54.9%), (Z)-β-farnesene (6.3%) (Z,Z)-matricaria ester (7.7%)	[51]
C. canadensis aerial parts EO (Kerman, Iran)	myrcene (8.9%), limonene (12.3%), (*E*)-β-farnesene (14.6%), *ar*-curcumene (7.8%), zingiberene (5.5%), spathulenol (14.1%), isospathulenol (7.7%), phytol (7.3%)	[52]
C. canadensis aerial parts EO (Athens, Greece)	β-pinene (9.5%), limonene (57.3%), matricaria ester (14.4%)	[43]
C. canadensis aerial parts EO (Korea)	limonene (68.3%), (*E*)-β-ocimene (15.9%) [b]	[53]
C. canadensis EO (China)	limonene (14.8%), *epi*-bicyclosesquiphellandrene (11.0%), $C_7H_{30}B_4Si$ (25.1%) [c], 1-phenyl-1-nonyne (7.3%)	[54]
C. canadensis aerial parts EO (Szeged, Hungary)	limonene (79.2%)	[55]
C. canadensis aerial parts EO (Manavgat, Antalya, Turkey)	β-pinene (9.7%), limonene (28.1%), spathulenol (16.3%)	[56]
C. canadensis aerial parts EO	β-pinene (8.8%), limonene (41.5%), (*Z*)-lachnophyllum ester (5.5%)	this work
C. sumatrensis aerial parts EO (Rondônia state, Brazil)	sabinene (5.3%), limonene (22.9%), (*E*)-β-ocimene (5.0%), (*E*)-β-farnesene (5.3%), (*Z*)-lachnophyllum ester (43.7%)	[57]
C. sumatrensis leaf EO (N'gorato village, Côte d'Ivoire)	limonene (13.0%), (*E*)-β-ocimene (6.5%), (*E*)-caryophyllene (10.5%), (*E*)-β-farnesene (17.0%), (*Z*)-lachnophyllum ester (5.9%), germacrene D (13.6%), bicyclogermacrene (5.2%)	[58]
C. sumatrensis leaf EO (Monastir, Tunisia)	matricaria ester (7.5%), spathulenol (13.8%), caryophyllene oxide (20.5%)	[59]
C. sumatrensis aerial parts EO	limonene (25.5%), (*E*)-caryophyllene (5.5%), (*E*)-β-farnesene (6.7%), (*Z*)-lachnophyllum ester (20.7%), spathulenol (5.2%), caryophyllene oxide (5.8%)	this work

[a] The identification of this compound is uncertain; it is not found in the *Dictionary of Natural Products* [60]. [b] This compound was listed as δ-3-carene, but the retention time is more consistent with (*E*)-β-ocimene rather than δ-3-carene. [c] The identification of this compound (2,3-μ-trimethylsilyl-C,C′-dimethyl-4,5-dicarba-*nido*-hexaborane) is not correct; the compound listed is not a natural product.

2.2. Mosquito Larvicidal Activity

The mosquito larvicidal activities of the *Conyza* essential oils are summarized in Table 3. The essential oil of *C. canadensis* showed the best larvicidal activity against both *Ae. aegypti* (24-h LC_{50} = 9.80 μg/mL) and *Ae. albopictus* (24-h LC_{50} = 18.0 μg/mL) and good larvicidal activity against *Cx. quinquefasciatus* (24-h LC_{50} = 39.4 μg/mL). *Conyza sumatrensis* essential oil also showed good larvicidal activity against the three mosquito species (24-h LC_{50} = 21.7, 19.1, and 26.7 μg/mL, respectively, for *Ae. aegypti*, *Ae. albopictus*, and *Cx. quinquefasciatus*). *Conyza bonariensis* essential oil was less active (24-h LC_{50} = 69.7, 81.1 and 130.0 μg/mL against *Ae. aegypti*, *Ae. albopictus*, and *Cx. quinquefasciatus*, respectively).

The larvicidal activities of *Conyza* essential oils roughly coincides with the concentration of limonene in the samples (41.5%, 25.5%, and 0.2%, respectively, for *C. canadensis*, *C. sumatrensis*, and *C. bonariensis*), and this relationship is borne out in a principle component analysis based on the major essential oil components (limonene, *allo*-aromadendrene, (*Z*)-lachnophyllum ester, caryophyllene oxide, β-caryophyllene, β-pinene, (*E*)-β-farnesene, spathulenol, and α-humulene, along with the 24-h larvicidal activities) (Figure 1). Limonene has shown excellent larvicidal activities against *Ae. aegypti* (24-h LC_{50} = 17.7 μg/mL) and *Cx. quinquefasciatus* (24-h LC_{50} = 31.6 μg/mL) (Table 3) as well as *Ae. albopictus* (LC_{50} 10.8-41.8 μg/mL) [34]. Consistent with these results, Zeng and co-workers found the larvicidal activity of *C. canadensis* from China (14.8% limonene) to be 56.9 μg/mL and 32.1 μg/mL against *Ae. albopictus* and *Cx. quinquefasciatus*, respectively [54]. These workers also appreciated the remarkable larvicidal activity and noted that *C. canadensis* essential oil has a potential for further development. Furthermore, *Citrus* peel oils, rich in limonene, have also shown remarkable larvicidal activities against *Ae. albopictus* [61] and *Cx. quinquefasciatus* [62].

Table 3. Mosquito larvicidal activity and insecticidal activity of *Conyza* essential oils.

Essential Oil or Major Compound	24 h LC_{50} (95% Limits), μg/mL	LC_{90} (95% Limits), μg/mL	χ^2	p	Slope
	Aedes aegypti				
C. bonariensis	69.71 (64.82–75.36)	88.61 (82.13–97.54)	9.39	0.009	9.45
C. canadensis	9.801 (8.730–10.986)	23.27 (19.93–28.36)	8.70	0.069	12.18
C. sumatrensis	21.74 (20.16–23.36)	31.02 (28.29–35.50)	0.131	0.988	7.98
β-Pinene	23.63 (22.16-25.33)	32.12 (29.47-36.00)	0.225	0.994	7.69
Limonene	17.66 (16.45–18.97)	23.62 (22.03–25.73)	0.784	0.941	10.68
(*E*)-Caryophyllene	70.80 (65.49–76.69	107.2 (98.4–118.6)	4.08	0.395	12.75
α-Humulene	53.05 (48.69–58.08)	82.78 (75.81–91.87)	15.9	0.003	12.79
Caryophyllene oxide	136.6 (129.2–143.9)	180.2 (171.4–191.2)	30.1	0.000	12.37
Permethrin control	0.000643 (0.000551–0.00753)	0.00246 (0.00192–0.00344)	12.5	0.006	11.57
	Aedes albopictus [a]				
C. bonariensis	81.13 (74.61–87.97)	127.1 (117.5–139.9)	0.395	0.821	11.44
C. canadensis	18.04 (16.71–19.52)	26.20 (24.22–28.82)	1.46	0.834	11.30
C. sumatrensis	19.13 (17.73–20.66)	27.49 (25.41–30.38)	3.19	0.364	9.97
Permethrin control	0.0024 (0.0021–0.0026)	0.0042 (0.0038–0.0049)	4.64	0.031	8.45
	Culex quinquefasciatus				
C. bonariensis	130.0 (122.5–138.8)	178.4 (165.6–197.2)	0.675	0.713	8.97
C. canadensis	39.37 (36.83–42.00)	52.29 (49.04–56.56)	0.493	0.974	10.49
C. sumatrensis	26.74 (24.80–29.20)	36.83 (33.56–41.92)	8.97	0.030	7.96
β-Pinene	30.46 (28.21–33.21)	41.58 (38.10–46.58)	0.399	0.983	9.38
Limonene	31.63 (29.37–34.50)	41.51 (38.03–46.78)	0.874	0.928	8.23
(*E*)-Caryophyllene	165.4 (157.5–174.0)	220.6 (207.8–238.5)	10.0	0.040	9.91
α-Humulene	108.3 (101.4–115.5)	158.2 (148.5–170.5)	1.0	0.910	13.32
Caryophyllene oxide	98.52 (90.70–108.68)	144.5 (129.6–165.7)	1.60	0.809	9.20
Permethrin control	0.0165 (0.0149–0.0181)	0.0305 (0.0266–0.0367)	5.24	0.073	10.12
	Diplonychus rusticus [a]				
C. canadensis	135.7 (129.3–142.8)	182.5 (172.6–195.5)	7.78	0.051	12.35
C. sumatrensis	111.0 (106.1–116.7)	137.0 (129.5–147.6)	16.1	0.001	9.85

Table 3. *Cont.*

Essential Oil or Major Compound	48 h LC$_{50}$ (95% Limits), µg/mL	LC$_{90}$ (95% Limits), µg/mL	χ^2	*p*	Slope
	Aedes aegypti				
C. bonariensis	63.85 (59.07–70.75)	81.84 (74.16–94.79)	3.43	0.180	6.89
C. canadensis	7.091 (6.099–8.141)	22.46 (18.63–28.59)	5.98	0.201	11.63
C. sumatrensis	22.52 (21.18–23.87)	29.00 (27.23–31.68)	0.0488	0.997	10.12
β-Pinene	22.91 (21.29–24.85)	31.37 (29.03–35.03)	0.323	0.988	9.08
Limonene	17.43 (16.24–18.74)	23.17 (21.58–25.28)	0.664	0.956	10.48
(*E*)-Caryophyllene	65.92 (60.45–72.08)	106.4 (98.4–116.7)	14.2	0.007	13.10
α-Humulene	46.25 (42.27–50.94)	74.14 (67.47–82.99)	19.2	0.001	12.21
Caryophyllene oxide	120.2 (112.7–127.5)	165.4 (156.4–176.6)	19.8	0.001	12.34
Permethrin control	0.000575 (0.000483–0.00688)	0.00281 (0.00208–0.00423)	5.29	0.152	10.93
	Aedes albopictus [a]				
C. bonariensis	69.42 (63.20–75.93)	113.2 (103.8–125.8)	3.10	0.212	10.72
C. canadensis	15.12 (13.93–16.47)	22.67 (20.84–25.09)	7.23	0.124	12.22
C. sumatrensis	18.43 (17.05–19.93)	26.76 (24.71–29.58)	4.25	0.236	8.44
	Culex quinquefasciatus				
C. bonariensis	108.1 (101.4–115.1)	152.1 (142.4–165.1)	2.32	0.313	10.84
C. canadensis	29.81 (27.33–32.68)	47.06 (43.03–52.39)	14.5	0.006	12.17
C. sumatrensis	22.95 (21.22-25.08)	33.06 (30.07-37.60)	2.38	0.498	9.37
β-Pinene	28.36 (26.20–31.19)	39.01 (35.41–44.50)	2.41	0.661	8.39
Limonene	29.15 (26.89–31.98)	40.83 (37.19–46.07)	7.05	0.133	9.50
(*E*)-Caryophyllene	138.5 (129.3–148.5)	215.3 (200.1–234.9)	13.5	0.009	13.11
α-Humulene	87.81 (81.14–94.89)	140.0 (130.0–152.7)	9.80	0.044	13.50
Caryophyllene oxide	95.19 (86.69–106.26)	141.0 (127.6–160.8)	4.01	0.405	10.12
	Diplonychus rusticus [a]				
C. canadensis	124.0 (118.0–130.4)	165.0 (156.1–176.6)	1.17	0.760	12.17
C. sumatrensis	107.8 (103.1–113.4)	133.6 (126.1–144.4)	8.07	0.045	9.37

[a] *Aedes albopictus* and *Diplonychus rusticus* were obtained from the wild; the limited numbers of organisms available precluded screening of individual components on these two insect species.

Other components in the *Conyza* essential oils likely contribute to the mosquito larvicidal effects. *Conyza bonariensis* was rich in (*E*)-caryophyllene (13.3%) and caryophyllene oxide (12.2%), but both of these compounds have been found to have weak larvicidal activities against *Ae. aegypti* (24-h LC$_{50}$ = 70.8 and 137 µg/mL, respectively (Table 3). On the other hand, β-pinene, a major component of *C. canadensis* essential oil (8.8%), has shown larvicidal activity against *Ae. aegypti* (24-h LC$_{50}$ = 23.6 µg/mL), *Cx. quinquefasciatus* (24-h LC$_{50}$ = 30.5 µg/mL) (Table 3), and *Ae. albopictus* [61]. In addition, synergy between essential oil components may also be important [63,64]. Scalerandi and coworkers have found that the housefly (*Musca domestica*) metabolizes the major components in an essential oil, but leaves the minor components to act as toxicants [65].

In order to assess the potential detrimental impact of the *Conyza* essential oils on beneficial aquatic species, the insecticidal activity was assessed against the water bug, *Diplonychus rusticus*, an insect predator of mosquito larvae [66]. Both *C. canadensis* and *C. sumatrensis* essential oils were substantially less toxic to *D. rusticus* than they were to the mosquito larvae.

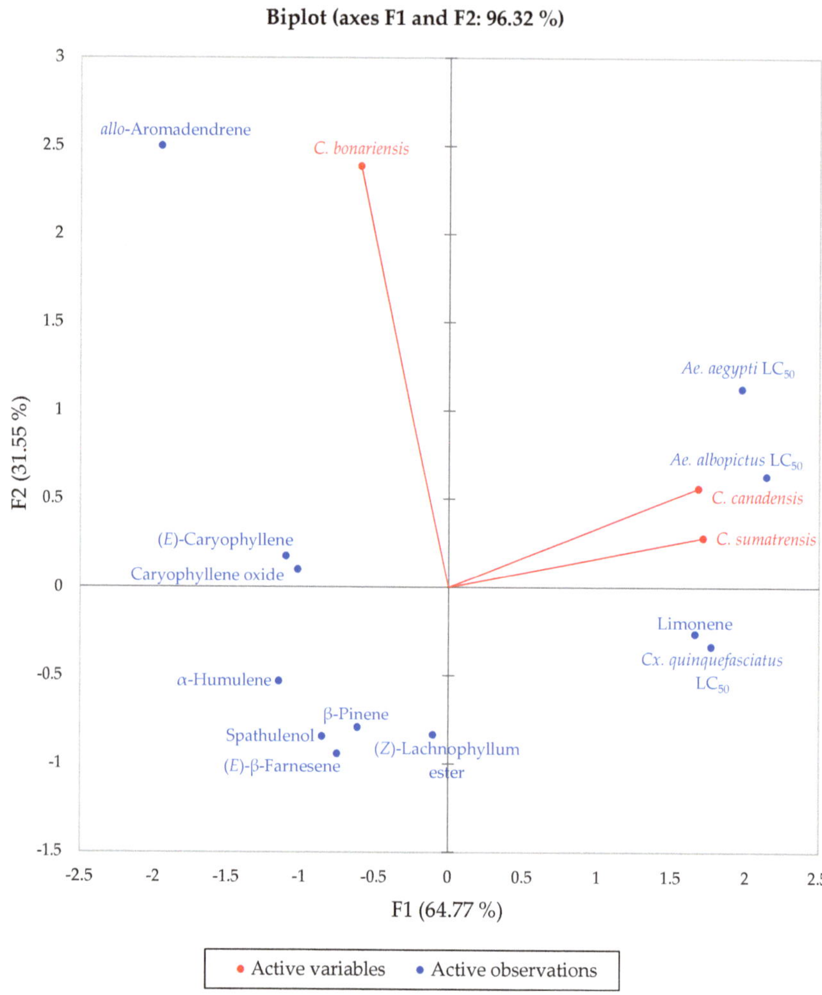

Figure 1. Principal component biplot of PC1 and PC2 scores and loadings demonstrating the relationships between *Conyza* essential oil major components and larvicidal activities.

3. Materials and Methods

3.1. Chemicals

Chemicals used for this study, dimethylsulfoxide (DMSO), β-pinene, limonene, (*E*)-caryophyllene, α-humulene, caryophyllene oxide, dichloromethane, and permethrin, were obtained from Sigma-Aldrich (St. Louis, MO, USA) and used as received without further purification.

3.2. Plant Material

The three *Conyza* species were collected from Bach Ma National Park, Thue Thien Hue province (16° 11′ 34″ N, 107° 51′ 12″ E) in April 2020. The plants were identified by Dr. Do Ngoc Dai and Dr. Le Thi Huong. Voucher specimens, LTH129 (*Conyza canadensis*), LTH130 (*Conyza sumatrensis*), and LTH131 (*Conyza bonariensis*) have been deposited in the Pedagogical Institute of Science, Vinh University. Four-kg samples of fresh aerial parts (leaves, stems, and flowers) of each of the plants were shredded and hydrodistilled for 4 h using a Clevenger-type apparatus.

3.3. *Gas Chromatography–Mass Spectrometry*

The *Conyza* essential oils were analyzed by GC-MS as previously described [67]: Shimadzu GCMS-QP2010 Ultra, electron impact (EI) mode, electron energy = 70 eV, scan range = 40–400 atomic mass units, scan rate = 3.0 scans/s, ZB-5 fused silica capillary column (30 m × 0.25 mm, 0.25 μm film thickness), He carrier gas, 552 kPa column head pressure, and 1.37 mL/min flow rate. Injector temperature was 250 °C and the ion source temperature was 200 °C. The GC oven temperature program was programmed for 50 °C initial temperature, temperature increased at a rate of 2 °C/min to 260 °C. A 5% *w*/*v* solution of the sample in CH_2Cl_2 was prepared and 0.1 μL was injected with a splitting mode (30:1). Identification of the oil components was based on their retention indices determined by reference to a homologous series of *n*-alkanes, and by comparison of their mass spectral fragmentation patterns with those reported in the databases [36–39].

3.4. *Mosquito Larvicidal Assay*

Mosquito larvicidal activity was carried out on *Ae. aegypti*, *Ae. albopictus*, and *Cx. quinquefasciatus* as previously described [67]: For the assay, 1% stock solutions of each essential oil in dimethylsulfoxide (DMSO) were prepared, and aliquots of the stock solutions were placed in 500-mL beakers and added to water that contained 20 larvae (fourth instar). With each experiment, a set of controls using DMSO was also run for comparison. Mortality was recorded after 24 h and again after 48 h of exposure during which no nutritional supplement was added. The experiments were carried out 25 ± 2°C. Each test was conducted with four replicates with three concentrations (50, 25, and 12.5, μg/mL for *C. canadensis* and *C. sumatrensis*; 150, 100, and 50 μg/mL for *C. bonariensis*). Permethrin was used as a positive control.

3.5. *Non-Target Insecticidal Assay*

The *Diplonychus rusticus* adults were collected in the field and maintained in glass tanks (60 cm long × 50 cm wide) containing water at 25 °C with a water depth of 20 cm. The essential oils were tested at concentrations of 200, 150, 100, 75, 50, and 25 μg/mL. Four replicates were performed for each concentration. Twenty *D. rusticus* adults were introduced into each solution. The non-target organism was observed for mortality after 24 h and 48 h exposure.

3.6. *Data Analysis*

The mortalities were recorded 24 h and 48 h after treatment. The data obtained were subjected to log-probit analysis [68] to obtain LC_{50} values, LC_{90} values, 95% confidence limits, and chi square values using Minitab® 18 (Minitab Inc., State College, PA, USA). For the principal component analysis (PCA), the 9 major components (limonene, *allo*-aromadendrene, (*Z*)-lachnophyllum ester, caryophyllene oxide, (*E*)-caryophyllene, β-pinene, (*E*)-β-farnesene, spathulenol, and α-humulene), and the 24-h larvicidal activities against *Ae. aegypti*, *Ae. albopictus*, and *Cx. quinquefasciatus* were taken as variables using a Pearson correlation matrix using XLSTAT Premium, version 2018.5 (Addinsoft, Paris, France). A total of 33 data (11 variables × 3 samples) were used for the PCA.

4. Conclusions

Invasive plant species are generally considered to be ecologically and detrimental with potential economic impacts, and the control or eradication of invasive plant species can be prohibitively costly. However, identification of beneficial uses of invasive plants could be economically advantageous and aid in the control of the species. *Conyza* spp., as well as *Erechtites* spp. [34], *Crassocephalum crepidioides* [35], and *Severinia monophylla* [33], are invasive weeds in Vietnam, and essential oils from these plants have demonstrated promising mosquito larvicidal activities. The plant materials are readily available and harvesting of these weeds may provide economically valuable "cash crops" as well as serve as a means for ecological remediation. Note that *C. bonariensis* [69], *C. canadensis* [70], and *C. sumatrensis* [71] have all shown resistance to the commonly used herbicide glyphosate, so herbicidal control of these weeds is impractical as well as environmentally detrimental. Further research on potential formulations (e.g., nanoemulsions or essential oil-loaded nanoparticles) [72] for field use of these promising essential oils is warranted.

Author Contributions: Conceptualization, N.H.H.; methodology, T.M.H., P.S., and W.N.S.; software, P.S.; validation, T.M.H., D.N.D., P.S., and W.N.S.; formal analysis, P.S. and W.N.S.; investigation, T.M.H., D.N.D., H.V.C., D.V.H., T.A.T., L.T.H., N.H.H., and V.T.H.; resources, T.M.H.; data curation, W.N.S.; writing—original draft preparation, W.N.S.; writing—review and editing, T.M.H., D.N.D., N.H.H., P.S., and W.N.S.; supervision, N.H.H.; project administration, T.M.H.; funding acquisition, T.M.H. All authors have read and agreed to the published version of the manuscript.

Funding: Please add: This research is funded by Vietnam National Foundation for Science and Technology Development (NAFOSTED) under grant number: 106.03-2019.315.

Acknowledgments: P.S. and W.N.S. participated in this work as part of the activities of the Aromatic Plant Research Center (APRC, https://aromaticplant.org/).

Conflicts of Interest: The authors declare no conflict of interest.

References

1. Kim Lien, P.T.; Briant, L.; Tang, T.B.; Trang, B.M.; Gavotte, L.; Cornillot, E.; Duoc, V.T.; Duong, T.N.; Frutos, R.; Nga, P.T. Surveillance of dengue and chikungunya infection in Dong Thap, Vietnam: A 13-month study. *Asian Pac. J. Trop. Med.* **2016**, *9*, 39–43. [CrossRef] [PubMed]
2. Pham Thi, K.L.; Briant, L.; Gavotte, L.; Labbe, P.; Perriat-Sanguinet, M.; Cornillot, E.; Vu, T.D.; Nguyen, T.Y.; Tran, V.P.; Nguyen, V.S.; et al. Incidence of dengue and chikungunya viruses in mosquitoes and human patients in border provinces of Vietnam. *Parasit. Vectors* **2017**, *10*, 556. [CrossRef] [PubMed]
3. Powell, J.R. Mosquito-borne human viral diseases: Why *Aedes aegypti*? *Am. J. Trop. Med. Hyg.* **2018**, *98*, 1563–1565. [CrossRef] [PubMed]
4. Lourenço de Oliveira, R.; Vazeille, M.; de Filippis, A.M.B.; Failloux, A.B. Large genetic differentiation and low variation in vector competence for dengue and yellow fever viruses of *Aedes albopictus* from Brazil, the United States, and the Cayman Islands. *Am. J. Trop. Med. Hyg.* **2003**, *69*, 105–114. [CrossRef] [PubMed]
5. Lambrechts, L.; Scott, T.W.; Gubler, D.J. Consequences of the expanding global distribution of *Aedes albopictus* for dengue virus transmission. *PLoS Negl. Trop. Dis.* **2010**, *4*, e646. [CrossRef]
6. Vazeille, M.; Moutailler, S.; Coudrier, D.; Rousseaux, C.; Khun, H.; Huerre, M.; Thiria, J.; Dehecq, J.S.; Fontenille, D.; Schuffenecker, I.; et al. Two Chikungunya isolates from the outbreak of La Reunion (Indian Ocean) exhibit different patterns of infection in the mosquito, *Aedes albopictus*. *PLoS ONE* **2007**, *2*, e1168. [CrossRef]
7. Wong, P.-S.J.; Li, M.I.; Chong, C.-S.; Ng, L.-C.; Tan, C.-H. *Aedes* (*Stegomyia*) *albopictus* (Skuse): A potential vector of Zika virus in Singapore. *PLoS Negl. Trop. Dis.* **2013**, *7*, e2348. [CrossRef]
8. Albuquerque, C.M.R.; Cavalcanti, V.M.S.; Melo, M.A.V.; Verçosa, P.; Regis, L.N.; Hurd, H. Bloodmeal microfilariae density and the uptake and establishment of *Wuchereria bancrofti* infections in *Culex quinquefasciatus* and *Aedes aegypti*. *Mem. Inst. Oswaldo Cruz* **1999**, *94*, 591–596. [CrossRef]
9. Turell, M.J. Members of the *Culex pipiens* complex as vectors of viruses. *J. Am. Mosq. Control Assoc.* **2012**, *28*, 123–127. [CrossRef]
10. van den Hurk, A.F.; Hall-Mendelin, S.; Jansen, C.C.; Higgs, S. Zika virus and *Culex quinquefasciatus* mosquitoes: A tenuous link. *Lancet Infect. Dis.* **2017**, *17*, 1014–1016. [CrossRef]
11. Thebaud, C.; Abbott, R.J. Characterization of invasive *Conyza* species (Asteraceae) in Europe: Quantitative trait and isozyme analysis. *Am. J. Bot.* **1995**, *82*, 360–368. [CrossRef]
12. Prieur-Richard, A.-H.; Lavorel, S.; Grigulis, K.; Dos Santos, A. Plant community diversity and invasibility by exotics: Invasion of Mediterranean old fields by *Conyza bonariensis* and *Conyza canadensis*. *Ecol. Lett.* **2000**, *3*, 412–422. [CrossRef]

13. Bajwa, A.A.; Sadia, S.; Ali, H.H.; Jabran, K.; Peerzada, A.M.; Chauhan, B.S. Biology and management of two important *Conyza* weeds: A global review. *Environ. Sci. Pollut. Res.* **2016**, *23*, 24694–24710. [CrossRef] [PubMed]
14. Wu, H.; Walker, S.; Rollin, M.J.; Tan, D.K.Y.; Robinson, G.; Werth, J. Germination, persistence, and emergence of flaxleaf fleabane (*Conyza bonariensis* [L.] Cronquist). *Weed Biol. Manag.* **2007**, *7*, 192–199. [CrossRef]
15. Wu, H. The biology of Australian weeds. *Plant Prot. Quart.* **2007**, *22*, 122–131.
16. Pruski, J.F.; Sancho, G. *Conyza sumatrensis* var. *leiotheca* (Compositae: Astereae), a new combination for a common Neotropical weed. *Novon* **2006**, *16*, 96–101. [CrossRef]
17. Raghubanshi, A.S.; Rai, L.C.; Gaur, J.P.; Singh, J.S. Invasive alien species and biodiversity in India. *Curr. Sci.* **2005**, *88*, 539–540.
18. Weidenhamer, J.D.; Callaway, R.M. Direct and indirect effects of invasive plants on soil chemistry and ecosystem function. *J. Chem. Ecol.* **2010**, *36*, 59–69. [CrossRef]
19. Pimentel, D.; Zuniga, R.; Morrison, D. Update on the environmental and economic costs associated with alien-invasive species in the United States. *Ecol. Econ.* **2005**, *52*, 273–288. [CrossRef]
20. Kettenring, K.M.; Adams, C.R. Lessons learned from invasive plant control experiments: A systematic review and meta-analysis. *J. Appl. Ecol.* **2011**, *48*, 970–979. [CrossRef]
21. Pasko, S.; Goldberg, J. Review of harvest incentives to control invasive species. *Manag. Biol. Invasions* **2014**, *5*, 263–277. [CrossRef]
22. Stocker, R.K. Mechanical harvesting of *Melaleuca quinquenervia* in Lake Okeechobee, Florida. *Ecol. Eng.* **1999**, *12*, 373–386. [CrossRef]
23. Carson, B.D.; Lishawa, S.C.; Tuchman, N.C.; Monks, A.M.; Lawrence, B.A.; Albert, D.A. Harvesting invasive plants to reduce nutrient loads and produce bioenergy: An assessment of Great Lakes coastal wetlands. *Ecosphere* **2018**, *9*, e02320. [CrossRef]
24. Benelli, G.; Pavela, R.; Cianfaglione, K.; Nagy, D.U.; Canale, A.; Maggi, F. Evaluation of two invasive plant invaders in Europe (*Solidago canadensis* and *Solidago gigantea*) as possible sources of botanical insecticides. *J. Pest Sci.* **2019**, *92*, 805–821. [CrossRef]
25. Kamrin, M.A. *Pesticide Profiles: Toxicity, Environmental Impact, and Fate*; CRC Press: Boca Raton, FL, USA, 1997; ISBN 0-56670-190-2.
26. Goulson, D. An overview of the environmental risks posed by neonicotinoid insecticides. *J. Appl. Ecol.* **2013**, *50*, 977–987. [CrossRef]
27. Cuervo-Para, J.A.; Romero Cortés, T.; Ramirez-Lepe, M. Mosquito-borne Diseases, Pesticides Used for Mosquito Control, and Development of Resistance to Insecticides. In *Insecticides Resistance*; Trdan, S., Ed.; IntechOpen: London, UK, 2016; pp. 111–134. ISBN 978-953-51-2258-6.
28. Silva, W.J.; Dória, G.A.A.; Maia, R.T.; Nunes, R.S.; Carvalho, G.A.; Blank, A.F.; Alves, P.B.; Marçal, R.M.; Cavalcanti, S.C.H. Effects of essential oils on *Aedes aegypti* larvae: Alternatives to environmentally safe insecticides. *Bioresource Technol.* **2008**, *99*, 3251–3255. [CrossRef]
29. Benelli, G. Research in mosquito control: Current challenges for a brighter future. *Parasitol. Res.* **2015**, *114*, 2801–2805. [CrossRef]
30. Masetti, A. The potential use of essential oils against mosquito larvae: A short review. *Bull. Insectol.* **2016**, *69*, 307–310.
31. Pavela, R.; Benelli, G. Essential oils as ecofriendly biopesticides? Challenges and constraints. *Trends Plant Sci.* **2016**, *21*, 1000–1007. [CrossRef]
32. Ntalli, N.; Koliopoulos, G.; Giatropoulos, A.; Menkissoglu-Spiroudi, U. Plant secondary metabolites against arthropods of medical importance. *Phytochem. Rev.* **2019**, *18*, 1255–1275. [CrossRef]
33. Satyal, P.; Hieu, H.V.; Chuong, N.T.H.; Hung, N.H.; Sinh, L.H.; Van The, P.; Tai, T.A.; Hien, V.T.; Setzer, W.N. Chemical composition, *Aedes* mosquito larvicidal activity, and repellent activity against *Triatoma rubrofasciata* of *Severinia monophylla* leaf essential oil. *Parasitol. Res.* **2019**, *118*, 733–742. [CrossRef]
34. Hung, N.H.; Satyal, P.; Hieu, H.V.; Chuong, N.T.H.; Dai, D.N.; Huong, L.T.; Tai, T.A.; Setzer, W.N. Mosquito larvicidal activity of the essential oils of *Erechtites* species growing wild in Vietnam. *Insects* **2019**, *10*, 47. [CrossRef] [PubMed]

35. Hung, N.H.; Satyal, P.; Do, N.D.; Tai, T.A.; Huong, L.T.; Chuong, N.T.H.; Hieu, H.V.; Tuan, P.A.; Vuong, P.; Van Setzer, W.N. Chemical compositions of *Crassocephalum crepidioides* essential oils and larvicidal activities against *Aedes aegypti*, *Aedes albopictus*, and *Culex quinquefasciatus*. *Nat. Prod. Commun.* **2019**, *14*. [CrossRef]
36. Adams, R.P. *Identification of Essential Oil Components by Gas Chromatography/Mass Spectrometry*, 4th ed.; Allured Publishing: Carol Stream, IL, USA, 2007; ISBN 978-1-932633-21-4.
37. Mondello, L. *FFNSC 3*; Shimadzu Scientific Instruments: Columbia, MD, USA, 2016.
38. *NIST17*; National Institute of Standards and Technology: Gaithersburg, MD, USA, 2017.
39. Satyal, P. Development of GC-MS Database of Essential Oil Components by the Analysis of Natural Essential Oils and Synthetic Compounds and Discovery of Biologically Active Novel Chemotypes in Essential Oils. Ph.D. Thesis, University of Alabama in Huntsville, Huntsville, AL, USA, 2015.
40. Maia, J.G.S.; da Silva, M.H.L.; das Gracas, M.B.Z.; Andrade, E. Composition of the essential oils of *Conyza bonariensis* (L.) Cronquist. *J. Essent. Oil Res.* **2002**, *14*, 325–326. [CrossRef]
41. Souza, M.C.; Siani, A.C.; Ramos, M.F.S.; Menezes-de-lima, O., Jr.; Henriques, M.G.M.O. Evaluation of anti-inflammatory activity of essential oils from two Asteraceae species. *Pharmazie* **2003**, *58*, 582–586. [PubMed]
42. Barbosa, L.C.A.; Paula, V.F.; Azevedo, A.S.; Silva, E.A.M.; Nascimento, E.A. Essential oil composition from some plant parts of *Conyza bonariensis* (L.) Cronquist. *Flavour Fragr. J.* **2005**, *20*, 39–41. [CrossRef]
43. Tzakou, O.; Vagias, C.; Gani, A.; Yannitsaros, A. Volatile constituents of essential oils isolated at different growth stages from three *Conyza* species growing in Greece. *Flavour Fragr. J.* **2005**, *20*, 425–428. [CrossRef]
44. Urdampilleta, J.D.; Amat, A.G.; Bidau, C.J.; Koslobsky, N.K. Biosystematic and chemosystematic studies in five South American species of *Conyza* (Asteraceae). *Bol. Soc. Argent. Bot.* **2005**, *40*, 101–107.
45. Mabrouk, S.; Elaissi, A.; Ben Jannet, H.; Harzallah-Skhiri, F. Chemical composition of essential oils from leaves, stems, flower heads and roots of *Conyza bonariensis* L. from Tunisia. *Nat. Prod. Res.* **2011**, *25*, 77–84. [CrossRef]
46. Benzarti, A.; Hammami, S.; Piras, A.; Falconieri, D.; El Mokni, R.; M'Henni, M.F.; Marongiu, B.; Mighri, Z. Effects of different ecological conditions and extraction techniques on the quality of volatile oils from flaxleaf fleabane (*Erigeron bonariensis* L.). *J. Med. Plant Res.* **2013**, *7*, 3059–3065.
47. Araujo, L.; Moujir, L.M.; Rojas, J.; Carmona, J.; Rondón, M. Chemical composition and biological activity of *Conyza bonariensis* essential oil collected in Mérida, Venezuela. *Nat. Prod. Commun.* **2013**, *8*, 1175–1178. [CrossRef] [PubMed]
48. Musembei, R.; Joyce, K.J. Chemical composition and antibacterial activity of essential oil from Kenyan *Conyza bonariensis* (L.) Cronquist. *Sci. Lett.* **2017**, *5*, 180–185.
49. do Amaral, W.; Deschamps, C.; Biasi, L.A.; Bizzo, H.R.; Machado, M.P.; da Silva, L.E. Yield and chemical composition of the essential oil of species of the Asteraceae family from Atlantic Forest, South of Brazil. *J. Essent. Oil Res.* **2018**, *30*, 278–284. [CrossRef]
50. Stoyanova, A.; Georgiev, E.; Kermedchieva, D.; Lis, A.; Gora, J. Changes in the essential oil of *Conyza canadensis* (L.) Cronquist. during its vegetation. *J. Essent. Oil Res.* **2003**, *15*, 44–45. [CrossRef]
51. Lis, A.; Piggott, J.R.; Góra, J. Chemical composition variability of the essential oil of *Conyza canadensis* Cronq. *Flavour Fragr. J.* **2003**, *18*, 364–367. [CrossRef]
52. Rustaiyan, A.; Azar, P.A.; Moradalizadeh, M.; Masoudi, S.; Ameri, N. Volatile constituents of three Compositae herbs: *Anthemis altissima* L. var. *altissima*, *Conyza canadensis* (L.) Cronq. and *Grantina aucheri* Boiss. growing wild in Iran. *J. Essent. Oil Res.* **2004**, *16*, 579–581. [CrossRef]
53. Choi, H.-J.; Wang, H.-Y.; Kim, Y.-N.; Heo, S.-J.; Kim, N.-K.; Jeong, M.-S.; Park, Y.-H.; Kim, S. Composition and cytotoxicity of essential oil extracted by steam distillation from horseweed (*Erigeron canadensis* L.) in Korea. *J. Korean Soc. Appl. Biol. Chem.* **2008**, *51*, 55–59.
54. Zeng, D.-Q.; Peng, Y.-H.; Chen, F.-F.; Zhang, Y.; Liu, M. Insecticidal activity of essential oil derived from horseweed *Conyza canadensis* (L.) Cronq.against two mosquitoes and its volatile components. *Acta Entomol. Sin.* **2014**, *57*, 204–211.
55. Veres, K.; Csupor-Löffler, B.; Lázár, A.; Hohmann, J. Antifungal activity and composition of essential oils of *Conyza canadensis* herbs and roots. *Sci. World J.* **2012**, *2012*, 489646. [CrossRef]
56. Ayaz, F.; Küçükboyacı, N.; Demirci, B. Chemical composition and antimicrobial activity of the essential oil of *Conyza canadensis* (L.) Cronquist from Turkey. *J. Essent. Oil Res.* **2017**, *29*, 336–343. [CrossRef]

57. Machado, S.M.F.; Militão, J.S.L.T.; Facundo, V.A.; Ribeiro, A.; de Morais, S.M.; de Alencar, J.W.; Braz Filho, R. Essential oil of *Conyza sumatrensis* (Retz) Walk. *J. Essent. Oil Res.* **1995**, *7*, 83–84. [CrossRef]
58. Boti, J.B.; Koukoua, G.; N'Guessan, T.Y.; Casanova, J. Chemical variability of *Conyza sumatrensis* and Microglossa pyrifolia from Côte d'Ivoire. *Flavour Fragr. J.* **2007**, *22*, 27–31. [CrossRef]
59. Mabrouk, S.; Salah, K.B.H.; Elaissi, A.; Jlaiel, L.; Ben Jannet, H.; Aouni, M.; Harzallah-Skhiri, F. Chemical composition and antimicrobial and allelopathic activity of Tunisian *Conyza sumatrensis* (Retz.) E. Walker essential oils. *Chem. Biodivers.* **2013**, *10*, 209–223. [CrossRef] [PubMed]
60. *Dictionary of Natural Products on DVD*; CRC Press: Boca Raton, FL, USA, 2019; ISBN 0-412-49150-8.
61. Giatropoulos, A.; Papachristos, D.P.; Kimbaris, A.; Koliopoulos, G.; Polissiou, M.G.; Emmanouel, N.; Michaelakis, A. Evaluation of bioefficacy of three *Citrus* essential oils against the dengue vector *Aedes albopictus* (Diptera: Culicidae) in correlation to their components enantiomeric distribution. *Parasitol. Res.* **2012**, *111*, 2253–2263. [CrossRef]
62. Manimaran, A.; Cruz, M.M.J.J.; Muthu, C.; Vincent, S.; Ignacimuthu, S. Larvicidal and knockdown effects of some essential oils against *Culex quinquefasciatus* Say, *Aedes aegypti* (L.) and *Anopheles stephensi* (Liston). *Adv. Biosci. Biotechnol.* **2012**, *3*, 855–862. [CrossRef]
63. Pavela, R. Acute toxicity and synergistic and antagonistic effects of the aromatic compounds of some essential oils against *Culex quinquefasciatus* Say larvae. *Parasitol. Res.* **2015**, *114*, 3835–3853. [CrossRef]
64. Tak, J.H.; Isman, M.B. Penetration-enhancement underlies synergy of plant essential oil terpenoids as insecticides in the cabbage looper, *Trichoplusia ni*. *Sci. Rep.* **2017**, *7*, 42432. [CrossRef]
65. Scalerandi, E.; Flores, G.A.; Palacio, M.; Defagó, M.T.; Carpinella, M.C.; Valladares, G.; Bertoni, A.; Palacios, S.M. Understanding synergistic toxicity of terpenes as insecticides: Contribution of metabolic detoxification in *Musca domestica*. *Front. Plant Sci.* **2018**, *9*, 1579. [CrossRef]
66. Saha, N.; Aditya, G.; Bal, A.; Saha, G.K. A comparative study of predation of three aquatic heteropteran bugs on *Culex quinquefasciatus* larvae. *Limnology* **2007**, *8*, 73–80. [CrossRef]
67. Huong, L.T.; Hung, N.H.; Dai, D.N.; Tai, T.A.; Hien, V.T.; Satyal, P.; Setzer, W.N. Chemical compositions and mosquito larvicidal activities of essential oils from *Piper* species. *Molecules* **2019**, *24*, 3871. [CrossRef]
68. Finney, D. *Probit Analysis*; Reissue, Ed.; Cambridge University Press: Cambridge, UK, 2009; ISBN 978-0521135900.
69. Travlos, I.S.; Chachalis, D. Glyphosate-resistant hairy fleabane (*Conyza bonariensis*) is reported in Greece. *Weed Technol.* **2010**, *24*, 569–573. [CrossRef]
70. Koger, C.H.; Poston, D.H.; Hayes, R.M.; Montgomery, R.F. Glyphosate-resistant horseweed (*Conyza canadensis*) in Mississippi. *Weed Technol.* **2004**, *18*, 820–825. [CrossRef]
71. Santos, G.; Oliveira, R.S., Jr.; Constantin, J.; Francischini, A.C.; Machado, M.F.P.S.; Mangolin, C.A.; Nakajima, J.N. *Conyza sumatrensis*: A new weed species resistant to glyphosate in the Americas. *Weed Biol. Manag.* **2014**, *14*, 106–114. [CrossRef]
72. Pavela, R.; Maggi, F.; Iannarelli, R.; Benelli, G. Plant extracts for developing mosquito larvicides: From laboratory to the field, with insights on the modes of action. *Acta Trop.* **2019**, *193*, 236–271. [CrossRef] [PubMed]

Sample Availability: Samples of the Conyza essential oils are no longer available.

molecules

MDPI

Article

Larvicidal Enzyme Inhibition and Repellent Activity of Red Mangrove *Rhizophora mucronata* (Lam.) Leaf Extracts and Their Biomolecules against Three Medically Challenging Arthropod Vectors

Sengodan Karthi [1], Karthic Uthirarajan [1,†], Vinothkumar Manohar [1,†], Manigandan Venkatesan [2,3,†], Kamaraj Chinnaperumal [4], Prabhakaran Vasantha-Srinivasan [5] and Patcharin Krutmuang [6,7,*]

1 Department of Biochemistry, Centre for Biological Sciences, K.S. Rangasamy College of Arts and Science (Autonomous), Namakkal, Tiruchengode Tamil Nadu 637 215, India; karthientomology@gmail.com (S.K.); uk2147karthic@gmail.com (K.U.); vinothvk02@gmail.com (V.M.)
2 Division of Biopesticides and Environmental Toxicology, Sri Paramakalyani Centre for Excellence in Environmental Sciences, Manonmaniam Sundaranar University, Alwarkurichi, Tirunelveli Tamil Nadu 627 412, India
3 Department of Biomedical Engineering, College of Engineering, Michigan State University, East Lansing, MI 48824, USA; manisscience@gmail.com
4 Chettinad Academy of Research and Education, Kelambakkam, Chennai Tamil Nadu 603 103, India
5 Department of Biotechnology, Periyar University, Salem Tamil Nadu 636 011, India; mullaikamaraj@gmail.com
6 Department of Biotechnology, St. Peter's Institute of Higher Education and Research, Avadi, Chennai Tamil Nadu 600 054, India; vasanth.bmg@gmail.com
7 Department of Entomology and Plant Pathology, Faculty of Agriculture, Chiang Mai University, Muang Chiang Mai 50200, Thailand
8 Innovative Agriculture Research Center, Faculty of Agriculture, Chiang Mai University, Chiang Mai 50200, Thailand
* Correspondence: patcharink26@gmail.com
† These authors contributed equally to this work.

Academic Editor: Giovanni Benelli
Received: 3 August 2020; Accepted: 21 August 2020; Published: 24 August 2020

Abstract: The larvicidal potential of crude leaf extracts of *Rhizophora mucronata*, the red mangrove, using diverse solvent extracts of the plant against the early fourth instar larvae of *Anopheles stephensi*, *Culex quinquefasciatus* and *Aedes aegypti* mosquito vectors was analyzed. The acetone extract of *R. mucronata* showed the greatest efficacy: for *Cx. quinquefasciatus* (LC_{50} = 0.13 mg/mL; LC_{90} = 2.84 mg/mL), *An. stephensi* (LC_{50} = 0.34 mg/mL; LC_{90} = 6.03 mg/mL), and *Ae. aegypti* (LC_{50} = 0.11 mg/mL; LC_{90} = 1.35 mg/mL). The acetone extract was further fractionated into four fractions and tested for its larvicidal activity. Fraction 3 showed stronger larvicidal activity against all the three mosquito larvae. Chemical characterization of the acetone extract displayed the existence of several identifiable compounds like phytol, 3,7,11,15-tetramethyl-2-hexadecen-1-ol, 1-hexyl-2-nitrocyclohexane, eicosanoic acid etc. Enzyme assay displayed that *R. mucronata* active F3-fractions exert divergent effects on all three mosquitos' biochemical defensive mechanisms. The plant fractions displayed significant repellent activity against all the three mosquito vectors up to the maximum repellent time of 210 min. Thus, the bioactive molecules in the acetone extract of *R. murconata* leaves showed significant larvicidal and enzyme inhibitory activity and displayed novel eco-friendly tool for mosquito control.

Keywords: mangrove; larvicidal activity; enzyme inhibition; *Rhizophora mucronata*; repellent; mosquitoes

1. Introduction

Mosquitoes are central to the spreading of many infections including dengue fever, malaria, yellow fever and lymphatic filariasis, especially in areas with ecosystems that favor their breeding [1]. Mosquitoes are the primary arthropod vectors of different blood borne illness that cause millions of mortalities per year in humans [2]. Mosquitoes can also cause allergic reactions in human beings such as angioedema [3]. In India alone, there were 1.5 million cases and more than 1500 deaths caused due to malaria cases in the past decades due to mosquito vectors [4].

Malaria is chiefly spread by six mosquito species in India, with *Anopheles stephensi* the most prevalent in urban society [5]. *Aedes aegypti* are the major vectors for dengue and dengue hemorrhagic fever, which are prevalent in Africa, the Americas and India [6]. *Culex quinquefasciatus*, which often breeds in contaminated water, is the major domestic mosquito in many tropical countries and a significant lymphatic filariasis vector, with lymphatic filariasis being the highest growing vector borne illness in the tropical countries affecting more than 146 million people [7]. Larval mosquitoes are particularly striking targets for major insecticides since their breeding site is located in water, an accessible habitat [4]. Among the approaches to reduce mosquito populations and entomological inoculation rates, larvicidal represent an attractive tool to be developed for mosquito control [8,9].

Phytochemicals with higher mosquitocidal actions are now recognized as effective natural pesticides due to their exceptional larvicidal, repellent, pupicidal, and adulticidal actions [7]. Mangrove plants are generally suggested as an alternative source of bioactive ingredients, because they are thought to contain bioactive substances that are considered safe and environmentally friendly to both humans and animals [10]. The red mangrove plant, *Rhizophora mucronata* Lam. (Malpighiales: Rhizophoraceae) is indigenous to East Africa, India, Indonesia, wet tropical regions of Australia and other prominent countries of Asia [11]. The major advantage of choosing this plant is it can be collected easily and cost effective for preparing mosquitocides as compared to commercial pesticides, is easily degradable and, most importantly, has active blends of bio-active chemicals with diverse pharmacological activity. The chief tactic of the present research is to target the developmental stages, especially the Laval stage, of blood sucking pests which keen to block the adult emergence and blocks spreading dreadful diseases. Furthermore, an aim is to detect the repellent activity of the bio-active fractions of *R. mucronata* since any green based pesticides delivering higher repellent activity has higher commercial value in the market for developing better insecticides. It is crucial to investigate the toxic range of any botanical extracts or its derivatives, just as estimated for commercial pesticides, to provide safer dealings to non-targets, especially humans [3]. Thus, the present research was to investigate the chemical composition of leaf extracts of the mangrove *R. mucronata* and larvicidal, enzyme inhibitory activity and repellent against the major medically challenging pest larvae of *An. stephensi*, *Ae. aegypti*, and *Cx. quinquefasciatus*.

2. Results and Discussion

2.1. Larvicidal Activity of *R. mucronata* Crude Extract

Extracts isolated from the plant leaves and flowers deliver novel visions into bio-rational mosquitocidal development. Previous research displayed that botanicals are a dynamic resource with significant biological activities, such as antiviral, antifungal, phytotoxic and, most importantly, larvicidal actions [2,3,12]. An increment number of plant molecules were stated in the previous research, suggesting that there is the atmosphere possible to provide a fertile source of compounds for medicinal or therapeutic use. The synthetic chemical usage has been an essential agent for arthropod management, but they deliver harmful outcomes, including higher tenacity in the surrounding, non-target impact to other well beings, polluting the natural sources, insecticide residues being found on food, the development of insecticide resistance by the targeted pests and impact on non-targets [13]. The aquatic regions are a unique reservoir of bio-active compounds, most of them delivers inimitable structures [13]. Plant-derived chemicals are likely to be part of the future arsenal of mosquito control programs, as they can act as common toxins, reproduction and growth inhibitors or as active repellents and also oviposition deterrent [3]. Investigating the natural botanicals and their active deliverables against the mosquito larvae may ultimately prime to their practicing green based pesticides [14]. Larval management using bio-active compounds is a major vector for the effective management of blood sucking arthropods. Plants are considered to be a viable and preferred substitute to commercial larvicides for managing mosquitoes at the community level [7,14]. For these reasons, we decided to explore our local environment for possible sources of botanic larvicides that might provide effective and safe alternatives to synthetic anti-mosquito products. Here, the larvicidal action of mangrove leaf extracts was determined against three important mosquito vectors. Ethyl acetate, acetone, benzene and methanol leaf extracts of *R. mucronata* (Figure S1) were assayed against the fourth instars. Among them, acetone extracts against *Cx. quinquefasciatus* (LC_{50} = 0.129; LC_{90} = 2.8417 mg/mL), *An. stephensi* (LC_{50} = 0.378; LC_{90} = 6.035 mg/mL), and *Ae. aegypti* (LC_{50} = 0.113; LC_{90} = 1.334 mg/mL) produced the highest larval mortality (Table 1). Similarly, the lethal dosage (1.5 mg) of crude volatile oil derived from the *Piper betle* leaf exhibited a significant mortality rate against the dengue vector with more than 94% of larval mortality and the LC_{50} was observed at 0.63 mg/L [15].

2.2. Mortality Bioassays of Acetone Extract Fractions

The efficacy of plant extracts varies from species to species and plant parts [16]. The variability in toxic concentrations of the various plant extracts, i.e., to achieve mosquitocidal activity, may be due to differences in concentration levels between the insecticidal components of each plant; moreover, the effect of each plant extract will vary with the time of collection and season [13]. It has been reported [15] that the seed methanol extract of *Clitoria ternatea* was effective against *An. stephensi*, *Ae. aegypti*, and *Cx. quinquefasciatus* larvae with LC_{50} values of 65.2, 154.5, and 54.4 ppm, respectively. In our study, the acetone extracts of *R. mucronata* leaves were fractioned and four different fractions were collected. These four fractions were tested for larvicidal activity against three important mosquitoes in public health terms (Table 2). When testing Fraction 3, we observed the greatest larvicidal activity against *Ae. aegypti* (LC_{50} = 0.1037; LC_{90} = 1.0025 mg/mL), *An. stephensi* (LC_{50} = 0.1480; LC_{90} = 4.6480 mg/mL) and *Cx. quinquefasciatus* (LC_{50} = 0.174348; LC_{90} = 16.73929 mg/mL), respectively. Earlier studies reported that methanolic extracts of plants are larvicidal against *An. stephensi* and *Cx. quinquefasciatus* [16,17]. Kamaraj and colleagues reported the highest larval mortality in leaf petroleum ether and flower methanol extracts of *Cryptocoryne auriculata*, flower methanol extracts of *Leucas aspera*, leaf and seed methanol extracts of *Solanum torvum*, and leaf hexane extracts of *Vitex negundo* against *An. subpictus* larvae (LC_{50} = 44.21, 44.69, 53.16, 41.07, 35.32, 28.90 and 44.40 ppm; LC_{90} = 187.31, 188.29, 233.18, 142.66, 151.60, 121.05 and 192.11 ppm, respectively) and *Cx. tritaeniorhynchus* larvae (LC_{50} = 69.83, 51.29, 81.24, 71.79, 44.42, 84.47 and 65.35 ppm; LC_{90} = 335.26, 245.63, 300.45, 361.83, 185.09, 351.41 and 302.42 ppm, respectively). Ansari and colleagues examined the larvicidal activity of *Pinus longifolia* oil against three vector mosquitoes, namely, *Aedes aegypti* (LC_{50} = 82.1 ppm), *Cx. quinquefasciatus* (LC_{50} = 85.7 ppm) and *An. stephensi* (LC_{50} = 112.6 ppm).

2.3. Chemical Characterization of *R. mucronata* Extract

GC-MS analysis of acetone solvent extracts of *R. mucronata* leaves (Figure 1) showed 13 peaks representing 13 compounds. The 13 compounds were characterized and identified as shown in Table 3. phytol, 3,7,11,15-tetramethyl-2-hexadecen-1-ol, 1-hexyl-2-nitrocyclohexane, eicosanoic acid, estra-1,3,5(10)-trien-17, betaol, sulfurous acid, octadecyl 2-propyl ester, 2-heptadecenal, 1-hexyl-2-nitrocyclohexane, 17-pentatriacontene, tritetracontane, urs-12-en-28-ol and squalene. The peak area percentage was prominent in Eicosanoic Acid (38.24%) with retention time at 19.205. Since the Peak area percentage was prominent in Eicosanoic Acid and this might have played a key role in mosquitocidal activity against all the three arthropod vectors. Preliminary phytochemical screening of whole plant extracts revealed the presence of phenol, flavonoids, alkaloids, saponins, tannins, glycosides, amino acids, quinones and carbohydrates in the plant extracts (Table 4).

Table 1. Larvicidal activity of mangrove plant extracts of *Rhizosphora murconata* against *Ae. aegypti*, *An. stephensi* and *Cx. quinquefasciatus*. LC_{50} Lethal concentration 50% mortality, LC_{90} Lethal concentration 90% mortality, LCL: lower confidence limits, UCL: upper confidence limits, χ^2: chi square, df: degrees of freedom.

Species	Solvents	LC_{50} mg/mL (95% Confidence Limit)	LC_{90} mg/mL (95% Confidence Limit)	χ^2	df	*p*-Value
Aedes aegypti	Acetone	0.113 (0.1935–1.147)	1.334 (0.886–2.9626)	1.274	3	0.763
	Ethyl Acetate	0.305 (0.125–0.556)	1.037 (1.413–8.009)	3.078	3	0.486
	Methanol	0.154 (0.210–0.307)	1.1453 (1.258–11.809)	4.524	3	0.527
	Petroleum benzene	0.502 (0.271–1.026)	1.3725 (1.803–8.368)	5.164	3	0.498
Anopheles stephensi	Acetone	0.378 (0.119–0.481)	6.035 (1.1045–11.930)	3.500	3	0.003
	Ethyl Acetate	0.427 (0.389–2.358)	4.418 (2.902–5.972)	2.410	3	0.395
	Methanol	0.415 (0.269–1.243)	2.088 (1.202–7.169)	4.319	3	0.375
	Petroleum benzene	0.504 (0.304–0.895)	5.8592 (1.2245–3.2839)	2.311	3	0.269
Culex quinquefasciatus	Acetone	0.129 (0.030–0.239)	2.8417 (2.700–6.302)	1.346	3	0.865
	Ethyl Acetate	0.378 (0.165–0.751)	1.7374 (1.861–5.499)	2.746	3	0.468
	Methanol	0.295 (0.116–0.539)	1.0615 (1.4216–9.4711)	3.092	3	0.037
	Petroleum benzene	0.584 (0.324–1.302)	1.6477 (2.0227–13.219)	2.275	3	0.284

Table 2. LC_{50}, LC_{90}, and chi square analysis of larvicidal activity of *Rhizosphora murconata* acetone extract column fraction against *Ae. aegypti*, *An. stephensi* and *Cx. quinquefasciatus*. LC_{50} Lethal concentration 50% mortality, LC_{90} Lethal concentration 90% mortality, LCL: lower confidence limits, UCL: upper confidence limits, χ^2: chi square, df: degrees of freedom. Significance at $p < 0.05$. * denotes the predominant lethal concentration dosage of Fraction F3.

Species	Column Fraction	LC_{50} mg/mL (95% Confidence Limit)	LC_{90} mg/mL (95% Confidence Limit)	χ^2	df	*p*-Value
Culex quinquefasciatus	F1	0.245333 (0.159992–2.886752)	2.322 (5.002817–7.2119)	1.77430	3	0.187
	F2	0.341783 (0.212881–0.493197)	12.58869 (4.985508–86.38909)	3.647	3	0.476
	F3	0.174348 * (0.061645–0.291515)	16.73929 (5.101697–366.6364)	3.919	3	0.528
	F4	0.217996 (0.115948–0.323425)	8.073846 (3.511051–45.48615)	1.82923	3	0.098
Aedes aegypti	F1	0.314289 (0.148417–0.511897)	39.94815 (9.04682–2474.55)	3.41649	3	0.461
	F2	0.130 (0.119–0.1561)	1.1158 (0.9422–2.2120)	65.1	3	0.782
	F3	0.1037 * (0.069–0.112)	1.0025 (0.8871–2.1147)	4.84	3	0.521
	F4	0.20831 (0.48172–1.19820)	22.581621 (20.1682–24.8216)	3.082	3	0.391
Anopheles stephensi	F1	0.266881967 (0.0021–0.469229)	3.1525 (2.9900–3.1211)	1.938	3	0.207
	F2	0.31175 (0.01451–0.22655)	3.3851 (2.44364–7.5935)	0.992	3	0.034
	F3	0.1480 * (0.2957–0.767397)	4.6480 (3.2585–9.4680)	3.147	3	0.218
	F4	1.358 (0.0198–3.2210)	5.8546 (4.3215–7.5842)	0.678	3	0.004

Table 3. Chemical characterization of acetone leaf extract of *R. murconata* through GC-MS analysis.

S.No.	Name of the Compounds	RI Polar Column Exp	Lit	RI Polar Column Exp	Lit	Peak Area %	Formula	Structure
1	Phytol	925	919	2622	2617	9.704	$C_{20}H_{40}O$	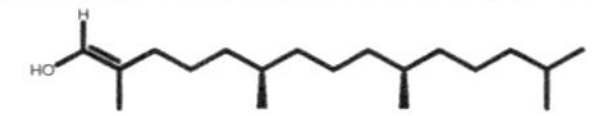
2	3,7,11,15-Tetramethyl-2-Hexadecen-1-ol	957	900	2114	2116	3.738	$C_{20}H_{40}O$	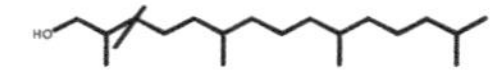
3	1-Hexyl-2-Nitrocyclohexane	817	814	1054	1060	1.338	$C_{12}H_{23}O_2N$	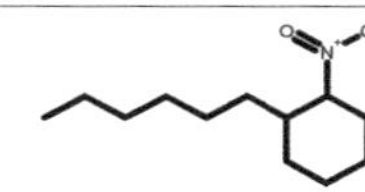
4	Eicosanoic Acid	913	912	2442	2445	38.246	$C_{20}H_{40}O_2$	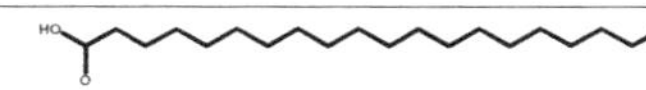
5	Estra-1,3,5(10)-Trien-17-Beta-ol	871	869	1145	1152	15.447	$C_{18}H_{24}O$	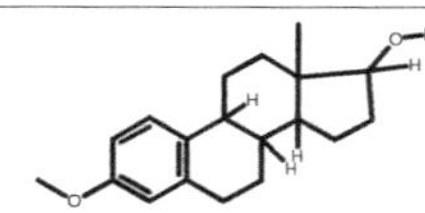
6	Sulfurous acid, Octadecyl 2-Propyl Ester	915	911	1231	1237	3.108	$C_{21}H_{44}O_3S$	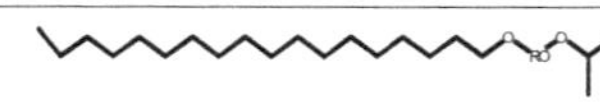
7	2-Heptadecenal	918	909	1174	1183	3.406	$C_{17}H_{32}O$	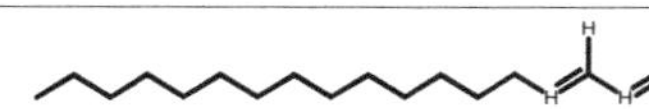
8	1-Hexyl-2-Nitrocyclohexane	923	916	1214	1217	5.675	$C_{12}H_{23}O_2N$	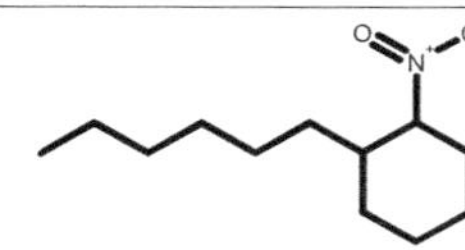
9	17-Pentatriacontene	929	921	1063	1066	2.450	$C_{35}H_{70}$	
10	Sulfurous acid, Octadecyl 2-Propyl Ester	952	947	1118	1120	1.310	$C_{21}H_{44}O_3S$	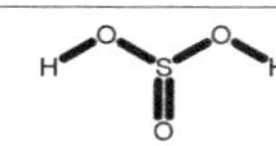

Table 3. *Cont.*

S.No.	Name of the Compounds	RI Polar Column Exp	Lit	RI Polar Column Exp	Lit	Peak Area %	Formula	Structure
11	Tritetracontane	943	939	4297	4300	1.433	$C_{43}H_{88}$	
12	2,6,10,14,18,22-tetracosahexaene, 2,6,10,15,19,23-Hexamethyl-, (all-e)-	983	975	2814	2819	12.222	$C_{30}H_{50}$	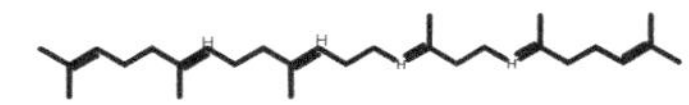
13	Urs-12-En-28-ol	749	748	987	992	1.923	$C_{30}H_{50}O$	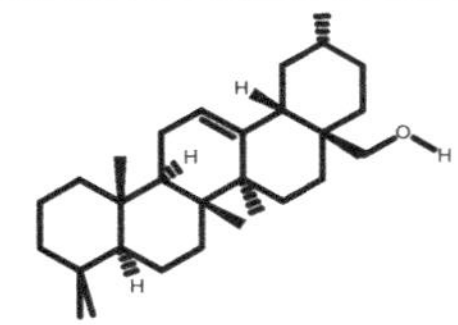

Table 4. Phytochemical constituent's analysis of different solvent extracts of *R. murconata*. Where + indicates presence, - indicates absence.

S. No.	Phytochemical Test	Petroleum Benzene	Ethyl Acetate Extract	Acetone Extract	Methanol Extract
1	Phenols	+	+	+	+
2	Flavonoids	+	+	+	+
3	Alkaloids	-	+	+	+
4	Saponins	+	+	+	+
5	Tannins	+	+	+	+
6	Glycosides	+	+	+	+
7	Proteins	-	-	-	-
8	Amino Acid	-	+	+	-
9	Quinones	+	+	+	+
10	Carbohydrates	-	-	-	+

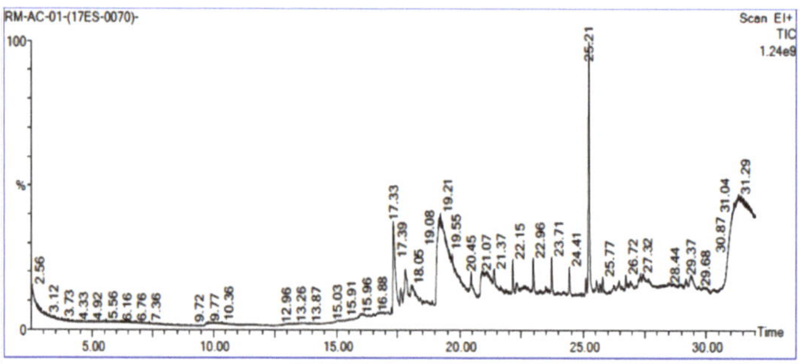

Figure 1. Chemical composition of GC-MS analysis using acetone leaf extract of *R. murconata*.

Plant-based compounds are known to be eco-friendly and can be used in mosquito larval control safely. Furthermore, these natural products are readily biodegradable and safe to other organisms [18]. The potent larvicidal activity of *R. murconata* could be attributed to the presence of tannins, phenols, flavonoids, saponins, glycosides and quinones (Figure 2). The isolation of these compounds, as defined here, could provide the basis for developing natural mosquitocidal products as a substitute for synthetic insecticides- or for developing mosquito repellents. The fractions obtained from the *Adhatoda vasica* were an effective larvicidal agent against the *Cx. quinquefasciatus* and *Ae. aegypti* larvae [19]. It was highly toxic to mosquito larvae and also inhibited the development of pupae. The high rate of larval mortality of mosquitoes observed at higher concentrations (250 ppm of *A. vasica*) within a 24 h exposure indicates the high toxicity of the product. The plants tested in the present study are reported to be eco-friendly and toxic to agriculturally important insect pest [20]. For example, essential oils from Citronella and Eucalyptus are often used in repellents sold under several brand names [21]. Recent studies have indicated that phytol, a precursor of synthetic vitamin E and vitamin K (which was identified in our analysis as well), exhibits antioxidant and antinociceptive effects and was cytotoxic against the MCF−7 breast cancer cell line [22].

The FT-IR spectrum analysis of acetone solvent extract of *R. mucronata* (Figure 3) showed 13 peaks indicating the presence of functional groups. The functional groups of the detected peak values were identified (Table 5). Alkane, Amine amd Alkyl were invariably present in acetone extracts. The column separation was eluted the different fractions of *R. mucronata*. Besides the active fraction of F3 was compared with a standard of squalene. The retention time of fraction F3 got eluted in 4.8 min at 220 nm and it was comparable to the standard peak of squalene (Figure 4). Correspondingly, Octacosane, derived from *Couroupita guianensis*, showed the highest percentage, at 31.86% peak area in the active fraction F6 and was previously shown to have toxicity to early third instars of *Cx. quinquefasciatus* (Say.) [23]. Octacosane was also shown to be present in the unsaponifiable phase of common chicory, *Cichorium intybus L.* at 1.34% peak area, exhibiting larvicidal activity against *Anopheles pharoensis* (Theobald) with LC_{50} value of 13.62 mg/kg_1 [23]. Spirostan-3,15-diol,3-(4-methylbenzenesulfo-nate) at 5.20% of peak area, was isolated by Soule et al. [24] as a spirostane glycoside from the *Solanum laxum* (Steud.) aerial parts were reported to have toxicity against the aphid, *Schizaphisgra minum* (Rondani) [25]. FractionF6 also contains hexacosane at 17.54% peak area. Hexacosane, in the leaf oil of *Solanum sarrachoides* (Sendt.), exhibits oviposition deterrence against red spidermites, (*Tetranychus evansi*) [26,27]. The sensitivity of adults of *Ae. aegypti* has been shown as resulting from the presence of 1,8-cineole, α-pinene and *p*-cymene and is correlated to the amount of 1,8-cineole in the tested extract [28]. The insecticidal activity of the essential oil of *Eucalyptus tereticornis* observed in the current work, might be explained by the presence of one of its major components (*p*-cymene) but also by a minor compound (1,8-cineole) in our extracts, which both have demonstrated insecticidal activity against *An. gambiae* [27].

Figure 2. Phytochemical analysis of *R. murconata* using different solvent extracts. (**A**) Phenol; (**B**) flavonoids; (**C**) alkaloids; (**D**) saponins; (**E**) tannins; (**F**) glycosides; (**G**) protein; (**H**) amino acids; (**I**) quinones; (**J**) carbohydrates.

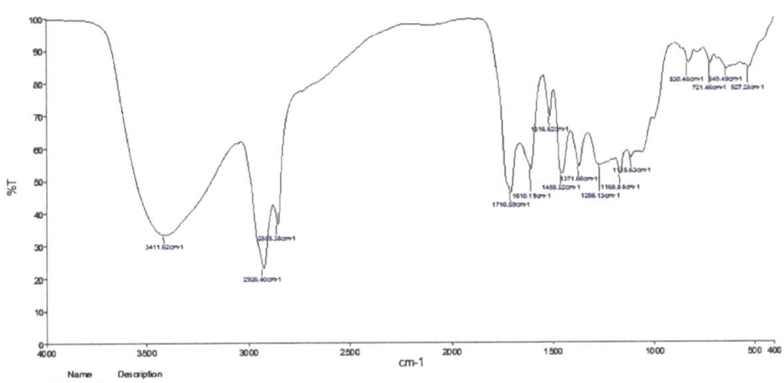

Figure 3. FT-IR spectrum analysis of acetone extract of *R. murconata*.

2.4. Enzyme Assay

The active acetone extract F3-fractions of *R. murconata* inhibited α-carboxylesterase activity of fourth instar larvae tested at 24 h. The activity decreased in dose dependent manner across *Cx. quinquefasciatus* (Figure 5A, $F_{4,20} = 41.62$; $p < 0.0001$), *Ae. aegypti* (Figure 5B, $F_{4,20} = 32.12$; $p < 0.0001$) and *An. stephensi*

(Figure 5C, $F_{4,20} = 27.22$; $p < 0.0001$), respectively. Similarly, β carboxylesterase activity decreased significantly to the sub-lethal dosages of F3-fractions against all the three mosquito vectors. However, the reduction rate was prominent in *Cx. quinquefasciatus* (Figure 5D, $F_{4,20} = 27.22$; $p < 0.0001$) as compared to *Ae. aegypti* (Figure 5E) and *An. stephensi* (Figure 5F). Likewise, the level of SOD activity was also declined at the maximum sub-lethal dosages of all the three mosquito vectors (Figure 5G–I). However, the reduction rate was minimal at the dosage of 0.6 mg/mL in *Cx. quinquefasciatus* ($F_{4,20} = 13.11$; $p < 0.0001$), *Ae. aegypti* ($F_{4,20} = 15.11$; $p < 0.0001$) and *An. stephensi* ($F_{4,20} = 17.26$; $p < 0.0001$), respectively. It has been well known that many bio-insecticides affects the insect metabolism by inhibiting or stimulating the activity of digestive enzymes. Generally, SOD (superdioxide) expressed in several regions especially in anal gills of mosquitoes to catalyzes the superoxide radical dismutation into hydrogen peroxide [28].

Table 5. FT-IR analysis of peak values of *R. mucronata* acetone extract.

S.No.	Peak (Wave Number cm^{-1})	Intensity	Bond	Functional Group Assignment
1	3411.62	39.09	N-H Stretch	Amine
2	2926.40	23.06	C-H Stretch	Alkyl
3	2855.38	36.55	C-H Stretch	Alkyl
4	1710.59	42.02	C=O Stretch	Aldehyde
5	1610.19	50.93	C=O Stretch	Amide
6	1516.62	59.60	C=C Bending	Aromatic
7	1458.22	49.94	C-H Bending	Alkane
8	1371.66	53.83	C-H Bending	Alkane
9	1266.13	47.76	C-N Stretch	Amine
10	1168.88	49.26	C-N Stretch	Amine
11	1115.63	53.74	C-N Stretch	Amine
12	830.48	80.77	C-H Bending	Aromatic
13	721.46	85.12	C-Cl Stretch	Alkyl Halide

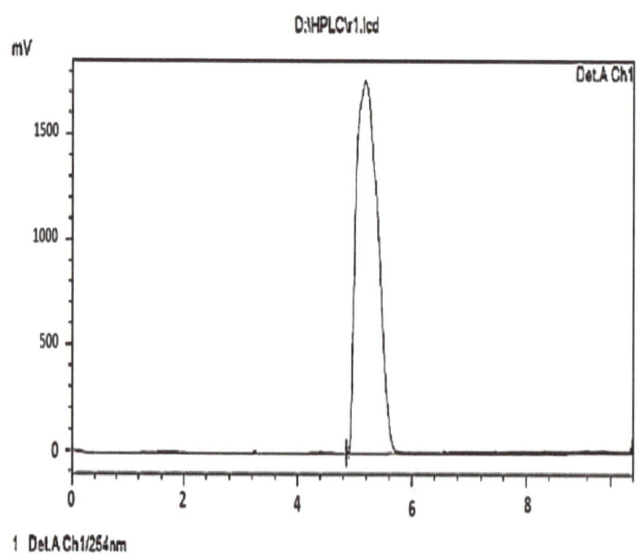

Figure 4. HPLC chromatogram analysis for acetone extract active F3-fractions of *R. murconata*.

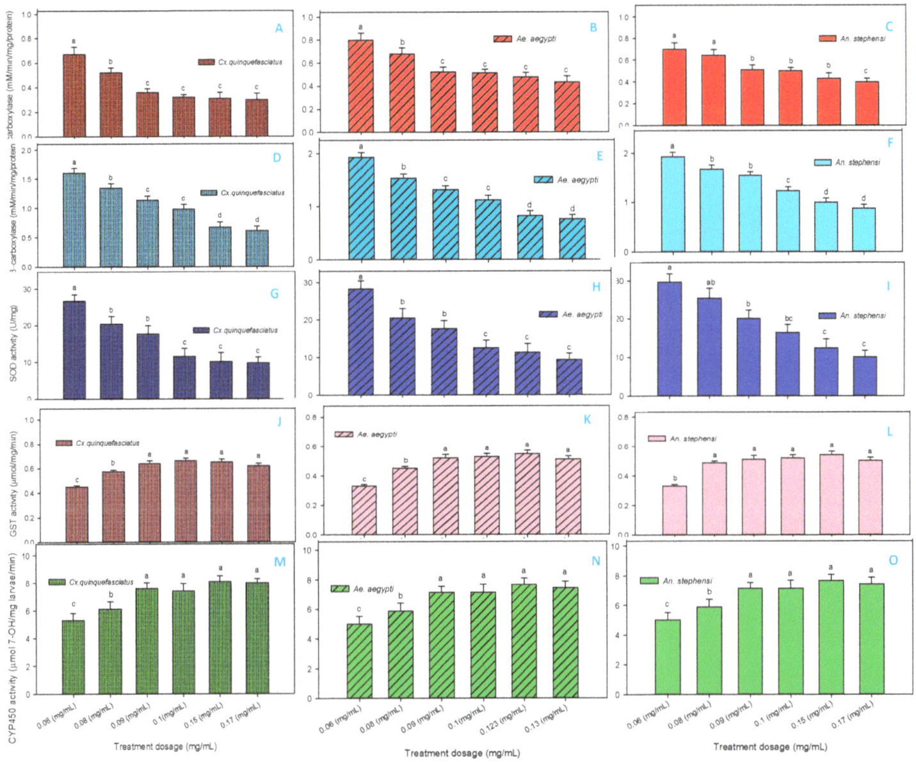

Figure 5. (**A–C**) α- carboxylestrase (**D–F**) β carboxylestrase (**G–I**) SOD activity (**J–L**) Glutathione S-transferase (**M–O**) CYP450 enzyme activity of *Cx. quinquefasciatus*, *Ae. aegypti*, *An. stephensi* fourth instar larvae after treatment with active acetone extract F3- fractions of *R. murconata*. Mean (± SEM) followed by the same letter in the above bars indicate no significant difference ($p < 0.05$) in a Tukey's test.

In contrast to the above enzyme levels, GST and CYP450 enzyme regulations upregulated significantly in dose dependent manner. The enzyme ratio increased significantly initially in the lower doses (0.06 mg/mL) and relics constant at the higher sub-lethal dosages across all the mosquito vectors. Besides, the GST (Figure 5J–L) and CYP450 (Figure 5M–O) enzyme regulations uplifted significantly in *Cx. quinquefasciatus* ($F_{4,20} = 11.21$; $p < 0.0001$) as compared to *Ae. aegypti* ($F_{4,20} = 16.71$; $p < 0.0001$) and *An. stephensi* ($F_{4,20} = 14.55$; $p < 0.0001$). Our enzyme analysis data was interpreted to display that *R. mucronata* active F3-fractions exert divergent effects on all three mosquitos' biochemical defensive mechanisms.

In addition to their potential for mosquito control, the *R. mucronata* extracts may exhibit enzyme inhibitory activity at currently unknown bioactivities and further studies are required to investigate the possibility of the herein identified constituents of *R. mucronata* extracts for the development of environmentally-friendly applications. Indeed, phytochemicals from plant sources may also act as insect growth regulators, repellents, ovipositor attractants, among others [26,27].

2.5. Repellent Activity

The active fractions of *R. mucronata* showed significant repellent activity against *Cx. quinquefasciatus* (96.4%-$F_{4,20} = 24.22$, $p \leq 0.0001$), *A. aegypti* (94.32%-$F_{4,20} = 18.82$, $p \leq 0.0001$) and *An. stephensi* (97.2%-$F_{4,20} = 21.88$, $p \leq 0.0001$), respectively, at the maximum repellent dosage of 0.1 (mg/mL) at the maximum repellent time of 210 min (Figure 6. Similarly, the sub-lethal dosage (100 ppm) of crude seed extracts of *Terminalia chebula Retz.* displayed significant protection time at the maximum time of 210 min [29].

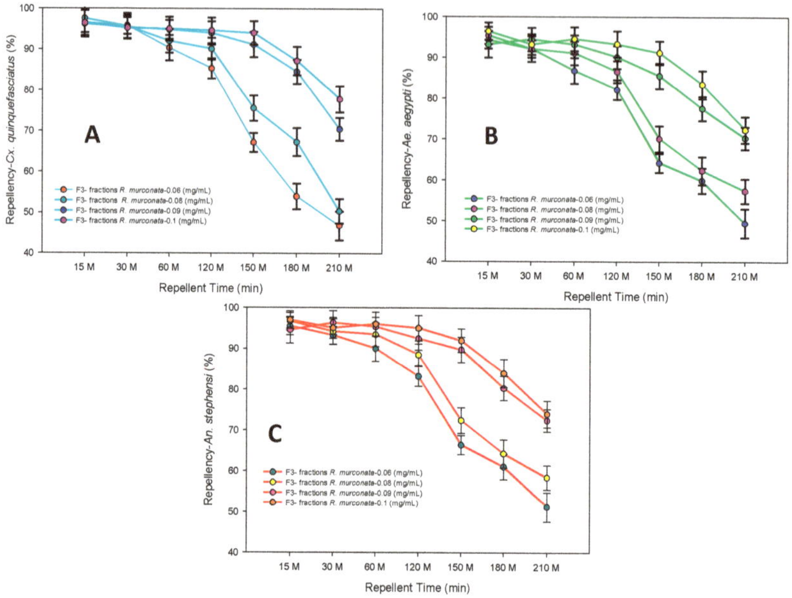

Figure 6. Repellency of acetone extract F3- fractions of *R. murconata* against *Cx. Quinquefasciatus* (**A**), *Ae. Aegypti* (**B**), and *An. Stephensi* (**C**). Mean (± SEM) followed by the same letter in the above bars indicate no significant difference ($p < 0.05$) in a Tukey's test.

3. Materials and Methods

3.1. Plant Harvesting

The fresh leaves of *R. mucronata* were harvested from the region of Pichavaram, Rameshwaram (Latitude: 13.129991; Longitude: 79°18′46.54″ E), Tamil Nadu, India, with taxonomy confirmed by an expert (Prof. Kathiresan, Department of Marine Biology, Annamalai University, Tamil Nadu) (Figure S1). The specimen voucher was preserved in the herbarium (Ref. No. BC¬/2016/Rm- 07) for further assays.

3.2. Crude Extract Preparation

Collected fresh leaves were air dried under shadow at room temperature for 7–10 days. The leaves (250 g) were dried and mechanically powdered using a mixer and crushed (Mixer Grinders Stylo 750) to well particle size. Leaf powdered extracts of *R. mucronata* (250 g) was prepared by using soxhlet device utilizing diverse solvents including acetone, ethyl acetate, petroleum benzene, and methanol. Further the solvents were evaporated by using rotary evaporator and the remaining were preserved under 4 °C for further assays. The total yields were observed 2.15, 1.76, 1.95 and 2.02 g, respectively.

3.3. Mosquitoes

Ae. aegypti, *An. stephensi*, and *Cx. quinquefasciatus* larvae were collected from the local areas of Pallipalayam (Latitude: 11.3450° N; Longitude: 77.7309° E), Tiruchengode, Tamil Nadu, India. Furthermore, the culture were maintained in the laboratory were kept in sterile vessels filled with water, maintained at 28 ± 2 °C and 75%–80% relative humidity (RH) under a fixed photoperiod (14:10 L/D). The emerged adult mosquitoes were maintained under the same conditions as the larvae.

3.4. Larval Mortality Assay

Larval bioassays were executed on the fourth instars of *Cx. quinquefasciatus*, *An. stephensi* and *Ae. aegypti* with diversified dosages (0.5, 0.10, 0.15, 0.20 and 0.25 mg/mL) of *R. mucronata* leaf extracts. Minimal of 20 larvae/each concentration were utilized for all the assays, and the procedure were three times replicated. The lethal dosage (LC_{50} and LC_{90}) was calculated based on Probit analysis [30]. Twenty larvae of fourth instars were presented to a 200 mL glass jar supplemented with discriminating dosage of leaf extracts along with 50 mg/L of yeast extract. A group of control was also kept alone treated with methanol. Three times were replicated along with control in each replication. Mortality percentage in the treatments was corrected whenever required using Abbott formula [31].

3.5. Preparation of Whole-Body Homogenates for Enzyme Assay

The control and plant extract exposed fourth instar larvae were rinsed with sterile distilled water, and the adhering water was distant by using blotting tissue paper from the surface. The larval homogenized were distinct using a handy homogenizer in Eppendorf tubes containing ice-cold sodium phosphate buffer (500 µL–20 mM, pH 7.0). Homogenates were further spin for 20 min at 8000× *g* at 4 °C) and the supernatants separated were utilized for enzyme experiments. The homogenates separated were reserved at −4 °C for further experiments.

3.6. Carboxyl Esterase Assays

The α- and β- carboxylesterase activity was estimated using the larval extracts in phosphate potassium buffer (0.2 M; pH = 7.1) were set to (20 µL; 84 µg protein). Further, it was assorted with 500 µL buffer (0.3 mMα- or β-naphthyl acetate in 0.1 M phosphate potassium at pH 7.2 containing 1% acetone). Enzyme activity of one unit was definite as the enzyme amount essential to generate 1 µmol of α- or β-naphthol/minute.

3.7. Superoxide Dismutase Activity

The Superoxide Dismutase (SOD) assays were carried out by using the Superoxide dismutase determination kit (Sigma-Aldrich, Bangalore, India). Activity of SOD per 1 unit was resolute as the required enzyme to slab the increment absorbance by 50% at 440 nm.

3.8. Glutathione-S-Transferase Activity

A total of 250 µL of fourth instars were homogenized in sodium phosphate solution (50 mM; pH = 7.2) and spin at 10,000× *g* at 4 °C for 20 min. The Glutathione-S-Transferase (GST) assay Kit (Sigma-Aldrich, Catalog 0410, Bangalore, India) was utilized to investigate the conjugation of the thiol group of glutathione to the 1-chloro-2,4-dinotrobenzene (CDNB) substrate. Further assays were performed based on the standard protocol prescribed in the Kit. The GST activity was stated as µmol/mg protein/min substrate conjugated.

3.9. Cytochrome P450 Activity

Fourth instar visible to bioactive fractions3 were rinsed in sterile water and separated. Further the larvae were kept in 40 mM sodium phosphate solution with maintained pH: 7.2 and cooled earlier separation. The abdominal segments, heads, and digestive tissues were extracted. For determining the enzyme activity, carcasses were taken. The enzymes were measured by using ethoxycoumarin-*O*-deethylase existing in the body walls and it was stated as mol 7-OH/mg larvae/min.

3.10. Phytochemical Analysis

Chemical testing for the presence of carbohydrates, alkaloids, saponins, phenolics, tannins, terpenes and flavonoids was evaluated in the mangrove extracts and the most active extract using the standard procedure of Harborne [32].

3.10.1. Phenols: Ferric Chloride Test

Extracts were treated with 3–4 drops of ferric chloride solution. Formation of bluish black color indicates the presence of phenols.

3.10.2. Flavonoids Test

Ammonia solution (5 mL) was added to a portion of the crude extract followed by addition of concentrated H_2SO_4. Formation of a yellow coloration in the extract indicates the presence of flavonoids. The yellow coloration disappears after some time.

3.10.3. Alkaloids: Wagner Reagent

Extracts were dissolved individually in dilute Hydrochloric acid and filtered. Filtrates were treated with Wagner's reagent (1.27 g Iodine in 2 g of Potassium Iodide). The formation of a brown/reddish precipitate indicates the presence of alkaloids.

3.10.4. Saponin

Half mg of the extract was shaken with 2 mL of distilled water. If foam produced persists for ten min, it indicates the presence of saponins.

3.10.5. Tannin Test: Gelatin Test

To the extract, 1% gelatin solution containing sodium chloride was added. Formation of white precipitate indicates the presence of tannins.

3.10.6. Glycosides Test

Minimum quantities of the extracts were hydrolyzed with hydrochloric acid for a few minutes on a water bath and the hydrolysate was subjected to the following tests.

The extracts were treated with chloroform and evaporate it to dryness. Separately, 0.4 mL of glacial acetic acid containing a trace amount of ferric chloride was added and transferred to a small test tube added with carefully 0.5 mL of concentrated sulfuric acid by the side of the test tube; blue color appeared in the acetic acid layer, indicating the presence of glycosides.

3.10.7. Ninhydrin Test

To the extract, 0.25% *w/v* Ninhydrin reagent was added and boiled for few minutes. Formation of blue or blue to violet indicates the presence of amino acid.

3.10.8. Benedict Test

Two mL of crude extracts were dissolved individually in 5 mL distilled water and filtered. Filtrates were treated with Benedict's reagent and heated gently. Orange-red precipitate indicates the presence of reducing sugars.

3.10.9. Starch Test

To 5 mL of the extract, a few drops of iodine was added. The presence of starch was indicated by the formation of blue color.

3.10.10. Flavonoids Test

Ammonia solution (5 mL) was added to a portion of the crude extract followed by addition of concentrated H_2SO_4. Formation of a yellow coloration in the extract indicates the presence of flavonoids. The yellow coloration disappears after some time.

3.11. Repellent Assay

The active fractions F3 derived from *R. mucronata* extracts evaluated for its repellency against all the three mosquito vectors with sub-lethal dosages (0.06, 0.08, 0.09 and 0.1 mg/mL). Further repellent assay experiments were adapted from our previous research protocol Thanigaivel et al. [29].

3.12. GC–MS Analysis and Compound Identification

Among the different extracts (acetone, petroleum benzene, ethyl acetate and methanol) which were tested for their larvicidal toxic, and based on our preliminary tests, the highest toxicity was observed in acetone extracts of *R. mucronata*. Therefore, the acetone extract was used for further experiments. The isolated *R. mucronata* acetone extract was dissolved in (1:1) ratio with ethyl alcohol. From this, 2 μL of crude solution was dissolved in HPLC-grade methanol and subjected to GC and MS JEOL GC (Agilent Technologies 6890N, PerkinElmer, Bangalore, India) mate equipped with secondary electron multiplier. The downstream procedure of chemical characterization was carried out by our previous methodology (Agilent Technologies 6890N, PerkinElmer, Bangalore, India) [29]. The molecular weight, molecular formula and structure of the compounds of tested materials were ascertained by interpretation on mass spectrum of GC-MS using the database of the National Institute Standard and Technology (NIST).

3.13. FT-IR Analysis

Dried acetone extract was used for FT-IR analysis according to the combined methods reported in [33,34]. A total of 2 mg sample were mixed with 100 mg KBr (FT-IR grade) and then compressed to prepare a translucent sample disc (3 mm diameter), which was immediately kept in the sample holder. The sample was scanned and the FT-IR spectra recorded in the absorption range of 400 and 4000 cm^{-1}. FT-IR analysis was performed using a Perkin-Elmer spectrophotometer (Perkin-Elmer FT-IR, Spectrum 2 Singapore, L160000A) to detect characteristic peaks, types of chemical bonds, and probable functional groups present in the sample. FT-IR peak values were recorded and the analysis was repeated twice for spectrum confirmation.

3.14. High Performance Liquid Chromatography (HPLC) Analysis

The HPLC (Flexar HPL, Perkin Elmer, Chennai, India) analysis was carried out with a C18 column (220 nm) (250 × 4.6 mm). Acetonitrile and water were used as mobile phase at flow rate of 1 mL/min. Peaks obtained were detected at 220 nm. Fraction F3 was analyzed and compared with a purchased standard of squalene (Sigma-Aldrich, Analytical Standard). The yield of the collected fraction dry weight is 0.57 mg.

3.15. Statistical Analysis

Average larval mortality data were analyzed by probit analysis to calculate the LC_{50}, LC_{90}, and other statistics with 95% confidence intervals of upper confidence limit (UCL) and lower confidence limit (LCL) standards and the chi-squared test. SPSS 14.0 (IBM Inc., Chicago, IL, USA) was used to analyze the data.

4. Conclusions

As an endnote, the phyto-chemical characterization of acetone extracts of *R. mucronata* through GC-MS analysis displayed 13 major phyto-chemicals and the individual bio-active fractions were purified and characterized through HPLC and FT-IR assays which signified that active fraction-F3 was prominent as compared to other fractions and the larvicidal activity of active fraction F3 delivers that they are lethal against larvae of all three mosquito vectors. Moreover, the enzyme assays signify that the acetone extract F3-fractions significantly drifted the enzyme pattern in all five major digestive and detoxifying enzymes. Furthermore, the sub-lethal dosage of active fraction F3 also showed significant repellent activity in a dose dependent manner across all the three arthropod vectors. From this research, it is suggested that the mangrove leaf extract of *R. murconata* contains several bioactive molecules that might be useful as eco-friendly larvicides and repellent for managing blood sucking pests of medical importance. Further functional and mechanistic studies on how these compounds exert larvicidal activity may pave the way for environmentally safe botanical insecticides to control mosquito populations.

Supplementary Materials: The following are available online at http://www.mdpi.com/1420-3049/25/17/3844/s1, Figure S1: *Rhizophora mucronata* (A) whole plant and (B) stem and leaf.

Author Contributions: S.K., M.V., K.U., V.M. and P.K. designed the research plan and drafted the manuscript. M.V., K.U., V.M., P.V.-S., K.C., performed the experimental works and data compilation. M.V., K.U., P.K., and S.K. coordinated the work and discussed the results. S.K. and P.K. revise the manuscript. All authors read and approved the final manuscript.

Funding: This research was partially funded by Post-Doctoral Program, Chiang Mai University, Thailand.

Acknowledgments: We would like to thank the Department of Biochemistry, K.S. Rangasamy College of Arts and Science (Autonomous), Tiruchengode, Namakkal, Tamil Nadu, India for providing the infrastructural facility for carrying out this research work and also, this work was partially supported Post-Doctoral Program, Chiang Mai University, Thailand. We also thank the anonymous reviewers for their critical and constructive comments which helped a lot in improving the draft.

Conflicts of Interest: The authors declare that no competing interests exist.

References

1. Narendhiran, S.; Mohanasundaram, S.; Arun, J.; Saravanan, L.; Catherine, L.; Subathra, M.; Rannjith, R. Comparative study in larvicidal efficacy of medicinal plant extracts against *Culex quinquefasciatus*. *Int. J. Res. Plant Sci.* **2014**, *4*, 22–25.
2. Chellappandian, M.; Thanigaivel, A.; Vasantha-Srinivasan, P.; Edwin, E.; Ponsankar, A.; Selin-Rani, S.; Kalaivani, K.; Senthil-Nathan, S.; Benelli, G. Toxicological effects of Sphaeranthus indicus Linn. (Asteraceae) leaf essential oil against human disease vectors, Culex quinquefasciatus Say and Aedes aegypti Linn. and impacts on a beneficial mosquito predator. *Environ. Sci. Pollut. Res. Int.* **2017**, *1110*, 294–10306. [CrossRef]
3. Chellappandian, M.; Vasantha-Srinivasan, P.; Senthil-Nathan, S.; Karthi, S.; Thanigaivel, A.; Ponsankar, A.; Kalaivani, K.; Hunter, W.B. Botanical essential oils and uses as mosquitocides and repellents against dengue. *Environ. Int.* **2018**, *113*, 214–230. [CrossRef] [PubMed]
4. Govindarajan, M. Chemical composition and larvicidal activity of leaf essential oil from *Clausena anisata* (Willd.) Hook. f. ex Benth (Rutaceae) against three mosquito species. *Asian Pac. J. Trop. Med.* **2018**, *3*, 74–877. [CrossRef]
5. Panneerselvam, C.; Murugan, K.; Kovendan, K.; Kumar, P.M.; Ponarulselvam, S.; Amerasan, D. Larvicidal efficacy of *Catharanthus roseus* Linn. (Family: Apocynaceae) leaf extract and bacterial insecticide *Bacillus thuringiensis* against *Anopheles stephensi* Liston. *Asian Pac. J. Trop. Biomed.* **2013**, *6*, 847–853. [CrossRef]
6. Edwin, E.; Vasantha-Srinivasan, P.; Senthil-Nathan, S.; Thanigaivel, A.; Ponsankar, A.; Pradeepa, V.; Selin-Rani, S.; Kalaivani, K.; Hunter, W.B.; Abdel-Megeed, A.; et al. Anti-dengue efficacy of bioactive andrographolide from *Andrographis paniculata* (Lamiales: Acanthaceae) against the primary dengue vector *Aedes aegypti* (Diptera: Culicidae). *Acta Trop.* **2016**, *163*, 167–178. [CrossRef] [PubMed]
7. Pavela, R.; Maggi, F.; Iannarelli, R.; Benelli, G. Plant extracts for developing mosquito larvicides: From laboratory to the field, with insights on the modes of action. *Acta Trop.* **2019**, *193*, 236–271. [CrossRef] [PubMed]
8. Elumalai, D.; Hemalatha, P.; Kaleena, P. Larvicidal activity and GC–MS analysis of *Leucas aspera* against *Aedes aegypti, Anopheles stephensi* and *Culex quinquefasciatus*. *J. Saudi Soc. Agric. Sci.* **2017**, *16*, 306–313. [CrossRef]
9. Kalaivani, K.; Senthil-Nathan, S.; Murugesan, A.G. Biological activity of selected Lamiaceae and Zingiberaceae plant essential oils against the dengue vector *Aedes aegypti* L. (Diptera: Culicidae). *Parasitol. Res.* **2012**, *110*, 1261–1268. [CrossRef]
10. Sakulku, U.; Nuchuchua, O.; Uawongyart, N.; Puttipipatkhachorn, S.; Soottitantawat, A.; Ruktanonchai, U. Characterization and mosquito repellent activity of citronella oil nanoemulsion. *Int. J. Pharm.* **2009**, *372*, 105–111. [CrossRef]

11. Senthilkumar, A.; Venkatesalu, V. Larvicidal potential of *Acorus calamus* L. essential oil against filarial vector mosquito *Culex quinquefasciatus* (Diptera: Culicidae). *Asian Pac. J. Trop. Dis.* **2012**, *2*, 324–326. [CrossRef]
12. Skenderidis, P.; Mitsagga, C.; Giavasis, I.; Petrotos, K.; Lampakis, D.; Leontopoulos, S.; Hadjichristodoulou, C.; Tsakalof, A. The in vitro antimicrobial activity assessment of ultrasound assisted *Lycium barbarum* fruit extracts and pomegranate fruit peels. *J. Food Measur. Charact.* **2019**, *13*, 2017–2031. [CrossRef]
13. Senthil-Nathan, S. Physiological and biochemical effect of Neem and other Meliaceae plants secondary metabolites against Lepidopteran insects. *Front. Physiol.* **2013**, *4*, 1–17. [CrossRef] [PubMed]
14. Senthil-Nathan, S. A review of bio pesticides and their mode of action against insect pests. In *Environmental Sustainability-Role of Green Technologies*; Springer: Berlin, Germany, 2015; pp. 49–63.
15. Mathew, N.; Anitha, M.; Bala, T.; Sivakumar, S.; Narmadha, R.; Kalyanasundaram, M. Larvicidal activity of *Saraca indica, Nyctanthes arbortristis*, and *Clitoria ternatea* extracts against three mosquito vector species. *Parasitol. Res.* **2009**, *104*, 1017–1025. [CrossRef]
16. Manjari, S.; Karthi, S.; Ramkumar, G.; Muthusamy, R.; Natarajan, D.; Shivakumar, M.S. Chemical composition and larvicidal activity of plant extracts from *Clausena dentata* (Willd) (Rutaceae) against dengue, malaria, and filariasis vectors. *Parasitol. Res.* **2014**, *113*, 2475–2481. [CrossRef]
17. Senthil-Nathan, S.; Kalaivani, K. Efficacy of nucleopolyhedrovirus and azadirachtin on *Spodoptera litura* fabricius (Lepidoptera: Noctuidae). *Biol. Control.* **2005**, *34*, 93–98. [CrossRef]
18. Senthil-Nathan, S.; Kalaivani, K.; Murugan, K. Effects of neem limonoids on the malaria vector *Anopheles stephensi* Liston (Diptera: Culicidae). *Acta Trop.* **2005**, *96*, 47–55. [CrossRef]
19. Sadek, M.M. Antifeedant and toxic activity of *Adhatoda vasica* leaf extract against *Spodoptera littoralis* (Lepidoptera: Noctuidae). *J. Appl. Entomol.* **2003**, *127*, 396–404. [CrossRef]
20. Vasantha-Srinivasan, P.; Senthil-Nathan, S.; Ponsankar, A.; Thanigaivel, A.; Edwin, E.; Selin-Rani, S.; Chellappandian, M.; Pradeepa, V.; Lija-Escaline, J.; Kalaivani, K. Comparative analysis of mosquito (Diptera: Culicidae: *Aedes aegypti* Liston) responses to the insecticide Temephos and plant derived essential oil derived from *Piper betle* L. *Ecotoxicol. Environ. Saf.* **2017**, *139*, 439–446. [CrossRef]
21. Vasantha-Srinivasan, P.; Thanigaivel, A.; Edwin, E.; Ponsankar, A.; Senthil-Nathan, S.; Selin-Rani, S.; Kalaivani, K.; Hunter, W.B.; Duraipandiyan, V.; Al-Dhabi, N.A. Toxicological effects of chemical constituents from Piper against the environmental burden *Aedes aegypti* Liston and their impact on non-target toxicity evaluation against biomonitoring aquatic insects. *Ecotoxicol. Environ. Saf.* **2018**, *25*, 10434–10446. [CrossRef]
22. Rajkumar, S.; Jebanesan, A. Mosquitocidal activities of octacosane from *Moschosma polystachyum* Linn. (lamiaceae). *J. Ethnopharmacol.* **2004**, *90*, 87–89. [CrossRef] [PubMed]
23. Mansour, S.A.; Ibrahim, R.M.; El-Gengaihi, S.E. Insecticidal activity of chicory (*Cichorium intybus* L.) extracts against two dipterous insect-disease vectors: Mosquito and housefly. *Ind. Crops Prod.* **2014**, *54*, 192–202. [CrossRef]
24. Soule, S.; Guntner, C.; Vazquez, A.; Argandona, V.; Moyna, P.; Ferreira, F. An aphid repellent glycoside from *Solanum laxum*. *Phytochemistry* **2000**, *55*, 217–222. [CrossRef]
25. Murungi, L.; Kirwa, H.; Torto, B. Differences in essential Oil Content of berries and Leaves of *Solanum sarrachoides* (Solanaceae) and the effects on oviposition of the tomato spider mite (*Tetranychus evansi*). *Ind. Crops Prod.* **2013**, *46*, 73–79. [CrossRef]
26. Lucia, A.; Licastro, S.; Zerba, E.; Audino, P.G.; Masuh, H. Sensitivity of *Aedes aegypti* adults (Diptera: Culicidae) to the vapors of Eucalyptus essential oils. *Bioresour. Technol.* **2009**, *100*, 6083–6087. [CrossRef]
27. Bossou, D.A.; Mangelinckx, S.; Yedomonhan, H.; Boko, P.M.; Akogbeto, M.C.; Kimpe, N. Chemical composition and insecticidal activity of plant essential oils from Benin against *Anopheles gambiae* (Giles). *Parasit. Vectors.* **2013**, *6*, 337. [CrossRef]
28. Nivsarkar, M.; Kumar, G.P.; Laloraya, M.; Laloraya, M.M. Superoxide dismutase in the anal gills of the mosquito larvae of *Aedes aegypti*: Its inhibition by alpha-terthienyl. *Arch. Insect Biochem. Physiol.* **1991**, *16*, 249–255. [CrossRef]
29. Thanigaivel, A.; Vasantha-Srinivasan, P.; Senthil-Nathan, S.; Edwin, E.; Ponsankar, A. Chellappandian, M.; Impact of *Terminalia chebula* Retz. against *Aedes aegypti* L. and non-target aquatic predatory insects. *Ecotoxicol. Environ. Saf.* **2017**, *137*, 210–217. [CrossRef]
30. Perumalsamy, H.; Jang, M.J.; Kim, J.R.; Kadarkarai, M.; Ahn, Y.J. Larvicidal activity and possible mode of action of four flavonoids and two fatty acids identified in *Millettia pinnata* seed toward three mosquito species. *Parasit. Vectors.* **2015**, *8*, 237. [CrossRef]

31. Abbott, W. A method of computing the effectiveness of an insecticide. *J. Econ. Entomol.* **1925**, *18*, 265–267. [CrossRef]
32. Harborne, J. Phenolic compounds. In *Phytochemical Methods*; Springer: Berlin, Germany, 1973; pp. 33–88. ISBN1 978-0-412-23050-9. (Print); ISBN2 978-94-009-5921-7. (Online).
33. Al-Tameme, H.J.; Hameed, I.H.; Idan, S.A.; Hadi, M.Y. Biochemical analysis of *Origanum vulgare* seeds by fourier-transform infrared (FT-IR) spectroscopy and gas chromatography-mass spectrometry (GC-MS). *J. Pharmacogn. Phytother.* **2015**, *7*, 221–237.
34. Hussein, A.O.; Mohammed, G.J.; Hadi, M.Y.; Hameed, I.H. Phytochemical screening of methanolic dried galls extract of *Quercus infectoria* using gas chromatography-mass spectrometry (GC-MS) and Fourier transform-infrared (FT-IR). *J. Pharmacogn. Phytother.* **2016**, *8*, 49–59.

Sample Availability: Samples of the compounds are not available.

molecules

MDPI

Article

Fractionation of Biomolecules in *Withania coagulans* Extract for Bioreductive Nanoparticle Synthesis, Antifungal and Biofilm Activity

Murtaza Hasan [1,2,*,†], Ayesha Zafar [2,†], Irum Shahzadi [2], Fan Luo [1], Shahbaz Gul Hassan [3], Tuba Tariq [2], Sadaf Zehra [4], Tauseef Munawar [5], Faisal Iqbal [5] and Xugang Shu [1,*]

1 College of Chemistry and Chemical Engineering, Zhongkai University of Agriculture and Engineering, Guangzhou 510225, China; luofan01@21cn.com
2 Department of Biochemistry and Biotechnology (Baghdad-ul-Jadeed Campus), The Islamia University of Bahawalpur, Bahawalpur 63100, Pakistan; ayeshazafar510@gmail.com (A.Z.); s.irum66@gmail.com (I.S.); tubatariq222@gmail.com (T.T.)
3 College of Information Science and Engineering, Zhongkai University of Agriculture and Engineering, Guangzhou 510225, China; mhasan387@gmail.com
4 Department of Botany, The Islamia University of Bahawalpur, Bahawalpur 63100, Pakistan; sadaf.zahra@iub.edu.pk
5 Department of Physics, The Islamia University of Bahawalpur, Bahawalpur 63100, Pakistan; rana.tuseefmunawar@gmail.com (T.M.); faisal.iqbal@iub.edu.pk (F.I.)
* Correspondence: murtaza@zhku.edu.cn (M.H.); xgshu@21cn.com (X.S.); Tel.: +86-020-8900-3114 (X.S.)
† These authors contributed equally to this work.

Academic Editor: Giovanni Benelli
Received: 5 June 2020; Accepted: 20 July 2020; Published: 31 July 2020

Abstract: *Withania coagulans* contains a complex mixture of various bioactive compounds. In order to reduce the complexity of the plant extract to purify its phytochemical biomolecules, a novel fractionation strategy using different solvent combination ratios was applied to isolate twelve bioactive fractions. These fractions were tested for activity in the biogenic synthesis of cobalt oxide nanoparticles, biofilm and antifungal activities. The results revealed that plant extract with bioactive fractions in 30% ratio for all solvent combinations showed more potent bioreducing power, according to the observed color changes and the appearance of representative absorption peaks at 500–510 nm in the UV-visible spectra which confirm the synthesis of cobalt oxide nanoparticles (Co_3O_4 NPs). XRD diffraction was used to define the crystal structure, size and phase composition of the products. The fractions obtained using 90% methanol/hexane and 30% methanol/hexane showed more effectiveness against biofilm formation by *Pseudomonas aeruginosa* and *Staphylococcus aureus* so these fractions could potentially be used to treat bacterial infections. The 90% hexane/H_2O fraction showed excellent antifungal activity against *Aspergillus niger* and *Candida albicans*, while the 70% methanol/hexane fraction showed good antifungal activity for *C. albicans*, so these fractions are potentially useful for the treatment of various fungal infections. On the whole it was concluded that fractionation based on effective combinations of methanol/hexane was useful to investigate and study bioactive compounds, and the active compounds from these fractions may be further purified and tested in various clinical trials.

Keywords: fractionation; reducing activity; biomolecules; antibiofilm; microbial infection

1. Introduction

Adverse increases in the rates of microbial, fungal and viral infections worldwide prompted by compromised and human immunity are due in part to the indiscriminate use of antibiotics that enhances resistance in microbial communities against the corresponding antigens [1]. The generation of biofilms by microbes, which root in a self-produced matrix on living and non-living surfaces [2], is a peculiar behavior of microbes in inducing and producing resistance. Biofilm affinity is associated to a firm attachment of the microbe and biofilm-forming microbes have a great tendency to stick permanently to the large variety of surfaces [3]. These tiny creatures' biofilms are protected by a layer of exopolysaccharides, which can be up to 1000 times more resistant to antimicrobials, which has increased exponentially the rate of chronic infections caused by increased resistance against the host immune system and antibiotics [4,5]. Among such microbes is *Candida albicans*, a well-known resistant nosocomial bacterium primarily known for being the main cause of infectious diseases [6] such as oral thrush [7], vaginitis [8] organ transplant recipients [9] and forms of cancer in HIV/AIDS patients. Besides resistance, the limited availability of commercial drugs effective against bacteria and the resulting toxicity has increased the global rate and effects of infections in people. This severe problem has driven the interest of researchers in developing less toxic, herbal bioactive compounds that could work against such strains of microbes. Similarly, the commonly known resistant fungus, *Aspergillus* species, responsible for pulmonary diseases, has also acquired resistance to many common drugs [10]. In order to overcome these biofilm-producers alternative treatments include the use of antibiofilm agents produced by medicinal plants as this mode of action reduces the resistance susceptibility [11]. Plants, being an enriched source of naturally occurring biologically active components, play a vital role in the prevention and treatment of diseases by boosting immunity and reducing toxicity [12,13]. Ancient plants like *W. coagulans* contain many useful bioactive molecules such as withanolide, withaferin, withacoagin [14], etc., that have been used to synthesize therapeutic drugs for the prevention and treatment of various diseases due to their reduced side effects [15,16]. *W. coagulans* belongs to the Solanaceae, a family of common traditional therapeutic plants with wide range of pharmacological applications [17], including antimicrobial, anti-inflammatory [18], antitumor [19], antihyperglycemic [20], cardiovascular, and immunosuppressive properties [21]. The constituents of *W. coagulans* include free amino acids, essential oils, steroidal lactones and esterases, widely used for their pharmacological activities [22]. A few studies have also recommended the use of withanolide, withaferin and other biological entities found in *W. coagulans* for their bioreducing potential in the synthesis of nanoparticles [23,24], and studies have reported the eco-friendly and less toxic preparation of nanoparticles and pharmacological studies using *W. coagulans* components [25].

So far, all these biological activities were tested using crude extracts containing complex mixtures of active biomolecules and the solvents-based screening, fractionation and functionalization of bioactive compounds has not been previously reported. The development of antibiofilm strategies is a major interest and also the basis of an important field of investigation that is the development of premium, environmentally friendly antibiofilm biomolecules. The present work was focused on investigate the functional role of fractions obtained using methanol and hexane with water and mixtures of methanol and hexane to purified active biomolecules from *W. coagulans* extract. For this purpose, *W. coagulans* fractions were extracted with mixtures containing different ratios of methanol and *n*-hexane and water and methanol and *n*-hexane mixtures in order to evaluate the bioactivities such as bioreducing potential for the synthesis of cobalt nanoparticles, and antioxidant, anti-biofilm and antifungal activities.

2. Results

The increasing resistance of microbes against antibiotics calls for the urgent discovery of unique biomolecules from extracts of plants like *W. coagulans* that are of potential interest for their antibiofilm and antifungal activity and as bioreducing agents for the synthesis of cobalt oxide nanoparticles (Co_3O_4 NPs). The species *W. coagulans* is highly acclaimed in the Indian ayurvedic system of medicine, where it is known for its medicinal significance in promoting physical and mental health [26,27]. Its active components include alkaloids, steroidal compounds, lactones, withaferin a [28], withanoloids [29], withanone [30], etc. that act as anti-inflammatory, anticancer, chemoprotective, hepatoprotective, immune modulatory, antifungal, antibacterial, hypocholestroemic, and radical scavenging agents [31].

The complex bioactive extract of *W. coagulans* contains potent and functional molecules that must be fractionated to simplify the complexity and provide separate bioactive molecules that can exhibit their functionalities efficiently. Different fractions of plant extract obtained using different solvents and mixtures of solvents were used to resolve the complexity of the biological entities of *W. coagulans* used as bioreducing, antibacterial, antifungal agents [32]. This fractionation route provided a means to separate, simplify and unveil the hidden active molecules in the complex. Initially using a *W. coagulans* extract, 12 different methanol, hexane and their mixture fractions in ratios of 30%, 50%, 70%, 90% were made (Figure S1) and their bio-reducing, antibiofilm and antifungal potential in vitro evaluated (Figure 1).

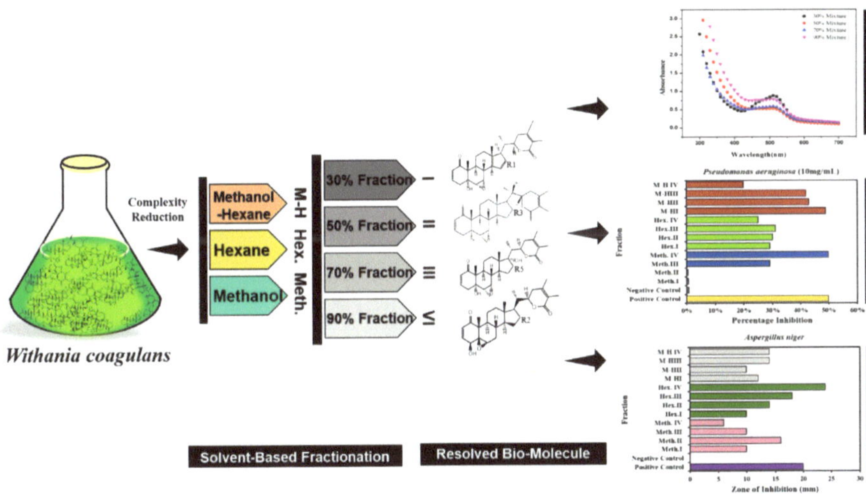

Figure 1. Schematic illustration of *W. coagulans* biomolecules and their applications.

2.1. Green Synthesis of Co_3O_4 NPs

Pink coloured cobalt chloride solution was mixed individually with all 12 different solvent-based plant extract fractions that turned to a dark brown color upon addition and continuous magnetic stirring at 90 °C for five h. As the chemical reaction proceeded the color changed from dark brown to light brown indicating the synthesis of Co_3O_4 NPs (Figure S2).

2.1.1. Characterization of Green Synthesized Co_3O_4 NPs

Monitoring the reduction potential of synthesized Co_3O_4 NPs by UV spectroscopy using the *Withania*-based fractions showed different peaks within the 500–510 nm range for different solvent fractions [33]. The methanol and water ratio results conclusively indicated that 30% methanol/H_2O (3:1) showed the highest peak, indicating that the 30% fraction was a more active bioreducing fraction than 50% methanol/H_2O (5:5), 70% methanol/H_2O (7:3), or 90% methanol/H_2O (9:1), as they all showed less bioreducing activity [34,35] (Figure 2a). Among the next four fractions based on hexane and water ratio 30% hexane/H_2O (3:1) and 90% hexane/H_2O (9:1) showed almost same highest peak which indicated that these fractions have more bioreducing potential than 50% hexane/H_2O (5:5) and 70% hexane/H_2O (7:3). Furthermore 70% hexane/H_2O showed a much lower peak with no bio-reducing potential [36,37] (Figure 2b). Similarly, the four methanol/hexane-based fractions with different ratios (30%, 50%, 70%, 90%) were evaluated next for bioreducing potential and was indicated that 30% fraction mixture of methanol/hexane (3:1) showed a much sharper peak indicating better bioreducing potential than 50%, 70%, 90% methanol/hexane fraction mixtures [23,38] (Figure 2c).

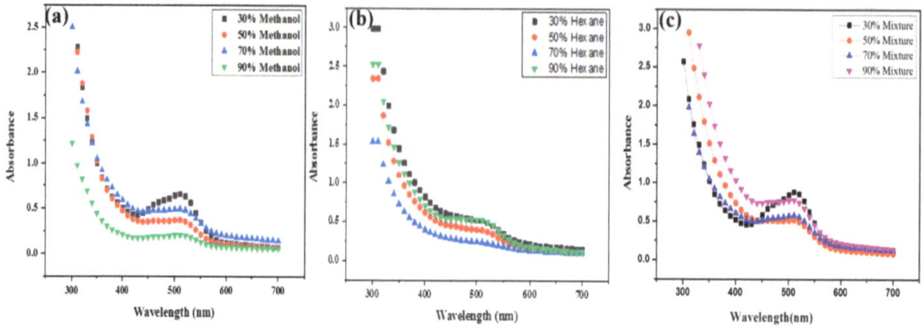

Figure 2. Bioreducing potential of *W. coagulans* based on: (**a**) methanol (**b**) hexane (**c**) methanol/hexane (mixture) fractions for Co_3O_4 NPs synthesis.

For optimizing the results a comparative analysis was done between 30% fraction of methanol/H_2O, 30% hexane/H_2O and 30% methanol/hexane and the results demonstrated that out of all mixtures the 30% methanol/hexane (3:7) fraction mixture showed a much sharper peak. meaning it had a higher bioreducing ability than 30% methanol/H_2O and 30% hexane/H_2O fraction, as seen in Figure 3a.

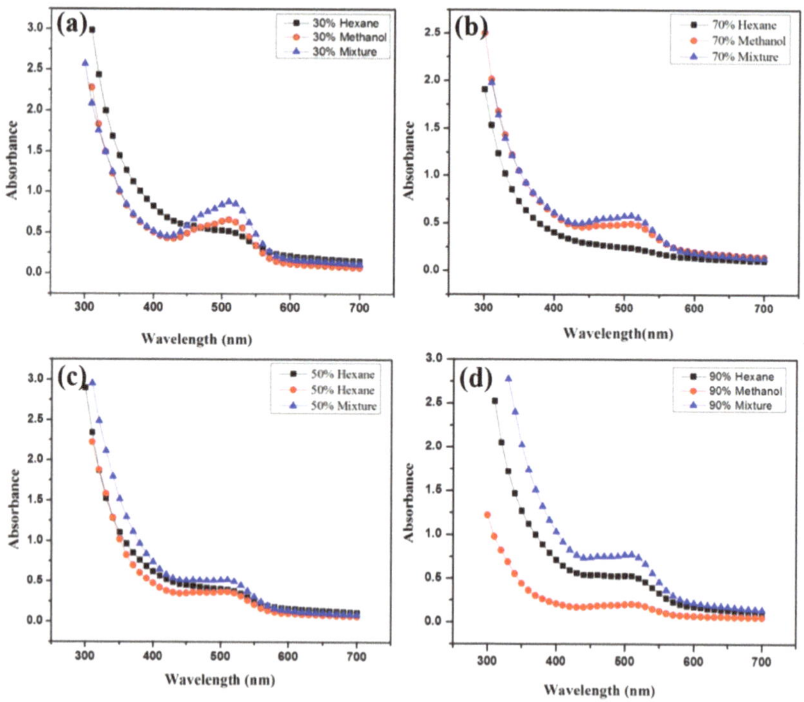

Figure 3. Bioreducing potential of *Withania coagulans* based on methanol, hexane and methanol/hexane (mixtures) using (**a**) 30% fraction, (**b**) 50% fraction, (**c**) 70% fraction, (**d**) 90% fraction.

Among the 50% fractions, 50% methanol/hexane (5:5) fraction mixture showed higher peaks corresponding to a higher bioreducing potential than 50% methanol/H_2O and 50% hexane/H_2O, but 50% hexane/H2O and 50% methanol/H_2O showed almost the same peak and almost the same bioreducing potential (Figure 3b).

Next, among the different 70% fractions of *W. coagulans*, the 70% methanol/hexane (7:3) mixture fraction showed a high peak with higher bioreducing potential than 70% methanol/H_2O and 70% hexane/H_2O. Here 70% methanol/H_2O showed a much sharper peak (indicating better bioreducing potential) than the 70% hexane/H_2O fraction, as illustrated in Figure 3c. Finally, out of all the 90% fractions of *W. coagulans*, 90% methanol/hexane (9:1) fraction mixture showed the highest peak indicating a higher bioreducing potential than 90% methanol/H_2O and 90% hexane/H_2O. Here different results were observed because 90% hexane/H_2O shows a much sharper peak than 70% methanol/H_2O meaning that 70% hexane/H_2O fraction has higher bioreducing ability than 70% methanol/H_2O (Figure 3d).

2.1.2. XRD Analysis of Co_3O_4 NPs

XRD diffraction was used to define the crystal structure and phase composition of the produced NPs. The XRD patterns of the samples obtained with different solvent fraction ratios are presented in Figure 4a–c. The observable diffraction pattern of materials obtained using methanol (fraction (a)), hexane (fraction (b)) and methanol/hexane (fraction (c)) were well-matched with Co_3O_4. The diffraction patterns of the methanol fraction were thus consistent with JCPDS Card No. 01-080-1534, hexane fraction (b) with JCPDS Card No. 01-074-1657, and methanol/hexane fraction (c) with JCPDS Card No. 01-076-1802, respectively. The peaks and related planes are indicated in Figure 4. The XRD results show that none of the samples have any characteristic peaks due to impurities, which shows that the grown samples have outstanding crystalline nature. The lattice parameters (*a*) and unit cell volume (*v*) of the samples were calculated using the following formula:

$$\frac{1}{d^2} = \frac{h^2 + k^2 + l^2}{a^2} \tag{1}$$

$$v = a^3 \tag{2}$$

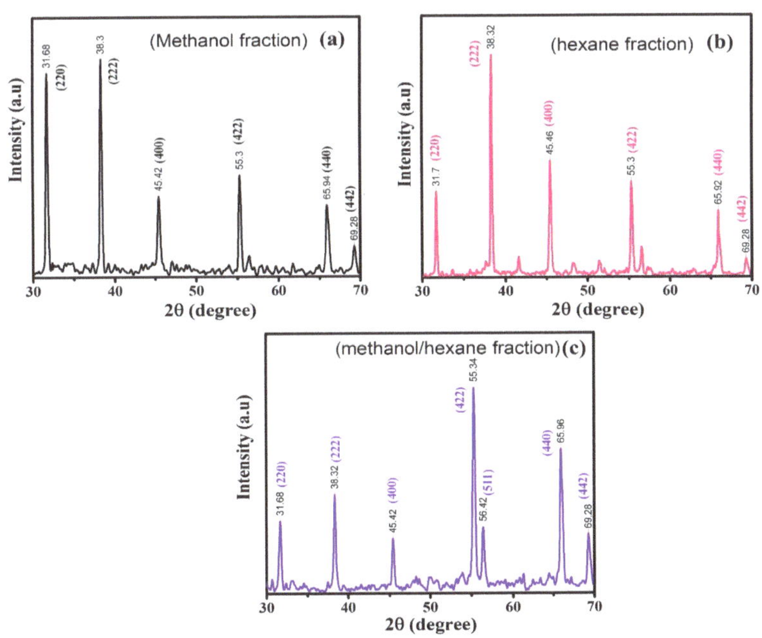

Figure 4. XRD analysis of Co_3O_4 nanoparticles based on solvent fractions: (**a**) methanol fraction (**b**) hexane fraction, (**c**) methanol/hexane fraction.

Where (hkl) are the miller index, 'd' is d-spacing, and 'a' is lattice constant. The calculated values are listed in Table 1. The average crystallite size (D) of all synthesized samples was determined by using the well-known Debye–Scherer Formula [39,40]:

$$D = \frac{K\lambda}{\beta \cos \theta} \tag{3}$$

Table 1. Structural parameters of grown samples.

Samples	a(Å)	b(Å)	c(Å)	d-Spacing	Volume (Å^3)	Crystallite Size (nm)	Dislocation Density δ (nm)$^{-2}$ × 10^{-4}
A	8.04702	-	-	1.8365	521.0805	50	4.000
B	8.07016	-	-	1.9657	525.5884	59	2.870
C	8.06895	-	-	1.8362	525.3527	49	4.160

In these equations K is the shape factor having value (0.94), λ is the wavelength of X-ray (1.5406 Å), β is the full width at half maxima. From the results, it can be concluded that the crystallite size follows the trend b (59 nm) > a (50 nm) > c (49 nm) (Table 1). The dislocation density (δ) and d-spacing can be calculated by:

$$\delta = 1/D^2 \tag{4}$$

$$2d\sin\theta = n\lambda \tag{5}$$

where 'λ' is the wavelength of X-rays in Å, 'θ' is the diffraction angle (Bragg angle) in degrees, n is the order of diffraction which is the spacing between adjacent crystal planes. The calculated values are listed in Table 1. The results show that d-spacing varies directly with crystallite size while dislocation density varies as square inverse of crystallite size.

Furthermore, compound microscopy results (Figure 5a–c) show that changing the nature of the solvent influenced on the shape of Co_3O_4 NPs. Figure 5a shows bead-shaped Co_3O_4 NPs obtained using methanol solvent extract as reducing agent [41] while in Figure 5b the shape of Co_3O_4 NPs obtained with hexane was different because of the different biomolecules present as compared to methanol solvent [42]. In the case of a mixed ratio of methanol and hexane solvents (Figure 5c), the Co_3O_4 NPs were cube-shaped, most probably because of the action of different active biomolecules in this fraction when they reduce the cobalt nanoparticles [27,43].

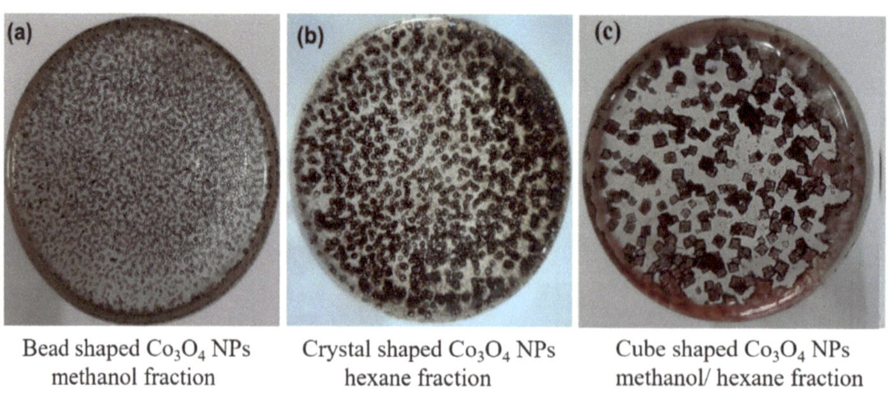

Figure 5. Variation in Co_3O_4 nanoparticle shape based on solvent fractions: (**a**) methanol fraction (**b**) hexane fraction, (**c**) methanol/hexane fraction.

From the above results, we can conclude that the active biomolecules exhibiting reducing potential found in the methanol/hexane fraction were proven to have the best bioreducing potential in the synthesis of Co_3O_4 NPs. Among the different fractional concentrations of similar solvents 30% fraction showed the best bioreducing efficiency. This means that when preparing fractions with these three solvents, and running a separate solvent fraction-based reaction, the 30% fraction will provide more significant results as previously reported [44,45]. It shows a well-defined sharp peak for every solvent containing a 30% solvent faction. The exposed binding sites for the binding of cobalt precursors and saturating the metal by biochemical agents in order to provide stability was done by solvent-based fractionation of *Withania* extract as reported earlier [46].

The scheme (Figure 1) shows a double dip strategy where the nature and concentration of a solvent reduce the complexity, provide active sites and finally highlight the functional activity of the biomolecules. This solvent fractionation actually works similarly to an enzyme substrate reaction, as active sites are provided as product gets generated. Here the fractionation helps expose and present the active sites by reducing the complexity and generating Co_3O_4 NPs. In the next level of optimization, the concentration was kept constant and the solvent was altered. The results showed that the mixture of methanol/hexane was a hybrid solvent that reinforced the characteristic properties of each solvents. Conclusively in order to optimize our study, mixtures of methanol/hexane, at all concentrations provide the best reduction capacity. Thus, to reduce complexity, unlocking the bioactive molecules in methanol/hexane mixtures of 30% fraction should provide an excellent lead for identifying compounds good at reducing cobalt to Co_3O_4 NPs.

2.2. Biofilm Activity of Prepared W. coagulans Fractions

Bioactive fractions from *W. coagulans* (12 different fractions) were evaluated for antibiofilm activity against the drug sensitive strains *Pseudomonas aeruginosa* and *Staphylococcus aureus* in 96 well micro-titer plates. The purpose was to evaluate the potential of the 12 different fractions to inhibit the growth of a preformed biofilm already established in the wells of the micro-titer plate [47]. In anti-biofilm assay biofilm was induced to grow on 96 well micro-titer plates by adding 100 µL nutrient broth, 100 µL plant extract and 20,100 µL bacterial culture in each well and incubating for 24 h at 37 °C then staining the next day with crystal violet (dye) give a dark blue color to the well where biofilm formation took place (Figure S3). Crystal violet is a dye that binds non-specifically to negatively charged surface molecules such as the polysaccharide matrix of biofilms and stains them with a blue color so it is generally used to estimate biofilm biomass [48], so a reduction in blue color indicates biofilm inhibition by different tested plant fractions.

2.3. Antibiotic Selectivity

First an effective positive control for *P. aeruginosa* and *S. aureus* (drug sensitive strains) was established by treating with four different antibiotics (clindamycin, moxifloxacin, penicillin and ciprofloxacin). The results showed that moxifloxacin and ciprofloxacin were more active drugs against the *P. aeruginosa* strain as indicated by a larger zone of inhibition shown by the drugs (Figure 6a,b) but ciprofloxacin was a more effective antibiotic against *S. aureus* as shown by its larger inhibition zone (Figure 6c). Thus, the strong antibiotic ciprofloxacin was selected to test the *W. coagulans*-based 12 different fractions of methanol and hexane and their mixtures to evaluate the biofilm inhibition potential against *P. aeruginosa* and *S. aureus* at concentrations of 5 mg/mL and 100 mg/mL.

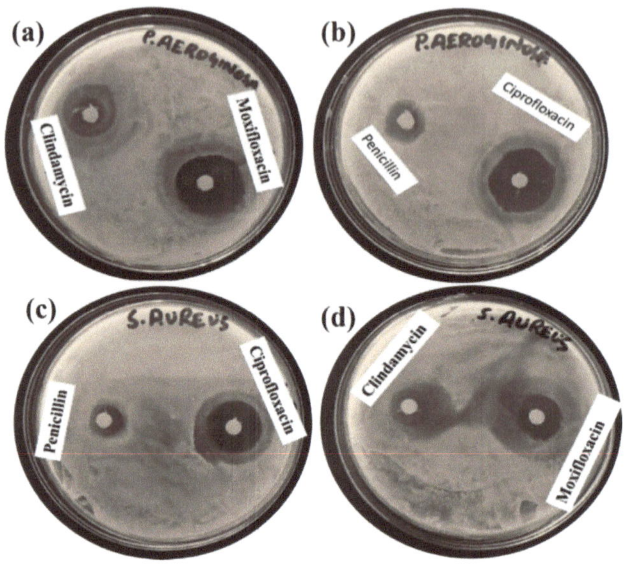

Figure 6. Antibiotic selectivity of (**a**) clindamycin, moxifloxacin (**b**) penicillin, ciprofloxacin against *P. aeruginosa* (**c**) penicillin, ciprofloxacin (**d**) clindamycin, moxifloxacin against *S. aureus*.

2.3.1. Biofilm Inhibition Potential of *W. coagulans* Fraction against *P. aeruginosa*

Ciprofloxacin, being a positive control against *P. aeruginosa*, shows a reduction of dark blue color of the dye (crystal violet) in the first well and solvent blank without bacterial strain marked as first negative control that does not contain bacteria so no biofilm formation occurred there, thus no crystal violet dye staining was observed (Figure S3), leaving a colorless well indicating the absence of bio-film formation. As a second negative control a well was loaded with 55 mg/mL of *P. aeruginosa* without the plant extract and blue colored biofilm was observed. Color reduction of the dark blue dye in the micro-titer plate well gave a rapid qualitative analysis of biofilm inhibition potential by the crystal violet staining technique that was measured as a percentage inhibition of biofilm formation. With the positive control, ciprofloxacin, the percentage inhibition against *P. aeruginosa* was found to be 50%, and it was 0.7% with the negative control.

After running the successful controls, the *Withania*-derived solvent-based fractions were assessed. For 30% methanol (Meth.^{I}) the inhibition was 0.6%, for 50% methanol (Meth.^{II}) it was 0.5%, for 70% methanol (Meth.^{III}) it was 29% and for 90% methanol (Meth.^{IV}) the inhibition reached 50%. Hexane was next and 30% hexane (Hex.^{I}) exhibited 29% inhibition, 50% hexane (Hex.^{II}) showed 30% inhibition, 70% hexane (Hex.^{III}) showed 31% and 90% hexane (Hex.^{IV}) gave about 24% inhibition.

The third series includes mixtures of methanol and hexane, among which 30% methanol-hexane (M–H^{I}) showed 49% inhibition, 50% methanol-hexane (M–H^{II}) 43%, 70% methanol-hexane (M–H^{III}) 42% and 90% methanol-hexane (M–H^{IV}) showed only 20% inhibition of biofilm formation. Overall Meth.^{IV} exhibited a 100% percentage inhibition of biofilm production with respect to control. On average Meth. inhibited 40%, Hex. inhibited 57% and methanol-hexane mixture inhibited 77% with respect to control. Hence the solvent mixture super-combination showed superior results on average at all concentrations by decoding the complexity with the hybrid mixture of solvents. Biofilm formation by dye degradation and calculated inhibitions are shown in Figure 7a. These results are relevant to previous work done using plant extracts against the biofilm activity [49].

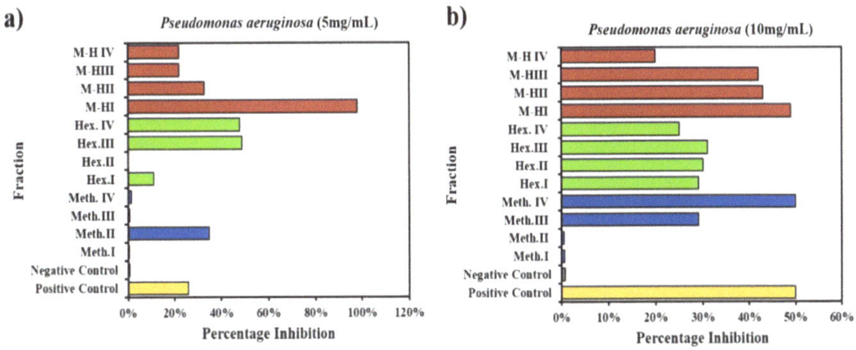

Figure 7. Biofilm activity of *W. coagulans* 12 fraction against *P. aeruginosa* with concentration (**a**) 5 mg/mL (**b**) 10 mg/mL.

Similarly, when using the 10 mg/mL extract against *P. aeruginosa* where the positive control showed 26% inhibition of biofilm and 0.7% of inhibition for the negative control, 0.5% > 35% > 0.6%> 1.5% inhibition was seen for Meth.^{I} > Meth.^{II} > Meth.^{III} > Meth. ^{IV}. Moving to the next solvent fraction Hex.^{I} > Hex.^{II} > Hex.^{III} > Hex.^{IV} (11% > 0.3%> 49%> 48%) and lastly, for the mixture fraction M−H^{I} > M−H^{II} > M−H^{III} > M−H^{IV} (98% >33% > 22%) as depicted in Figure 6b along with biofilm formation (Figure S4). The best fractions M−H^{I}, Hex.^{III}, Hex.^{IV}, Meth.^{II} and M−H^{II} provided an outstanding inhibition representing 277%, 88%, 85%, 34% and 26% more than the control. As a result, Meth provided 36% inhibition with respect to control, Hex exhibited 4% more inhibition with respect to control whereas the excelling M−H mixture exhibited 68% more inhibition with respect to the control on average. Some fractions had previously shown significant inhibition with 5 mg/mL *Withania* solution against *P. aeruginosa* [50] but changing the concentration to 100 mg/mL the bio-film percentage inhibition increased even above the control level, showing higher antibacterial activity as shown in Figure 7b.

2.3.2. Biofilm Inhibition Potential of *W. coagulans* Fractions against *S. aureus*

The activity of concentrations of each fraction up to 10 mg/mL against *S. aureus* was observed. Biofilm formation against *S. aureus* strain was done with ciprofloxacin as positive control which was found to be active against the drug sensitive *S. aureus* strain as shown by the white colour of wells. A negative control was also added (Figure S5).

The controls gave 55% and 0.4% inhibition, respectively. For the other 12 fractions a concentration of 55 g/mL was used that provided no significant or results as shown in Figure 8a where the positive control inhibition was 55% and that of the negative control was 0.4%. Meth.^{I} > Meth.^{II} > Meth.^{III} > Meth.^{IV} values were 1.2% > 1.8% > 1.9% > 48%. For hexane, i.e., Hex.^{I} > Hex.^{II} > Hex.^{III} > Hex.^{IV} the inhibition was 3.1% > 36% > 2.1% > 3.2% and for mixtures M−H^{I} > M−H^{II} > M−H^{III} > M−H^{IV}, percentage inhibitions of 0.8% > 20% > 17.5% > 12.5% were exhibited which were quite insignificant against such a resistant strain and at a such minute concentration.

Next the change in concentration up to 10 mg/mL against *S. aureus* showed significant results, whereby the positive control showed 40% inhibition and the negative one showed 0% inhibition. In the first fraction series Meth.^{I} > Meth.^{II} > Meth.^{III} > Meth.^{IV} the inhibition was 71% > 65% > 28% > 24%. For hexane fractions, i.e., Hex.^{I} > Hex.^{II} > Hex.^{III} > Hex.^{IV} the results showed 3.5% > 20% > 19% > 3.7% inhibition and the percentage inhibition was calculated as 2.7% > 62%> 72% > 72% for M−H^{I} > M−H^{II} > M−H^{III} > M−H^{IV} as shown in Figure 8b. Compared to the control M−H^{III} >M−H^{IV} > Meth.^{I} > Meth.^{II} > M−H^{II} exhibited 80% > 80% > 78% > 62% > 55% more biofilm formation indicating an outstanding result at the particular dilutions that revealed the presence of antibacterial biomolecules in such fractions. On average Meth. showed 18% more inhibition,

Hex. showed only 28% inhibition with respect to control and M–H was superior, exhibiting more film formation with 30% inhibition. The present observations regarding bacterial biofilm formation match the work rewported by previous researchers [51].

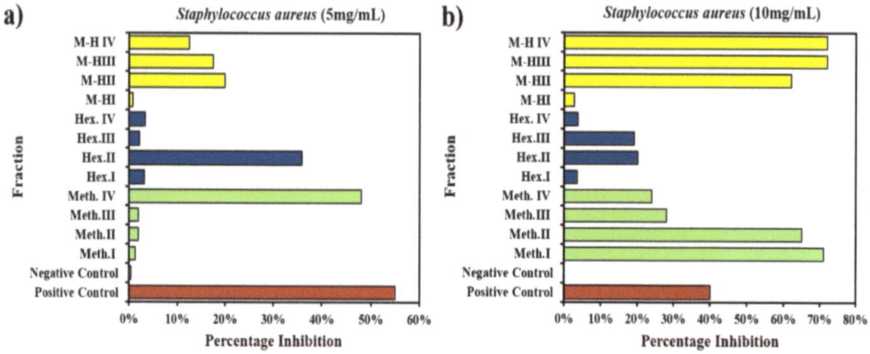

Figure 8. Biofilm activity of *W. coagulans* 12 fraction against *S. aureus* with concentration (**a**) 5 mg/mL (**b**) 10 mg/mL.

2.4. Antifungal Activity of Prepared W. coagulans Fractions

2.4.1. Antifungal Activity of Prepared *W. coagulans* against *A. Niger*

The antifungal activity was evaluated using all 12 different fractions extracts of methanol and hexane and their mixtures using plant extract of *W. coagulans* against *A. niger* and *C. albicans* by the disc method [52]. The active biomolecules were resolved into simple *W. coagulans* plant molecules that exhibited antifungal activity. The active principal molecules were measure and made visible by the zone of inhibition produced by the fraction molecules against the specific strains (Figure 9a–h).

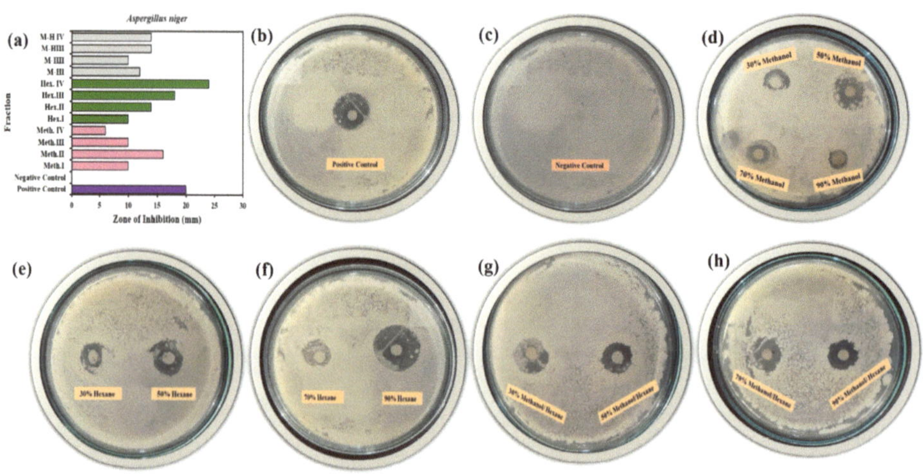

Figure 9. Antifungal activity of 12 *W. coagulans* fractions against *A. niger* measured by zone of inhibition: (**a**) Positive control; (**b**) negative control; (**c**) methanol fractions; (**d**) hexane fraction 30%, 50%; (**e**) hexane fraction 70%, 90%; (**f**) methanol/hexane fraction 30%, 50% (**g**); methanol/hexane fraction 70%, 90% (**h**).

The tested fractions provided significant results as follows: amphotericin B at concentration (10 mg/mL) was used as standard for both fungus strains that were pathogenic [53]. The positive control shows good antifungal activity against *A. niger* as indicated by the large clear inhibition zone (20 mm) whereas the negative control exhibited no clear zone of inhibition as shown in Figure 9b,c. The tested concentrations beginning with Meth.^{I} > Meth.^{II} > Meth.^{III} > Meth.^{IV} exhibited 10 mm > 16 mm > 10 mm > 6 mm inhibition, with an average of 50% antifungal activity compared to the control (Figure 9d). Next is the hexane fractions, Hex.^{I} > Hex.^{II} > Hex.^{III} > Hex.^{IV} showing 10 mm >14 mm >18 mm > 24 mm zone of inhibition with 83% agreement with the control (Figure 9e,f). Finally M–H^{I} > M–H^{II} > M–H^{III} > M–H^{IV} where the zone of inhibition provided 62% similar result on average with control and 12 mm > 10 mm > 14 mm > 14 mm inhibition zones, respectively (Figure 9g,h). Surprisingly Hex.^{IV} showed 20% more antifungal activity than the control against *A. niger*.

2.4.2. Antifungal Activity of Prepared *W. coagulans* against *C. albicans*

C. albicans showed a 20 mm zone of inhibition with the positive control amphotericin B, an effective drug against this strain (Figure 10a–h). The negative control provided no zone of inhibition indicating no antifungal activity (Figure 10c). On further treatment the 12 fractions provided significant results, where Meth. provided 6% antifungal activity, Hex. 70% activity and M–H 83% activity with respect to the control.

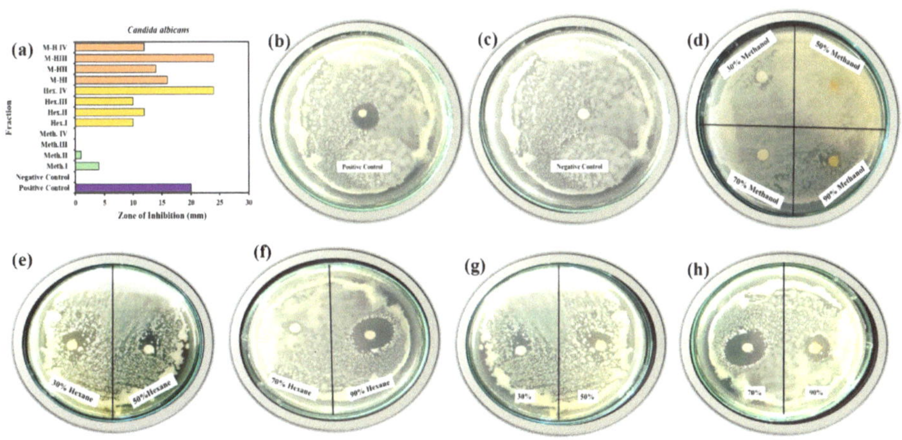

Figure 10. Antifungal activity of 12 *W. coagulans* fractions against *C. albicans* (zone of inhibition): (**a**) Positive control (**b**) negative control (**c**) methanol fractions (**d**) hexane fraction 30%, 50% (**e**) hexane fraction 70%, 90% (**f**) methanol/hexane fraction 30%, 50% (**g**) methanol/hexane fraction 70%, 90% (**h**).

Individually Meth.^{I} > Meth.^{II} > Meth.^{III} > Meth.^{IV} provided 4 mm > 1 mm > 0 mm > 0 mm of inhibition zone (Figure 10d), Hex.^{I} > Hex.^{II} > Hex.^{III} > Hex.^{IV} had 10 mm > 12mm > 10 mm > 24 mm inhibition (Figure 10e,f). Finally 16 mm > 14 mm > 24 mm > 12 mm zones of inhibition were measured for M–H^{I} > M–H^{II} > M–H^{III} > M–H^{IV} fractions (Figure 10g,h). With *C. albicans* Hex.^{IV} and M–H^{III} exhibited 20% more antifungal activity that the control.

The antibacterial and antifungal activity using *W. coagulans* was proven to be significant because of the biomolecules initially present in complex form that were resolved into simple and more functionally active groups by the solvent-based fractionation method. Owing to such a strategy and the significant activity this set of optimizations can be incorporated in the medicinal field in order to combat bacterial and fungal infections. Plant extracts have shown a variety of potentials such as reducing, antioxidant, synthetic, and medicinal activities due to the presence of numerous bio-molecules that exist in different parts of the plant. Depending upon the nature each show different extents of variation in their capabilities due to the presence of some additional biomolecules and the varying concentrations of those biomolecules. Considering *Withani*, it is truly rich in phenols, flavonoids, alkaloids, steroids and other complex structures that provide reducing, antibacterial and antifungal activities. The results of this work show that the separation of these components using different solvents such as water, hexane, methanol, acetone, etc., enhanced the activities by aiding in resolving the complexity, dissolving components of different nature according to their solubility in different solvents, combining the biomolecules for effective interaction and thus showing their potentials at their maximum level. Similarly, *Withania* had shown antibacterial activity against *Salmonella typhi, Klebsiella pneumonia, S. aureus* with percentage antibacterial activities as 43%, 0%, 73% respectively. *Withania*-decorated iron rods enhanced the activity up to 30% for *S. aureus* and *P. aeruginosa,* whereas *Withania* showed less inhibition against a *Brucella* strain. Multiple examples have shown that the bacterial inhibition of crude extracts was not so high as that achieved by using solvent- based fractionation methods that enhance the values and activity to a significant level. Antifungal activity was exhibited against various strains such as *A. flavus, A. niger, Penicillium* and *Alternaria alternate,* where a significant 6–10 mm zone was measured. Thus, the addition of solvents, mixtures of solvents, and the concentration help in simplifying the complex structure of the plant extracts that displayed much higher activities, including bioreduction, antibacterial and antifungal properties.

3. Materials and Methods

3.1. Plant Material

Plant of *W. coagulans* was obtained from a local market in Bahawalpur, Pakistan in September 2018. Fresh plant was washed three time with distilled H_2O and kept in the shade until it was completely dried, then it was crushed into powder form for further use.

3.2. Preparation of Plant Extract

Whole plant was dried and crushed using a pestle and mortar to obtain a fine powder, then 10 g of extract powder was dipped in different concentrations of methanol and hexane to make 12 different fractions with ratios of 90%, 70%, 50%, 30% (final volume 200 mL). After overnight incubation the extracts were filtered and the filtrates were dried in an incubator at 37 °C. These powder extracts then used to check the bioreducing, antifungal and biofilm activities.

3.3. Synthesis of Cobalt Oxide Nanoparticles

For the synthesis of cobalt oxide nanoparticles, a 0.5 M solution of cobalt chloride was prepared Flasks containing 40 mL cobalt chloride solution and 10 mL plant extract (90%, 70%, 50%, 30%) were prepared, put on a magnetic stirrer (150 rpm) and kept there for 4 h at 90 °C as a reaction occurred indicated by a change in color confirming the synthesis of nanoparticles. After this the mixture was centrifuged at 6000 rpm for 10 min., the pellet was separated and dried for characterization.

3.4. UV-Vis Spectroscopy

All 12 fractions were subjected to UV-Vis spectroscopy (Instrument model VT05404-0998, Biotek, Winooski, VT, USA) at predetermined time intervals to confirm the formation of cobalt nanoparticles and the wavelength was noted. Peaks between 550–510 nm give a positive indication of nanoparticle synthesis. Also, the color changes of reaction mixtures were used as evidence of cobalt oxide nanoparticle formation

3.5. Morphology Analysis of via Compound Microscope

The dried form of the cobalt oxide nanoparticles was uniformly distributed in Petri plates with relevant solvent and allowed to dry. Later compound microscopy (model IM-850, IRMECO GmbH, Hamburg, Germany) was used to observe the morphological variations in all three fractions.

3.6. Biofilm Assay of W. coagulans Fraction

Biofilm assays were performed by a crystal violet staining assay. The effect of extracts on biofilm formation was evaluated in 96-well polystyrene plates. Firstly, the 96-well micro-titer plates were washed with sterile distilled water, air dried and then oven-dried at 60 °C for 45 min. Briefly, nutrient broth, standard drug (ciprofloxacin) and bacterial culture were used as positive control while nutrient broth, distilled water and bacterial culture were used as negative control. Nutrient broth, plant fractions and bacterial culture were added to each micro-plate and incubated at 37 °C for 24 h. After that staining with 0.1% crystal violet was performed and the OD was recorded at 630 nm using an ELISA reader (model IM-850, IRMECO GmbH, Hamburg, Germany) and % inhibition was calculated by following formula:

$$\% \text{ inhibition} = (A_0 - A_1)/A_0 \times 100 \tag{6}$$

where A_0 is absorbance of negative control and A_1 is the absorbance of the plant fractions

3.7. Antifungal Activity of W. coagulans Fraction

Fresh plant was washed two times with distilled water and allowed to dry at room temperature for 3 to 4 days. The dried material was ground and extracted separately by making different methanol and hexane fractions. The extracts were filtered and the filtrate was dried. All extracts fractions were stored at 4 °C and used for the bioassays. The plant extracts were tested against two important fungal pathogens, *C. albicans* and *A. niger,* obtained from the laboratory of the Department of Biochemistry and Biotechnology (Islamia University Bahawalpur). All cultures were maintained on SDA agar at 37 °C. Overnight cultures on SGA slants at 37 °C were used to prepare the fungal inoculum to be used in the antimicrobial assays. The antifungal activity of *W. coagulans* methanolic and hexane extracts was measured according to the disc diffusion method. Sterile blank discs of 6 mm diameter were soaked with the prepared *W. coagulans* extracts to give a final concentration of 10 mg/mL, respectively. The discs were then placed firmly on a SDA surface which has been previously seeded with *C. albicans* strain suspension. The same steps were repeated for the A. *niger* strain. All plates were incubated overnight at 37 °C. Throughout this experiment, a blank disc impregnated with sterile distilled water represented as negative control while a disc soaked with 100 μL of amphotericin B was the positive control. The susceptibility of each *Candida* spp. was determined by the diameter of the growth inhibited zone surrounding the disc.

4. Conclusions

Twelve different *W. coagulans*-based fractions prepared using combinations of different solvents (methanol, hexane) and their mixture were used to study the effect of different solvent combinations on various biological activities. Plant fractions of different concentration (30%, 50%, 70%, 90%) were used. These fraction were used to investigate the bioreducing potential of the plant extracts containing complex biomolecule mixtures, it was found that collectively 30% fraction of methanol. hexane, and mixture of methanol-hexane provided the highest reducing potential for the synthesis of cobalt oxide nanoparticles. Results also showed that 90% methanol/hexane and 30% methanol/hexane were more active against biofilm formation of *P. aeruginosa* and *S. aureus* so these fractions could be used for treatment of various drug resistance-related bacterial infections. A 90% fraction of hexane/H_2O showed excellent antifungal activity against *P. niger* and *C. albicans*, while 70% methanol/hexane show good antifungal activity for *C. albicans,* so these fractions are potentially useful for the treatment of various fungal infections. This solvent-based fractionation method provides a direct means to reduce the complexity of the *W. coagulans* extracts and reveal the strong bioreducing, antifungal and antibiofilm activities and optimize the particular activity for practical applications. This provides a cost-effective, ecofriendly, non-toxic and effective source for medicinal and synthetic applications.

Supplementary Materials: The following are available online, Figure S1 Screening strategy for exploring bioactive fraction of *W. coagulans*; Table S1 Different solvent fractionation of *Withania coagulans*.

Author Contributions: Conceptualization, M.H., S.G.H. and X.S.; methodology, S.Z.; software, S.G.H.; validation, T.M., F.L. and F.I.; formal analysis, S.Z.; investigation, A.Z.; resources, T.T.; data curation, S.G.H.; writing—original draft preparation, M.H.; writing—review and editing, M.H.; visualization, X.S.; supervision, M.H.; project administration, X.S.; funding acquisition, X.S.; Experimentation, I.S. All authors have read and agreed to the published version of the manuscript.

Funding: This research received no special external funding.

Acknowledgments: The authors express their gratitude for the financial support from the Provincial Education Department Project (Natural Science, 2017KZDXM045), the Agriculture and Rural Department Project of Guangdong Province, the Guangzhou Foreign Cooperation Project (201907010033), the Graduate Technology Innovation Fund (KJCX2019004), and the Undergraduate Innovation and Entrepreneurship Training Program (S201911347028). The authors would also like to thank, The Islamia University Bahawalpur, Pakistan, National Research Program for University (NRPU) for Higher Education Commission (9458).

Conflicts of Interest: The authors declare no conflict of interest.

References

1. Blair, J.M.A.; A Webber, M.; Baylay, A.J.; Ogbolu, D.O.; Piddock, L.J.V. Molecular mechanisms of antibiotic resistance. *Nat. Rev. Genet.* **2014**, *13*, 42–51. [CrossRef] [PubMed]
2. De Kievit, T. Biofilms. *Compr. Biotechnol.* **2011**, *1*, 529–540. [CrossRef]
3. Hall-Stoodley, L.; Costerton, J.W.; Stoodley, P. Bacterial biofilms: From the Natural environment to infectious diseases. *Nat. Rev. Genet.* **2004**, *2*, 95–108. [CrossRef] [PubMed]
4. Abu Bakar, M.; McKimm, J.; Haque, S.Z.; Majumder, A.A.; Haque, M. Chronic tonsillitis and biofilms: A brief overview of treatment modalities. *J. Inflamm. Res.* **2018**, *11*, 329–337. [CrossRef]
5. Orazi, G.; Ruoff, K.L.; O'Toole, G.A.; Wright, G.; Gilmore, M. Pseudomonas aeruginosa Increases the Sensitivity of Biofilm-Grown Staphylococcus aureus to Membrane-Targeting Antiseptics and Antibiotics. *mBio* **2019**, *10*, e01501-19. [CrossRef] [PubMed]
6. Dadar, M.; Tiwari, R.; Karthik, K.; Chakraborty, S.; Shahali, Y.; Dhama, K. Candida albicans—Biology, molecular characterization, pathogenicity, and advances in diagnosis and control—An update. *Microb. Pathog.* **2018**, *117*, 128–138. [CrossRef] [PubMed]
7. Cannon, R.D.; Chaffin, W. Oral Colonization By Candida Albicans. *Crit. Rev. Oral Boil. Med.* **1999**, *10*, 359–383. [CrossRef]
8. Nyirjesy, P. Management of Persistent Vaginitis. *Obstet. Gynecol.* **2014**, *124*, 1135–1146. [CrossRef]
9. Gavaldà, J.; Meije, Y.; Fortun, J.; Roilides, E.; Saliba, F.; Lortholary, O.; Muñoz, P.; Grossi, P.; Cuenca-Estrella, M. Invasive fungal infections in solid organ transplant recipients. *Clin. Microbiol. Infect.* **2014**, *20*, 27–48. [CrossRef]
10. Kosmidis, C.; Denning, D. The clinical spectrum of pulmonary aspergillosis. *Thorax* **2014**, *70*, 270–277. [CrossRef]
11. Slobodníková, L.; Fialová, S.; Rendeková, K.; Kováč, J.; Mučaji, P. Antibiofilm Activity of Plant Polyphenols. *Molecules* **2016**, *21*, 1717. [CrossRef]
12. Clain, E.; Haddad, J.G.; Koishi, A.C.; Sinigaglia, L.; Rachidi, W.; Despres, P.; Dos Santos, C.N.D.; Guiraud, P.; Jouvenet, N.; El Kalamouni, C.; et al. The Polyphenol-Rich Extract from Psiloxylon mauritianum, an Endemic Medicinal Plant from Reunion Island, Inhibits the Early Stages of Dengue and Zika Virus Infection. *Int. J. Mol. Sci.* **2019**, *20*, 1860. [CrossRef] [PubMed]
13. Ren, Z.; Chen, L.; Li, J.; Li, Y. Inhibition ofStreptococcus mutanspolysaccharide synthesis by molecules targeting glycosyltransferase activity. *J. Oral Microbiol.* **2016**, *8*, 623. [CrossRef] [PubMed]
14. Ichikawa, H. Withanolides potentiate apoptosis, inhibit invasion, and abolish osteoclastogenesis through suppression of nuclear factor-B (NF-B) activation and NF-B-regulated gene expression. *Mol. Cancer Ther.* **2006**, *5*, 1434–1445. [CrossRef] [PubMed]
15. Salwaan, C.; Singh, A.; Mittal, A.; Singh, P. Investigation of the Pharmacognostical, Phytochemical and Antioxidant Studies of Plant Withania coagulans Dunal. *J. Pharmacogn. Phytochem.* **2012**, *1*, 32–39.
16. Gupta, P.C. Withania coagulans Dunal—An overview. *Int. J. Pharm. Sci. Rev. Res.* **2012**, *11*, 448–451.

17. Prasad, S.; Singh, P.; Wahi, A.; Hemalatha, S. Pharmacognostical Standardization of Withania coagulans Dunal. *Pharmacogn. J.* **2010**, *2*, 386–394. [CrossRef]
18. Ismail, H.; Rasheed, A.; Haq, I.-U.; Jafri, L.; Ullah, N.; Dilshad, E.; Sajid, M.; Mirza, B. Five Indigenous Plants of Pakistan with Antinociceptive, Anti-Inflammatory, Antidepressant, and Anticoagulant Properties in Sprague Dawley Rats. *Evid.-Based Complement Altern. Med.* **2017**, *2017*, 1–10. [CrossRef]
19. Ahmad, R.; Fatima, A.; Srivastava, A.; Khan, M.A. Evaluation of apoptotic activity of Withania coagulans methanolic extract against human breast cancer and Vero cell lines. *J. Ayurveda Integr. Med.* **2017**, *8*, 177–183. [CrossRef]
20. Maurya, R.; Akanksha; Jayendra; Singh, A.B.; Srivastava, A.K. Coagulanolide, a withanolide from Withania coagulans fruits and antihyperglycemic activity. *Bioorganic Med. Chem. Lett.* **2008**, *18*, 6534–6537. [CrossRef]
21. Gupta, V.; Keshari, B.B. Withania Coagulans Dunal. (Paneer Doda): A Review. *Int. J. Ayurvedic Herb. Med.* **2013**, *3*, 1330–1336.
22. Khodaei, M.; Jafari, M.; Noori, M. Remedial Use of Withanolides from Withania Coagolans (Stocks) Dunal. *Adv. Life Sci.* **2012**, *2*, 6–19. [CrossRef]
23. Qasim, S.; Zafar, A.; Saif, M.S.; Ali, Z.; Nazar, M.; Waqas, M.; Haq, A.U.; Tariq, T.; Hassan, S.G.; Iqbal, F.; et al. Green synthesis of iron oxide nanorods using Withania coagulans extract improved photocatalytic degradation and antimicrobial activity. *J. Photochem. Photobiol. B Biol.* **2020**, *204*, 111784. [CrossRef] [PubMed]
24. Tripathi, D.; Modi, A.; Narayan, G.; Rai, S.P. Green and cost effective synthesis of silver nanoparticles from endangered medicinal plant Withania coagulans and their potential biomedical properties. *Mater. Sci. Eng. C* **2019**, *100*, 152–164. [CrossRef] [PubMed]
25. Maurya, R.; And, A. Jayendra Chemistry and pharmacology of Withania coagulans: An Ayurvedic remedy. *J. Pharm. Pharmacol.* **2010**, *62*, 153–160. [CrossRef]
26. Yousaf, M.; Ahmad, M.; Bhatti, I.A.; Nasir, A.; Hasan, M.; Jian, X.; Kalantar-Zadeh, K.; Mahmood, N. In Vivo and In Vitro Monitoring of Amyloid Aggregation via BSA@FGQDs Multimodal Probe. *ACS Sens.* **2018**, *4*, 200–210. [CrossRef]
27. Hasan, M.; Zafar, A.; Yousaf, M.; Gulzar, H.; Mehmood, K.; Hassan, S.G.; Saeed, A.; Yousaf, A.; Mazher, A.; Rongji, D.; et al. Synthesis of Loureirin B-Loaded Nanoliposomes for Pharmacokinetics in Rat Plasma. *ACS Omega* **2019**, *4*, 6914–6922. [CrossRef]
28. Kim, S.; Yu, J.S.; Lee, J.Y.; Choi, S.U.; Lee, J.; Kim, K.H. Cytotoxic Withanolides from the Roots of Indian Ginseng (Withania somnifera). *J. Nat. Prod.* **2019**, *82*, 765–773. [CrossRef]
29. Kaur, A.; Singh, B.; Ohri, P.; Wang, J.; Wadhwa, R.; Kaul, S.C.; Pati, P.K.; Kaur, A. Organic cultivation of Ashwagandha with improved biomass and high content of active Withanolides: Use of Vermicompost. *PLoS ONE* **2018**, *13*, e0194314. [CrossRef]
30. Vaishnavi, K.; Saxena, N.; Shah, N.; Singh, R.; Manjunath, K.; Uthayakumar, M.; Kanaujia, S.P.; Kaul, S.C.; Sekar, K.; Wadhwa, R. Differential Activities of the Two Closely Related Withanolides, Withaferin A and Withanone: Bioinformatics and Experimental Evidences. *PLoS ONE* **2012**, *7*, e44419. [CrossRef]
31. Rai, M.K.; Jogee, P.S.; Agarkar, G.; Dos Santos, C.A. Anticancer activities of Withania somnifera: Current research, formulations, and future perspectives. *Pharm. Biol.* **2015**, *54*, 189–197. [CrossRef]
32. Yousaf, A.; Zafar, A.; Ali, M.; Bukhary, S.M.; Manzoor, Y.; Tariq, T.; Saeed, A.; Akram, M.; Bukhari, F.; Abdullah, M.; et al. Intrinsic Bio-Enhancer Entities of Fagonia cretica for Synthesis of Silver Nanoparticles Involves Anti-Urease, Anti-Oxidant and Anti-Tyosinase Activity. *Adv. Biosci. Biotechnol.* **2019**, *10*, 455–468. [CrossRef]
33. Zulfiqar, H.; Zafar, A.; Rasheed, M.N.; Ali, Z.; Mehmood, K.; Mazher, A.; Hasan, M.; Mahmood, N.; Ayesha, Z.; Rasheed, N. Synthesis of silver nanoparticles using Fagonia cretica and their antimicrobial activities. *Nanoscale Adv.* **2019**, *1*, 1707–1713. [CrossRef]
34. Hasan, M.; Teng, Z.; Iqbal, J.; Awan, U.; Meng, S.; Dai, R.; Qing, H.; Deng, Y. Assessment of Bioreducing and Stabilizing Potential of Dragon's Blood (Dracaena Cochinchinensis, Lour. S. C. Chen) Resin Extract in Synthesis of Silver Nanoparticles. *Nanosci. Nanotechnol. Lett.* **2013**, *5*, 780–784. [CrossRef]
35. Hasan, M.; Iqbal, J.; Awan, U.; Saeed, Y.; RanRan, Y.; Liang, Y.; Dai, R.; Deng, Y. Mechanistic Study of Silver Nanoparticle's Synthesis by Dragon's Blood Resin Ethanol Extract and Antiradiation Activity. *J. Nanosci. Nanotechnol.* **2015**, *15*, 1320–1326. [CrossRef] [PubMed]

36. Hasan, M.; Ullah, I.; Zulfiqar, H.; Naeem, K.; Iqbal, A.; Gul, H.; Ashfaq, M.; Mahmood, N. Biological entities as chemical reactors for synthesis of nanomaterials: Progress, challenges and future perspective. *Mater. Today Chem.* **2018**, *8*, 13–28. [CrossRef]
37. Akbar, S.; Haleem, K.S.; Tauseef, I.; Rehman, W.; Ali, N.; Hasan, M. Raphanus sativus Mediated Synthesis, Characterization and Biological Evaluation of Zinc Oxide Nanoparticles. *Nanosci. Nanotechnol. Lett.* **2017**, *9*, 2005–2012. [CrossRef]
38. Mahmood, S.; Fatima, T.; Zulfaqar, H.; Saher, R.; Rafiq, M.; Rehman, A.; Dai, R.; Ashfaq, M.; Hasan, M. Meta-analysis of dragon's blood resin extract as radio-protective agent. *J. Coast. Life Med.* **2017**, 409–416. [CrossRef]
39. Munawar, T.; Yasmeen, S.; Hasan, M.; Mahmood, K.; Hussain, A.; Ali, A.; Arshad, M.; Iqbal, F. Novel tri-phase heterostructured ZnO–Yb_2O_3–Pr_2O_3 nanocomposite; structural, optical, photocatalytic and antibacterial studies. *Ceram. Int.* **2020**, *46*, 11101–11114. [CrossRef]
40. Munawar, T.; Yasmeen, S.; Mukhtar, F.; Nadeem, M.S.; Mahmood, K.; Saif, M.S.; Hasan, M.; Ali, A.; Hussain, F.; Iqbal, F. $Zn_{0.9}Ce_{0.05}M_{0.05}O$ (M = Er, Y, V) nanocrystals: Structural and energy bandgap engineering of ZnO for enhancing photocatalytic and antibacterial activity. *Ceram. Int.* **2020**, *46*, 14369–14383. [CrossRef]
41. Fathalipour, S.; Pourbeyram, S.; Sharafian, A.; Tanomand, A.; Azam, P. Biomolecule-assisted synthesis of Ag/reduced graphene oxide nanocomposite with excellent electrocatalytic and antibacterial performance. *Mater. Sci. Eng. C* **2017**, *75*, 742–751. [CrossRef] [PubMed]
42. Somu, P.; Paul, S.; S, P. A biomolecule-assisted one-pot synthesis of zinc oxide nanoparticles and its bioconjugate with curcumin for potential multifaceted therapeutic applications. *New J. Chem.* **2019**, *43*, 11934–11948. [CrossRef]
43. Hasan, M.; Iqbal, J.; Awan, U.; Xin, N.; Dang, H.; Waryani, B.; Saeed, Y.; Ullah, K.; Rongji, D.; Deng, Y. LX loaded nanoliposomes synthesis, characterization and cellular uptake studies in H_2O_2 stressed SH-SY5Y cells. *J. Nanosci. Nanotechnol.* **2014**, *14*, 4066–4071. [CrossRef] [PubMed]
44. Ran, Y.; Wang, R.; Hasan, M.; Jia, Q.; Tang, B.; Shan, S.; Deng, Y.; Qing, H. Radioprotective effects of dragon's blood and its extracts on radiation-induced myelosuppressive mice. *J. Ethnopharmacol.* **2014**, *154*, 624–634. [CrossRef] [PubMed]
45. Ran, Y.; Wang, R.; Gao, Q.; Jia, Q.; Hasan, M.; Awan, M.U.F.; Tang, B.; Zhou, R.; Dong, Y.; Wang, X.; et al. Dragon's blood and its extracts attenuate radiation-induced oxidative stress in mice. *J. Radiat. Res.* **2014**, *55*, 699–706. [CrossRef]
46. Ran, Y.; Wang, R.; Lin, F.; Hasan, M.; Jia, Q.; Tang, B.; Xia, Y.; Shan, S.; Wang, X.; Li, Q.; et al. Radioprotective effects of Dragon's blood and its extract against gamma irradiation in mouse bone marrow cells. *Phys. Med.* **2014**, *30*, 427–431. [CrossRef] [PubMed]
47. Petrachi, T.; Resca, E.; Piccinno, M.S.; Biagi, F.; Strusi, V.; Dominici, M.; Veronesi, E. An Alternative Approach to Investigate Biofilm in Medical Devices: A Feasibility Study. *Int. J. Environ. Res. Public Health* **2017**, *14*, 1587. [CrossRef]
48. Good, M.; Wend, C.; Bond, L.; McLean, J.S.; Panetta, P.; Ahmed, S.; Crawford, S.; Daly, D. An estimate of biofilm properties using an acoustic microscope. *IEEE Trans. Ultrason. Ferroelectr. Freq. Control* **2006**, *53*, 1637–1648. [CrossRef]
49. Biswas, B.; Rogers, K.; McLaughlin, F.; Daniels, D.; Yadav, A. Antimicrobial Activities of Leaf Extracts of Guava (Psidium guajavaL.) on Two Gram-Negative and Gram-Positive Bacteria. *Int. J. Microbiol.* **2013**, *2013*, 1–7. [CrossRef]
50. Türeli, N.G.; Torge, A.; Juntke, J.; Schwarz, B.C.; Schneider-Daum, N.; Türeli, A.E.; Lehr, C.-M.; Schneider, M. Ciprofloxacin-loaded PLGA nanoparticles against cystic fibrosis P. aeruginosa lung infections. *Eur. J. Pharm. Biopharm.* **2017**, *117*, 363–371. [CrossRef]
51. Manandhar, S.; Luitel, S.; Dahal, R.K. In Vitro Antimicrobial Activity of Some Medicinal Plants against Human Pathogenic Bacteria. *J. Trop. Med.* **2019**, *2019*, 1895340-5. [CrossRef]

52. Samadi, F.M.; Suhail, S.; Sonam, M.; Sharma, N.; Singh, S.; Gupta, S.; Dobhal, A.; Pradhan, H. Antifungal efficacy of herbs. *J. Oral Biol. Craniofacial Res.* **2018**, *9*, 28–32. [CrossRef] [PubMed]
53. Hamill, R.J. Amphotericin B Formulations: A Comparative Review of Efficacy and Toxicity. *Drugs* **2013**, *73*, 919–934. [CrossRef] [PubMed]

Sample Availability: Samples of the compounds are available from the authors.

molecules

MDPI

Article

Effect of Naringenin and Its Derivatives on the Probing Behavior of *Myzus persicae* (Sulz.)

Katarzyna Stec [1], Joanna Kozłowska [2], Anna Wróblewska-Kurdyk [1], Bożena Kordan [3], Mirosław Anioł [2] and Beata Gabryś [1,*]

1 Department of Botany and Ecology, Institute of Biological Sciences, University of Zielona Góra, Szafrana 1, 65-516 Zielona Góra, Poland; katarzyna.rozycka1985@interia.pl (K.S.); a.wroblewska-kurdyk@wnb.uz.zgora.pl (A.W.-K.)

2 Department of Chemistry, Faculty of Biotechnology and Food Science, Wrocław University of Environmental and Life Sciences, Norwida 25, 50-375 Wrocław, Poland; joannakozlowska3@gmail.com (J.K.); miroslaw.aniol@upwr.edu.pl (M.A.)

3 Department of Entomology, Phytopathology and Molecular Diagnostics, University of Warmia and Mazury in Olsztyn, Prawocheńskiego 17, 10-719 Olsztyn, Poland; bozena.kordan@uwm.edu.pl

* Correspondence: b.gabrys@wnb.uz.zgora.pl

Academic Editors: George Grant and Angelo Canale
Received: 28 May 2020; Accepted: 8 July 2020; Published: 13 July 2020

Abstract: Substances that alter insect behavior have attracted a lot of attention as potential crop protection agents. Naringenin (5,7,4′-trihydroxyflavanone) is a naturally occurring bioactive flavanone. We evaluated the influence of naringenin on aphid activities during individual phases of probing and feeding and the effect of structural modifications of naringenin on its activity towards aphids. We monitored the probing behavior of *Myzus persicae* (Sulz.) (Hemiptera: Aphididae) using the Electrical Penetration Graph (EPG) technique. The chemical modifications were the substitution of hydrogen atoms with methyl, ethyl or pentyl groups and the replacement of the carbonyl group in naringenin and its derivatives with an oxime moiety. Depending on the substituents, the activity of naringenin-derived compounds varied in potency and mode of action. Naringenin was an attractant of moderate activity, which enhanced sap ingestion. The naringenin derivative with two methyl groups—7,4′-di-*O*-methylnaringenin—was a deterrent, which hindered aphid probing in non-phloem tissues. Naringenin oxime derivatives with methyl substituents—7,4′-di-*O*-methylnaringenin oxime, 7-*O*-methylnaringenin oxime, and 5,7,4′-tri-*O*-methylnaringenin oxime—and the derivative with a pentyl substituent—7-*O*-pentylnaringenin oxime—were strong attractants which stimulated aphid probing in non-phloem tissues and the ingestion of phloem sap.

Keywords: electrical penetration graph; peach potato aphid; antifeedants; attractants; structure-activity relationships

1. Introduction

Naringenin (5,7,4′-trihydroxyflavanone) is a natural flavonoid, most common in *Citrus* fruits, known to have bioactive effects on human health, such as antidiabetic, antidepressant, immunomodulatory, antitumor, anti-inflammatory, DNA protective, and antioxidant effects [1,2]. Various effects of naringenin on insect development and behavior were also reported; naringenin inhibited the feeding of adult Japanese beetles *Popillia japonica* (Newman) (Coleoptera: Scarabaeidae) [3], caused a reduction in larval growth and development in the common cutworm *Spodoptera litura* (Fabricius) (Lepidoptera: Noctuidae) [4], stimulated oviposition in the spotted pink ladybeetle *Coleomegilla maculata* De Geer (Coleoptera: Coccinellidae) [5], and impaired the learning abilities of the honey bee *Apis mellifera* L. (Hymenoptera: Apidae) [6].

Aphids (Hemiptera: Aphididae) are responsible for at least 2% of losses caused by insect feeding in the world's crops [7]. In addition to the removal of assimilates from plant phloem transporting vessels, aphids transfer viral diseases from infected to healthy plants. The extremely polyphagous peach potato aphid *Myzus persicae* (Sulz.) can transmit over 100 plant viruses among plants within over 40 families [8]. To our knowledge, there exists only one published study that reports the effect of naringenin on aphids. Goławska et al. [9] showed that the addition of naringenin into a sucrose–agarose diet caused an increase in the duration of the pre-reproductive period and mortality, as well as a decrease in fecundity and the intrinsic rate of natural increase in the pea aphid *Acyrthosiphon pisum* (Harris). In the same study, the authors demonstrated that high concentrations of naringenin inhibited the passive ingestion (analogous to passive ingestion of phloem sap on plants and represented by EPG waveform g-E2) of the naringenin-supplemented sucrose–agarose diet, but stimulated the active ingestion (analogous to active ingestion of xylem sap on plants and represented by EPG waveform g-G) of the diet [9].

Nowadays, aphid control relies mainly on neurotoxic insecticides. However, several aphid species, especially *M. persicae*, evolved diverse mechanisms of resistance to various insecticides [10,11]. At the same time, a global trend for the reduction in insecticide use is observed in response to environmental issues. In recent years, serious restrictions in neonicotinoid use have been established in the European Union [12]. Therefore, there is a growing demand for the replacement of traditional insecticides, at least in part, by natural product-based insect control agents. Specifically, behavior modifying substances (repellents, antifeedants, attractants, etc.), which may cause the withdrawal of the herbivore from the plant or other substrates, are searched for [13–15]. The exogenous application of xenobiotics may alter aphid response to otherwise acceptable host plants, which has been shown in studies on aphid antifeedants involving different chemical groups, including terpenoids, quassinoids, flavonoids, and cyanogenic glycosides [9,16–20]. At the same time, aphids may be attracted to other areas, such as trap crops or barrier crops, in 'push–pull' strategies [21–23]. Unfortunately, the application of natural compounds for the protection of plants is limited. The main constraints are the low content in natural sources and usually complicated structures, which make their synthesis complex and expensive. Therefore, the synthesis of natural compound analogues is one of the most promising ways leading to their practical use in insect pest population control [24]. Structural transformations of the natural molecule usually change the mode of action and potency of its activity [25,26]. The possibility of reducing aphid infestation of crop plants by naringenin and its analogues application has never been explored.

The aim of the present study was to assess the influence of naringenin on aphid activities during individual phases of probing and feeding and evaluate the effect of structural modifications of naringenin on its activity towards aphids. The chemical modifications were the substitution of hydrogen atoms with methyl, ethyl or pentyl groups and the replacement of the carbonyl group in naringenin and its derivatives with an oxime moiety. We monitored aphid probing with the Electrical Penetration Graph (EPG) technique, which visualizes the movements of aphid mouthparts within individual plant tissues. The values of parameters derived from EPG recordings are reliable and accurate indicators of aphid behavioral responses to alteration in plant suitability due to exogenous application of xenobiotics [19–24].

2. Results

The typical behavior of *M. persicae* on control untreated plants consisted of non-probing (11% time of the 8 h experiment), probing in non-phloem tissues (34%), and probing in phloem tissues (55%). Sap ingestion occupied 95% of the phloem phase. Aphid probing activities were divided into 19.8 (±10.0) probes on average, and these probes were approximately 0.6 (±0.6) hours long. Nearly 10% of these probes contained a phloem phase (Table 1). *M. persicae* needed approximately 2.0 (±1.3) hours to reach phloem vessels and commence sap ingestion. In that time, 25% were non-probing activities. Probing activities in non-phloem tissues before the first phloem phase were divided into 12.7 (±8.8) events (probes) (Table 2). The phloem phase in *M. persicae* consisted of 4.4 (±2.8) separate periods and almost all of these phloem phases included sap ingestion. Of these sap ingestion periods, 60% were longer than 10 min and were 2.1 (±2.2) hours long on average. The first contact with sieve elements (the first phloem phase) was 1.7 (±2.4) hours long (Table 3).

M. persicae probing behavior on plants treated with naringenin and its derivatives (**2**–**17**) (Figures 1 and 2) was significantly different from the aphid behavior on the control plants. Non-probing activities were significantly reduced on plants treated with (**8**) and (**17**). The total duration of the non-phloem phase was significantly longer on (**2**)- and (**12**)-treated plants, and the phloem phase was significantly shorter on (**2**)-treated plants than on the control plants. The total duration of sap ingestion was longer on (**13**)-treated plants than on the control. The number of probes was lower, and the probes were longer on (**3**)-, (**4**)-, (**8**)-, (**13**)-, (**15**)-, and (**16**)-treated plants (Table 1). During the pre-phloem phase, i.e., the period before the first phloem phase occurred, the total duration of non-probing was longer on (**7**)-treated plants but shorter on (**3**)-, (**8**)-, and (**13**)-treated plants than on the control. The duration of probing in non-phloem tissues, the total time to the first phloem phase and the first phloem sap ingestion phase from the onset of probing were shorter on (**13**)-treated plants than on the control. The total number of probes and the number of short probes before the first phloem phase were lower on (**3**)-, (**4**)-, (**8**)-, (**13**)-, (**15**)-, and (**16**)-treated plants (Table 2). The number of phloem phases and the number of sap ingestion phases were lower on naringenin, (**3**)-, (**7**)-, (**8**)-, (**10**)-, (**15**)-, and (**17**)-treated plants than on the control. On these plants, the first phloem phases were significantly longer than on the control plants. The mean duration of phloem sap ingestion phase was longer on naringenin-, (**8**)-, and (**17**)-treated plants than on the control (Table 3).

The comparison within the group of aphids on treated plants showed that the highest number of probes occurred in aphids on (**7**)-treated plants (16.7 ± 9.7 probes) in contrast to aphids on (**3**)- and (**16**)-treated plants (4.4 ± 4.2 and 5.2 ± 7.3 probes, respectively), the durations of periods before the first phloem phase and phloem sap ingestion phase were longest in aphids on (**7**)-treated plants (2.8±1.4 h) in contrast to aphids on (**13**)-treated plants (1.0 ± 0.8 h) (Tables 1 and 2). The mean duration of the phloem phase was the shortest in aphids on (**2**)-treated plants (1.1 ± 2.0 h) in contrast to aphids on (**15**)-treated plants (5.0 ± 2.7 h) (Table 3).

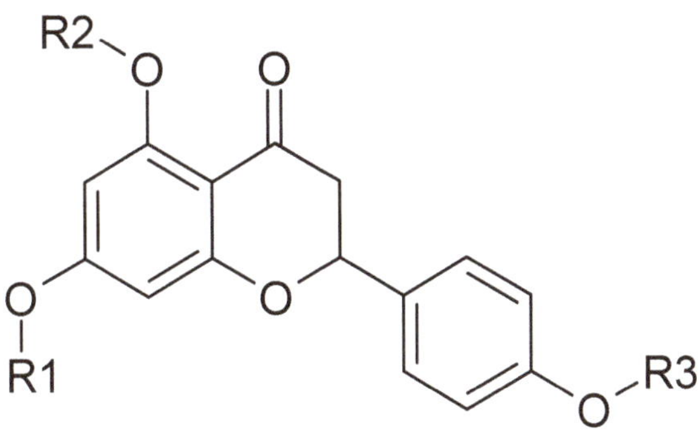

Figure 1. Naringenin and its derivatives. Naringenin: R1=H, R2=H, R3=H **(1)**; 5,7,4′-tri-*O*-methylnaringenin: R1=CH_3, R2=CH_3, R3=CH_3 **(2)**; 5,7,4′-tri-*O*-ethylnaringenin: R1=CH_3CH_2, R2=CH_3CH_2, R3=CH_3CH_2 **(3)**; 7-*O*-ethylnaringenin: R1=CH_3CH_2, R2=H, R3=H **(4)**; 7-*O*-pentylnaringenin: R1=$CH_3(CH_2)_4$, R2=H, R3=H **(5)**; 7,4′-di-*O*-ethylnaringenin: R1=CH_3CH_2, R2=H, R3=CH_3CH_2 **(6)**; 7,4′-di-*O*-methylnaringenin: R1=CH_3, R2=H, R3=CH_3 **(7)**; 7-*O*-methylnaringenin: R1=CH_3, R2=H, R3=H **(8)**.

Table 1. General aspects of *Myzus persicae* probing behavior on naringenin and naringenin derivatives (**1–17**, Figures 1 and 2)-treated plants (means ± SD). The mean and SD values given are a representation of non-Gaussian data, but the statistical analysis was done by non-parametric tests, in which all individual data were included; n—number of replications; C—pathway; E1—phloem salivation; E2—phloem sap ingestion; F—derailed stylet movements; G—xylem sap ingestion. Different small letters in columns show significant differences in the values of specific parameters among aphids on plants treated with individual naringenin derivatives (Kruskal–Wallis test, $p < 0.05$); different capital letters show significant differences in the values of specific parameters between aphids on plants treated with individual compounds and control (Mann–Whitney U-test, $p < 0.05$).

Compound/EPG Parameter	Sample Size	Total Duration of Non-Probing (h)	Total Duration of Probing in Non-Phloem Tissues C + F + G (h)	Total Duration of Phloem Phase E1 + E2 (h)	Total Duration of Sap Ingestion Phase E2 (h)	Number of Probes (#)	Mean Duration of a Probe (h)
Control	$n = 16$	0.9 ± 1.2 A	2.7 ± 1.3 A	4.4 ± 1.7 A	4.2 ± 1.8 A	19.8 ± 10.0 A	0.6 ± 0.6 A
1	$n = 14$	0.8 ± 0.5 aA	2.7 ± 1.8 aA	4.5 ± 2.1 aA	4.5 ± 2.1 aA	19.1 ± 13.7 aA	0.8 ± 1.0 aA
2	$n = 13$	1.3 ± 1.1 aA	4.0 ± 1.7 aB	2.7 ± 2.4 aB	2.6 ± 2.4 aA	24.2 ± 16.3 aA	1.2 ± 2.2 aA
3	$n = 13$	0.9 ± 1.6 aA	3.2 ± 1.4 aA	4.0 ± 1.8 aA	3.9 ± 1.8 aA	10.8 ± 8.4 aB	1.5 ± 2.0 aB
4	$n = 13$	0.5 ± 0.5 aA	3.5 ± 1.8 aA	3.9 ± 2.1 aA	3.9 ± 2.1 a A	12.9 ± 6.9 aB	0.8 ± 0.5 aB
5	$n = 14$	0.8 ± 0.8 aA	3.3 ± 2.3 aA	3.9 ± 2.8 aA	3.9 ± 2.8 aA	20.3 ± 15.1 aA	0.8 ± 0.8 aA
6	$n = 14$	0.6 ± 0.7 aA	2.3 ± 1.8 aA	5.1 ± 2.3 aA	5.1 ± 2.3 aA	13.1 ± 10.3 aA	1.5 ± 1.9 aA
7	$n = 16$	1.2 ± 1.2 aA	3.5 ± 1.9 aA	3.3 ± 2.1 aA	3.3 ± 2.2 aA	22.3 ± 12.3 aA	0.6 ± 0.6 aA
8	$n = 13$	0.3 ± 0.2 aB	2.9 ± 2.1 aA	4.8 ± 2.2 aA	4.7 ± 2.2 aA	11.5 ± 7.8 aB	1.3 ± 1.2 aB
9	$n = 16$	0.8 ± 0.6 aA	2.8 ± 1.6 aA	4.4 ± 1.9 aA	4.4 ± 1.9 aA	17.3 ± 12.3 aA	0.8 ± 0.7 aA
10	$n = 14$	0.6 ± 0.6 a A	3.7 ± 2.3 aA	3.7 ± 2.7 aA	3.7 ± 2.7 aA	14.9 ± 11.7 aA	1.4 ± 2.0 aA
11	$n = 13$	0.7 ± 0.6 aA	3.7 ± 2.0 aA	3.6 ± 2.2 aA	3.5 ± 2.2 aA	15.8 ± 7.0 aA	0.8 ± 1.0 aA
12	$n = 14$	0.7 ± 0.4 aA	4.3 ± 2.0 aB	3.0 ± 2.2 aA	3.0 ± 2.2 aA	15.5 ± 10.5 aA	0.7 ± 0.4 aA
13	$n = 14$	0.3 ± 0.3 a B	2.1 ± 1.5 aA	5.6 ± 1.7 aA	5.6 ± 1.7 aB	11.1 ± 8.8 a B	1.1 ± 0.7 aB
14	$n = 14$	1.0 ± 0.7 aA	2.9 ± 1.6 aA	4.1 ± 2.1 aA	4.1 ± 2.1 aA	19.7 ± 9.4 aA	0.5 ± 0.3 aA
15	$n = 14$	0.6 ± 0.8 aA	2.0 ± 1.8 aA	5.5 ± 2.4 aA	5.5 ± 2.4 aA	11.9 ± 12.1 aB	2.0 ± 2.5 aB
16	$n = 12$	0.7 ± 0.9 aA	2.3 ± 1.8 aA	4.9 ± 2.2 aA	4.9 ± 2.2 aA	9.8 ± 7.8 aB	2.2 ± 2.7 aB
17	$n = 13$	0.8 ± 0.7 aA	2.8 ± 1.9 aA	4.4 ± 2.5 aA	4.4 ± 2.5 aA	20.01 ± 4.2 aA	0.7 ± 0.6 aA

Table 2. *Myzus persicae* behavior in non-phloem tissues prior to the first phloem phase during probing on naringenin and naringenin derivatives (**1–17**, Figures 1 and 2)-treated plants (means ± SD); the mean and SD values given are a representation of non-Gaussian data, but the statistical analysis was done by non-parametric tests, in which all individual data were included; *n*—number of replications; only replications where phloem phase occurred were included; C—pathway; E1—phloem salivation; F—derailed stylet movements; G—xylem sap ingestion. Different small letters in columns show significant differences in the values of specific parameters among aphids on plants treated with individual naringenin derivatives (Kruskal–Wallis test, $p < 0.05$); different capital letters show significant differences in the values of specific parameters between aphids on plants treated with individual compounds and control (Mann–Whitney U-test, $p < 0.05$).

Compound/EPG Parameter	Sample Size	Total Duration of Non-Probing (h)	Total Duration of Probing in Non-Phloem Tissues C + F + G (h)	Number of Probes	Time from 1st Probe to 1st Phloem Phase E1 (h)
Control	$n = 16$	0.4 ± 0.2 A	1.6 ± 1.2 A	12.7 ± 8.8 A	2.0 ± 1.3 A
1	$n = 13$	0.7 ± 0.5 aA	1.8 ± 1.4 aA	15.8 ± 13.6 abA	2.4 ± 1.8 abA
2	$n = 13$	0.7 ± 0.7 aA	1.8 ± 1.1 aA	10.5 ± 8.7 abA	2.5 ± 1.5 abA
3	$n = 12$	0.2 ± 0.2 aB	2.0 ± 1.2 aA	4.4 ± 4.2 aB	2.1 ± 1.2 abA
4	$n = 13$	0.2 ± 0.2 aA	1.4 ± 1.2 aA	4.9 ± 4.0 abB	1.5 ± 1.3 abA
5	$n = 13$	0.3 ± 0.4 aA	1.0 ± 0.7 aA	8.6 ± 7.9 abA	1.3 ± 1.0 abA
6	$n = 14$	0.4 ± 0.3 aA	1.7 ± 1.5 aA	8.7 ± 7.6 abA	2.1 ± 1.8 abA
7	$n = 15$	0.7 ± 0.4 aB	2.2 ± 1.2 aA	16.7 ± 9.7 bA	2.8 ± 1.4 bA
8	$n = 13$	0.2 ± 0.2 aB	2.0 ± 1.6 aA	6.8 ± 7.3 abB	2.1 ± 1.7 abA
9	$n = 16$	0.4 ± 0.3 aA	1.5 ± 0.8 aA	8.4 ± 6.4 abA	1.8 ± 0.9 abA
10	$n = 13$	0.2 ± 0.2 aA	1.9 ± 1.7 aA	6.2 ± 4.1 abA	2.1 ± 1.8 abA
11	$n = 12$	0.4 ± 0.6 aA	2.3 ± 2.4 aA	8.3 ± 7.0 abA	2.7 ± 2.6 abA
12	$n = 14$	0.4 ± 0.3 aA	1.8 ± 1.6 aA	9.2 ± 8.1 abA	2.1 ± 1.9 abA
13	$n = 14$	0.2 ± 0.1 aB	0.9 ± 0.6 aB	5.4 ± 5.0 abB	1.0 ± 0.8 aB
14	$n = 12$	0.5 ± 0.4 aA	1.2 ± 0.8 aA	11.2 ± 7.4 abA	1.7 ± 1.1 abA
15	$n = 14$	0.2 ± 0.2 aA	1.0 ± 0.8 aA	5.2 ± 3.5 abB	1.2 ± 0.9 abA
16	$n = 12$	0.5 ± 0.8 aA	1.5 ± 1.5 aA	5.2 ± 7.3 aB	2.0 ± 2.0 abA
17	$n = 11$	0.4 ± 0.3 aA	1.5 ± 0.7 aA	10.9 ± 7.8 abA	1.8 ± 1.0 abA

Table 3. *Myzus persicae* behavior associated with probing in sieve elements on naringenin and naringenin derivatives (**1–17**, Figures 1 and 2)-treated plants (means ± SD); the mean and SD values given are a representation of non-Gaussian data, but the statistical analysis was done by non-parametric tests, in which all individual data were included; *n*—replication number; * all replications were included in analysis (the missing phloem phase was quantified as 0.0); ** aphids that did not reach phloem elements during 8 h experiment were excluded from this analysis; E1—phloem salivation; E2—phloem sap ingestion. Different small letters in columns show significant differences in the values of specific parameters among aphids on plants treated with individual naringenin derivatives (Kruskal–Wallis test, $p < 0.05$); different capital letters show significant differences in the values of specific parameters between aphids on plants treated with individual compounds and control (Mann–Whitney U-test, $p < 0.05$).

Compound/EPG Parameter	Sample Size *	Number of Phloem Phases E1 and E1 + E (#)	Number of Phloem Sap Ingestion Phases (#)	Number of Sustained Sap Ingestion Phases E2 > 10 min (#)	Sample Size **	Mean Duration of 1st Phloem Phase E1 + E2 (h)	Mean Duration of Phloem Sap Ingestion Phase E2 (h)
Control	*n* = 16	4.4 ± 2.8 A	4.4 ± 2.7 A	2.6 ± 1.5 A	*n* = 16	1.7 ± 2.4 A	2.1 ± 2.2 A
1	*n* = 14	1.9 ± 1.6 aB	1.7 ± 1.2 aB	1.4 ± 0.9 aA	*n* = 13	3.0 ± 2.3 abB	3.4 ± 2.0 aB
2	*n* = 13	3.1 ± 1.7 aA	3.1 ± 1.7 aA	1.6 ± 1.1 aA	*n* = 13	1.1 ± 2.0 aA	1.5 ± 2.0 aA
3	*n* = 13	2.4 ± 2.6 aB	2.3 ± 2.3 aB	1.5 ± 0.8 aA	*n* = 12	3.0 ± 2.4 abA	3.2 ± 2.1 aA
4	*n* = 13	2.6 ± 1.3 aA	2.6 ± 1.3 aA	1.5 ± 1.2 aA	*n* = 13	1.9 ± 2.3 abA	2.3 ± 2.0 aA
5	*n* = 14	3.1 ± 2.2 aA	2.9 ± 2.1 aA	1.8 ± 1.1 aA	*n* = 13	2.6 ± 3.1 abA	2.8 ± 3.0 aA
6	*n* = 14	2.8 ± 1.8 aA	2.7 ± 1.7 aA	1.8 ± 1.3 aA	*n* = 14	2.7 ± 3.0 abA	2.8 ± 2.7 aA
7	*n* = 16	1.9 ± 1.2 aB	1.9 ± 1.2 aB	1.6 ± 0.9 aA	*n* = 15	2.3 ± 2.2 abA	2.5 ± 2.2 aA
8	*n* = 13	2.2 ± 1.6 aB	2.2 ± 1.6 aB	1.6 ± 0.8 aA	*n* = 13	3.7 ± 2.5 abB	3.6 ± 2.4 aA
9	*n* = 16	2.9 ± 2.3 aA	3.0 ± 2.4 aA	1.6 ± 0.9 aA	*n* = 16	2.9 ± 2.5 abA	3.0 ± 2.5 aA
10	*n* = 14	2.1 ± 1.6 aB	2.1 ± 1.6 aB	1.6 ± 1.2 aA	*n* = 13	3.4 ± 2.8 abB	3.2 ± 2.8 aA
11	*n* = 13	2.9 ± 2.7 aA	2.9 ± 2.7 aA	2.0 ± 1.8 aA	*n* = 12	1.6 ± 2.2 abA	2.1 ± 2.0 aA
12	*n* = 14	3.1 ± 1.8 aA	2.8 ± 1.7 aA	1.6 ± 0.8 aA	*n* = 14	2.7 ± 2.2 abA	2.8 ± 2.2 aA
13	*n* = 14	3.1 ± 1.9 aA	3.1 ± 1.9 aA	2.0 ± 1.2 aA	*n* = 14	3.0 ± 3.1 abA	3.3 ± 2.8 aA
14	*n* = 14	2.9 ± 2.2 aA	2.8 ± 2.2 aA	2.1 ± 1.5 aA	*n* = 12	2.0 ± 2.3 abA	2.4 ± 2.0 aA
15	*n* = 14	1.8 ± 1.4 aB	1.8 ± 1.4 aB	1.4 ± 0.7 aB	*n* = 14	5.0 ± 2.7 bB	4.7 ± 3.0 aB
16	*n* = 12	2.6 ± 1.8 aA	2.5 ± 1.8 aA	1.9 ± 1.7 aA	*n* = 12	3.6 ± 2.9 abB	3.4 ± 2.9 aA
17	*n* = 13	1.7 ± 1.6 aB	1.7 ± 1.6 aB	1.5 ± 1.3 aA	*n* = 11	3.8 ± 2.7 abB	4.0 ± 2.5 aB

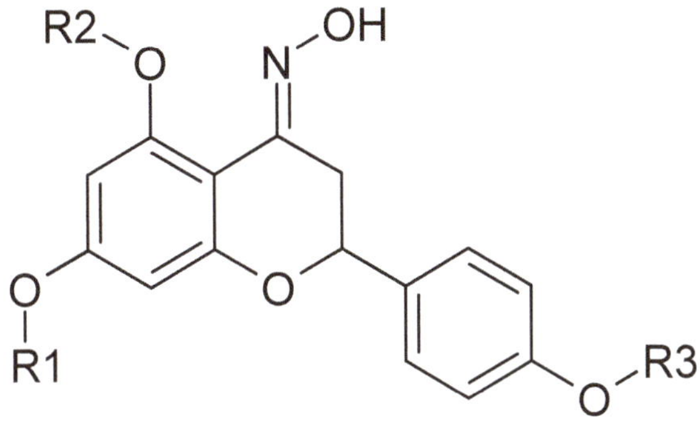

Figure 2. Naringenin oxime and its derivatives. Naringenin oxime: R1=H, R2=H, R3=H **(9)**; 5,7,4′-tri-*O*-methylnaringenin oxime: R1=CH_3, R2=CH_3, R3=CH_3 **(10)**; 5,7,4′-tri-*O*-ethylnaringenin oxime: R1=CH_3CH_2, R2=CH_3CH_2, R3=CH_3CH_2 **(11)**; 7-*O*-ethylnaringenin oxime: R1=CH_3CH_2, R2=H, R3=H **(12)**; 7-*O*-pentylnaringenin oxime: R1=$CH_3(CH_2)_4$, R2=H, R3=H **(13)**; 7,4′-di-*O*-ethylnaringenin oxime: R1=CH_3CH_2, R2=H, R3=CH_3CH_2 **(14)**; 7,4′-di-*O*-methylnaringenin oxime: R1=CH_3, R2=H, R3=CH_3 **(15)**; 7-*O*-methylnaringenin oxime: R1=CH_3, R2=H, R3=H **(16)**; 7,4′-di-*O*-pentylnaringenin oxime: R1=$CH_3(CH_2)_4$, R2=H, R3=$CH_3(CH_2)_4$ **(17)**.

3. Discussion

The pre-phloem and phloem phases of aphid probing in plant tissues are two crucial steps in the chemosensory-based host plant selection and host plant acceptance processes. Allelochemicals are the main cues used by aphids for host plant selection during either the pathway or phloem phase [16]. The long duration of probing time in non-phloem tissues as compared to total penetration time, the relatively long time to the 1st phloem phase within a probe, and a failure in finding sieve elements may be interpreted as pre-ingestive effects of antifeedants that restrain aphid probing at the level of non-phloem tissues [27]. In contrast, the reduction in the number of probes and the elongation of these probes indicates the attractant character of chemical factors. The long total and mean durations of phloem sap ingestion may point to the ingestive mode of feeding stimulatory activity [28,29]. The interpretation of aphid behavior in response to the chemical properties of plant tissues is based on studies on plant resistance mechanisms. Aphids feeding on susceptible plant genotypes have a significantly greater duration of sieve element phase than when feeding on resistant genotypes and the time taken to reach the first sieve element phase in resistant genotypes is significantly greater than in susceptible genotypes [30–33].

Based on the comparison of the EPG-monitored *M. persicae* behavior on naringenin and naringenin derivatives-treated plants to control and the overall trends for each compound, it is possible to group the studied compounds according to their potential to modify aphid probing activities: (i) strong attractants (**8**), (**13**), (**15**), and (**16**), which stimulated aphid activities in the non-phloem as well as in the phloem tissues. In comparison to control untreated plants, on treated plants, aphids rarely withdraw stylets from plant tissues (**8, 13, 15, 16**) and the non-probing time was significantly reduced (**8, 13**), which caused a significant reduction in time to reach phloem vessels from the onset of probing (**13**), the individual phloem sap ingestion periods were long and rarely interrupted (**8, 15, 16**), and the total duration of sap ingestion was longer (**13**); (ii) moderate attractants naringenin, (**3**), (**10**), and (**17**)—naringenin, (**10**), and (**17**) had no effect on aphid behavior during the pre-phloem phase but encouraged sap ingestion, and (**3**) caused a slight reduction in the number of phloem phases as well as the number of probes and a decrease in the duration of non-probing before the phloem phase; (iii) weak attractant (**4**) caused a slight reduction in the number and a slight increase in the duration of probes; (iv) weak deterrent (**7**) caused aphid restlessness by hindering pre-phloem pathway activities, which was manifested in the frequent withdrawals of stylets from plant tissues, an increased time of non-probing, and, in consequence, a delay in reaching phloem vessels; (v) inactive compounds (**2**), (**5**), (**6**), (**9**), (**11**), (**12**), (**14**). In comparison to control, aphid behavior was not altered on plants treated with these compounds.

The biological activity of a given compound is species-specific and depends on its structural characteristics. Variations, such as incorporation of functional groups, epoxidation, or lactonization, can produce radical changes in activity [29]. In our previous studies, we determined that chemical modifications of naturally occurring terpenoids, e.g., incorporation of functional groups, epoxidation or lactonization, evoked significant changes in their activity profiles. We have established that the potency and persistence of behavioral effects on aphid probing of piperitone-, β-damascone-, xanthohumol-, isoxanthohumol-, and *cis*-jasmone-derived compounds depended on their substituents. Certain modifications caused shifts from attractant to deterrent properties, or vice versa [24–26,29]. In the present study, we also revealed specific structure–activity relationships. Naringenin appeared to be an attractant of moderate activity. Three ethyl groups incorporated at positions 5,7, and 4′ in (**3**) did not significantly alter the naringenin activity. The compound with one methyl group in the position 7 (**4**) was a weak attractant. However, the compound with two methyl groups in the positions 7 and 4′ (**7**) was a weak deterrent. The incorporation of three methyl groups at positions 5,7, and 4′ (**2**), two ethyl groups at positions 7, and 4′ (**6**), or one pentyl group at position 7 (**5**) caused a loss of naringenin activity towards *M. persicae*. Naringenin oxime (**9**) was inactive towards *M. persicae*. However, the substitution of hydrogen atoms with methyl, ethyl or pentyl groups, in addition to oxime moiety, caused a significant rise in the activity of some of the derived compounds. All naringenin oxime (**9**) derivatives with methyl substituents at positions 7 (**15**), 7 and 4′ (**16**), and 5,7,4′ (**10**) and the derivative with a pentyl substituent at position 7 (**13**) appeared strong attractants. Pentyl groups at positions 7 and 4′ made the compound (**17**) a weak attractant. The ethyl group substituents did not improve the activity of naringenin oxime (**9**); all derivatives with one (**12**), two (**14**), or three (**11**) ethyl substituents remained inactive towards *M. persicae*.

The results of the experiments in the present work illustrate two major aspects of the biological activity of naringenin and its derivatives that depend on their substituents: (1) the variation in the potency of the behavioral effect and (2) a switch from attractant to deterrent properties. In summary, the most effective transformations of the naringenin molecule were the substitutions of hydrogen atom(s) in hydroxyl group(s) with methyl or pentyl group(s) in combination with the replacement of the carbonyl group with an oxime moiety. The behavioral effects of these transformations were manifested mainly in the stimulation of probing in non-phloem tissues as well as the ingestion of phloem sap.

The results of the present study could be applied towards modifying aphid attraction or deterrence to plants in the field using genetic modification or topical application of naringenin or naringenin-derived analogues [1,34]. This is especially important in the context of virus transmission. Aphids may acquire and inoculate viruses during various stages of plant penetration with sucking–piercing mouthparts. During brief intracellular probes in the epidermis and parenchyma (mesophyll in leaves) that precede feeding in phloem vessels, aphids may transmit nonpersistent and semi-persistent viruses. When aphid stylets reach sieve elements, persistent viruses may be transmitted [35–38]. It is crucial then, to deter aphid probing or at least prevent feeding, to protect plants from pathogen infection and limit the virus spread within the field crops. Besides the direct negative effect on aphid feeding, a deterrent that impedes activities during pre-phloem and phloem stylet penetration should also prevent the transmission of non-persistent and persistent viruses, respectively. Considering the activities of naringenin and its derivatives revealed here, the strong attractants (**8**), (**10**), (**13**), (**15**), and (**16**) have the highest potential for practical applications in 'push–pull' strategies. As the probing and feeding stimulants, they can be applied topically to any species of barrier plant to pull *M. persicae* out of the protected crop. By making barrier plants more attractive to aphids, virus spread within the crop may be reduced [22]. In addition, the weak deterrent (**7**) that makes aphids restless can be applied on the crop plant to push *M. persicae* out of the crop plant stand.

4. Materials and Methods

4.1. Naringenin and Naringenin Derivatives

Naringenin (5,7,4′-trihydroxyflavanone) was purchased from SIGMA (W530098). The naringenin derivatives (**2–17**) were prepared as described previously by Kozłowska et al. [39]. Briefly, mono- and di-*O*-alkyl compounds (**4–8**) were obtained by stirring anhydrous potassium carbonate and significant excess of appropriate alkyl iodide in the solution of naringenin in anhydrous acetone at room temperature for 24–96 h. After solvent evaporation and washing with a saturated brine, the products were extracted with diethyl ether, dried, concentrated and separated by column chromatography. Tri-*O*-alkyl derivatives of naringenin (**2**, **3**) were obtained similarly to the method mentioned above, but dimethylformamide was used instead of acetone; after stirring for 7–24 h at room temperature, the reaction mixtures were neutralized with 1 M HCl and extracted with methylene chloride. The reaction yields were in the range of 20–72%.

Syntheses of oximes (**9–17**) were performed by stirring the *O*-alkyl derivatives of naringenin (1.0 eq.) (**2–8**), hydroxylamine hydrochloride (1.5 eq.), and anhydrous sodium acetate (1.5 eq.) in anhydrous ethanol at 40–50 °C. The reaction mixtures were poured into ice water. The precipitated crystals were collected, dried in a vacuum, and purified by column chromatography. The reaction yields were in the range of 81–99%.

The purity of obtained compounds was monitored using thin layer chromatography, high-performance liquid chromatography and proton nuclear magnetic resonance.

4.2. Aphid and Plant Cultures

Laboratory culture of the peach potato aphid *Myzus persicae* (Sulz.), kept as a multiclonal colony (i.e., deriving from different parthenogenetically reproducing females), was maintained on *Brassica rapa* L. ssp. *pekinensis* L. in the laboratory at 21 °C, 65% r.h., and L16:8D photoperiod. *M. persicae* had originally been collected in the greenhouse and kept on *B. rapa* ssp. *pekinensis* in the laboratory since 2000. The plants for the EPG experiments were *B. rapa* ssp. *pekinensis* and were grown under similar laboratory conditions as aphid cultures, at 21 °C, 65% r.h., and L16:8D photoperiod. Plants were grown in plastic pots (0.33 L) filled with fine garden soil commonly used for greenhouse experiments. Plants were watered regularly, and no additional nutrients were supplied.

4.3. Preparation and Application of Compounds

To mimic the natural environment under laboratory conditions, naringenin and its analogues were offered to aphids by application through their host plants. Preparation and application of the compounds followed the procedure described by Polonsky et al. [40], later modified by Gabryś et al. [26]. Briefly, each compound was dissolved in 70% ethanol to obtain a 0.1% solution [40]. All compounds were applied on the adaxial and abaxial leaf surfaces by immersing one leaf of the experimental plant in the ethanolic solution of a given compound for 30 s. Leaves of similar size of the control plants were immersed in 70% ethanol that was used as a solvent for the studied compounds. There was no effect of ethanol application on aphid probing behavior and plant condition [41]. Treated and control leaves were allowed to dry for 1 h before the start of the experiment to permit the evaporation of the solvent.

4.4. Aphid Probing Behavior

Myzus persicae probing behavior was monitored by using the Electrical Penetration Graph (EPG) technique. The EPG technique provides a unique opportunity to reveal aphid mouthparts stylets activities in plant tissues [41–43]. The parameters describing aphid behavior during probing and feeding, such as total time of probing, duration and frequency of sap ingestion periods, number of probes, etc., are good indicators of plant suitability or interference of probing by chemical or physical factors in individual plant tissues [16–20]. In this experimental setup, aphid and plant are made parts of an electric circuit, which is completed when the aphid inserts its stylets into the plant. Weak voltage is supplied in the circuit, and all changing electric properties are recorded as EPG waveforms that have been correlated with aphid activities and stylet position in plant tissues [42,43].

In the present study, one- to seven day old adult apterous females of *M. persicae* and three week old plants with four to five fully developed leaves were used for all experiments according to the standard procedure applied in similar studies [17–20,24–27]. Aphids were attached to a golden wire electrode with conductive silver paint and starved for 1 h prior to the experiment. Probing behavior of apterous *M. persicae* was monitored for 8 h continuously with a Giga-8 DC EPG with 1 GΩ of input resistance (EPG Systems, Wageningen, The Netherlands) and Stylet+ software (www.epgsystems.eu). Each aphid was given access to a freshly prepared leaf of an unused plant, which means that each plant and each aphid were used only once. One aphid–plant combination was considered a replication. Two rounds of 8 replications ($n = 16$) were carried out for each studied substance and control. Giga-8 DC EPG allows the recording of 8 samples simultaneously. Incomplete EPG recordings, i.e., those that were prematurely ended due to the aphid falling off the plant or other incidents, were discarded from analysis. All experiments were carried out under the same conditions of temperature, relative humidity, and photoperiod, as described for the rearing of plants and aphids. The bioassays started at 10–11 a.m. MEST (Middle European Summer Time).

The following aphid behaviors related to mouthparts positions in or out of the plant tissues were distinguished: non-probing, which represents aphid stylets outside the plant tissues, pathway phase 'C', which represents the movement of aphid stylets within the epidermis and mesophyll; phase 'F', which represents unidentified ('derailed') stylet movements within apoplast; xylem phase 'G', which represents active xylem sap uptake; phloem phase consisting of watery salivation E1 and passive ingestion of phloem sap 'E2'. 'F' and 'G' occurred sporadically irrespective of a treatment, therefore, these activities were analyzed together with phase 'C' and referred to as the 'non-phloem' phase of probing.

4.5. Statistical Analysis

The EPG parameters describing aphid probing behavior were calculated manually and individually for every aphid and the means and standard deviations were subsequently calculated using the EPG analysis Excel worksheet created for this study. Two comparative analyses were carried out. First, aphid behavior on control plants was compared to aphid behavior on naringenin- and naringenin derivatives-treated plants individually for each compound/treatment. This comparison (Mann–Whitney U-test) was performed and the results were interpreted to reveal the mode of action of a given compound (deterrent, attractant, neutral), which allowed grouping of the studied compounds according to their similarity in the effect they had on aphid behavior. A second comparison was carried out to determine the effect of structural modifications in the naringenin molecule on the aphid behavior modifying activity. For this purpose, aphid behavior on only the treated plants was compared. Due to failure to meet the assumptions of analysis of variance, the obtained data were analyzed by the Kruskal–Wallis test and post hoc multiple comparisons of mean ranks for all groups (Dunn's test). The Kruskal–Wallis test is a non-parametric alternative to the one-factor ANOVA test for independent measures and it is commonly used to analyze data deriving from EPG recordings of aphid probing [32]. The mean and SD values given in Tables 1–3 are a representation of non-Gaussian data, but the statistical analysis was done by non-parametric tests, in which all individual data were included. All statistical calculations were performed using StatSoft, Inc. (2014) STATISTICA (data analysis software system, version 12, www.statsoft.com).

Author Contributions: Conceptualization, B.G. and M.A.; Methodology, B.G., M.A. and B.K.; Formal Analysis, A.W.-K.; Investigation, K.S.; A.W.-K.; B.K.; M.A. and J.K.; Data Curation, B.G. and M.A.; Writing-Original Draft Preparation, B.G. and M.A.; Writing-Review & Editing, B.G.; Visualization, K.S. and M.A.; Supervision, B.K.; B.G. and M.A.; Funding Acquisition, M.A. and B.K. All authors have read and agreed to the published version of the manuscript.

Funding: Publication was supported by the National Science Centre, Grant No. 2016/21/B/NZ9/01904. The project was financially supported by Minister of Science and Higher Education in the range of the program entitled "Regional Initiative of Excellence" for the years 2019-2022, Project No. 010/RID/2018/19, amount of funding 12.000.000 PLN.

Conflicts of Interest: The authors declare no conflict of interest. The funders had no role in the design of the study; in the collection, analyses, or interpretation of data; in the writing of the manuscript, and in the decision to publish the results.

References

1. Venkateswara, R.P.; SDVS, K.; Rohini, P.; Bhagyasree, P. Flavonoid: A review on naringenin. *J. Pharmacogn. Phytochem.* **2017**, *6*, 2778–2783.
2. Khan, M.K.; Zill-E-Huma, O.; Dangles, O. A comprehensive review on flavanones, the major citrus polyphenols. *J. Food Compos.* **2014**, *33*, 85–104. [CrossRef]
3. Patton, C.A.; Ranney, T.G.; Burton, J.D.; Walgenbach, J.F. Feeding responses of japanese beetle to naturally occurring metabolites found in rosaceous plants. *J. Environ. Hort.* **1997**, *15*, 222–227.
4. Su, Q.; Zhou, Z.; Zhang, J.; Shi, C.; Zhang, G.; Jin, Z.; Wang, W.; Li, C. Effect of plant secondary metabolites on common cutworm, *Spodoptera litura* (Lepidoptera: Noctuidae). *Entomol. Res.* **2017**, *48*, 18–26. [CrossRef]
5. Riddick, E.W.; Wu, Z.; Eller, F.J.; Berhow, M.A. Do bioflavonoids in *Juniperus virginiana* heartwood stimulate oviposition in the ladybird *Coleomegilla maculata*? *Int. J. Insect Sci.* **2018**, *10*, 1–13. [CrossRef] [PubMed]
6. Hernandez, I.G.; Palottini, F.; Macri, I.; Galmarini, C.R.; Farina, W.M. Appetitive behavior of the honeybee *Apis mellifera* in response to phenolic compounds naturally found in nectars. *J. Exp. Biol.* **2019**, *222*, 189910. [CrossRef]
7. Wellings, P.W.; Ward, S.A.; Dixon, A.F.G.; Rabbinge, R. Crop Loss Assessment. In *Aphids, Their Biology, Natural Enemies and Control*; Minks, A.K., Harrewijn, P., Eds.; Elsevier: Amsterdam, The Netherlands, 1989; Volume C, pp. 49–64.
8. Blackman, R.L.; Eastop, V.F. *Aphids on the World's Crops: An Identification Guide*; John Wiley and Sons: Chichester, UK; New York, NY, USA; Brisbane, Australia; Toronto, Japan; Singapore, 1985.
9. Goławska, S.; Sprawka, I.; Łukasik, I.; Goławski, A. Are naringenin and quercetin useful chemicals in pest-management strategies? *J. Pest. Sci.* **2014**, *87*, 173–180. [CrossRef]
10. Dedryver, C.A.; Le Ralec, A.; Fabre, F. The conflicting relationships between aphids and men: A review of aphid damage and control strategies. *C. R. Biol.* **2010**, *333*, 539–553. [CrossRef]
11. Bass, C.; Puinean, A.M.; Zimmer, C.T.; Denholm, I.; Field, L.M.; Foster, S.P.; Gutbrod, O.; Nauen, R.; Slater, R.; Williamson, M.S. The evolution of insecticide resistance in the peach potato aphid, *Myzus persicae*. *Insect Biochem. Mol. Biol.* **2014**, *51*, 41–51. [CrossRef]
12. Blake, R. Eu Neonicotinoid Ban Removes Vital Tools in Global Fight Against Pests. *Outlooks Pest Manag.* **2018**, *29*, 197–200. [CrossRef]
13. Pickett, J.A. Lower terpenoids as natural insect control agents. In *Ecological Chemistry and Biochemistry of Plant Terpenoids*; Harborne, J.B., Tomas-Barberan, F.A., Eds.; Clarendon Press: Oxford, UK, 1991; pp. 297–313.
14. Dayan, F.E.; Cantrell, C.l.; Duke, S.O. Natural products in crop protection. *Bioorg. Med. Chem.* **2009**, *17*, 4022–4034. [CrossRef] [PubMed]
15. Lattanzio, V. Phenolic compounds: Introduction. In *Natural Products*; Ramawat, K.G., Me'rillon, J.M., Eds.; Springer: Berlin/Heidelberg, Germany, 2013; pp. 1543–1580.
16. Powell, G.; Hardie, J.; Pickett, J. Effects of the antifeedant polygodial on plant penetration by aphids, assessed by video and electrical recording. *Entomol. Exp. Appl.* **1993**, *68*, 193–200. [CrossRef]
17. Wróblewska-Kurdyk, A.; Gniłka, R.; Dancewicz, K.; Grudniewska, A.; Wawrzeńczyk, C.; Gabryś, B. *β*-thujone and its derivatives modify the probing behavior of the peach potato aphid. *Molecules* **2019**, *24*, 1847. [CrossRef]
18. Wróblewska-Kurdyk, A.; Dancewicz, K.; Gliszczyńska, A.; Gabryś, B. New insight into the behaviour modifying activity of two natural sesquiterpenoids farnesol and nerolidol towards *Myzus persicae* (Sulzer) (Homoptera: Aphididae). *Bull. Entomol. Res.* **2020**, *110*, 249–258. [CrossRef]
19. Gabryś, B.; Dancewicz, K.; Gliszczyńska, A.; Kordan, B.; Wawrzeńczyk, C. Systemic deterrence of aphid probing and feeding by β-damascone analogues. *J. Pest. Sci.* **2015**, *88*, 507–516. [CrossRef] [PubMed]

20. Stompor, M.; Dancewicz, K.; Gabryś, B.; Anioł, M. Insect antifeedant potential of xanthohumol, isoxanthohumol, and their derivatives. *J. Agric. Food Chem.* **2015**, *63*, 6749–6756. [CrossRef]
21. Shelton, A.M.; Badenes-Perez, F.R. Concepts and applications of trap cropping in pest management. *Annu. Rev. Entomol.* **2006**, *51*, 285–308. [CrossRef]
22. Fereres, A. Barrier crops as a cultural control measure of non-persistently transmitted aphid-borne viruses. *Virus Res.* **2000**, *71*, 221–231. [CrossRef]
23. Cook, S.M.; Khan, Z.R.; Pickett, J.A. The use of push-pull strategies in integrated pest management. *Annu. Rev. Entomol.* **2007**, *52*, 375–400. [CrossRef]
24. Grudniewska, A.; Dancewicz, K.; Białońska, A.; Ciunik, Z.; Gabryś, B.; Wawrzeńczyk, C. Synthesis of piperitone-derived halogenated lactones and their effect on aphid probing, feeding, and settling behavior. *RSC Adv.* **2011**, *1*, 498–510. [CrossRef]
25. Paprocka, M.; Gliszczyńska, A.; Dancewicz, K.; Gabryś, B. Novel hydroxy- and epoxy-cis-jasmone and dihydrojasmone derivatives affect the foraging of the peach potato aphid *Myzus persicae* (Sulzer) (Homoptera: Aphididae). *Molecules* **2018**, *23*, 2362. [CrossRef] [PubMed]
26. Dancewicz, K.; Gabryś, B.; Dams, I.; Wawrzeńczyk, C. Enantiospecific effect of pulegone and pulegone–derived lactones on settling and feeding of *Myzus persicae* (Sulz.). *J. Chem. Ecol.* **2008**, *34*, 530–538. [CrossRef] [PubMed]
27. Dancewicz, K.; Sznajder, K.; Załuski, D., Kordan; Gabryś, B. Behavioral sensitivity of *Myzus persicae* to volatile isporenoids in plant tissues. *Entomol. Exp. Appl.* **2016**, *160*, 229–240. [CrossRef]
28. Gabryś, B.; Pawluk, M. Acceptability of different species of Brassicaceae as hosts for the cabbage aphid. *Entomol. Exp. Appl.* **1999**, *91*, 105–109. [CrossRef]
29. Grudniewska, A.; Dancewicz, K.; Białońska, A.; Wawrzeńczyk, C.; Gabryś, B. Piperitone-derived saturated lactones: Synthesis and aphid behavior-modifying activity. *J. Agric. Food Chem.* **2013**, *61*, 3364–3372. [CrossRef]
30. Tjallingii, W.F. Regulation of phloem sap feeding by aphids. In *Regulatory Mechanisms in Insect Feeding*; Chapman, R.F., de Boer, G., Eds.; Chapman & Hall: New York, NY, USA, 1995; pp. 190–209.
31. Mayoral, A.M.; Tjallingii, W.F.; Castanera, P. Probing behavior of Diuraphis noxia on five cereal species with different hydroxyamic acid levels. *Entomol. Exp. Appl.* **1996**, *78*, 341–348. [CrossRef]
32. Philippi, J.; Schliephake, E.; Jurgens, H.U.; Jansen, G.; Ordon, F. Feeding behavior of aphids on narrow-leafed lupin (*Lupinus angustifolius*) genotypes varying in the content of quinolizidine alkaloids. *Entomol. Exp. Appl.* **2005**, *156*, 37–51. [CrossRef]
33. Schwarzkopf, A.; Rosenberger, D.; Niebergall, M.; Gershenzon, J.; Kunert, G. To feed or not to feed: Plant factors located in the epidermis, mesophyll, and sieve elements influence pea aphid's ability to feed on legume species. *PLoS ONE* **2013**, *8*, e75298. [CrossRef]
34. Ververidis, F.; Trantas, E.; Douglas, C.; Vollmer, G.; Kretzschmar, G.; Panopoulos, N. Biotechnology of flavonoids and other phenylpropanoid-derived natural products. Part I: Chemical diversity, impacts on plant biology and human health. *Biotechnol. J.* **2007**, *2*, 1214–1234. [CrossRef]
35. Prado, E.; Tjallingii, W.F. Aphid activities during sieve element punctures. *Entomol. Exp. Appl.* **1994**, *72*, 157–165. [CrossRef]
36. Powell, G.; Pirone, T.; Hardie, J. Aphid stylet activities during potyvirus acquisition from plants and in vitro system that correlate with subsequent transmission. *Eur. J. Plant Pathol.* **1995**, *101*, 411–420. [CrossRef]
37. Martin, B.; Collar, J.L.; Tjallingii, W.F.; Fereres, A. Intracellular ingestion and salivation by aphids may cause the acquisition and inoculation of non-persistently transmitted plant viruses. *J. Gen. Virol.* **1997**, *78*, 2701–2705. [CrossRef] [PubMed]
38. Brault, V.; Uzest, M.; Monsion, B.; Jacquot, E.; Blanc, S. Aphids as transport devices for plant viruses. *C. R. Biol.* **2010**, *333*, 524–538. [CrossRef]
39. Kozłowska, J.; Potaniec, B.; Żarowska, B.; Anioł, M. Synthesis and biological activity of novel *O*-alkyl derivatives of naringenin and their oximes. *Molecules* **2017**, *22*, 1485. [CrossRef]
40. Polonsky, J.; Bhatnagar, S.C.; Griffiths, D.C.; Pickett, J.A.; Woodcock, C.M. Activity of qassinoids as antifeedants against aphids. *J. Chem. Ecol.* **1989**, *15*, 933–998. [CrossRef] [PubMed]
41. Halarewicz-Pacan, A.; Gabryś, B.; Dancewicz, K.; Wawrzeńczyk, C. Enantiospecific effect of limonene and limonene-derived bicyclic lactones on settling and probing behaviour of the peach-potato aphid *Myzus persicae* (Sulz.). *J. Plant Prot. Res.* **2003**, *43*, 133–142.

42. Pettersson, J.; Tjallingii, W.F.; Hardie, J. Host-plant selection and feeding. In *Aphids as Crop Pests*; van Emden, H.F., Harrington, R., Eds.; CABI: Wallingford, UK, 2017; pp. 173–195.
43. Tjallingii, W.F.; Hogen Esch, T.H. Fine-structure of aphid stylet routes in plant-tissues in correlation with EPG signals. *Physiol. Entomol.* **1993**, *18*, 317–328. [CrossRef]

Sample Availability: Not available.

molecules

MDPI

Article

Spinasterol, 22,23-Dihydrospinasterol and Fernenol from *Citrullus Colocynthis* L. with Aphicidal Activity against Cabbage Aphid *Brevicoryne Brassicae* L.

Maqsood Ahmed [1,2], Peiwen Qin [1,*], Mingshan Ji [1,*], Ran An [1], Hongxia Guo [1] and Jamil Shafi [3]

1 College of Plant Protection, Shenyang Agricultural University, Shenyang 110866, China; maqsoodahmed200@hotmail.com (M.A.); anan8u8@outlook.com (R.A.); ghx331682817@163.com (H.G.)

2 Department of Agriculture, Pest Warning & Quality Control of Pesticides, Gujrat 50700, Pakistan

3 Department of Plant Pathology, University of Agriculture Faisalabad, Sub-Campus Depalpur, Okara, Faisalabad 56300, Pakistan; jamil_shafi786@yahoo.com

* Correspondence: qinpeiwen@syau.edu.cn (P.Q.); jimingshan@syau.edu.cn (M.J.); Tel.: +86-24-8848-7148 (M.J.); Fax: +86-24-8834-2315 (M.J.)

Academic Editor: Giovanni Benelli
Received: 7 April 2020; Accepted: 29 April 2020; Published: 7 May 2020

Abstract: *Brevicoryne brassicae* is a problematic pest in cabbage and other field crops. Synthetic pesticides are used to control this pest, but they are injurious for human health and the environment. The present study aimed to purify and identify the active compounds from *Citrullus colocynthis* leaves with an appraisal of their efficacy against *B. brassicae*. Separation and purification were performed via different chromatographic techniques. Molecular analysis and chemical structures were recognized by mass spectrum (MS) and nuclear magnetic resonance (NMR), respectively. Moreover, in vitro and in vivo aphicidal activity was assessed using various concentrations, i.e., 6.25, 12.5, 25 and 50 μg/mL at 12, 24, 48 and 72 h exposure. The outcome shows that mass spectrum analyses of the purified compounds suggested the molecular formulae are $C_{30}H_{50}O$ and $C_{29}H_{50}O$, $C_{29}H_{48}O$. The compounds were characterized as fernenol and a mixture of spinasterol, 22,23-dihydrospinasterol by ^{1}H-NMR and ^{13}C-NMR spectrum analysis. The toxicity results showed that the mixture of spinasterol and 22,23-dihydrospinasterol showed LC_{50} values of 32.36, 44.49 and 37.50 μg/mL by contact, residual and greenhouse assay at 72 h exposure, respectively. In contrast, fernenol recorded LC_{50} values as 47.99, 57.46 and 58.67 μg/mL, respectively. On the other hand, spinasterol, 22,23-dihydrospinasterol showed the highest mortality, i.e., 66.67%, 53.33% and 60% while, 30%, 23.33% and 25% mortality was recorded by fernenol after 72 h at 50 μg/mL by contact, residual and greenhouse assay, respectively. This study suggests that spinasterol, 22,23-dihydrospinasterol are more effective against *B. brassicae* which may be introduced as an effective and suitable substitute of synthetic chemical pesticides.

Keywords: *Citrullus colocynthis*; spinasterol, 22,23-dihydrospinasterol; fernenol; insecticidal activity; LC_{50}

1. Introduction

The cabbage aphid, *Brevicoryne brassicae* L. belongs to (Hemiptera: Aphididae) is one of the key pest of vegetables and crops, commonly distributed in warm and moderate areas of the world [1,2]. Due to the heavy infestation of *B. brassicae* on cabbage, yield losses can be increased up to 70% and under-promising circumstances, it can cause complete loss of leafy vegetables of brassica [3,4]. Being an injurious pest, it negatively affects cabbage production by developing sooty mold on plants surface [2,5]. Continued feeding of this pest caused yellowing, wilting and stunting of plant growth, ultimately cause plant death and lead to economic losses [6]. Moreover, most of the species of aphid have gained resistance against many synthetic aphicidal agents [7].

Although synthetic chemicals are widely used to manage this pest but, intensive and continuous use of these chemicals has resulted in the development of pest resistance, resurgence to these chemicals and may also leave hazardous effects on humans and the environment [8]. In contrast, botanical insecticides are relatively safe to some extent and are effective substitutes of these chemicals which play a fundamental role in the field of biopesticides. However, plant derivatives play a vital role in their biologic tasks [9]. Numerous findings have specified that phenols, terpenoids and nitrogen-based constituents are imperative phytoalexins which afford a protective system to plants being attacked by other injurious insect-pests [9,10].

Citrullus colocynthis is an important plant from the pharmacological and pesticidal viewpoint. It belongs to the Cucurbitaceae family, mostly grows in desert areas and has attracted the consideration of researchers as a natural botanic pesticide. The insecticidal potential of this desert plant has been assessed against various insect species [11]. *Citrullus colocynthis* exerts carcinogenic, antidiabetic, antibacterial, antioxidant properties, and possesses toxic properties against several harmful insects [12–15]. Numerous active chemical compounds have been reported from *C. colocynthis*, including bitter materials (colocynthetin and colocynthin), various cucurbitacins such as A, B, C, D and E [16], other cucurbitacins like E, I, J, K and L [17], cucurbitacins glycosides [18,19], the cucurbitacins glucosides I and L [19], flavone glycosides and flavonoids [19,20]. However, the cucurbitacins tetracyclic triterpens possess extensive biologic activities. Numerous biologic compounds belong to this group being examined because of their cytotoxic, hepatoprotective, cardiovascular, antioxidant action of B and I cucurbitacins [21].

Song et al. [22] isolated two cucurbitacins from the ethyl acetate extract of *C. colocynthis* fruit, but these were not evaluated against insects. Similarly, Ding et al. [23] identified a mixture of spinasterol, 22,23-dihydrospinasterol from the roots of *Bermeuxia thibetica*, but its bioactivity was also not evaluated. However, Sinha et al. [24] reported that spinasterol, 22,23-dihydrospinasterol exhibited by *Melothria maderaspatana* showed biologic activities. It has been reported that Artemisia extracts contain valuable phytochemicals, possess insecticidal activity which were mainly attributed to the presence of fernenol and other phytoconstituents [25]. Furthermore, constituents of *Artemisia vulgaris* like psilostachy C, Maackiain, psilostachyin A and fernenol possess medicinal as well as antibacterial activities and also used by farmers for the preservation of crops and stored grains products [26].

Although some studies have been conducted on isolation and identification of various compounds including spinasterol, 22,23-dihydrospinasterol and fernenol from natural plant resources, their use as insecticidal purposes especially against *B. brassicae* has not been evaluated so far. Therefore, the present novel study was conducted in a comprehensive way for the isolation, characterization and evaluation of biochemical compounds from *C. colocynthis* leaves against *B. brassicae*.

2. Results

2.1. Preliminary Toxicity Evaluations

Interestingly, all solvents extract caused significant mortality of *B. brassicae*. However, maximum mortality (85%) was recorded at 96 h post exposure by methanol extract, followed by ethanol and chloroform extract (80%) and (76.67%), respectively. In contrast, at 72 h exposure, maximum mortality afforded by methanol extract was 66.67%, followed by chloroform and ethanol extract (58.33%) and (53.33%), respectively. However, minimum mortality was recorded via distilled water extract which was 35% and 18.33% at 96 and 72 h exposure, respectively. The results also revealed that mortality is dependent upon concentration and prolonged time exposure (Figure 1).

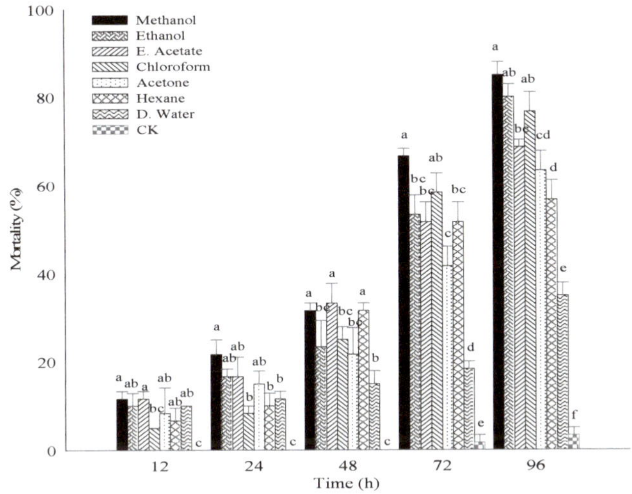

Figure 1. Mortality of *Brevicoryne brassicae* by *Citrullus colocynthis* leaves extracts. Values are represented as mean ± standard deviation followed by different superscripts (a, b, c, d, e, f, ab, bc, cd) were not significantly different according to Duncan multiple range test (DMRT) at $p > 0.05$. E. Acetate (ethyl acetate); D. Water (distilled water); CK (check in distilled water); Time (h).

2.2. Extraction, Separation and Purification of Extract

Powdered leaves of *C. colocynthis* were extracted with different polarity based solvents. Fractionation of the extract was performed by chromatographic techniques. Collected fractions followed by separation by silica gel column were concentrated and weighed following their Thin Layer Chromatography (TLC) analysis (Table 1). Out of 8 of the obtained fractions, D_2 fraction was further purified by repeated column chromatography on silica gel column, Sephadex gel column and preparative TLC and finally two compounds referred to as D2 (N) and D3A bearing weight 35 mg and 21 mg, respectively were purified.

Table 1. Obtained fractions by silica gel column.

Eluted Fractions	Collected Fractions	Weight (g)
1–10	A	0.293
11–13	B	0.355
14–23	C	0.796
24–27	D	0.662
28–30	E	0.710
31–34	F	0.270
35–40	G	0.447
41–43	H	0.428
44–49	I	2.841

2.3. Mass Spectrum and Elemental Analysis

The C, H and N elemental analysis is helpful for the calculation of empirical formula. The mass spectrometer examination recorded the molecular ion peak (M) at *m/z* 426, 414 and 412. These data suggested the expected molecular formula as $C_{30}H_{50}O$ (Figure 2) and a mixture of $C_{29}H_{50}O$, $C_{29}H_{48}O$ (Figure 3) for D3A and D2 (N), respectively. However, D2 (N) was a mixture of two compounds referred to as D2 (N1) and D2 (N2) and the ratio of these two compounds was recorded as 5:4 from the mass spectrum and nuclear magnetic resonance analysis.

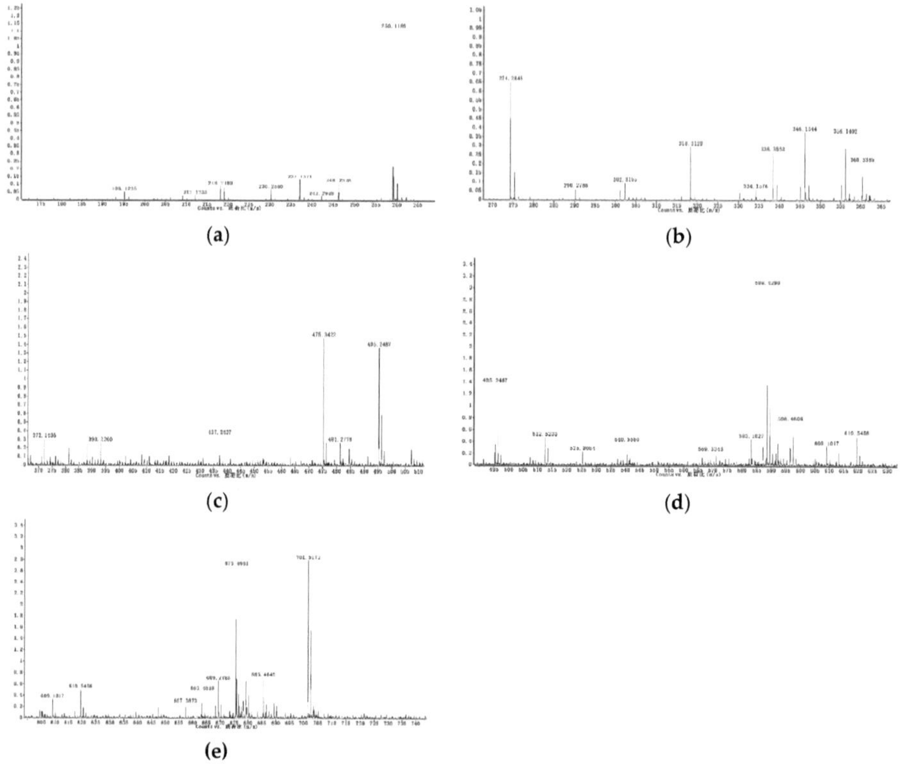

Figure 2. (**a**–**e**) The full mass spectrum for the purified compound, D3A from *Citrullus colocynthis* leaves.

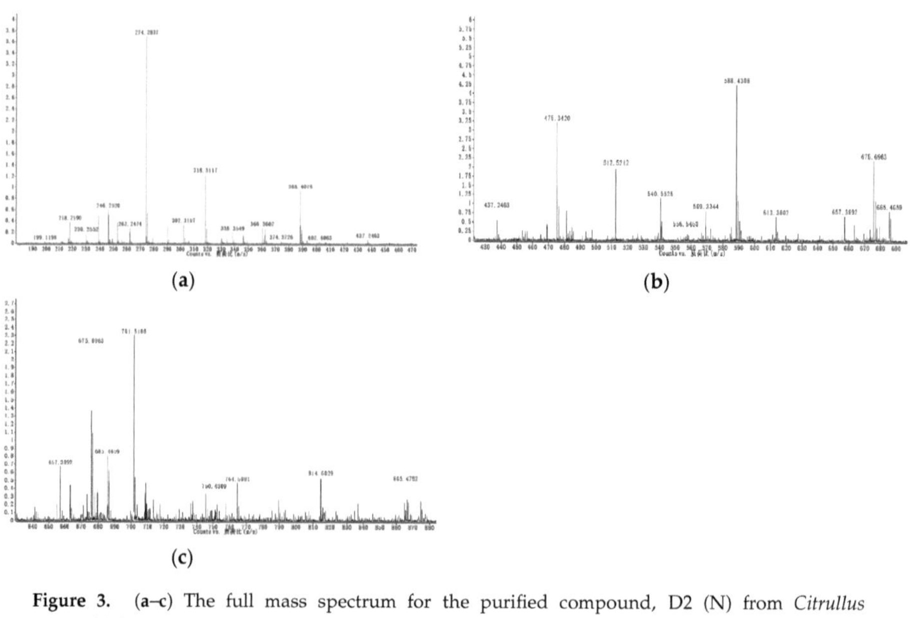

Figure 3. (**a**–**c**) The full mass spectrum for the purified compound, D2 (N) from *Citrullus colocynthis* leaves.

2.4. NMR Analysis (^{1}H-NMR and ^{13}C-NMR)

The chemical structures of the purified compounds D2 (N) and D3A were characterized by nuclear magnetic resonance (NMR), Bruker BioSpin, Billerica MA for 1D (^{1}H-NMR, ^{13}C-NMR), mass spectrum and elemental analysis. All NMR experiments were carried out at room temperature.

In the ^{1}H-NMR and ^{13}C-NMR spectrum of the compound D_2 (N 1) ^{1}H-NMR (600 MHz, $CDCl_3$) δ 5.16 (m, 1H, H-7), 3.60 (m, 1H, H-3) allylic protons, 0.93 (d, J = 7.0 Hz, 3H, -CH_3), 0.85 (t, J = 6.8 Hz, 3H, -CH_3), 0.82 (d, J = 6.9 Hz, 3H, -CH_3), 0.81 (d, J = 6.9 Hz, 3H, -CH_3), 0.80 (s, 3H, -CH_3), 0.55 (s, 3H, -CH_3) (Figure 4a). ^{13}C-NMR (151 MHz, $CDCl_3$) δ 11.8, 12.0, 13.0, 18.8, 19.0, 19.7, 21.5, 22.9, 23.0, 26.1, 27.9, 29.1, 29.6, 31.4, 33.8, 34.1, 36.5, 37.1,37.9, 39.5, 40.2, 43.3, 45.8, 49.4, 55.0, 56.0, 71.0, 117.3, 139.5, (Spinasterol, $C_{29}H_{50}O$; 414.0 g/mol) (Figure 4b). However, ^{1}H-NMR and ^{13}C-NMR spectra of the compound D_2 (N 2) ^{1}H-NMR (600 MHz, $CDCl_3$) δ 5.16 (overlap, 1H H-7), 5.16 (overlap, 1H, H-22), 5.03 (dd, J = 15.1, 8.8 Hz, 1H, H-23), 3.60 (m, 1H, H-3),) the allylic protons, 1.03 (d, J = 7.0 Hz, 3H, -CH_3), 0.85 (t, J = 6.8 Hz, 3H, -CH_3), 0.82 (d, J = 6.9 Hz, 3H, -CH_3), 0.81 (d, J = 6.9 Hz, 3H, -CH_3), 0.80 (s, 3H, -CH_3), 0.54 (s, 3H, -CH_3) (Figure 4). ^{13}C-NMR (151 MHz, $CDCl_3$) δ 11.9, 12.2, 13.0, 18.9, 21.0, 21.3, 21.5, 22.9, 25.3, 28.4, 29.6, 31.4, 31.8, 34.1, 37.1, 37.9, 39.4, 40.2, 40.8, 43.2, 49.4, 51.2, 55.0, 55.8, 71.0, 117.4, 129.4, 138.1, 139.5, (22,23-dihydrospinasterol, $C_{29}H_{48}O$; 412.0 g/mol) (Figure 4b).

In the ^{1}H-NMR and ^{13}C-NMR spectra of the compound D_3A, ^{1}H-NMR (600MHz, $CDCL_3$) δ 5.30 (s, 1H, C=CH), 3.26–3.17 (m, 1H, -OH), the allylic protons, 1.07 (s, 3H, -CH_3), 0.96 (s, 3H, -CH_3), 0.89 (d, J = 6.4 Hz, 3H, -CH_3), 0.87 (s, 3H, -CH_3), 0.84–0.80 (m, 6H, two -CH_3), 0.76 (s, 3H, -CH_3), 0.73 (s, 3H, -CH_3), 0.73–2.03 (m, 48H, including 8 CH_3) (Figure 4c). ^{13}C-NMR (151 MHz, $CDCL_3$) δ: 13.90, 15.77, 17.87, 19.07, 20.05, 22.04, 22.98, 25.13, 27.37, 28.13, 28.06, 29.20, 29.63, 30.69, 36.07, 36.65, 37.57, 37.70, 39.19, 39.27, 39.90, 40.96, 42.85, 44.20, 50.29, 51.88, 59.58, 79.07, 116.11, 150.98, (Fernenol; $C_{30}H_{50}O$; 426.7 g/mol) (Figure 4d).

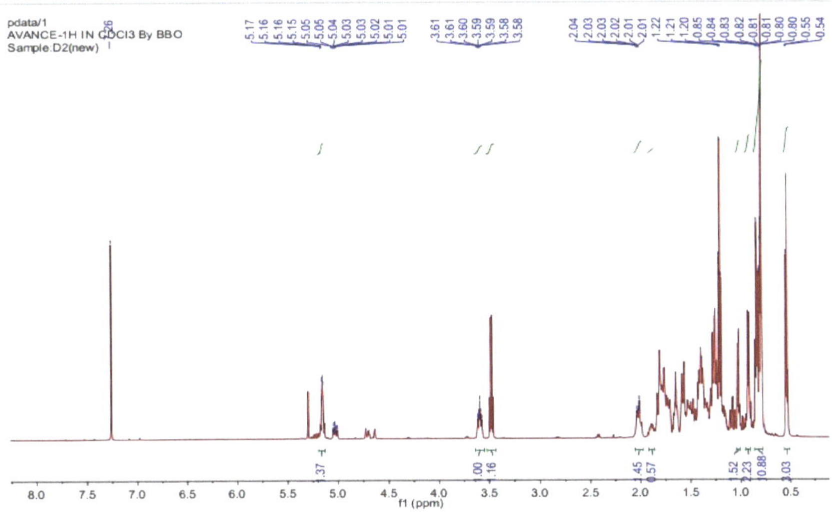

a

Figure 4. *Cont.*

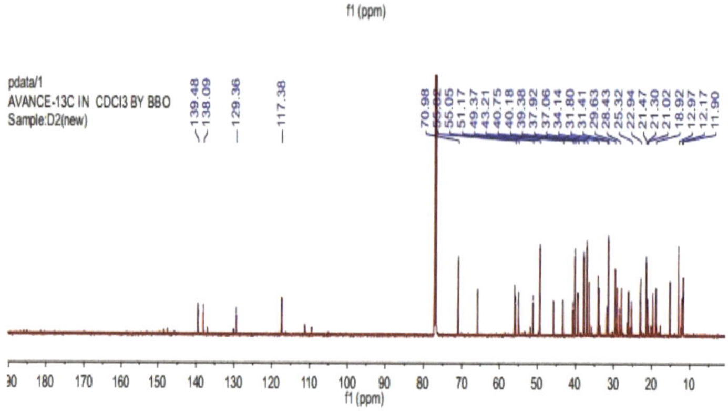

b

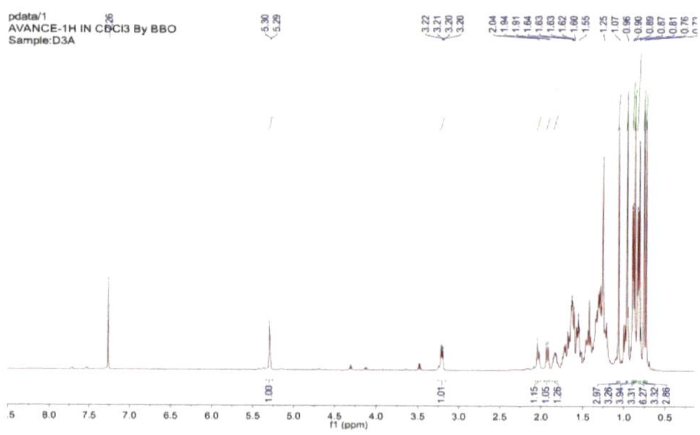

c

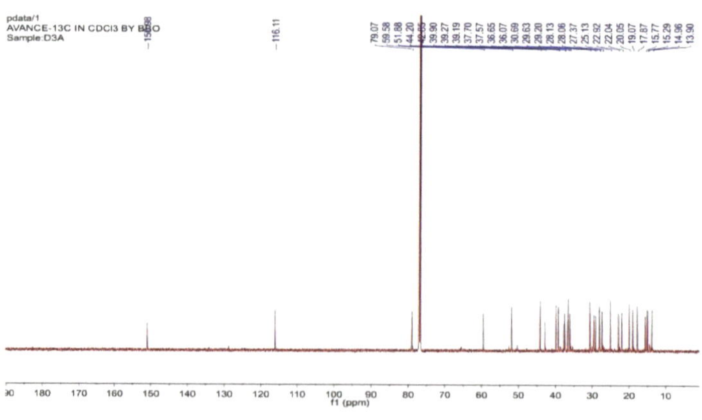

d

Figure 4. (**a**). ^{1}H-NMR spectrum of the compound D2 (N); (**b**). ^{13}C-NMR spectrum of compound D2 (N); (**c**). ^{1}H-NMR spectrum of the compound D3A; (**d**). ^{13}C-NMR spectrum of the compound D3A.

2.5. Structures Elucidation of the Compounds

Chemical structures of the identified fractions of D3A and D2 (N) were characterized by ^{1}H-NMR and ^{13}C-NMR spectroscopy and their molecular formulae were predicted by the mass spectrum. D3A compound was identified as fernenol (Figure 5). Another purified compound was a little bit impure and due to this impurity, the ^{1}H-NMR spectrum was overlapped and hence, mixture of two compounds was identified as Spinasterol and 22, 23-dihydrospinasterol from the fraction D2 (N) with ratio 5:4 in the mixture (Figure 6a,b).

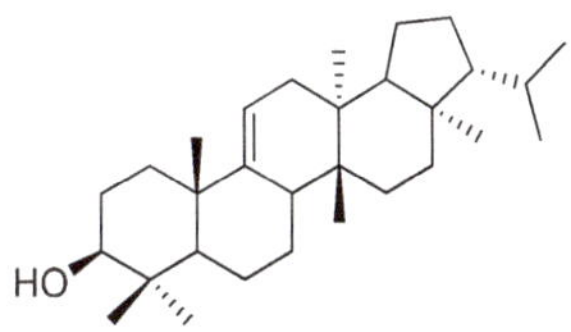

Figure 5. Chemical structure of compound; Fernenol from D3A.

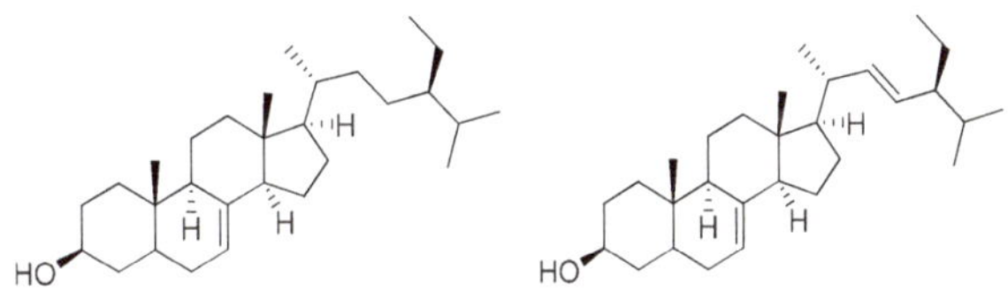

(a) (b)

Figure 6. Chemical structures of compound; (**a**). Spinasterol; (**b**) 22, 23-dihydrospinasterol from D2 (N).

2.6. Bioassay Study

Toxicity of the identified compounds was evaluated extensively against *B. brassicae* by contact, residual and greenhouse assay and data were collected after a specific period of time exposure.

2.6.1. In Vitro Contact Toxicity

The mortality data described in Table 2 exposed the contact efficacy of the spinasterol, 22,23-dihydrospinasterol and fernenol against *B. brassicae*. It was observed that the percent mortality rate of aphids was directly associated with the concentration and period of exposure. Results indicated that extreme mortality recorded was 66.67% and 56.66% at 72 and 48 h exposure, respectively by spinasterol, 22,23-dihydrospinasterol at 50 μg/mL. In contrast, moderate mortality was recorded by fernenol which was 30.00% and 21.67% at 72 and 48 h exposure, respectively, at the same concentration.

2.6.2. In Vitro Residual Toxicity

The mortality data described in Table 3 exposed the residual efficacy of spinasterol, 22,23-dihydrospinasterol and fernenol against *B. brassicae*. Results demonstrated that highest mortality (53.33%) was observed at exposure of 72 h followed by (43.33%) at 48 h by spinasterol, 22,23-dihydrospinasterol at the concentration of 50 μg/mL. On the other hand, fernenol afforded lower mortality, i.e., (23.33%) and (16.67%) at the same exposure time and concentration, respectively.

2.6.3. In Vivo Assay

The mortality data presented in Table 4 revealed the insecticidal activity of the purified compounds against *B. brassicae* under greenhouse conditions. Results showed that mortality rate of aphids under greenhouse conditions was 60.00% at 72 h exposure via spinasterol, 22,23-dihydrospinasterol at 50 μg/mL. In contrast, mortality recorded by fernenol was 25.00% at 50 μg/mL concentration and 72 h exposure.

Table 2. Mortality of *Brevicoryne brassicae* by spinasterol, 22,23-dihydrospinasterol and fernenol via contact assay.

Conc. (µg/mL)	Mean Mortality (%) with Time (h)							
	12		24		48		72	
	Spinasterol, 22,23-Dihydr-Spinasterol	Fernenol	Spinasterol, 22,23-Dihydr-Spinasterol	Fernenol	Spinasterol, 22,23-Dihydr-Spinasterol	Fernenol	Spinasterol, 22,23-Dihydr-Spinasterol	Fernenol
6.25	0.00 ± 0.00^{c}	0.00 ± 0.00^{a}	0.00 ± 0.00^{c}	0.00 ± 0.00^{c}	3.33 ± 3.33^{c}	1.67 ± 1.67^{c}	6.67 ± 3.33^{c}	3.33 ± 1.67^{c}
12.50	3.33 ± 3.33^{c}	0.00 ± 0.00^{a}	10.00 ± 0.00^{c}	5.00 ± 2.87^{bc}	13.3 ± 3.33^{c}	6.67 ± 1.67^{bc}	20.00 ± 5.77^{c}	10.0 ± 2.89^{bc}
25	16.67 ± 3.33^{b}	3.33 ± 1.67^{a}	23.33 ± 3.33^{b}	8.33 ± 1.67^{ab}	26.67 ± 3.33^{b}	11.67 ± 1.67^{b}	43.33 ± 6.67^{b}	15.0 ± 2.89^{b}
50	26.67 ± 3.33^{a}	3.3 ± 1.67^{a}	40.00 ± 0.00^{a}	13.3 ± 1.67^{a}	56.67 ± 3.33^{a}	21.67 ± 1.67^{a}	66.67 ± 3.33^{a}	30.002.89^{a}
CK	0.00 ± 0.00^{c}	0.00 ± 0.00^{a}	0.00 ± 0.00^{c}	0.00 ± 0.00^{c}	3.33 ± 3.33^{c}	1.67 ± 1.67^{c}	3.33 ± 3.33^{ed}	3.33 ± 1.67^{c}
	Statics Summary							
S.S	1693	40.00	3506	390.0	5960	840.0	8573	1460
df	4	4	4	4	4	4	4	4
M.S	423	10.00	876.6	97.50	1490	210.0	2143	365.0
F	21.17***	3.00^{NS}	131.50***	11.70***	44.70***	25.20***	32.150***	19.91***

Values are described as mean ± standard error followed by different superscripts were significantly different according to (Duncan multiple range test DMRT $p > 0.05$). S.S (Sum of square); df (Degree of freedom); M.S (Mean square); F (Significance); CK (Check); NS (Non-significant); *** (Highly significant).

Table 3. Mortality of *Brevicoryne brassicae* by spinasterol, 22,23-dihydrospinasterol and fernenol via residual assay.

Conc. (µg/mL)	Mean Mortality (%) with Time (h)							
	12		24		48		72	
	Spinasterol, 22,23-Dihydro-Spinasterol	Fernenol	Spinasterol, 22,23-Dihydro-Spinasterol	Fernenol	Spinasterol, 22,23-Dihydro-Spinasterol	Fernenol	Spinasterol, 22,23-Dihydro-Spinasterol	Fernenol
6.25	0.00 ± 0.00^{b}	0.00 ± 0.00^{a}	0.00 ± 0.00^{c}	0.00 ± 0.00^{b}	3.33 ± 3.33^{c}	0.00 ± 0.00^{c}	6.67 ± 3.33^{c}	1.67 ± 1.67^{c}
12.50	0.00 ± 0.00^{b}	0.00 ± 0.00^{a}	3.33 ± 3.33^{c}	1.67 ± 1.67^{b}	6.67 ± 3.33^{c}	3.33 ± 1.67^{bc}	13.33 ± 3.33^{c}	5.00 ± 0.00^{bc}
25	6.67 ± 3.33^{b}	1.67 ± 1.67^{a}	13.33 ± 3.33^{b}	5.00 ± 2.89^{b}	26.67 ± 3.33^{b}	6.67 ± 1.67^{b}	36.676 ± 3.33^{b}	10.00 ± 2.89^{b}
50	13.33 ± 3.33^{a}	1.67 ± 1.67^{a}	23.33 ± 3.33^{a}	11.67 ± 1.67^{a}	43.33 ± 3.33^{a}	16.67 ± 1.67^{a}	53.33 ± 3.33^{a}	23.33 ± 1.67^{a}
CK	0.00 ± 0.00^{b}	0.00 ± 0.00^{a}	0.00 ± 0.00^{c}	0.00 ± 0.00^{b}	0.00 ± 0.00^{c}	3.33 ± 1.67^{bc}	3.33 ± 3.33^{c}	3.33 ± 3.33^{c}
	Statics summary							
S.S	426.6	10.00	1240	290.0	4093	493.3	5560.0	923.3
df	4	4	4	4	4	4	4	4
M.S	106.67	2.50	310.0	72.50	1023	123.3	1390	230.8
F	8.00**	0.750*	15.00***	8.70***	38.37***	18.50***	41.70***	23.08***

Values are described as mean ± standard error followed by different superscripts were significantly different according to (Duncan multiple range test DMRT $p > 0.05$). S.S (Sum of square); df (Degree of freedom); M.S (Mean square); F (Significance); CK (Check); *** (Highly significant); ** (Highly significant); * (Significant).

Table 4. Mortality of *Brevicoryne brassicae* by spinasterol, 22,23-dihydrospinasterol and fernenol via greenhouse assay.

Conc. (μg/mL)	Mean Mortality (%) with Time (h)							
	12		24		48		72	
	Spinasterol, 22,23-Dihydro-Spinasterol	Fernenol	Spinasterol, 22,23-Dihydro-Spinasterol	Fernenol	Spinasterol, 22,23-Dihydro-Spinasterol	Fernenol	Spinasterol, 22,23-Dihydro-Spinasterol	Fernenol
6.25	0.00 ± 0.00^{b}	0.00 ± 0.00^{a}	0.00 ± 0.00^{b}	0.00 ± 0.00^{c}	0.00 ± 0.00^{c}	0.00 ± 0.00^{d}	3.33 ± 3.33^{cd}	5.00 ± 2.89^{c}
12.50	3.33 ± 3.33^{b}	0.00 ± 0.00^{a}	6.67 ± 3.33^{b}	3.33 ± 1.67^{bc}	10.00 ± 0.00^{c}	5.00 ± 0.00^{c}	13.33 ± 3.33^{c}	6.67 ± 1.67^{c}
25	10.00 ± 0.00^{b}	3.33 ± 1.67^{a}	20.00 ± 5.77^{a}	6.67 ± 1.67^{ab}	23.33 ± 3.33^{b}	10.00 ± 0.00^{b}	40.00 ± 0.00^{b}	15.00 ± 2.89^{b}
50	20.00 ± 5.77^{a}	3.33 ± 1.67^{a}	30.00 ± 5.77^{a}	10.00 ± 2.89^{a}	50.00 ± 5.77^{a}	18.33 ± 1.67^{a}	60.00 ± 5.77^{a}	25.00 ± 2.89^{a}
CK	0.00 ± 0.00^{b}	0.00 ± 0.00^{a}	0.00 ± 0.00^{b}	0.00 ± 0.00^{c}	3.33 ± 3.33^{c}	1.67 ± 1.67^{d}	3.33 ± 3.33^{c}	3.33 ± 1.67^{c}
	Statics summary							
S.S	866.6	40.00	2106.6	226.6	4960.0	656.6	8000	976.6
df	4	4	4	4	4	4	4	4
M.S	216.6	10.00	526.67	56.67	1240.0	164.1	2000.0	244.1
F	8.13**	3.00*	11.29**	6.80***	37.20***	49.25***	60.00***	13.32***

Values are described as mean ± standard error followed by different superscripts were significantly different according to (Duncan multiple range test DMRT $p > 0.05$). S.S (Sum of square); df (Degree of freedom); M.S (Mean square); F (Significance); CK (Check); *** (Highly significant); ** (Highly significant); * (Significant).

Probit analysis exposed the LC_{50}, slope value, Chi-square and fiducial limits at 95% confidence interval. Lowest LC_{50} values recorded by spinasterol, 22,23-dihydrospinasterol were 32.36 and 44.79 μg/mL after 72 h and 48 h of exposure period, respectively via contact assay. In contrast, at the same exposure period, LC_{50} values by residual assay were 44.57 and 58.38 μg/mL, respectively. On the other hand, LC_{50} values recorded by greenhouse assay were 37.50 and 48.90 μg/mL at 72 and 48 h exposure, respectively (Table 5). Similarly, LC_{50} values recorded via fernenol were 47.99 and 63.08 μg/mL; 57.46 and 104.46 μg/mL; 58.67 and 76.52 μg/mL at 72 and 48 h by contact, residual and greenhouse assay, respectively (Table 6).

Table 5. Probit analysis of the effects of spinasterol, 22,23-dihydrospinasterol on *Brevicoryne brassicae*.

Bioassay	Time (h)	LC_{50} (μg/mL)	95% F.L		Slope ± SE	$\chi^{(2)}$
			Lower	Upper		
Contact	12	87.07	52.59	437.6	2.039 ± 0.60	1.75
	24	61.04	41.42	151.1	2.089 ± 0.52	1.06
	48	44.79	32.68	82.15	2.478 ± 0.72	0.31
	72	32.36	23.66	48.39	2.322 ± 0.54	0.12
Residual	12	143.5	69.10	1634	1.272 ± 1.01	0.77
	24	104.1	57.74	1003	1.927 ± 0.62	1.24
	48	58.38	39.46	143.7	1.981 ± 0.49	0.49
	72	44.57	30.99	90.95	2.078 ± 0.59	0.42
G. house	12	133.4	65.40	6044	1.870 ± 0.68	0.33
	24	83.23	50.25	387.1	1.920 ± 0.55	1.09
	48	48.90	36.29	86.10	2.462 ± 0.54	0.78
	72	37.50	27.86	57.92	2.502 ± 0.63	0.71

F.L; Fiducial limits. $\chi^{(2)}$; Chi-square.

Table 6. Probit analysis of the effects of fernenol on *Brevicoryne brassicae*.

Bioassay	Time (h)	LC_{50} (μg/mL)	95% F.L		Slope ± SE	$\chi^{(2)}$
			Lower	Upper		
Contact	12	345.2	89.21	1145	1.89 ± 0.73	1.62
	24	135.0	61.44	6531	0.91 ± 1.42	1.65
	48	63.08	40.63	296.3	1.47 ± 1.02	0.39
	72	47.99	32.59	94.62	1.27 ± 1.11	0.66
Residual	12	775.9	79.10	1634	1.43 ± 1.19	0.82
	24	109.7	60.35	1456	2.05 ± 0.69	0.24
	48	104.4	57.08	1031	1.82 ± 0.58	0.71
	72	57.46	44.78	221.1	3.12 ± 1.19	0.24
G. house	12	345.2	98.43	1316	1.61 ± 0.94	0.61
	24	158.0	67.56	2353	1.52 ± 0.57	0.99
	48	76.72	49.39	1781	2.33 ± 0.95	2.39
	72	58.67	38.68	282.9	2.13 ± 0.89	0.22

F.L; Fiducial limits. $\chi^{(2)}$; Chi-square.

3. Discussion

Because of the problems associated with the use of chemical pesticides for pest administration, the introduction of natural products particularly from plants source is pressing. Essential oils or extracts of plant origin are commonly used for plant protection measures because of their effectiveness against different life stages of insect pests. However, screening of suitable candidate plants for isolation, purification and identification of active ingredients is very crucial [27]. This technique was adopted to obtain active compounds from *C. colocynthis* leaves. Additionally, different techniques are used in the extraction, separation and purification of bioactive compounds from natural plant resources.

In our results for the preliminary toxicity evaluations of the solvents extract showed significant mortality of *B. brassicae.* Among the solvents extracts, methanol extract afforded high mortality which is in accordance with [28–30] who reported maximum mortality of *Aphis craccivora* by methanolic extract of *C. colocynthis* followed by ethyl acetate and petroleum ether extract. However, the activity of the crude extract can be attributed to the existence of specific chemical compounds of the plants like fatty acids (linoleic and oleic acid) glycosides (flavonoids, phenols, saponins), terpenoids and alkaloids, etc. [31]. Recently, Ahmed et al. [32] reported that solvents (methanol, ethanol, ethyl acetate, chloroform, acetone and hexane) extract of *C. colocynthis* leaves exhibited important phytochemicals such as alkaloids, glycosides, steroids, saponins, phenol, tannins and flavonoids along with potential antioxidant activities however, acetone and ethanol extract displayed as potent antioxidants. Further, these solvents extract also exhibited pronounced insecticidal activities [33]. Butanol fraction of the methanol extract from the *C. colocynthis* plant possessed insecticidal properties because of the presence of flavone glucosides and two cucurbitacins glucosides [19] Moreover, Yoshikawa et al. [34] described that alcoholic extract obtained from *C. colocynthis* fruits contains various compounds comprising cucurbitacins E 2-*O*-a-D-glucopyranoside, colocynthosides A and B, aglycon and cucurbitacins E. Among the isolated two triterpens glycosides and four cucurbitacins from *C. colocynthis* leaves, one was found effective against colon cancer cells HT29 and Caco-2 [35]. Plants possess a wide range of biologic compounds involved in their mechanism of chemical defense. These natural products contain significant prospective against variety of insects [36,37]. Activity of some of the natural compounds was evaluated against *Blattellea germanica* which showed LC_{50} values as 0.07 mg cm^{-1}, 0.06 mg cm^{-1} and 0.07 mg cm^{-1} via camphor, pulegone and verbenone, respectively [38]. Similarly, eugenol, isoeugenol, carvecol, thymol and p-cymene had shown anti-adulticidal activity at 1 mg $adult^{-1}$ against *B. germanica* [39].

As different techniques are employed in the extraction, separation, purification and identification of bioactive compounds thus, following these techniques, two compounds were identified as; a mixture of spinasterol and 22,23-dihydrospinasterol and other pure compound fernenol (d:c-friedo-b′:a′-neogammacer-9(11)-en-3alpha-olfern-9(11)-en-3alpha-o). Spinasterol is contained by Cucurbitaceae, Stegnospermaceae, Phytolaccaceae and Polygalaceae. In a study Meneses-Sagrero et al. [40] identified spinasterol form the methanol extract of *Stegnosperma halimifolium* (B.) and evaluated against cancer cell line. It was reported that spinasterol exhibited potential activity as antiproliferative against two cell lines of cervical cancer such as HeLa and RAW 264.7. Studies have also demonstrated that spinasterol exhibited different biologic activities including antidepressant [41], anti-ulcerogenic [42], anti-inflammatory [43], and antiproliferative activities [44]. Spinasterol, 22,23-dihydrospinasterol possess pharmacological and cytotoxic exertions likewise, it was isolated from *Bougainvillea spectabillis* and exhibited strong inhibition of xanthine oxidase being IC_{50} of 39.21 μM [45].

Fernenol belongs to the class of organic compounds known as triterpenoids containing six isoprene units. Xian-xue W et al. [46] isolated this compound from the whole plant extract of *Arenaria polytrichoides*, however, its activity was not evaluated against pests. Similarly, Li et al. [47] isolated three terpenoids including fernenol from *Ainsliaea yunnanensis* showed cytotoxic effects on human acute monocytic Leukemia cell line with IC_{50} being 1.73 μM. Some studies revealed that fernenol play an important role in the inhibition of mycelial development of *Colletotrichum gloeosporioides* being 47.5 mg mL^{-1} EC_{50}, and also significantly effective against anthracnose of mango when used at 100 mg L^{-1} and 200 mg L^{-1} concentration [48].

Our results demonstrated that spinasterol, 22,23-dihydrospinasterol exhibited aphicidal activity which caused significant mortality of this pest via different bioassays whereas, fernenol exhibited moderate aphicidal activity. Similar findings on extraction, purification and activity of purified compound 2-*O*-β-D-glucapyranosylcucurbitacin E was evaluated by Torkey et al. [49] against *Aphis craccivora* showed significant mortality with LC_{50} being 11,003 ppm. Moreover, aphicidal activity of *Eupatorium adenophorum* isolated compound 9-oxo-10, 11-dehydroageraphorone was evaluated against *Pseudoregma bambucicola* displayed 73.33% mortality at 2 mg mL^{-1} with 6 h exposure. Similarly, Nong et al. [50] recorded complete control of this pest at the same concentration after 30 days in a field experiment.

Thus, this unique and innovative research was performed first time for isolation, purification and characterization of bioactive compounds from *C. colocynthis* leaves and to investigate their insecticidal potential against cabbage aphid *B. brassicae*, which is an injurious pest of cabbage and other crops. It is worth mention that these compounds were isolated and evaluated from *C. colocynthis* for the first time against this pest.

4. Material and Methods

4.1. Collection of Plants and Aphids

C. colocynthis (Colocynth), locally recognized as Tumba, was the study plant. Leaves were collected from natural habitat of desert one climate of District Bahawalnagar, Pakistan with latitudinal and longitudinal gradients (29°59′34′′N, 73°15′13′′E) from March to April 2018. This area contains dry climate with an average precipitation of 204 mm annually and temperature ranges from 12.7–45 °C. The samples were authenticated at Entomological Research Institute Faisalabad, Pakistan.

Cabbage aphids were collected from wild cabbage plants from the surrounding field area of Shenyang Agricultural University, Shenyang China. During the collection of aphids it was insured that no pesticides was applied on the plants. The population of the aphids was maintained on cabbage plants grown in the greenhouse at 20 ± 5 °C and 45% ± 5% relative humidity (RH), along with a photoperiod of 16:8 (light: dark).

4.2. Extraction of Plant Material

Extraction was performed thrice at room temperature for three days by using the cold extraction/solvent extraction method. Different organic solvents were used for extraction purposes bearing varying polarity. Then, filter the contents and volume was reduced by concentrating the filtered material through the rotary evaporator (Model R-210, Buchi Switzerland, Flawil, Switzerland). Next, the filtrate was allowed to dry for 12 h in a fume hood at 28 °C and then, the dry extract was preserved in glass bottles at 4 °C for further experimentation.

4.3. Sample Preparations and Preliminary Toxicity Evaluations

Each of the extract obtained by using different solvents was evaluated against *B. brassicae* at 50 µg/mL concentration at an exposure of 6, 12, 24, 48, 72 and 96 h. For toxicity tests, dried extracts were dissolved in acetone (1 mL for each solvent extract) and mixed with 1% Tween-20 (prepared in distilled water) for preparing concentration. For control treatment, check (CK) was also prepared in 1% Tween-20 including acetone but excluding extract to prepare a 50 µg/mL concentration. Thus, to obtain homogenous mixture for separation and purification purposes, a 2-g sample from each solvent extract was taken in porcelain dish (14 g in total), dissolved in appropriate amount of dichloromethane and then added equal quantity of silica gel into it. The resulting mixture was left for overnight in a fume hood at 28 °C to evaporate the solvent and on drying; the residue was ground to fine powder by mortar and pestle and stored in glass stopper vials at 4 °C for further use.

4.4. Separation and Purification of Extract

The prepared sample was separated and purified by column chromatographic techniques. The dried sample was chromatographed on silica gel column (200–300 mesh). Gradient eluent of 300 mL each of the eluted fraction was collected using a gradient ratio of petroleum ether and ethyl acetate. The mobile phases used for silica gel column were PE:EA of 100:0, 100:2, 100:4, 100:8, 100:16, 100:32,

100:64, 100:100 and at the end methanol was used and finally, total 49 fractions were collected. Lastly, eight sub-fractions (A-H) were obtained followed by the mixing of the same polarity fractions. The mobile phases used for gel column Sephadex, LH-20 (40–120 μm) were methanol and dichloromethane with 1:1 ratio. For further purification preparative TLC plates were used and target compounds were scrapped, dissolved in ethyl acetate and concentrated on a rotary evaporator (Buchi Switzerland R-210, Flawil, Switzerland) for obtaining pure compounds with weight calculation.

4.5. Mass Spectrum and Elemental Analysis

Mass spectrum data were calculated on the Triple Quad GC-MS 7000C and Triple Quad LC/MS 6440 mass Spectrometer (Agilent Technologies, Santa Clara, CA, USA). Vario EL III element analyzer was used for elemental analysis (Elementar Analyssensysteme GmbH, Frankfurt, Germany).

4.6. Nuclear Magnetic Resonance (NMR) ^{1}H-NMR and ^{13}C-NMR Spectrum

To determine ^{1}H-NMR spectra, Deuterated chloroform ($CDCL_3$) was used as a solvent while for an internal standard, tetramethylsilane (TMS) was used. ^{1}H-NMR spectra were evaluated on 300 MHz and 600 MHz spectrometer whereas, ^{13}C-NMR spectra were assessed on a 151-MHz spectrometer (Bruker, Karlsruhe, Germany). Chemical shift values were recorded in parts per million (δ, ppm). The coupling constants value (J) was described in Hz. The splitting patterns of proton signals were also designated as follows: singlet (s), doublet (d), a doublet of doublets (dd), a doublet of the doublet of doublets (ddd), triplet (t), the quartet (q) and the multiplet (m)

4.7. Bioassay Study

Activity of the compound against *B. brassicae* was evaluated In vitro and In vivo. For In vitro assay, contact and residual toxicity methods were adopted while; In vivo experiment was conducted under greenhouse condition of Shenyang Agricultural University. Serial concentrations such as 6.25, 12.50, 25 and 50 μg/mL were prepared in Tween-20 (1%) solutions and replicated thrice.

4.7.1. In Vitro Contact Toxicity

For contact assay, 10 adult wingless aphids were dipped in respective concentrations for 5 s and released on freshly cut cabbage leaves placed in glass petri dishes. Check (CK) was prepared in 1% Tween-20 solution, but without adding a purified compound. All of the prepared petri dishes including control were placed in an incubator at 20 ± 5 °C, 65% RH for three days along with 16:8 (light: dark) photoperiod.

4.7.2. In Vitro Residual Toxicity

Fresh cabbage leaves were cut off and dipped for 10 s in respective concentration and dried in air for half an hour. Next, 10 adult wingless aphids were released on these leaves contained in glass petri dishes. Check (CK) was prepared in 1% Tween-20 solution, but without adding purified compound. All of the prepared petri dishes including check were placed in an incubator at 20 ± 5 °C, 65% RH for three days along with 16:8 (light: dark) photoperiod.

4.7.3. In Vivo Toxicity

Cabbage plants were grown and maintained in the greenhouse for the rearing of aphids and bioassay study. Prior to experiment, plants were sprayed with water to remove impurities and left for half an hour to dry in the open environment. Then, 10 wingless adult aphids were released on cabbage plants at 6–8 true leaf stage. After one hour of releasing aphids, plants were sprayed with respective concentrations (1–2 showers) with hand pump/sprayer. A control treatment was sprayed with distilled water solution in 1% Tween-20. Treated plants along with control were left for three days in greenhouse conditions.

4.8. Data Collection and Statistical Analysis

Data on mortality at a specific time period of 12, 24, 48 and 72 h were recorded regularly by examination under the microscope. Those individuals who presented no response on probing by needle were recorded as dead.

Analysis of the mortality data were performed by using analysis of variance (ANOVA). The mean mortality difference among treatments was intended at $p = 0.05$ by Duncan multiple range test DMRT with IBM-SPSS statistics 25.0 version software. Probit analysis was performed using EPA Probit analysis program version 1.5.

5. Conclusions

The present investigations indicated that *C. colocynthis* possess potential botanical agents. Results also demonstrated that *B. brassicae* showed reasonable sensitivity to isolated spinasterol, 22,23-dihydrospinasterol via all bioassay. In contrast, fernenol displayed moderate toxicity against this pest. Additionally, the contact bioassay produced higher mortality than the residual bioassay. Therefore, these chemical compounds may be introduced as alternatives to synthetic chemical insecticides. However, additional research is necessary for the purification and characterization of more bioactive constituents and their appraisals against *B. brassicae* and other field crops insect pests.

Author Contributions: Formal analysis, M.A., R.A. and H.G.; methodology, M.A; project administration, M.J.; software, J.S.; supervision, M.J.; Visualization, P.Q.; Writing—original draft, M.A.; Writing—review & editing, M.J., P.Q. and J.S. All authors have read and agreed to the published version of the manuscript.

Funding: The financial support provided by the National Key Research & Development Program of China (2016YFD0200500) is greatly acknowledged.

Conflicts of Interest: The authors have declared no conflict of interest.

References

1. CABI. Crop Protection Compendium. Available online: http://www.cabi.org/cpc (accessed on 3 March 2020).
2. Munthali, D. Evaluation of cabbage varieties for resistance to the cabbage aphid. *Afr. Entomol.* **2009**, *17*, 1–7. [CrossRef]
3. Swiader, J.M.; Ware, G.W.; McCollum, J.P. *Producing Vegetable Crops*; Interstate Printers and Publishers Inc.: Danville, IL, USA, 1992.
4. Parugrug, M.L.; Roxas, A.C. Insecticidal action of five plants against maize weevil, *Sitophilus zeamais* Motsch.(Coleoptera: Curculionidae). *Curr. Appl. Sci. Technol.* **2008**, *8*, 24–38.
5. Bhatia, R.; Verma, A. Seasonal incidence of major insect pests associated with winter crop of cabbage in Himachal Pradesh. *Ann. Agric. Res.* **1995**, *16*, 278–281.
6. Opfer, P.; McGranth, D. *Oregon Vegetables, Cabbage Aphid and Green Peach Aphid*; Oregon State University: Corvallis, OR, USA, 2013.
7. Rashid, T.; Chen, J.; McLeod, P. Toxicity of newly isolated piperideine alkaloids from the red imported fire ant, *Solenopsis invicta* Buren, against the green peach aphid, *Myzus persicae* (Sulzer). *Adv. Entomol.* **2013**, *2013*. [CrossRef]
8. Mvumi, C.; Marais, D.; Ngadze, E.; du Toit, E.S.; Tsindi, A. Effect of moringa extract on the leaf anatomy and yield potential of tomato infected by *Alternaria solani*. *South Afr. J. Plant Soil* **2018**, *35*, 389–392. [CrossRef]
9. Springob, K.; Kutchan, T.M. Introduction to the different classes of natural products. In *Plant-Derived Natural Products*; Donald Danforth Plant Science Center: St. Louis, MO, USA, 2009; pp. 3–50.
10. Fukuta, M.; Dang Xuan, T.; Deba, F.; Tawata, S.; Dang Khanh, T.; Min Chung, I. Comparative efficacies in vitro of antibacterial, fungicidal, antioxidant, and herbicidal activities of momilatones A and B. *J. Plant Interact.* **2007**, *2*, 245–251. [CrossRef]
11. Soam, P.S.; Singh, T.; Vijayvergia, R. *Citrullus colocynthis* (LINN.) and *Luffa acutangula* (L.) roxb, schrad. source of bioinsecticides and their contribution in managing climate change. *IJABPT* **2013**, *4*, 7–9.
12. Memon, U.; Brohi, A.; Ahmed, S.W.; Azhar, I.; Bano, H. Antibacterial screening of *Citrullus colocynthis*. *Pak. J. Pharm. Sci.* **2003**, *16*, 1–6.

13. Mohammad, D.; Al-Khateeb, M.; Riyadh, E.; Al-Hashem, F.; Nabil, B.; Mohammad, K. In vivo, acute, normo-hypoglycemic, antihyperglycemic, insulinotropic actions of orally administered ethanol extract of *Citrullus colocynthis* (L.) Schrab pulp. *Am. J. Biochem. Biotechnol.* **2009**, *5*, 119–126.
14. Kumar, S.; Kumar, D.; Jusha, M.; Saroha, K.; Singh, N.; Vashishta, B. Antioxidant and free radical scavenging potential of *Citrullus colocynthis* (L.) Schrad. methanolic fruit extract. *Acta Pharm.* **2008**, *58*, 215–220. [CrossRef]
15. Dehghani, F.; Panjeh, S.M.; Talaei, K.T.; Mesbah, A.S.; Azizi, M. Toxic effects of hydroalcoholic extract of *Citrullus colocynthis* on pregnant mice. *Iran. J. Vet. Res. Shiraz Univ.* **2008**, *9*, 42–45.
16. Adam, S.; Al-Yahya, M.; Al-Farhan, A. Response of Najdi sheep to oral administration of *Citrullus colocynthis* fruits, Nerium oleander leaves or their mixture. *Small Rumin. Res.* **2001**, *40*, 239–244. [CrossRef]
17. Sturm, S.; Schneider, P.; Seger, C.; Stuppner, H. Analysis of *Citrullus colocynthis* cucurbitacine derivatives with HPLC-SPE-NMR. *Sci. Pharm.* **2009**, *77*, 254. [CrossRef]
18. Seger, C.; Sturm, S.; Mair, M.E.; Ellmerer, E.P.; Stuppner, H. 1H and 13C NMR signal assignment of cucurbitacin derivatives from *Citrullus colocynthis* (L.) Schrader and *Ecballium elaterium* L.(Cucurbitaceae). *Magn. Reson. Chem.* **2005**, *43*, 489–491. [CrossRef] [PubMed]
19. Delazar, A.; Gibbons, S.; Kosari, A.R.; Nazemiyeh, H.; Modarresi, M.; Nahar, L.; Sarker, S.D. Flavone C-glycosides and cucurbitacin glycosides from *Citrullus colocynthis*. *DARU J. Pharm. Sci.* **2006**, *14*, 109–114.
20. Maatooq, G.T.; El-Sharkawy, S.H.; Afifi, M.; Rosazza, J.P. Cp-hydroxybenzoylglycoflavones from *Citrullus colocynthis*. *Phytochemistry* **1997**, *44*, 187–190. [CrossRef]
21. Jayaprakasam, B.; Seeram, N.P.; Nair, M.G. Anticancer and antiinflammatory activities of cucurbitacins from *Cucurbita andreana*. *Cancer Lett.* **2003**, *189*, 11–16. [CrossRef]
22. Song, F.; Dai, B.; Zhang, H.-Y.; Xie, J.-W.; Gu, C.-Z.; Zhang, J. Two new cucurbitane-type triterpenoid saponins isolated from ethyl acetate extract of *Citrullus colocynthis* fruit. *J. Asian Nat. Prod. Res.* **2015**, *17*, 813–818. [CrossRef]
23. Ding, L.; Chen, Y.; Wu, F. Constituents of the root of Berneuxia thibetica Decne. *Zhongguo Zhong Yao Za Zhi= Zhongguo Zhongyao Zazhi= China J. Chin. Mater. Med.* **1991**, *16*, 289–290, 318.
24. Sinha, B.; Sasmal, D.; Basu, S. Pharmacological studies on *Melothria maderaspatana*. *Fitoter (Milano)* **1997**, *68*, 75–78.
25. Glasby, J.S. *Directory of Plants Containing Secondary Metabolites*; Taylor & Francis group CRC Press: Boca Raton, FL, USA, 1991.
26. Rai, P.; Paudel, N.; Shakya, S. Cytological Effects of Leaf Extract of *Artemisia vulgaris* L. on Meristematic Cells of *Allium cepa* L. *Our Nat.* **2012**, *10*, 242–248. [CrossRef]
27. Chermenskaya, T.D.; Stepanycheva, E.A.; Shchenikova, A.V.; Savelieva, E.I.; Chakaeva, A.S. Insecticidal effects of *Ungernia severtzovii* bulb extracts against the grain aphid *Schizaphis graminum* (Rondani). *Ind. Crop. Prod.* **2012**, *36*, 122–126. [CrossRef]
28. Shehawy, A.A.; Khalil, A.-H.E.; Maklad, A.M.; Qari, S.H. Toxicity Effects and Biochemical Changes of Insecticide Alternatives on Cowpea Aphid,(*Aphis craccivora*)(Homoptera: Aphididae). *Egypt. Acad. J. Biol. Sci. F. Toxicol. Pest Control* **2019**, *11*, 129–138. [CrossRef]
29. Shehawy, A.; Alshehri, A. Toxicity and biochemical efficacy of Novel pesti-cides against *Aphis craccivora* (Hemiptera: Aphididae) in relation to enzymatic activity. *J. Plant Prot. Path. Mansoura Univ.* **2015**, *6*, 1507–1516.
30. Ibrahim, H.Y.E. Characterization of chemical constituents of *Citrullus colocynthis* (L.) extracts and their relation to toxicity against cowpea aphid, aphis craccivora koch. *JPPP* **2015**, *6*, 1323–1336. [CrossRef]
31. Yaniv, Z.; Shabelsky, E.; Schafferman, D. Colocynth: Potential arid land oilseed from an ancient cucurbit. *Perspect. New Crop. New Uses* **1999**, *3*, 257–261.
32. Ahmed, M.; Ji, M.; Peiwen, Q.; Liu, Y.; Gu, Z.; Sikandar, A.; Iqbal, M.; Javeed, A. Phytochemical screening, total phenolics and flavonoids content and antilxidant activities of *Citrullus colocynthis* L. and *Cannabis sativa* L. *Appl. Ecol. Environ. Res.* **2019**, *17*, 6961–6979. [CrossRef]
33. Ahmed, M.; Peiwen, Q.; Gu, Z.; Liu, Y.; Sikandar, A.; Hussain, D.; Javeed, A.; Shafi, J.; Iqbal, M.F.; An, R. Insecticidal activity and biochemical composition of *Citrullus colocynthis*, *Cannabis indica* and *Artemisia argyi* extracts against cabbage aphid (*Brevicoryne brassicae* L.). *Sci. Rep.* **2020**, *10*, 1–10. [CrossRef]

34. Yoshikawa, M.; Morikawa, T.; Kobayashi, H.; Nakamura, A.; Matsuhira, K.; Nakamura, S.; Matsuda, H. Bioactive saponins and glycosides. XXVII. Structures of new cucurbitane-type triterpene glycosides and antiallergic constituents from *Citrullus colocynthis*. *Chem. Pharm. Bull.* **2007**, *55*, 428–434. [CrossRef]
35. Chawech, R.; Jarraya, R.; Girardi, C.; Vansteelandt, M.; Marti, G.; Nasri, I.; Racaud-Sultan, C.; Fabre, N. Cucurbitacins from the leaves of *Citrullus colocynthis* (L.) Schrad. *Molecules* **2015**, *20*, 18001–18015. [CrossRef]
36. Michaelakis, A.; Papachristos, D.; Kimbaris, A.; Koliopoulos, G.; Giatropoulos, A.; Polissiou, M.G. Citrus essential oils and four enantiomeric pinenes against *Culex pipiens* (Diptera: Culicidae). *Parasitol. Res.* **2009**, *105*, 769. [CrossRef] [PubMed]
37. Lardeux, F.; Depickère, S.; Duchon, S.; Chavez, T. Insecticide resistance of *Triatoma infestans* (Hemiptera, Reduviidae) vector of Chagas disease in Bolivia. *Trop. Med. Int. Health* **2010**, *15*, 1037–1048. [CrossRef]
38. Jung, W.-C.; Jang, Y.-S.; Hieu, T.T.; Lee, C.-K.; Ahn, Y.-J. Toxicity of Myristica fragrans seed compounds against *Blattella germanica* (Dictyoptera: Blattellidae). *J. Med. Entomol.* **2007**, *44*, 524–529. [CrossRef]
39. Yeom, H.-J.; Kang, J.S.; Kim, G.-H.; Park, I.-K. Insecticidal and acetylcholine esterase inhibition activity of Apiaceae plant essential oils and their constituents against adults of German cockroach (*Blattella germanica*). *J. Agric. Food Chem.* **2012**, *60*, 7194–7203. [CrossRef] [PubMed]
40. Meneses-Sagrero, S.E.; Navarro-Navarro, M.; Ruiz-Bustos, E.; Del-Toro-Sánchez, C.L.; Jiménez-Estrada, M.; Robles-Zepeda, R.E. Antiproliferative activity of spinasterol isolated of *Stegnosperma halimifolium* (Benth, 1844). *Saudi Pharm. J.* **2017**, *25*, 1137–1143. [CrossRef] [PubMed]
41. Socała, K.; Wlaź, P. Evaluation of the antidepressant-and anxiolytic-like activity of α-spinasterol, a plant derivative with TRPV1 antagonistic effects, in mice. *Behav. Brain Res.* **2016**, *303*, 19–25. [CrossRef]
42. Klein, L.C.; Gandolfi, R.B.; Santin, J.R.; Lemos, M.; Cechinel Filho, V.; de Andrade, S.F. Antiulcerogenic activity of extract, fractions, and some compounds obtained from *Polygala cyparissias* St. Hillaire & Moquin (Polygalaceae). *Naunyn-Schmiedeberg's Arch. Pharmacol.* **2010**, *381*, 121–126.
43. Borges, F.R.; Silva, M.D.; Córdova, M.M.; Schambach, T.R.; Pizzolatti, M.G.; Santos, A.R. Anti-inflammatory action of hydroalcoholic extract, dichloromethane fraction and steroid α-spinasterol from *Polygala sabulosa* in LPS-induced peritonitis in mice. *J. Ethnopharmacol.* **2014**, *151*, 144–150. [CrossRef]
44. Ntie-Kang, F.; Yong, J.N. The chemistry and biological activities of natural products from Northern African plant families: From Aloaceae to Cupressaceae. *RSC Adv.* **2014**, *4*, 61975–61991. [CrossRef]
45. Chang, W.; Chang, Y.; Lu, F.; Chiang, H.-C. Inhibitory effects of phenolics on xanthine oxidase. *Anticancer Res.* **1994**, *14*, 501–506.
46. Wu, X.X.; Li, M.; Zhang, Q.; Cheng, L. Chemical constituents from *Arenaria polytrichoides* (II). *Chem. Res. Appl.* **2013**, *25*, 333–335.
47. Li, J.; Zhang, B.; Liu, H.; Zhang, X.; Shang, X.; Zhao, C. Triterpenoids from Ainsliaea yunnanensis Franch. and their biological activities. *Molecules* **2016**, *21*, 1481. [CrossRef] [PubMed]
48. Yenjit, P.; Issarakraisila, M.; Intana, W.; Chantrapromma, K. Fungicidal activity of compounds extracted from the pericarp of Areca catechu against *Colletotrichum gloeosporioides* in vitro and in mango fruit. *Postharvest Biol. Technol.* **2010**, *55*, 129–132. [CrossRef]
49. Torkey, H.; Abou-Yousef, H.; Abdel Azeiz, A.; Hoda, E. Insecticidal effect of cucurbitacin E glycoside isolated from *Citrullus colocynthis* against *Aphis Craccivora*. *Aust. J. Basic Appl. Sci.* **2009**, *3*, 4060–4066.
50. Nong, X.; Fang, C.-L.; Wang, J.-H.; Gu, X.-B.; Yang, D.-Y.; Liu, T.-F.; Fu, Y.; Zhang, R.-H.; Zheng, W.-P.; Peng, X.-R. Acaricidal activity of extract from *Eupatorium adenophorum* against the *Psoroptes cuniculi* and *Sarcoptes scabiei* in vitro. *Vet. Parasitol.* **2012**, *187*, 345–349. [CrossRef]

Sample Availability: Samples of the compounds are available from the authors.

molecules

MDPI

Article

Potential Synergy between Spores of *Metarhizium anisopliae* and Plant Secondary Metabolite, 1-Chlorooctadecane for Effective Natural Acaricide Development

Abid Hussain [1,2,3] and Ahmed Mohammed AlJabr [1,*]

[1] Laboratory of Bio-Control and Molecular Biology, Department of Arid Land Agriculture, College of Agricultural and Food Sciences, King Faisal University, Hofuf 31982, Al-Ahsa, Saudi Arabia; solvia_aah@yahoo.com

[2] Research and Consulting Institute, King Faisal University, Hofuf 31982, Al-Ahsa, Saudi Arabia

[3] Ministry of Environment, Water and Agriculture, Riyadh 11442, Saudi Arabia

* Correspondence: aljabr@kfu.edu.sa; Tel.: +9-6650-691-4442

Academic Editor: Giovanni Benelli

Received: 4 April 2020; Accepted: 18 April 2020; Published: 20 April 2020

Abstract: Date palm dust mites are important pests severely infesting valuable nutritious fruits (dates) of date palm. In search of an alternative to acaricides, joint action of *Metarhizium anisopliae* EBCL 02049 spores and 1-Chlorooctadecane was evaluated as a potential candidate for the management of *Oligonychus afrasiaticus* through natural products. In this regard, in vitro tests were performed to evaluate the interaction of *M. anisopliae* spores with multiple doses of 1-Chlorooctadecane (0.8, 1.6, 2.4, 3.2, and 4.0 mg/mL). Compatibility bioassay results evidenced from vegetative growth (77.7–84.40 mm), sporulation (5.50–7.30 $\times 10^6$ spores/mL), and germination (96.70–98.20%), revealed that all the tested doses are compatible (biological index > 82) with the spores of *M. anisopliae*. The impact of combined treatment of spores with 1-Chlorooctadecane in different proportions (Scheme I, II, III, and IV) compared to their sole application against *O. afrasiaticus* was evaluated by concentration–mortality response bioassays. Results showed that all the combined treatments revealed high mortality compared to the sole application, which showed relatively slow mortality response over time. Toxicity recorded from Scheme IV combinations (80% 1-Chlorooctadecane: 20% Spores), exhibited strong synergistic interaction (joint toxicity = 713). Furthermore, potent interactions have overcome the host antioxidant defense at the final stage of infection by tremendously reducing catalase, and superoxide dismutase activities. These experiments demonstrated fungal–toxin joint synergistic interaction as a promising date palm dust mite management option.

Keywords: antioxidants; compatibility; date palm dust mites; host defense; *Oligonychus afrasiaticus*; *Metarhizium anisopliae*; natural acaricide; synergism; 1-Chlorooctadecane

1. Introduction

Date palm (*Phoenix dactylifera* L.) is one of the oldest domesticated fruit trees with unique nutritional characteristics and great socioeconomic importance for the Middle East and North African countries. This cash crop, in addition to commercial and nutritional values, can successfully be grown in the deserts because of minimum water requirement and highly tolerance to salinity and harsh weather conditions [1]. In the Kingdom of Saudi Arabia, date palms are grown on vast areas of land, accounting for about 25% of the World's date palms (30 million). According to an estimate, approximately 450 different cultivars out of 2000 are grown here in Saudi Arabia [2].

The economic importance of date palms is mainly because of valuable nutritious fruit (dates), which contains carbohydrates, dietary fibers, fat, lipids, minerals and protein [3]. Saudi Arabia stands second (1,302,859 tons), after the major producer Egypt (1,562,171 tons), in the world in the production of dates [4]. However, the annual yield of dates in Saudi Arabia compared to the other countries is still low might because of pest infestations. Date palm dust mites, *O. afrasiaticus* (Acari: Tetranychidae) is considered a major pest of date palms in Saudi Arabia. The weather conditions prevailing in the Kingdom and neighboring countries including Qatar, Sultanate of Oman, United Arab Emirates, Yemen, Libya, Iraq, and Egypt greatly favor *O. afrasiaticus* growth and development.

The infestation of *O. afrasiaticus* mainly depends on the moisture contents of the dates. The attack of mites starts at the *Kimri* stage (greenish), during this stage these greenish dates tremendously increase in sugar contents and moisture level that ultimately led to an increase in size and weight. At this stage, mites start spinning the web around the bunches of dates. Dust particles and exuviae from different stages are trapped into these webs resulting in the dusty appearance of bunches. Under these conditions, mites grow rapidly and multiply logarithmically [5].

Currently, management of date palm dust mites heavily relied on the use of acaricides. The use of synthetic pesticides to manage mites on dates may cause serious health hazards because the fruit contains pesticide residues. Furthermore, the use of pesticides is discouraged because of (1) environmental pollution, (2) applicator safety issues, (3) detrimental effects on non-target animals and (4) decrease in biodiversity. These shortcomings prompted the policymakers and researchers to focus on the use of alternative environmentally friendly management strategies to manage target pest species. Among alternative methods, plant-based pesticides and naturally occurring bio-control agents attained considerable importance because of their high degree of host specificity and high searching ability, respectively. The use of biopesticides to manage other mite species have been described in numerous investigations [6–12]. The spores of the strain of *M. anisopliae* selected for the current study had shown virulence against various pest species including *Ocinara varians* Walker [13], *Coptotermes formosanus* Shiraki [14,15], and *O. afrasiaticus* (McGregor) [16]. On the other hand, plant secondary metabolite, 1-Chlorooctadecane selected for the current study is a promising pest management candidate known to extract from numerous plant species including *Albertisia papuana* Becc [17], *Syzygium cumini* (L.) [18], *Arisaema amurense* Maxim [19], and fungus, *Trichoderma harzianum* [20]. 1-Chlorooctadecane (CH_3 (CH_2) $16CH_2$ Cl) was screened from our preliminary study on the chemical fractionation of *Cucumis sativus* that showed toxicity against date palm dust mites. The successful implementation of these agents can yield numerous benefits such as no side effects, cost-effectiveness, reduced reliance on conventional pesticides, minimum environmental disturbance and self-perpetuation. However, few studies have explored the potential of biopesticides to manage *O. afrasiaticus* populations [16,21–23]. The current study is primarily aimed to fully exploit for the first time the biocontrol potential of plant secondary metabolite, 1-Chlorooctadecane and pathogenic fungus, *M. anisopliae* EBCL 02049 by a series of experimentation including compatibility assays, concentration–mortality response bioassays, physiological enzymatic regulations, and joint toxicity index analysis in order to facilitate a knowledge-based eco-friendly date palm dust mites management approach.

2. Results

2.1. Compatibility Bioassays

The spores of *M. anisopliae* EBCL 02049 were found to be compatible because they showed a very high value of biological index (BI > 82), against all the tested concentrations of 1-Chlorooctadecane. However, we found concentration-dependent inversely proportional relationship between concentrations of 1-Chlorooctadecane and BI. All the tested concentrations of 1-Chlorooctadecane revealed significant differences in the vegetative growth ($F = 3.32$; $df = 5, 54$; $p = 0.011$), and sporulation ($F = 3.00$; $df = 5, 54$; $p = 0.019$) of *M. anisopliae* EBCL 02049. We recorded a similar concentration-dependent response. The lower concentrations failed to inhibit the fungal growth resulting in significantly wider vegetative mycelial growth and high sporulation of *M. anisopliae* EBCL 02049 (Table 1). On the other hand, spores germination of *M. anisopliae* EBCL 02049 against all the tested concentrations revealed non-significant interaction ($F = 0.93$; $df = 5, 54$; $p = 0.468$), and resulting very high percent germination (> 96%).

Table 1. Compatibility of 1-Chlorooctadecane with the spores of *M. anisopliae*.

Treatments	Vegetative Growth (mm) [1]	Germination (%) [1]	Sporulation ($\times 10^6$ Spores/mL) [1]	Biological Index	Classification [2]
Control	85.70 ± 2.25^{a}	98.70 ± 0.79^{a}	7.80 ± 0.63^{a}	-	-
0.8 mg/mL	84.40 ± 2.28^{ab}	98.20 ± 0.99^{a}	7.30 ± 0.60^{ab}	96.48	Compatible
1.6 mg/mL	84.20 ± 21.69^{ab}	97.90 ± 1.03^{a}	6.40 ± 0.54^{abc}	91.38	Compatible
2.4 mg/mL	79.30 ± 1.82^{bc}	97.30 ± 0.71^{a}	6.20 ± 0.61^{bc}	87.53	Compatible
3.2 mg/mL	78.10 ± 2.11^{c}	96.70 ± 0.93^{a}	5.60 ± 0.40^{c}	83.50	Compatible
4.0 mg/mL	77.70 ± 1.54^{c}	97.20 ± 0.81^{a}	5.50 ± 0.34^{c}	82.78	Compatible

[1] Numerical values (means ± SE) are the means of ten replicates. Different lower-case letter(s) superscript followed by means ± SE within a column are significantly different (Fisher's LSD test; $\alpha = 0.05$). [2] Biological index-based classification criterion: BI more than 66 classified as Compatible, BI between 42 to 66 classified as Moderately toxic; BI less than or equal to 42 classified as Toxic interaction.

2.2. Screening Biossays

Date palm dust mites exposed with different concentrations of 1-Chlorooctadecane, and *M. anisopliae* EBCL 02049 spores alone or in different proportions revealed concentration-dependent mortality response (Figure 1). The sole application of *M. anisopliae* EBCL 02049 spore suspensions at different concentrations exhibited the lowest mortality response resulting the greater value for LC_{50} (21.66 mg/mL). However, all the tested concentrations of the spores ($F = 920.10$; $df = 4, 48$; $p < 0.0001$), at different time intervals ($F = 194.27$; $df = 2, 48$; < 0.0001), and their interaction ($F = 121.91$; $df = 8, 48$; < 0.0001), revealed significant differences in the mortality of date palm dust mites (Figure 1b). On the other hand, sole application of 1-Chlorooctadecane has shown comparatively lower value for LC_{50} (3.42 mg/mL). The mortality response of date palm dust mites recorded at different time intervals ($F = 138.93$; $df = 2, 48$; < 0.0001), after different concentrations of 1-Chlorooctadecane ($F = 1181.20$; $df = 4, 48$; $p < 0.0001$), and their interaction ($F = 140.63$; $df = 8, 48$; < 0.0001) exhibited significant differences (Figure 1a).

Different bioassay schemes designed to screen the most potent interaction revealed variable interaction response in terms of the mortality of second nymphal stage of date palm dust mites. The combination of 1-Chlorooctadecane with spores mentioned in Table 2 as Scheme I (20% 1-Chlorooctadecane: 80% Spores) led to an antagonistic interaction by revealing the least joint toxicity (joint toxicity = 47). However, mortality response of date palm dust mites exposed to different concentrations of these combinations ($F = 819.31$; $df = 4, 48$; $p < 0.0001$), recorded at different time intervals ($F = 179.31$; $df = 2, 48$; < 0.0001), and their interaction ($F = 75.94$; $df = 8, 48$; < 0.0001) revealed significant differences (Figure 1I). However, the combination Scheme II in which different concentrations were prepared to finalize the final concentrations with combination strength of 40% 1-Chlorooctadecane: 60% Spores led to a synergistic interaction (joint toxicity = 112). The mortality responses of date palm dust mites displayed in the Figure 1II revealed significant differences at different concentrations ($F = 439.53$; $df = 4, 48$; $p < 0.0001$), time intervals ($F = 265.19$; $df = 2, 48$; < 0.0001), and their interaction ($F = 26.81$; $df = 8, 48$; < 0.0001).

The combined application of 1-Chlorooctadecane and spores (Scheme III: 60% 1-Chlorooctadecane: 40% Spores) imparted significant differences in the mortality of date palm dust mites against different concentrations ($F = 1345.03$; $df = 4, 48$; $p < 0.0001$), time intervals ($F = 365.97$; $df = 2, 48$; < 0.0001), and their interaction ($F = 77.38$; $df = 8, 48$; < 0.0001) as shown in Figure 1III. On the other hand, joint toxicity recorded as a result of Scheme III combinations exhibited bit higher value (joint toxicity = 289) as shown in Table 2. Similarly, strong synergistic interaction (joint toxicity = 713) was recorded from the Scheme IV bioassays in which different concentrations were prepared to finalize the final concentrations with combination strength of 80% 1-Chlorooctadecane: 20% Spores (Table 2). The mortality response of date palm dust mites exposed to different concentrations of these combinations ($F = 1275.55$; $df = 4, 48$; $p < 0.0001$), recorded at different time intervals ($F = 394.27$; $df = 2, 48$; < 0.0001), and their interaction ($F = 112.46$; $df = 8, 48$; < 0.0001), revealed significant differences (Figure 1IV).

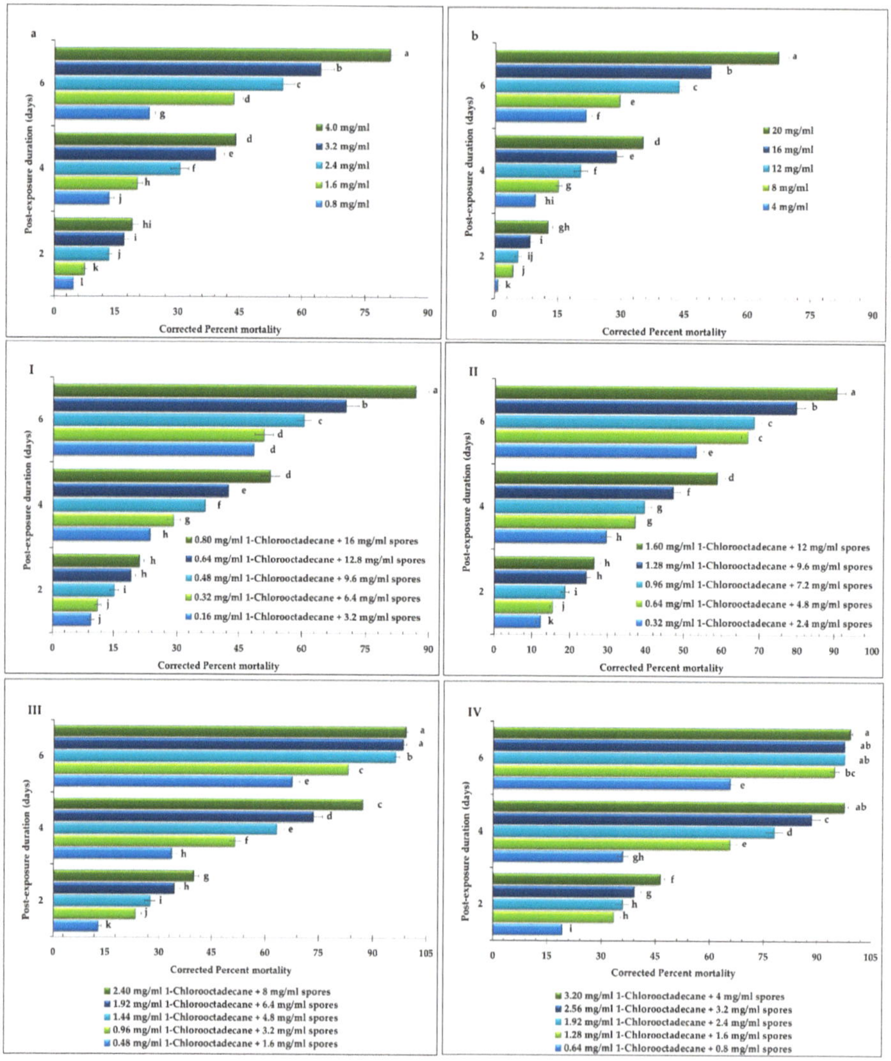

Figure 1. Impact of various concentrations of (**a**) 1-Chlorooctadecane; (**b**) suspensions of *M. anisopliae;* and their various interactions (**I**) 20% 1-Chlorooctadecane: 80% Spores; (**II**) 40% 1-Chlorooctadecane: 80% Spores; (**III**) 60% 1-Chlorooctadecane: 40% Spores; (**IV**) 80% 1-Chlorooctadecane: 20% Spores; on the mortality of second nymphal stage date palm dust mites. The comparisons of mortalities of *O. afrasiaticus* among different treatments were evaluated by repeated measures ANOVA (Fisher's LSD test; $\alpha = 0.05$).

Table 2. Toxicity of the interactions between 1-Chlorooctadecane and the spores of *M. anisopliae* against date palm dust mites.

Combinations	LC_{50} (mg/mL)	Joint Toxicity	Interaction [1]
Scheme I: 20% 1-Chlorooctadecane: 80% Spores			
0.16 mg/mL 1-Chlorooctadecane + 3.2 mg/mL spores 0.32 mg/mL 1-Chlorooctadecane + 6.4 mg/mL spores 0.48 mg/mL 1-Chlorooctadecane + 9.6 mg/mL spores 0.64 mg/mL 1-Chlorooctadecane + 12.8 mg/mL spores 0.80 mg/mL 1-Chlorooctadecane + 16.0 mg/mL spores	8.75 (7.10–10.75)	47	Antagonistic
Scheme II: 40% 1-Chlorooctadecane: 60% Spores			
0.32 mg/mL 1-Chlorooctadecane + 2.4 mg/mL spores 0.64 mg/mL 1-Chlorooctadecane + 4.8 mg/mL spores 0.96 mg/mL 1-Chlorooctadecane + 7.2 mg/mL spores 1.28 mg/mL 1-Chlorooctadecane + 9.6 mg/mL spores 1.60 mg/mL 1-Chlorooctadecane + 12 mg/mL spores	4.62 (3.41–5.65)	112	Synergistic
Scheme III: 60% 1-Chlorooctadecane: 40% Spores			
0.48 mg/mL 1-Chlorooctadecane + 1.6 mg/mL spores 0.96 mg/mL 1-Chlorooctadecane + 3.2 mg/mL spores 1.44 mg/mL 1-Chlorooctadecane + 4.8 mg/mL spores 1.92 mg/mL 1-Chlorooctadecane + 6.4 mg/mL spores 2.40 mg/mL 1-Chlorooctadecane + 8.0 mg/mL spores	2.39 (2.00–2.72)	289	Synergistic
Scheme IV: 80% 1-Chlorooctadecane: 20% Spores			
0.64 mg/mL 1-Chlorooctadecane + 0.8 mg/mL spores 1.28 mg/mL 1-Chlorooctadecane + 1.6 mg/mL spores 1.92 mg/mL 1-Chlorooctadecane + 2.4 mg/mL spores 2.56 mg/mL 1-Chlorooctadecane + 3.2 mg/mL spores 3.20 mg/mL 1-Chlorooctadecane + 4.0 mg/mL spores	1.47 (1.24–1.67)	713	Synergistic

[1] Interaction criterion: Joint toxicity less than 100 is classified as Antagonistic.

2.3. Host Antioxidant Defense Response

The exposure of different treatments sole or in different combination schemes produced various levels of catalase (CAT), and superoxide dismutase (SOD) activities relative to control treatment. The relative CAT activities of *O. afrasiaticus* analyzed after 24 h post-exposure revealed significant differences among different treatments ($F = 256.60$; $df = 5, 120$; $p < 0.0001$), concentrations ($F = 228.63$; $df = 4, 120$; $p < 0.0001$), and their interaction ($F = 2.35$; $df = 20, 120$; $p = 0.0031$). Overall, the most potent treatment especially the treatments from the higher concentrations of Scheme IV (80% 1-Chlorooctadecane: 20% Spores) tremendously induced the enzymatic activities of CAT, and remained statistically at the highest level. On the other hand, the lowest concentration of 1-Chlorooctadecane (0.8 mg/mL) after 24 h of post-exposure induced the lowest CAT activities among date palm dust mites (Table 3). In contrast, 0.16 mg/mL of 1-Chlorooctadecane + 3.2 mg/mL of spores induced the lowest CAT activities among date palm dust mites exposed for 72 h. However, all the treatments ($F = 208.53$; $df = 5, 120$; $p < 0.0001$), with different concentrations ($F = 464.81$; $df = 4, 120$; $p < 0.0001$), and their interaction ($F = 103.21$; $df = 20, 120$; $p < 0.0001$), revealed significant differences in the relative CAT activities from 72 h post-exposure date palm dust mites (Table 3). In addition, similar to 24 h post-exposure, date palm dust mites exposed for 72 h also tremendously induced CAT activities in response to the most potent treatment (3.20 mg/mL 1-Chlorooctadecane + 4.0 mg/mL spores). Unlike to relative CAT activities recorded 72 h post-exposure, the most potent treatments including the highest concentrations of Scheme IV (80% 1-Chlorooctadecane: 20% Spores) failed to induce the enzymatic activities of CAT after 120 h post-exposure, and remained statistically at the lowest level (Table 3). At this time interval, *M. anisopliae* EBCL 02049 spores at a concentration of 8 mg/mL induced the highest level of relative CAT activities of infected date palm dust mites, and remained statistically at the highest level. Overall, relative CAT activities recorded after 120 h showed significant differences against different treatments ($F = 3621.36$; $df = 5, 120$; $p < 0.0001$), concentrations ($F = 1536.63$; $df = 4, 120$; $p < 0.0001$), and their interaction ($F = 105.06$; $df = 20, 120$; $p < 0.0001$).

Relative enzymatic activities of SOD recorded from date palm dust mites after 24 h of exposure against different treatments ($F = 331.28$; $df = 5, 120$; $p < 0.0001$), concentrations ($F = 398.14$; $df = 4, 120$; $p < 0.0001$), and their interaction ($F = 5.44$; $df = 20, 120$; $p < 0.0001$), revealed significant differences (Table 4). The infection of the least potent treatment (4 mg/mL spores of *M. anisopliae* EBCL02049) failed to induce SOD activities and remained at the lowest level. On the other hand, 3.20 mg/mL 1-Chlorooctadecane + 4.0 mg/mL spores triggered the SOD activities. This trend continues to increase until 72 h post-exposure, and resulted the highest relative SOD activities from this potent treatment. Furthermore, significant differences in relative SOD activities calculated against different treatments ($F = 340.46$; $df = 5, 120$; $p < 0.0001$), concentrations ($F = 759.10$; $df = 4, 120$; $p < 0.0001$), and their interaction ($F = 164.03$; $df = 20, 120$; $p < 0.0001$), were recorded in this study.

Table 3. Relative catalase (CAT) activities of date palm dust mites fed for different time intervals on date palm leaf-discs treated solely and in different combinations of 1-Chlorooctadecane with *M. anisopliae* spores.

Treatments	Post-Exposure Duration		
	24 h [1]	72 h [1]	120 h [1]
	(%)	(%)	(%)
1-Chlorooctadecane			
0.8 mg/mL	13.35 ± 0.55^{p}	32.18 ± 0.34^{m}	46.03 ± 0.80^{g}
1.6 mg/mL	20.32 ± 1.02^{lm}	35.56 ± 0.49^{l}	39.59 ± 0.88^{i}
2.4 mg/mL	21.17 ± 1.03^{kl}	40.09 ± 0.69^{ij}	23.27 ± 0.44^{l}
3.2 mg/mL	24.38 ± 1.18^{hi}	46.16 ± 0.97^{g}	17.13 ± 0.34^{op}
4.0 mg/mL	27.14 ± 1.32^{fg}	52.86 ± 0.92^{d}	15.14 ± 0.25^{q}
Scheme I: 20% 1-Chlorooctadecane: 80% Spores			
0.16 mg/mL 1-Chlorooctadecane + 3.2 mg/mL spores	15.04 ± 0.21^{op}	31.52 ± 0.37^{m}	60.78 ± 0.94^{b}
0.32 mg/mL 1-Chlorooctadecane + 6.4 mg/mL spores	16.38 ± 0.24^{no}	39.64 ± 0.59^{j}	53.88 ± 0.98^{e}
0.48 mg/mL 1-Chlorooctadecane + 9.6 mg/mL spores	20.22 ± 0.54^{lm}	44.17 ± 0.72^{gh}	56.42 ± 0.93^{d}
0.64 mg/mL 1-Chlorooctadecane + 12.8 mg/mL spores	21.19 ± 0.78^{kl}	50.11 ± 0.95^{ef}	52.78 ± 0.95^{e}
0.80 mg/mL 1-Chlorooctadecane + 16.0 mg/mL spores	25.15 ± 0.92^{ghi}	43.11 ± 0.84^{h}	41.46 ± 0.76^{h}
Scheme II: 40% 1-Chlorooctadecane: 60% Spores			
0.32 mg/mL 1-Chlorooctadecane + 2.4 mg/mL spores	20.32 ± 0.31^{lm}	35.30 ± 0.55^{l}	40.03 ± 0.64^{hi}
0.64 mg/mL 1-Chlorooctadecane + 4.8 mg/mL spores	22.06 ± 0.70^{jkl}	36.50 ± 0.57^{l}	29.74 ± 0.55^{j}
0.96 mg/mL 1-Chlorooctadecane + 7.2 mg/mL spores	27.15 ± 0.72^{fg}	39.68 ± 0.78^{j}	24.09 ± 0.50^{l}
1.28 mg/mL 1-Chlorooctadecane + 9.6 mg/mL spores	30.01 ± 0.90^{de}	48.45 ± 0.91^{f}	20.10 ± 0.24^{m}
1.60 mg/mL 1-Chlorooctadecane + 12 mg/mL spores	34.18 ± 0.92^{bc}	52.93 ± 0.93^{d}	15.20 ± 0.20^{q}
Scheme III: 60% 1-Chlorooctadecane: 40% Spores			
0.48 mg/mL 1-Chlorooctadecane + 1.6 mg/mL spores	20.10 ± 0.24^{lm}	35.69 ± 0.46^{l}	26.54 ± 0.72^{k}
0.96 mg/mL 1-Chlorooctadecane + 3.2 mg/mL spores	24.19 ± 0.60^{hij}	37.43 ± 0.56^{kl}	24.19 ± 0.60^{l}
1.44 mg/mL 1-Chlorooctadecane + 4.8 mg/mL spores	29.12 ± 0.78^{ef}	42.02 ± 0.70^{hi}	20.29 ± 0.39^{m}
1.92 mg/mL 1-Chlorooctadecane + 6.4 mg/mL spores	32.22 ± 0.87^{cd}	55.90 ± 0.76^{c}	17.44 ± 0.32^{no}
2.40 mg/mL 1-Chlorooctadecane + 8.0 mg/mL spores	33.22 ± 0.90^{bc}	63.85 ± 0.93^{b}	09.79 ± 0.26^{r}
Scheme IV: 80% 1-Chlorooctadecane: 20% Spores			
0.64 mg/mL 1-Chlorooctadecane + 0.8 mg/mL spores	23.08 ± 0.57^{ijk}	38.92 ± 0.57^{jk}	15.58 ± 0.28^{pq}
1.28 mg/mL 1-Chlorooctadecane + 1.6 mg/mL spores	29.37 ± 0.64^{ef}	43.35 ± 0.69^{h}	09.67 ± 0.25^{r}
1.92 mg/mL 1-Chlorooctadecane + 2.4 mg/mL spores	33.02 ± 0.81^{bc}	50.16 ± 0.79^{ef}	03.58 ± 0.24^{s}
2.56 mg/mL 1-Chlorooctadecane + 3.2 mg/mL spores	35.12 ± 0.93^{b}	63.96 ± 0.91^{b}	01.79 ± 0.15^{t}
3.20 mg/mL 1-Chlorooctadecane + 4.0 mg/mL spores	39.32 ± 1.22^{a}	71.64 ± 1.05^{a}	01.24 ± 0.13^{t}
***M. anisopliae* EBCL02049 spores**			
4 mg/mL	15.34 ± 0.43^{op}	49.13 ± 0.81^{f}	58.19 ± 0.81^{c}
8 mg/mL	18.27 ± 0.56^{mn}	57.16 ± 0.98^{c}	63.83 ± 0.77^{a}
12 mg/mL	21.08 ± 0.78^{kl}	51.35 ± 0.97^{de}	48.38 ± 0.97^{f}
16 mg/mL	23.26 ± 0.88^{ijk}	43.67 ± 0.93^{h}	28.07 ± 0.62^{jk}
20 mg/mL	26.17 ± 0.91^{gh}	36.26 ± 0.68^{l}	19.13 ± 0.60^{mn}

[1] Numerical values (means ± SE) are the means of five replicates. Different lower-case letter(s) superscript followed by means ± SE of relative CAT activities of *O. afrasiaticus* within the column are significantly different (Fisher´s LSD test; α = 0.05).

Table 4. Relative superoxide dismutase (SOD) activities of date palm dust mites fed for different time intervals on date palm leaf-discs treated solely and in different combinations of 1-Chlorooctadecane with *M. anisopliae* spores.

Treatments	Post-Exposure Duration		
	24 h [1]	72 h [1]	120 h [1]
1-Chlorooctadecane	(%)	(%)	(%)
0.8 mg/mL	12.89 ± 0.53^{kl}	30.38 ± 0.85^{q}	66.27 ± 1.19^{bc}
1.6 mg/mL	20.42 ± 0.78^{h}	36.15 ± 0.90^{op}	57.13 ± 1.21^{d}
2.4 mg/mL	21.20 ± 0.87^{h}	43.85 ± 0.93^{l}	37.31 ± 0.65^{h}
3.2 mg/mL	26.87 ± 1.05^{d}	54.04 ± 1.27^{g}	31.36 ± 0.45^{ij}
4.0 mg/mL	29.65 ± 1.29^{c}	66.10 ± 1.35^{c}	19.77 ± 0.36^{lm}
Scheme I: 20% 1-Chlorooctadecane: 80% Spores			
0.16 mg/mL 1-Chlorooctadecane + 3.2 mg/mL spores	11.17 ± 0.27^{l}	53.13 ± 0.97^{gh}	76.13 ± 1.01^{a}
0.32 mg/mL 1-Chlorooctadecane + 6.4 mg/mL spores	13.33 ± 0.42^{k}	61.18 ± 0.99^{e}	68.07 ± 0.98^{b}
0.48 mg/mL 1-Chlorooctadecane + 9.6 mg/mL spores	18.13 ± 0.50^{j}	52.54 ± 1.04^{gh}	57.30 ± 1.11^{d}
0.64 mg/mL 1-Chlorooctadecane + 12.8 mg/mL spores	20.11 ± 0.76^{hi}	46.57 ± 0.95^{j}	40.43 ± 0.69^{g}
0.80 mg/mL 1-Chlorooctadecane + 16.0 mg/mL spores	21.32 ± 0.85^{gh}	29.46 ± 0.50^{q}	25.30 ± 0.58^{k}
Scheme II: 40% 1-Chlorooctadecane: 60% Spores			
0.32 mg/mL 1-Chlorooctadecane + 2.4 mg/mL spores	14.17 ± 0.61^{k}	37.27 ± 0.61^{o}	43.08 ± 0.82^{f}
0.64 mg/mL 1-Chlorooctadecane + 4.8 mg/mL spores	21.20 ± 0.63^{h}	49.65 ± 0.81^{i}	44.12 ± 0.96^{f}
0.96 mg/mL 1-Chlorooctadecane + 7.2 mg/mL spores	25.42 ± 0.91^{de}	58.40 ± 1.18^{f}	39.80 ± 0.88^{g}
1.28 mg/mL 1-Chlorooctadecane + 9.6 mg/mL spores	27.07 ± 0.93^{d}	61.44 ± 1.20^{e}	25.46 ± 0.51^{k}
1.60 mg/mL 1-Chlorooctadecane + 12.0 mg/mL spores	30.11 ± 1.02c	64.55 ± 1.25^{cd}	21.42 ± 0.50^{l}
Scheme III: 60% 1-Chlorooctadecane: 40% Spores			
0.48 mg/mL 1-Chlorooctadecane + 1.6 mg/mL spores	17.21 ± 0.39^{j}	41.14 ± 0.77^{mn}	36.26 ± 0.68^{h}
0.96 mg/mL 1-Chlorooctadecane + 3.2 mg/mL spores	23.10 ± 0.65^{fg}	44.39 ± 0.85^{kl}	29.32 ± 0.63^{j}
1.44 mg/mL 1-Chlorooctadecane + 4.8 mg/mL spores	24.27 ± 0.70^{ef}	54.25 ± 1.03^{g}	25.37 ± 0.53^{k}
1.92 mg/mL 1-Chlorooctadecane + 6.4 mg/mL spores	29.22 ± 0.95^{c}	63.76 ± 1.24^{d}	20.54 ± 0.43^{l}
2.40 mg/mL 1-Chlorooctadecane + 8.0 mg/mL spores	34.20 ± 1.02^{a}	72.26 ± 1.25^{b}	16.28 ± 0.25^{n}
Scheme IV: 80% 1-Chlorooctadecane: 20% Spores			
0.64 mg/mL 1-Chlorooctadecane + 0.8 mg/mL spores	23.38 ± 0.52^{f}	43.15 ± 0.81^{lm}	32.41 ± 0.52^{i}
1.28 mg/mL 1-Chlorooctadecane + 1.6 mg/mL spores	25.32 ± 0.70^{de}	53.19 ± 0.91^{gh}	26.42 ± 0.43^{k}
1.92 mg/mL 1-Chlorooctadecane + 2.4 mg/mL spores	26.19 ± 0.98^{d}	60.45 ± 0.99^{ef}	21.24 ± 0.37^{l}
2.56 mg/mL 1-Chlorooctadecane + 3.2 mg/mL spores	32.34 ± 1.12^{b}	66.23 ± 1.10^{c}	18.03 ± 0.52^{mn}
3.20 mg/mL 1-Chlorooctadecane + 4.0 mg/mL spores	35.41 ± 1.18^{a}	81.28 ± 1.26^{a}	11.80 ± 0.37^{o}
***M. anisopliae* EBCL02049 spores**			
4 mg/mL	09.28 ± 0.34^{m}	34.20 ± 0.79^{p}	56.53 ± 0.93^{d}
8 mg/mL	17.11 ± 0.48^{j}	40.14 ± 0.88^{n}	65.87 ± 1.03^{c}
12 mg/mL	18.41 ± 0.58^{ij}	46.43 ± 0.92^{jk}	54.14 ± 0.97^{e}
16 mg/mL	20.13 ± 0.66^{hi}	51.24 ± 0.98^{hi}	43.75 ± 0.99^{f}
20 mg/mL	23.16 ± 0.89^{fg}	60.24 ± 1.14^{ef}	25.46 ± 0.72^{k}

[1] Numerical values (means ± SE) are the means of five replicates. Different lower-case letter(s) superscript followed by means ± SE of relative SOD activities of *O. afrasiaticus* within the column are significantly different (Fisher's LSD test; $\alpha = 0.05$).

Overall, the synergistic interaction combinations at their highest concentrations comparatively showed higher relative SOD activities compared with antagonistic interaction combinations (Table 4). On the contrary, relative SOD activities tend to decline after 120 h post-exposure among synergistic interactions compared with antagonistic interaction (Table 4). However, date palm dust mites showed significant differences among relative SOD activities in response to different treatments ($F = 2230.32$; $df = 5, 120$; $p < 0.0001$), concentrations ($F = 1658.66$; $df = 4, 120$; $p < 0.0001$), and their interaction ($F = 71.62$; $df = 20, 120$; $p < 0.0001$).

3. Discussion

Plant secondary metabolite 1-Chlorooctadecane improved the efficiency of *M. anisopliae* EBCL 02049 against date palm dust mites by reducing the killing time and increasing killing capacity. However, the extent of pathogenicity that was calculated through joint toxicity varied with the combining proportion of spores with 1-Chlorooctadecane. The simultaneous pairing of *M. anisopliae* EBCL 02049 spores with 1-Chlorooctadecane in a compatible manner as depicted from the biological index in the current study is ideally an interesting alternative approach against such pests, which are difficult to manage with a sole application of fungal spores.

The mortality data of the date palm dust mites in the present study indicated that the sole application of *M. anisopliae* EBCL 02049 produced relatively slow mortality and failed to kill all the tested populations. Slow mortality response of date palm dust mites is in line with what was previously reported against this strain of *M. anisopliae* EBCL 02049 [16]. These findings were further strengthened from previous studies on the virulence of entomopathogenic fungi against date palm dust mites [23,24]. The results of these studies revealed 100% mortality after 15-days post-exposure with *M anisopliae*. Such a long time to impart mortality by the entomopathogenic fungi reported previously [16,24,25], and shown in the current study in the form of low mortality even at the highest concentration (20 mg spores/mL) is demanding an alternative approach to overcome the difficulties in the way of eco-friendly pest management tactics.

Generally, a mixture of plant secondary metabolite with mycopathogen provides eco-friendly stable pest management system compared to environment-deteriorating chemical pesticides. However, the first step to develop a compatible mixture mainly depends on the interaction of these products that might lead to compatible or toxic interaction. The in vitro tests to evaluate the interaction of *M. anisopliae* EBCL 02049 spores with 1-Chlorooctadecane showed encouraging results in laboratory assays. Different concentrations of 1-Chlorooctadecane had little or no effect on the vegetative growth, germination, and sporulation of *M. anisopliae* EBCL 02049. These minor impacts of 1-Chlorooctadecane on growth parameters, which are mainly responsible for the propagation of fungus [26], lead to the value of the biological index under the range of compatible interaction. These compatible interactions at all tested doses of 1-Chlorooctadecane enabled us to support the choice for their simultaneous use in different proportions against date palm dust mites. The search on such a pairing remained a novel option against numerous pest species in order to circumvent slow and low mortality criticism of mycopathogen [23,26–30].

It is well-known that both the fungi and plant secondary metabolites demonstrated an entirely different mode of actions to overcome the target host defense mechanisms to cause mortality. Fungal infection evades the host defense mechanism by triggering complex biochemical interactions in the form of series of events through cuticle adhesion, penetration, proliferation, and toxin production, which ultimately lead to the host mortality [29,31–33]. On the other hand, plant secondary metabolites are known antagonists that act by interfering with the signaling of the nervous and cellular systems to overcome the target host defense mechanism [34–38]. Therefore, the combined application is considered as a promising approach due to the differences in their mechanism of actions. Consequently, the mixture of fungal spores and toxin in a compatible manner were evaluated to develop synergistic interaction against various pest species [23,26,28,39,40]. The mixture of *M. anisopliae* EBCL 02049 and 1-Chlorooctadecane evaluated in the current study showed substantial synergistic interaction, which varies with their proportions. The cumulative corrected mortality of date palm dust mites infected with all the mixtures except 0.64 mg/mL 1-Chlorooctadecane + 0.8 mg/mL spores enlisted in Scheme IV (80% 1-Chlorooctadecane: 20% Spores) was more than 95%. Furthermore, LC_{50} results (1.47 mg/mL) upon exposure with the range of mixtures from Scheme IV showed that these combinations more quickly killed the date palm dust mites compared to *M. anisopliae* EBCL 02049 spores and 1-Chlorooctadecane alone. These findings are in line with the previous study on the synergistic interaction of *Beauveria bassiana* spores with Phytol [23]. However, current results are more promising compared to a previous study (Joint toxicity = 691), due to comparatively high toxicity (joint toxicity = 713). Furthermore, the findings of Zou et al., [40] on the joint action of fungal spores and toxin also strengthened our findings, which demonstrated that the synergistic interaction led to a quick mortality response of the combined treatment compared to their sole application. Similarly, the combined treatment enhanced the treatment effect revealed by Hernández et al., [28] against *Tetranychus urticae* further strengthened our findings. Their findings suggested that the compatibility of azadiractin with fungal spores against *Tetranychus urticae* revealed an additive effect, which contributes to improving their control by enhancing the treatment effect in the form of very high mortality. The outcome as a result of this study in the form of synergistic interaction undoubtedly will be useful for targeted management of date palm dust mites using natural products.

The application of the mixture of *M. anisopliae* EBCL 02049 spores and 1-Chlorooctadecane gave a varying antagonistic interaction especially from the bioassays of Scheme I (20% 1-Chlorooctadecane: 80% Spores). There might be a number of explanations for this interaction. Akbar et al., [41], illustrated in their findings that putative antifungal activity of the toxin might contribute towards antagonistic interaction. The studies conducted against *T. urticae* population to find a compatible interaction of fungal spores and toxin revealed that adjustment of toxin concentration is very important to avoid antagonistic interaction [28]. These findings are in line with our results and enabled us to suggest that treatment efficacy should be regulated through optimization of pest management products to develop pest management synergistic interaction.

The antioxidant defense mechanism of the target host is promptly activated to remove the reactive oxygen species produced during stressful situations. The response of antioxidant enzymes calculated from the activities of SOD, and CAT, have shown treatment-specific and time-specific different patterns. The enzymatic activities of the front-line antioxidant enzyme, SOD against oxidative stress revealed distinct patterns at all the tested time intervals. The most potent treatment established in the current study significantly induced SOD activities until 72 h post-exposure. However, the lateral stage nutrient-deficient date palm dust mites failed to induce relative SOD activities and remained significantly at the lowest level. The induction of SOD activities under stressful situations has already been reported from several mite species. For instance, SOD activities in *Tetranychus cinnabarinus* fed on transgenic cassava lines resistant to this mite species were significantly induced compared with control mites [42]. The recent findings on the antioxidant response of date palm dust mites upon exposure with fungal infections with variable virulence range strengthened our findings by revealing similar patterns of SOD activities from potent treatments as depicted in the current study [24]. The initial increase in SOD activities established here and elsewhere mainly aimed to minimize the cellular damage by regulating the traffic of ROS generated as a result of oxidative stress by converting ROS into H_2O_2. In the meanwhile, CAT is activated as described too in the current study, which detoxifies hydrogen peroxide by transforming into oxygen and water [43]. The least potent treatment in this study revealed high CAT activities over time for the safe removal of ROS corroborates with Hussain et al., [16]. On the other hand, enhanced CAT activities from the most potent treatment at the initial and middle staged infection, while negligible CAT activities from the late-staged nutrient-deficient infected mites are in agreement with previous studies conducted on different mite species on the exploration of antioxidant defense mechanism [16,23,24,42]. Such a pattern of defense enzymes may suggest that toxicity of the treatments regulates the host antioxidant enzyme activities.

4. Materials and Methods

4.1. Date Palm Dust Mites

Populations of *O. afrasiaticus* for laboratory experimentation were directly collected from NCPD (National Center for Palms & Dates), Al-Ahsa, Kingdom of Saudi Arabia. These populations were kept at 25 ± 1 °C; 62.5 ± 12.5% RH, as mentioned in the previous study [24].

4.2. M. anisopliae

The strain EBCL 02049 of *M. anisopliae* isolated from *Coptotermes formosanus* Shiraki during 2002 from Jiangxi, China was grown on potato dextrose agar for twenty-four days under controlled conditions (25 ± 0.5 °C; 70 ± 5% RH). This particular strain was selected because their spores suspension has shown virulence against *Ocinara varians* Walker [13,44], *Coptotermes formosanus* Shiraki [14,15], and *O. afrasiaticus* (McGregor) [16]. The spore suspensions of required concentrations were prepared using 0.05% Tween 80 (Sigma–Aldrich, London, UK) by Neubauer hemocytometer (Wertheim, Germany).

4.3. 1-Chlorooctadecane

1-Chlorooctadecane (CH_3 (CH_2) 16CH_2 Cl) was screened from our preliminary study (data not shown) on the chemical fractionation of *Cucumis sativus* that showed toxicity against date palm dust mites (data not shown). The identified fraction, 1-Chlorooctadecane, selected for the current study was purchased from Sigma–Aldrich in pure form (Cat # 238368; CAS Number 3386-33-2) to perform compatibility, toxicity, and synergistic studies. The stock solution of the 1-Chlorooctadecane was prepared by dissolving in ethanol.

4.4. Compatibility of M. anisopliae Spores with 1-Chlorooctadecane

The potato dextrose agar plates (90-mm × 15 mm) with different concentrations (0.80, 1.60, 2.40, 3.20, and 4.00 mg/mL) of 1-Chlorooctadecane were prepared as described in detail by a previous study [45] Ten microliter spores suspension of *M. anisopliae* EBCL 02049 at a single concentration (1 × 10^6 spores/mL) were pipetted in the center of each PDA-plate. The plates after inoculation were sealed with parafilm. All the plates were incubated in complete darkness at 25 ± 0.5 °C; 70 ± 5% RH. The values of three parameters including sporulation (spores/mL), germination (%), and vegetative growth (mm), were recorded in order to calculate biological index (BI) [23]. Ten replicates were prepared likewise separately for each parameter using ten PDA-plates.

In the case of percent germination, petri plates were incubated for 12 h post-inoculation. Percent germination (GR) of *M. anisopliae* EBCL 02049 spores was calculated by counting a hundred spores under seven different fields of vision. On the other hand, vegetative growth (VG) and sporulation (SP) was determined after 12-days post-inoculation. Seven perpendicular radial growths (mm) of each fungal culture were measured with the help of a transparent ruler for vegetative growth (VG) determination. Furthermore, a 15 mm diameter sporulating culture of *M. anisopliae* EBCL 02049 was taken using 15 mm sterilized cork borer in order to calculate sporulation with the help of Neubauer hemocytometer using 0.05% Tween 80 solution, under a compound microscope. All the parameters including sporulation, percent germination, and vegetative growth were separately analyzed by One-way Analysis of Variance, and their means by Fisher's LSD test ($\alpha = 0.05$).

The BI was calculated by adding the values of above-mentioned parameters into the standard formula (BI = [47 × VG + 43 × SP + 10 × GR]/100). As per standard criterion, BI > 66 ranked as Compatible interaction, while BI < 42 ranked as a toxic interaction. The BI between 42 to 66 is ranked as Moderately Toxic interaction [46].

4.5. Laboratory Evaluation of M. anisopliae Spores with 1-Chlorooctadecane in Different Proportions against Date Palm Dust Mites

Concentration–mortality response bioassays were performed in order to find the most potent fungal–toxin interaction. The treatments with different proportions used for leaf-dip bioassays are mentioned in Table 5. Each treatment (50 mL) sole or in different combinations mentioned in Scheme I, II, III, and IV was prepared separately in a sterilized glass beaker. Pesticide-free leaf-discs of date palm (7.5 cm length × 4 cm width) were dipped with the help of a forceps. After air drying, fifty deutonymphs (second nymphal stage) date palm dust mites were transferred with the help of a camel hair brush on these exposed leaf-discs surrounded with wet cotton in a petri dish (150 mm × 20 mm). Control leaf-discs were dipped in their respective solvent. In case of sole application of spores, 0.05% tween 80 was used as control treatment, while control treatment for 1-Chlorooctadecane was prepared using ethanol with strength used to prepare stock solution of 1-Chlorooctdecane. Control treatment for mixtures was separately prepared using Tween 80 and ethanol with their respective strengths to dissolve them. All the experimental units were maintained at 28.0 ± 0.5 °C with a photoperiod of 16:8 h (L:D). Five replicates were prepared likewise. Mortality data were recorded daily for 10 days post-exposure. Dead mites were transferred into the petri dishes lined with dampened sterile filter paper. Mycosis of the inoculating fungal isolate was confirmed by microscopic examination of the corpses. The Abbott formula was applied in order to correct the treatment mortality data from their respective control mortality data [47]. While each treatment mortality data were used to calculate LC_{50} (Lethal Concentration to impart 50% mortality of date palm dust mites) by Probit analysis [48]. Furthermore, mortality data after angular transformation were analyzed by Repeated Measures ANOVA, and means by Fisher's LSD test [49]. Interaction of different proportions with multiple concentrations as mentioned in Table 5 was designed mainly to calculate joint toxicity as described by Sun et al. [50].

Table 5. Various bioassay schemes evaluated against date palm dust mites.

1-Chlorooctadecane (mg/mL)	Interaction Schemes								*M. anisopliae* (mg/mL)
	Scheme I		Scheme II		Scheme III		Scheme IV		
	1-Chlorooctadecane (20%): Spores (80%)		1-Chlorooctadecane (40%): Spores (60%)		1-Chlorooctadecane (60%): Spores (40%)		1-Chlorooctadecane (80%): Spores (20%)		
	Toxin mg/mL	Spores mg/mL	Toxin mg/mL	Spores mg/mL	Toxin mg/mL	Spores mg/mL	Toxin mg/mL	Spores mg/mL	
0.8	0.16	3.2	0.32	2.4	0.48	1.6	0.64	0.8	4
1.6	0.32	6.4	0.64	4.8	0.96	3.2	1.28	1.6	8
2.4	0.48	9.6	0.96	7.2	1.44	4.8	1.92	2.4	12
3.2	0.64	12.8	1.28	9.6	1.92	6.4	2.56	3.2	16
4	0.8	16	1.6	12	2.4	8	3.2	4	20

4.6. Exploration of Host Antioxidant Defense Response

Second nymphal stage date palm dust mites infected separately with different treatments mentioned in Table 5 were allowed to feed on date palm leaf-discs at 28.0 ± 0.5 °C with a photoperiod of 16:8 h (L:D). Five replicates were prepared. Samples of live date palm dust mites were taken after 24 h, 72 h, and 120 h in order to calculate CAT and SOD activities. Each sample was individually homogenized in ice-cold potassium phosphate buffer saline. The tissues were crushed by a glass homogenizer. All the samples were centrifuged (12,000× *g*) for 15 min. The supernatants were served as enzyme source and taken for protein quantification by a standard method [51]. The enzymatic responses of CAT (Cat # CAT100-1KT, Sigma–Aldrich, London, UK), and SOD (Cat # 19160-1KT-F, Sigma–Aldrich, London, UK), upon exposure with different treatments were calculated using the standard protocols provided by the manufacturer of above-mentioned kits. The enzymatic activities in each treatment were expressed in percentages, which are relative to the enzymatic activities data obtained from their respective control treatments. Two-factor factorial analysis was used to analyze each treatment relative to control treatment enzymatic activities of date palm dust mites, and their significant differences among different treatments by Fisher's LSD test [52].

5. Conclusions

In conclusion, bioassay results of the present study demonstrated that *M. anisopliae* EBCL 02049 was found to be compatible with all tested concentrations of 1-Chlorooctadecane by showing a very high value of biological index (BI > 82). Our results demonstrated that combined application of *M. anisopliae* spores with 1-Chlorooctadecane resulted in enhanced treatment efficacy against second nymphal stage of date palm dust mites. However, we found that higher proportion of 1-Chlorooctadecane (80%) and lower proportion of *M. anisopliae* spores (20%) greatly improved the synergism (joint toxicity = 713). Furthermore, their synergism led to the most potent interaction, which reduced the time and concentration to cause high mortality among date palm dust mites by altering their antioxidant defense mechanism. Hence, the joint action of *M. anisopliae* spores and 1-Chlorooctadecae could be a promising component of integrated management of date palm dust mites.

Author Contributions: A.H. and A.M.A. conceived the idea. A.H. and A.M.A designed the experiments. A.M.A. provided the resources. A.H. performed the experiments. A.H. and A.M.A. analyzed the data. A.H. wrote the original draft. A.H. and A.M.A. revised and edited the manuscript. A.H. and A.M.A. approved the manuscript for publication. All authors have read and agreed to the published version of the manuscript.

Funding: This research was funded by King Abdul Aziz City for Science and Technology, Kingdom of Saudi Arabia, grant number AT-34-293.

Acknowledgments: We greatly appreciate the assistance provided by some of our colleagues.

Conflicts of Interest: The authors declare no conflict of interest. The funders had no role in the design of the study; in the collection, analyses, or interpretation of data; in the writing of the manuscript, or in the decision to publish the results.

References

1. Diallo, H. The role of date palm in combat desertification. In *The Date Palm: From Traditional Resource to Green Wealth*; Emirates Centre for Strategic Studies and Research: Abu Dhabi, UAE, 2005; pp. 13–19.
2. Bashah, M. *Date Varieties in the Kingdom of Saudi Arabia. Guidance Booklet: Palms and Dates*; King Abdulaziz University Press: Riyadh, Saudi Arabia, 1996.
3. El-Hadrami, I.; El-Hadrami, A. Breeding Date Palm. In *Breeding Plantation Tree Crops: Tropical Species*; Springer: New York, NY, USA, 2009; pp. 191–216.
4. FAOSTAT Food and Agricultural Commodities Production. Available online: http://www.fao.org/faostat/en/#data/QC/visualize (accessed on 26 February 2020).
5. Chaaban, S.B.; Chermiti, B.; Kreiter, S. Comparative demography of the spider mite, *Oligonychus afrasiaticus*, on four date palm varieties in Southwestern Tunisia. *J. Insect Sci.* **2011**, *11*, Article 136. [CrossRef]
6. Amer, S.A.A.; Refaat, A.M.; Momen, F.M. Repellent and oviposition-deterring activity of Rosemary and Sweet Marjoram on the spider mites *Tetranychus urticae* and *Eutetranychus orientalis* (Acari: Tetranychidae). *Acta Phytopathol. Entomol. Hungarica* **2001**, *36*, 155–164.
7. Reddy, G.; Srinivasa, N.; Muralidhara, M.S. Potentiality of *Cinnamomum* extracts to two spotted spider mite, *Tetranychus urticae* Koch and its predator *Neoseiulus longispinosus* (Evans). *J. Biopestic.* **2014**, *7*, 1–11.
8. Chiasson, H.; Bélanger, A.; Bostanian, N.; Vincent, C.; Poliquin, A. Acaricidal properties of *Artemisia absinthium* and *Tanacetum vulgare* (Asteraceae) essential oils obtained by three methods of extraction. *J. Econ. Entomol.* **2001**, *94*, 167–171. [CrossRef]
9. Idrees, A.; Qasim, M.; Ali, H.; Qadir, Z.A.; Idrees, A.; Qing, J. Acaricidal potential of some botanicals against the stored grain mites, *Rhizoglyphus tritici*. *J. Entomol. Zool. Stud.* **2016**, *4*, 611–617.
10. Elkertati, M.; Blenzar, A.; Jotei, A.B.; Belkoura, I.; Tazi, B. Acaricide effect of some extracts and fractions on *Tetranychus urticae* Koch (Acari: Tetranychidae). *African J. Agric. Res.* **2013**, *8*, 2970–2976.
11. Bussaman, P.; Sa-uth, C.; Rattanasena, P.; Chandrapatya, A. Effect of crude plant extracts on Mushroom mite, *Luciaphorus* sp. (Acari: Pygmephoridae). *Psyche A J. Entomol.* **2012**, *2012*, Article ID 150958. [CrossRef]

12. Miresmailli, S.; Bradbury, R.; Isman, M.B. Comparative toxicity of *Rosmarinus officinalis* L. essential oil and blends of its major constituents against *Tetranychus urticae* Koch (Acari: Tetranychidae) on two different host plants. *Pest Manag. Sci.* **2006**, *62*, 366–371. [CrossRef]
13. Hussain, A.; Tian, M.Y.; He, Y.R.; Ahmed, S. Entomopathogenic fungi disturbed the larval growth and feeding performance of *Ocinara varians* (Lepidoptera: Bombycidae) larvae. *Insect Sci.* **2009**, *16*, 511–517. [CrossRef]
14. Hussain, A.; Tian, M.Y.; He, Y.R.; Lei, Y.Y. Differential fluctuation in virulence and VOC profiles among different cultures of entomopathogenic fungi. *J. Invertebr. Pathol.* **2010**, *104*, 166–171. [CrossRef]
15. Hussain, A.; Tian, M.Y.; Wen, S.-Y. Proteomic analysis of Formosan Subterranean Termites during exposure to entomopathogenic fungi. *Curr. Proteomics* **2018**, *15*, 229–240. [CrossRef]
16. Hussain, A.; Rizwan-ul-haq, M.; AlJabr, A.M.; Al-Ayedh, H. Host-pathogen interaction for screening potential of *Metarhizium anisopliae* isolates against the date-palm dust mite, *Oligonychus afrasiaticus* (McGregor) (Acari: Tetranychidae). *Egypt. J. Biol. Pest Control* **2019**, *29*, 63. [CrossRef]
17. Kristiani, E.B.E.; Nugroho, L.H.; Moeljopawiro, S.; Widyarini, S. Characterization of volatile compounds of *Albertisia papuana* Becc root extracts and cytotoxic activity in breast cancer cell line T47D. *Trop. J. Pharm. Res.* **2016**, *15*, 959–964. [CrossRef]
18. Swami, S.B.; Thakor, N.S.J.; Patil, M.M.; Haldankar, P.M. Jamun (*Syzygium cumini* (L.)): A review of its food and medicinal uses. *Food Nutr. Sci.* **2012**, *3*, 1100–1117.
19. Li, G.; Jiang, Y.; Li, Y.; He, T.; Wang, Y.; Ji, T.; Zhai, W.; Zhao, L.; Zhou, X. Analysis and biological evaluation of *Arisaema amuremse* Maxim essential oil. *Open Chem.* **2019**, *17*, 647–654. [CrossRef]
20. Siddiquee, S.; Cheong, B.E.; Taslima, K.; Kausar, H.; Hasan, M.M. Separation and identification of volatile compounds from liquid cultures of *Trichoderma harzianum* by GC-MS using three different Capillary Columns. *J. Chromatogr. Sci.* **2012**, *50*, 358–367. [CrossRef]
21. Fetoh, B.; Al-Shammery, K. Acaricidal ovicial and repellent activities of some plant extracts on the date palm dust mite, *Oligonychus afrasiaticus* Meg. (Acari: Tetranychidae). *Int. J. Environ. Sci. Eng.* **2011**, *2*, 45–52.
22. Lakhdari, W.; Dehliz, A.; Acheuk, F.; Soud, A.; Hammi, H.; Mlik, R.; Doumandji-Mitiche, B. Acaricidal activity of aqueous extracts against the mite of date palm *Oligonychus afrasiaticus* Meg (Acari: Tetranychidae). *J. Med. Plants* **2015**, *3*, 113–117.
23. AlJabr, A.; Hussain, A.; Rizwan-ul-haq, M. Toxin-Pathogen synergy reshaping detoxification and antioxidant defense mechanism of *Oligonychus afrasiaticus* (McGregor). *Molecules* **2018**, *23*, 1978. [CrossRef]
24. Hussain, A.; Rizwan-ul-Haq, M.; AlJabr, A.M.; Al-Ayedh, H. Evaluation of host–pathogen interactions for selection of entomopathogenic fungal isolates against *Oligonychus afrasiaticus* (McGregor). *BioControl* **2020**, *65*, 185–195. [CrossRef]
25. Shi, W.-B.; Zhang, L.; Feng, M.-G. Time-concentration-mortality responses of carmine spider mite (Acari: Tetranychidae) females to three hypocrealean fungi as biocontrol agents. *Biol. Control* **2008**, *46*, 495–501. [CrossRef]
26. Gonçalves Diniz, A.; Barbosa, L.F.S.; da Silva Santos, A.C.; de Oliveira, N.T.; da Costa, A.F.; Carneiro-Leão, M.P.; Tiago, P.V. Bio-insecticide effect of isolates of *Fusarium caatingaense* (Sordariomycetes: Hypocreales) combined to botanical extracts against *Dactylopius opuntiae* (Hemiptera: Dactylopiidae). *Biocontrol Sci. Technol.* **2020**, *30*, 1–12.
27. Wraight, S.P.; Filotas, M.J.; Sanderson, J.P. Comparative efficacy of emulsifiable-oil, wettable-powder, and unformulated-powder preparations of *Beauveria bassiana* against the melon aphid *Aphis gossypii*. *Biocontrol Sci. Technol.* **2016**, *26*, 894–914. [CrossRef]
28. Hernández, M.M.; Martínez-Villar, E.; Peace, C.; Pérez-Moreno, I.; Marco, V. Compatibility of the entomopathogenic fungus *Beauveria bassiana* with flufenoxuron and azadirachtin against *Tetranychus urticae*. *Exp. Appl. Acarol.* **2012**, *58*, 395–405. [CrossRef]
29. Hussain, A.; Rizwan-ul-Haq, M.; Al-Ayedh, H.; AlJabr, A. Susceptibility and immune defence mechanisms of *Rhynchophorus ferrugineus* (Olivier) (Coleoptera: Curculionidae) against entomopathogenic fungal infections. *Int. J. Mol. Sci.* **2016**, *17*, 1518. [CrossRef]
30. Hussain, A.; Ahmed, S.; Shahid, M. Laboratory and field evaluation of *Metarhizium anisopliae* var. *anisopliae* for controlling subterranean termites. *Neotrop. Entomol.* **2011**, *40*, 244–250.
31. Hussain, A. Reprogramming the virulence: Insect defense molecules navigating the epigenetic landscape of *Metarhizium robertsii*. *Virulence* **2018**, *9*, 447–449. [CrossRef]

32. Hussain, A.; Rizwan-ul-Haq, M.; Al-Ayedh, H.; Al-Jabr, A. Mycoinsecticides: Potential and future perspective. *Recent Pat. Food. Nutr. Agric.* **2014**, *6*, 45–53. [CrossRef]
33. Hussain, A.; Tian, M.Y. Germination pattern and inoculum transfer of entomopathogenic fungi and their role in disease resistance among *Coptotermes formosanus* (Isoptera: Rhinotermitidae). *Int. J. Agric. Biol.* **2013**, *15*, 319–324.
34. Wink, M. Plant secondary metabolites modulate insect behavior-steps toward addiction? *Front. Physiol.* **2018**, *9*, 364. [CrossRef]
35. Wink, M. Interference of alkaloids with neuroreceptors and ion channels. *Stud. Nat. Prod. Chem.* **2000**, *21*, 3–122.
36. AlJabr, A.M.; Hussain, A.; Rizwan-ul-Haq, M.; Al-Ayedh, H. Toxicity of Plant Secondary Metabolites Modulating Detoxification Genes Expression for Natural Red Palm Weevil Pesticide Development. *Molecules* **2017**, *22*, 169. [CrossRef] [PubMed]
37. Hussain, A.; Rizwan-ul-haq, M.; AlJabr, A.M.; Al-Ayedh, H. Lethality of Sesquiterpenes Reprogramming Red Palm Weevil Detoxification Mechanism for Natural Novel Biopesticide Development. *Molecules* **2019**, *24*, 1648. [CrossRef] [PubMed]
38. Hussain, A.; Rizwan-ul-Haq, M.; Al-Ayedh, H.; Aljabr, A.M. Toxicity and detoxification mechanism of black pepper and its major constituent in controlling *Rhynchophorus ferrugineus* Olivier (Curculionidae: Coleoptera). *Neotrop. Entomol.* **2017**, *46*, 685–693. [CrossRef]
39. Sáenz-de-Cabezón Irigaray, F.J.; Marco-Mancebón, V.; Pérez-Moreno, I. The entomopathogenic fungus *Beauveria bassiana* and its compatibility with triflumuron: Effects on the twospotted spider mite *Tetranychus urticae*. *Biol. Control* **2003**, *26*, 168–173. [CrossRef]
40. Zou, C.; Li, L.; Dong, T.; Zhang, B.; Hu, Q. Joint action of the entomopathogenic fungus *Isaria fumosorosea* and four chemical insecticides against the whitefly *Bemisia tabaci*. *Biocontrol Sci. Technol.* **2014**, *24*, 315–324. [CrossRef]
41. Akbar, W.; Lord, J.C.; Nechols, J.R.; Loughin, T.M. Efficacy of *Beauveria bassiana* for red flour beetle when applied with plant essential oils or in mineral oil and organosilicone carriers. *J. Econ. Entomol.* **2005**, *98*, 683–688. [CrossRef]
42. Lu, F.; Liang, X.; Lu, H.; Li, Q.; Chen, Q.; Zhang, P.; Li, K.; Liu, G.; Yan, W.; Song, J.; et al. Overproduction of superoxide dismutase and catalase confers cassava resistance to *Tetranychus cinnabarinus*. *Sci. Rep.* **2017**, *7*, 40179. [CrossRef]
43. Deisseroth, A.; Dounce, A.L. Catalase: Physical and chemical properties, mechanism of catalysis, and physiological role. *Physiol. Rev.* **1970**, *50*, 319–375. [CrossRef]
44. Hussain, A.; Tian, M.Y.; He, Y.R.; Ruan, L.; Ahmed, S. *In vitro* and *in vivo* culturing impacts on the virulence characteristics of serially passed entomopathogenic fungi. *J. Food Agric. Environ.* **2010**, *8*, 481–487.
45. Hussain, A.; Tian, M.-Y.; Wen, S.-Y. Exploring the caste-specific multi-layer defense mechanism of Formosan Subterranean Termites, *Coptotermes formosanus* Shiraki. *Int. J. Mol. Sci.* **2017**, *18*, 2694. [CrossRef] [PubMed]
46. Schumacher, V.; Poehling, H.-M. In vitro effect of pesticides on the germination, vegetative growth, and conidial production of two strains of *Metarhizium anisopliae*. *Fungal Biol.* **2012**, *116*, 121–132. [CrossRef]
47. Abbott, W.S. A method of computing the effectiveness of an insecticide. *J. Econ. Entomol.* **1925**, *18*, 265–267. [CrossRef]
48. StatsDirect StatsDirect Statistical Software. England. StatsDirect Ltd., 2013. Available online: http://www.statsdirect.com (accessed on 25 February 2020).
49. *Statistix Statistix 8.1*; Analytical Software: Tallahassee, FL, USA, 2003.
50. Sun, Y.-P.; Johnson, E.R. Analysis of joint action of insecticides against House flies. *J. Econ. Entomol.* **1960**, *53*, 887–892. [CrossRef]
51. Bradford, M.M. A rapid and sensitive method for the quantitation of microgram quantities of protein utilizing the principle of protein-dye binding. *Anal. Biochem.* **1976**, *72*, 248–254. [CrossRef]
52. *SAS Institute SAS User's Guide: Statistics*; SAS Institute: Cary, NC, USA, 2000.

Sample Availability: Samples of the compound, 1-Chlorooctadecane and *Metarhizium anisopliae* EBCL 02049 are available from the authors.

molecules

MDPI

Article

Chemical Composition, Antifungal and Insecticidal Activities of the Essential Oils from Tunisian *Clinopodium nepeta* subsp. *nepeta* and *Clinopodium nepeta* subsp. *glandulosum*

Haïfa Debbabi [1], Ridha El Mokni [2,3,4], Ikbal Chaieb [5], Simona Nardoni [6], Filippo Maggi [7], Giovanni Caprioli [7,*] and Saoussen Hammami [1]

1 Research Unit 13ES63, Applied Chemistry and Environment, Faculty of Sciences of Monastir, University of Monastir, 5000 Monastir, Tunisia; debbabi_haifa10@hotmail.com (H.D.); h_saoussen@yahoo.fr (S.H.)

2 Department Pharmaceutical Sciences "A", Laboratory of Botany, Cryptogamy and Plant Biology, Faculty of Pharmacy of Monastir BP 207, Avenue Avicenna, University of Monastir, 5000 Monastir, Tunisia; riridah@yahoo.fr

3 Department of Silvo-Pastoral Resources, Laboratory of Research in Silvo-Pastoral Resources, Silvo-Pastoral Institute of Tabarka, BP. 345, University of Jendouba, Tabarka 8110, Tunisia

4 IRESA, Laboratory of Forest Ecology, I.N.R.G.R.E.F, BP N°10, Ariana 2080, Tunisia

5 Research unit UR13AGR09, Regional Center for Research in Horticulture and Organic Agriculture, Chott Mariem, University of Sousse, TN-4042 Sousse, Tunisia; ikbal_c@yahoo.fr

6 Department of Veterinary Sciences, Università degli Studi di Pisa, 56124 Pisa, Italy; simona.nardoni@unipi.it

7 School of Pharmacy, University of Camerino, 62032 Camerino, Italy; filippo.maggi@unicam.it

* Correspondence: giovanni.caprioli@unicam.it

Academic Editor: Giovanni Benelli
Received: 9 April 2020; Accepted: 1 May 2020; Published: 2 May 2020

Abstract: The present investigation was focused on the study of the chemical composition variability and biological activities of the essential oils from *Clinopodium nepeta* subsp. *nepeta* and subsp. *glandulosum*. Essential oils extraction was performed using hydrodistillation and the separation of the constituents was carried out by gas chromatography coupled with mass spectrometry (GC-MS). Antifungal activities were tested against *Aspergillus flavus, Aspergillus terreus, Microsporum canis, Microsporum gypseum, Trichophyton mentagrophytes,* and *Candida albicans*. Toxicity and repellency were evaluated against the stored product pests *Tribolium confusum* and *Sitophilus zeamais*. Both essential oils were characterized by a high content of oxygenated monoterpenes. Piperitone ranks first in the subspecies *nepeta* and piperitenone oxide is the dominant constituent in the subspecies *glandulosum*. All tested samples displayed noteworthy antifungal properties, with the highest activity observed for the essential oil of *C. nepeta* subsp. *glandulosum*, collected in Béni-M'tir, against *T. mentagrophytes* (MIC = 40 μg/mL). The essential oil samples of *C. nepeta* subsp. *glandulosum* were strongly repellent to the insect species (PR > 80%, after 2h) and highly toxic to *S. zeamais* reaching 97.5%–100% mortality after 24 h of exposure. In conclusion, this study showed considerable intra-specific changes in the quality of *C. nepeta* essential oils, which is reflected in different rates of antifungal and insecticidal activity.

Keywords: Lamiaceae; *Clinopodium nepeta* subsp. *nepeta*; *Clinopodium nepeta* subsp. *glandulosum*; essential oils; chemical variability; biological activities

1. Introduction

All across the globe, environmental problems such as soil and water pollution and food contamination are continuously increasing, inducing many disasters and human tragedies.

The excessive and indiscriminate use of available pesticides to control the losses of stored crops and to reduce insect-borne diseases as malaria, filaria and trypanosomiasis induced disturbances in ecosystem functioning [1,2]. Moreover, synthetic fungicides and fungal-drugs utilization to treat environmental and animal molds are not in a lesser class, when we talk about the negative effects and the appearance of fungicide-resistant pathogens [3,4]. Therefore, there is an increasing concern to search for new highly selective and eco-friendly alternatives of beneficial pest control materials to feed the growing human population in a healthy environment. Medicinal plants produce an arsenal of chemical compounds that alleviate various illnesses and rebalance human health. Much attention has been focused on the study of plant extracts and essential oils due to their pivotal role as a source of phototherapeutics widely used to fight infectious diseases [5,6]. Natural insecticides, fungicides and herbicides with promising effects have the properties to supplant or replace synthetic organic pesticides and therefore to avoid environmental contamination. Hence, various applications of essential oils acting as safe fungicidal agents against a large number of molds and as environmentally friendly insecticides have been reported [7,8]. It is noteworthy that the biosynthesis and accumulation of phytochemicals by medicinal herbs are influenced by environmental circumstances including temperature, climate, light and the region altitude [9]. According to the harvesting time and the environment in which they are found, the same plant species or subspecies may present different chemical compositions of essential oils [10]. Importantly, the nature of the produced secondary metabolites and essential oils influences the mechanisms of actions and determines the medicinal and economic utility of various plants. In this context, differences in essential oils productivity, including an increase or decrease in the yields, and variation of the chemical composition depending on harvesting phase (vegetative, pre-flowering, flowering and fruiting), as well as the plant geographical origin have been previously investigated [11,12].

The genus *Clinopodium* (Lamiaceae family) consists of 135 perennial herbs, most of them being rich sources of essential oils, distributed widely in Southern Europe, western Asia and all around the Mediterranean area [13]. Many preclinical studies have demonstrated the inhibitory effects of *Clinopodium* essential oils towards various bacterial and fungal strains. Their insecticidal properties have also been reported [14–17]. Among various *Clinopodium* species, *Clinopodium nepeta* L. Kuntze (Syn. *Calamintha nepeta* L. Kuntze) is apolymorphic and a fragrant plant that has been used traditionally around the world as an important antispasmodic, diaphoretic, stimulant, and tonic medicinal herb; it is also considered as a mint-like spice and is used in various culinary recipes [18,19]. A literature survey of the chemical investigation of essential oils produced by *C. nepeta* revealed the high content in oxygenated monoterpenes. Most of the phytochemical and biological studies on *C. nepeta* did not concern the level of subspecies [20]. However, some authors take this parameter into consideration during analysis. As a consequence, the presence of remarkable variations among the major constituents of various subspecies was distinguished [21]. Their biological features have also been well-confirmed [22,23]. Thus, antimicrobial, anti-*Candida*, antioxidant and insecticidal activities of *C. nepeta* subsp. *glandulosum* (Req.) Govaerts essential oils from Montenegro, Turkey, Italy, and Croatia have been reported [24–28].

Since antiquity, the essential oils from *C. nepeta* subsp. *nepeta* leaves have been used as a fragrance and insect repellent [29]. The essential oils from the subspecies collected in Portugal, Serbia and Italy have been characterized for antifungal, antiproliferative, antioxidant, and antimicrobial effects [30–33]. In Tunisia only two subspecies, namely *C. nepeta* subsp. *nepeta* and *C. n.* subsp. *glandulosum* have been reported up to now; they have a very attractive smell and are often visited by insect pollinators. As far as we know, nothing is reported on the chemical profile and biological efficacy of their essential oils.

The aim of this paper was to compare the phytochemical profile and to test the biological efficiency of the essential oils extracted from *C. nepeta* subsp. *nepeta* and *C. nepeta* subsp. *glandulosum* growing in different localities in North and North-western Tunisia (Table 1). A comprehensive evaluation of the antifungal and insecticidal activities was investigated considering the chemical variability depending on the subspecies and the geographical origin. The main objective of this study was to search for environmentally friendly insecticides and antifungals that are readily biodegradable, with minimal toxic effects on health and environment and which can be marketed at premium prices.

Table 1. Locality/Harvesting place, harvesting period and voucher specimen reference of both *Clinopodium nepeta* subspecies.

Taxon	Species Abbreviation	Harvesting Place	Harvesting Period (2016)	Voucher Specimen
Clinopodium nepeta (L.) Kuntze subsp. *nepeta*	CNN	Béni-M'tir	October	[LAM./Cal.n.n./ Kroumiria/BM.13/ 27102016]
Clinopodium nepeta subsp. *glandulosum* (Req.) Govaerts	CNG_1	Béni-M'tir	October	[LAM./Cal.n.g./ Kroumiria/BM.25/ 27102016]
Clinopodium nepeta subsp. *glandulosum* (Req.) Govaerts	CNG_2	Bizerte	July	[LAM./Cal.n.g./ NE/Bizerta. 03/ 10082016]

2. Results

2.1. Chemical Profiles of Essential Oils

The hydrodistillation of dry plant materials yielded 1.21%, 0.93% and 0.84% of essential oils, for *Clinopodium nepeta* subsp. *nepeta* (CNN), *Clinopodium nepeta* subsp. *glandulosum* from Béni-M'tir (CNG_1), and *Clinopodium nepeta* subsp. *glandulosum* from Bizerta (CNG_2), respectively. The percent occurrence of the oil phytochemicals, elucidated through GC-MS, is summarized in Table 2.

Table 2. Chemical profiles of essential oils obtained from *Clinopodium nepeta* subspecies harvested from different localities.

No.	RI Calc	RI LIT	Compounds	Content (%)		
				CNN	CNG_1	CNG_2
1	932	932	α-Pinene	0.23	0.2	0.1
2	975	974	β-Pinene	0.3	0.3	0.1
3	991	988	Myrcene	-	0.1	tr
4	995	994	3-Octanol	0.6	0.7	0.7
5	1000	1000	Decane	tr	-	
6	1024	1024	*o*-Cymene			tr
7	1024	1022	*p*-Cymene	-	0.3	-
8	1028	1024	Limonene	1.9	4.2	1.4
9	1030	1026	1,8-cineole	0.4	0.2	0.1
10	1058	1054	γ-Terpinene	tr	0.2	tr
11	1066	1070	*cis*-4-Thujanol	-	0.1	-
12	1100	1095	Linalool	0.7	0.6	0.5
13	1120	1119	*trans*-*p*-Mentha-2,8-dien-1-ol	-	0.1	tr
14	1164	1165	Borneol	-	0.3	0.4
15	1176	1174	Terpinen-4-ol	-	1.1	0.1
16	1184	1179	*p*-Cymen-8-ol	-	0.6	0.2
17	1189	1186	α-Terpineol	0.4	0.4	0.4
18	1197	1196	Methyl chavicol	-	0.3	-
19	1211	1220	4,7-dimethylbenzofuran	-	0.2	tr
20	1215	1221	8,9-Dehydrothymol	-	0.3	0.4
21	1239	1238	Cumin aldehyde	-	2.0	0.1
22	1243	1239	Carvone	-	0.2	tr
23	1253	1249	Piperitone	-	-	19.5
24	1255	1253	Piperitone oxide	51.7	23.5	16.3
25	1268	1274	Pseudodiosphenol	-	-	0.2
26	1271	1277 [a]	*p*-Mentha-1,8-dien-3-one	-	0.5	0.6
27	1286	1287	Bornyl acetate	0.3	0.3	0.2
28	1289	1298	*p*-Mentha-1,4-dien-7-al	-	1.0	0.2
29	1292	1289	Thymol	3.6	1.6	4.0
30	1299	1305	Diosphenol	0.6	-	1.1

Table 2. *Cont.*

No.	RI Calc	RI LIT	Compounds	Content (%)		
				CNN	CNG_1	CNG_2
31	1302	1308	6-Hydroxycarvotanacetone	0.7	5.1	1.2
32	1340	1340	Piperitenone	0.2	0.4	0.5
33	1366	1366	Piperitenone oxide	23.4	39.3	27.8
34	1376	1374	α-Copaene	-	0.6	0.3
35	1385	1387	β-Bourbonene	-	0.4	0.2
36	1400	1400	Tetradecane	0.2	0.9	1.8
37	1419	1417	(*E*)-Caryophyllene	0.3	1.4	0.6
38	1454	1452	α-Humulene	tr	0.1	tr
39	1458	1454	(*E*)-β-Farnesene	-	0.1	tr
40	1481	1484	Germacrene D	-	0.3	tr
41	1524	1522	δ-Cadinene	-	0.2	tr
42	1578	1577	Spathulenol	tr	-	0.2
43	1583	1582	Caryophyllene oxide	2.0	2.7	2.3
44	1600	1600	Hexadecane	-	tr	-
45	1613	1608	Humulene epoxide II	tr	0.2	0.2
46	1689	1687	Eudesma-4(15),7-dien-1β-ol	-	0.3	0.2
47	1848	1844	Phytone	tr	-	tr
Oxygenated monoterpenes				82.0	77.6	73.4
Monoterpene hydrocarbons				2.5	5.3	2.3
Oxygenated sesquiterpenes				2.1	3.1	2.8
Sesquiterpene hydrocarbons				0.4	3.1	1.2
Others				1.0	2.1	2.7
Total identified components				88.0	91.2	82.4

RI Calc: linear retention index calculated against homologue series of C_8–C_{30} alkanes. RI LIT: RI taken from Adams (2007) or NIST 17 (2017). Tr: Traces, % < 0.1. [a] RI value taken from [34].

As shown, 24, 40 and 42 components were identified and quantified in CNN, CNG_1 and CNG_2 essential oils accounting for 88.0%, 91.2% and 82.4% of the total compositions, respectively. Oxygenated monoterpenes constituted the main groups in CNN (82.0%), CNG_1 (77.6%) and CNG_2 (73.4%) essential oils of the selected Lamiaceae plants. It is worth noting that piperitone oxide and piperitenone oxide were identified as the most abundant components in the volatile oils of CNN (51.7% and 23.4%), CNG_1 (23.5% and 39.3%) and CNG_2 (16.3% and 27.8%). While the oxygenated monoterpene piperitone was found in high levels (19.5%) only in CNG_2 (Table 2).

2.2. *In Vitro Antifungal Activity of Essential Oils*

The anti-fungal activity of the selected essential oils was screened against potentially pathogenic fungi in humans and animals, namely *A. flavus, A. terreus, C. albicans, M. canis, M. gypseum,* and *T. mentagrophytes*, as shown in Table 3.

Table 3. Antifungal properties of essential oils produced by hydrodistillation of *Clinopodium nepeta* subsp. *nepeta* and *C. nepeta* subsp. *glandulosum* (MIC, mg·mL^{-1}).

Fungal Strains	Essential Oils		
	CNN	CNG_1	CNG_2
Aspergillus flavus	2	2	>2
Aspergillus terreus	0.4	0.4	0.4
Candida albicans	0.2	0.2	0.4
Microsporum canis	0.4	0.4	0.4
Microsporum gypseum	0.2	0.4	0.4
Trichophyton mentagrophytes	0.2	0.04	0.4

In general terms, the tested fungal isolates showed a variable degree of sensitivity patterns to the selected essential oils. The Minimal Inhibitory Concentrations (MIC) ranged from 0.04 to 2 mg·mL^{-1}. It was found that *A. terreus* and *M. canis* molds were moderately sensitive to all tested samples exhibiting a MIC value of 0.4 mg·mL^{-1}. It is relevant to note that the CNG_1 essential oil was considerably able to inhibit the fungal growth rate of *T. mentagrophytes* dermatophyte (MIC = 0.04 mg·mL^{-1}).

Moreover, the results showed that the CNN essential oil was a little bit more effective on *M. gypseum* when compared to CNG_1 and CNG_2 samples (MIC equal to 0.2 vs. 0.4 mg·mL^{-1}, respectively). On the other hand, CNG_2 essential oil was less effective on the yeast *C. albicans*, with a MIC value of 0.4 mg·mL^{-1}.

2.3. Insecticidal Activity of the Test Oils

The repellent properties of the three *C. nepeta* essential oils against *T. confusum* and *S. zeamais* adults were tested using the McDonald method. Table 4 gives the average repellency values of the essential oils tested at 2% concentration and for different exposure times.

Table 4. Repellency and toxicity of *Tribolium confusum* and *Sitophilus zeamais* exposed to the *C. nepeta subspecies* essential oils (2% and 5% concentrations for repellency and mortality tests, respectively) [A].

Essential Oils Plant Source	Exposure Duration (min)	Repellency (%)		Mortality (%)	
		T. confusum	*S. zeamais*	*T. confusum*	*S. zeamais*
CNN	15	52.5 ± 9.57 [a]	22.5 ± 9.57 [a]	35 ± 5.00 [b]	32.5 ± 5.00 [a]
	30	52.5 ± 5.00 [a]	40 ± 8.16 [a]		
	60	55 ± 5.77 [a]	57.5 ± 9.57 [a]		
	120	57.5 ± 18.92 [a]	57.5 ± 9.57 [a]		
CNG_1	15	82.5 ± 9.57 [b]	60 ± 11.54 [b]	17.5 ± 8.00 [a]	100 ± 0.00 [b]
	30	85 ± 10.00 [b]	62.5 ± 9.57 [b]		
	60	92.5 ± 5.00 [b]	92.5 ± 9.57 [b]		
	120	95 ± 5.77 [b]	92.5 ± 5.00 [b]		
CNG_2	15	80 ± 18.25 [b]	87.5 ± 18.92 [b]	17.5 ± 5.00 [a]	97.5 ± 5.00 [b]
	30	80 ± 8.16 [b]	87.5 ± 5.00 [c]		
	60	82.5 ± 9.57 [b]	90 ± 0.00 [b]		
	120	87.5 ± 5.00 [b]	92.5 ± 9.57 [b]		

[A] Data are mean ± SE ($n = 4$). Means with same alphabetic letters are not significantly different at $p < 0.01$ using Tukey's HSD test between essential oil plant sources for the same exposure duration and same insect species.

We noticed that the repellent activity increased with a prolonged exposure time and varied depending on the tested sample. Thus, both populations of *Clinopodium nepeta* subsp. *glandulosum* (CNG_1 and CNG_2) had a prominent repellency effect against both insects (PR > 80%, after 2 h), while *C. nepeta* subsp. *nepeta* displayed a significantly low activity (PR < 60%, after 2 h), ($F_{df2,11} = 11.3$; $p = 0.003$ for *T. confusum* and $F_{df2,11} = 23.5$; $p < 0.00$ for *S. zeamais*).

Data presented in Table 4 showed that the tested oils exhibited, at 5% concentration, various degrees of toxicity against *T. confusum* and *S. zeamais* after 24 h exposure.

Using topical application bioassay, CNG_1 and CNG_2 essential oils were highly toxic to *S. zeamais*, inducing a mortality rate ranging from 97.5% for CNG_2 to 100% for CNG_1, respectively. However, they were weakly toxic to *T. confusum* (17.5% for both essential oils) ($F_{df2,11} = 8.16$; $p = 0.009$).

On the other hand, the CNN essential oil induced moderate toxicity towards both adult species, with mortality rate values of 35% and 32.5% on *T. confusum* and *S. zeamais*, respectively ($F_{df2,11} = 351.5$; $p < 0.001$).

3. Discussion

The main purpose of this study was to investigate for the first time the phytochemical variability in Tunisian *C. nepeta* essential oils taking into consideration the subspecies *nepeta* and *glandulosum* and their environmental origin (Béni-M'tir and Bizerta), and consequently its effect on the biological properties. As shown in Table 1, *C. nepeta* subsp. *nepeta* and subsp. *glandulosum* essential oils were mostly characterized by high contents of piperitone oxide and piperitenone oxide. Though the type oil resemblance between the subspecies, some quantitative and qualitative differences have been highlighted. Importantly, piperitone oxide which ranks first in CNN essential oil (51.7%), was identified as the second most abundant phytochemical (23.5%) in CNG_1 after piperitenone oxide (39.3%). Nonetheless, the monoterpene piperitone ranks second in frequency in CNG_2 population (19.5%) and was totally absent in the remaining analyzed samples. Some reports stated a strong relationship between the variability of *C. nepeta* essential oil chemical compositions and the geographical origin, environmental conditions and the vegetative state of the plants [31,35,36].

Thus, the observed variability in essential oil compositions is probably due to the various intrinsic genetic factors between the subspecies and different environmental aspects such as altitude and climate for both *glandulosum* subspecies collected in Bizerta and Béni-M'tir [37].

The literature data emphasized a great intraspecific variability of the natural volatile constituents of *C. nepeta* subsp. *nepeta* and subsp. *glandulosum* [30,32]. However, according to some authors, the chemical composition is independent of the subspecies *nepeta* or *glandulosum* as both constitute common sources of terpenoids and may produce the same major volatiles with the C-3 oxygenated *p*-menthane skeleton [33,38]. Others demonstrated a taxonomic understanding of the subspecies [39].

Some interesting observations were made on the diversity of the secondary metabolism of *C. nepeta* essential oils. Almost three types of volatile oils can be distinguished with some exceptions [40]. The first and most popular one consists mainly of pulegone associated with other cyclohexanones as menthone and piperitenone, along with piperitone oxide and piperitenoneoxide. The second kind of essential oil obtained from *Clinopodium* taxa is characterized by the predominance of piperitenone and/or of piperitone oxides, with the last one correlated with the presence of *iso*-pulegone and 1,8-cineole [32].

Thus, it is interesting to note that the essential oils of *C. nepeta* subsp. *nepeta* and subsp. *glandulosum* growing in Tunisia belong to the second type reported in literature as they were rich in piperitone oxide and piperitenone oxide. The compositions of the studied samples are almost similar to those of *C. nepeta* collected in Corsica [21].

Due to their safety characteristics and various aromatherapeutic effects, essential oils have got a lot of attention in several fields of modern chemistry to treat patients and as environmentally friendly preservatives. In this regard, we evaluated the antifungal activities of the aforementioned essential oils, taking into consideration the effect of the chemical polymorphism on the pharmacological effects of essential oils.

Using the micro-dilution assay, the screening revealed that the antifungal effectiveness of all tested oils differs depending on the chemical composition as well as the dissimilarity of target fungus.

The most important effect was recorded by CNG_1 essential oil on *T. mentagrophytes* (MIC = 0.04 mg/mL). In contrast, the CNG_2 oil was less effective on this dermatophyte (MIC = 0.4 mg/mL), along with *A. flavus* (MIC > 2 mg/mL) and *C. albicans* (MIC = 0.4 mg/mL).

The differences may be related to the diversity of their chemical composition and the good antifungal effect of CNG_1 essential oil against *T. mentagrophytes* may be due to the high amount of piperitenone oxide and piperitone oxide. Moreover, we suggest that the lower action of CNG_2 oil against the same dermatophyte is due to the antagonist effect of piperitone quantified at a percentage of 19.5% in this volatile oil.

In the present study, *A. flavus* was the most resistant for all tested oils. Similarly and accordingly with previous reports, this mold presented the highest MIC value for *C. nepeta* subsp. *nepeta* essential oil from Portugal (MIC = 10 $\mu L{\cdot}mL^{-1}$) [31].

Moreover, CNN essential oil appeared to be the most effective on the dermatophyte *M. gypseum* showing the presence of compounds with known antifungal activity as the highly detected piperitone oxide (51.7%).

Similarly, previous studies confirmed the high antifungal activity of *C. nepeta* subsp. *nepeta* essential oil, especially against *Aspergillus* and dermatophyte molds [31]. After screening, the Italian *C. nepeta* volatile oil gave significant MIC values (0.32–1.25 $\mu L{\cdot}mL^{-1}$) in comparison with the less effective oils extracted in Portugal.

Herbivorous insects constitute the most exciting targets that facilitate chemical communication and adaptation to the environment. Repellents are substances with an offensive smell or taste, produced to fight off arthropod insect and to prevent attack from phytophagous. The uses of aromatic herbs and essential oils as insect repellents, has a long history in the herbal folklore [41].

This study represents the first report on the insecticidal efficacy of *C. nepeta* subspecies against *T. confusum* and *S. zeamais*.

The results showed that the repellent action was dependent upon the subspecies *nepeta* or *glandulosum*, while no significant differences were detected between the ability of both samples of *glandulosum* subspecies collected in Bizerta and Béni-M'tir regions to repel or kill the insects. Using the McDonald method, both essential oils at a concentration of 2% showed almost the same repellent activity against *T. confusum* and *S. zeamais* adults after 2 h of exposure. The percentages of repellency were 95 and 87.5% for CNG_1 and CNG_2, respectively, on *T. confusum* and 92.5% for both essential oils on *S. zeamais*. These oils fall under repellency class V (Highly Repulsive) according to Jilani and Su. [42]. However, 57.5% repellency of *C. nepeta* subsp. *nepeta* essential oil was observed against both adult insects, thus the volatile oil falls under class III repellency (Moderately Repulsive).

Our results are in line with the repellent efficacy of *C. nepeta* essential oil towards *Aedes aegypti* mosquito. For instance, the essential oil gave promising scores for both space repellent properties and olfactory studies carried out on human volunteers [43].

Using the topical application bioassay, the application of these essential oils, at a concentration of 5%, resulted in mortality of *T. confusum* and *S. zeamais* within 24 h of exposure.

CNN essential oil was moderately toxic against both *T. confusum* and *S. zeamais*, whereas CNG_1 and CNG_2 essential oils were extremely toxic against *S. zeamais*. The mortality rate was 100 and 97.5%, respectively, after 24 h of exposure. For *T. confusum*, the mortality rate caused by *glandulosum* subspecies was 17.5%. Thus, the CNG_1 and CNG_2 essential oils presented a weaker activity against *T. castaneum* compared to the oil from the Montenegro accession of *C. nepeta* subsp. *glandulosum*, which showed mortality rates of 56.7 and 96.7% after 24 and 96 h of treatment, respectively. [27].

Previous literature data attributed the insecticidal properties of many essential oils to monoterpenoids, mainly the oxygenated ones, which are typically volatile compounds that can penetrate rapidly into the insects and interfere with their physiological functions [41,44].

Actually, the antifungal, repellent and insecticidal assays put in evidence that CNG_1 essential oil was the most active one. This oil was characterized by the highest percentages of piperitenone oxide (39.3 vs. 23.4 and 27.8% of CNN and CNG_2, respectively). This monoterpenoid, endowed with an epoxide group, is typical of several *Mentha* L. essential oils, for instance *M. longifolia* L., *M. suaveolens* Ehrh. and *M. microphylla* K. Koch, which have demonstrated significant toxicity against stored product insects and fungal species [45–47]. Its biological power is given by the epoxide ring, which is interacting with proteins, neurotransmitters and nucleic acids [48]. Furthermore, piperitone oxide has been reported as a toxic repellent and reproduction retardant secondary metabolite towards the malarial vector *Anopheles stephensi* [49] as well as an effective antimicrobial agent [45]. In the present study, the *C. nepeta* subspecies oils showed differences in the activity against *T. confusum* and *S. zeamais*. Thus, repellent and toxic effects of the tested essential oils depend on the chemical composition variability as well as the insect susceptibility.

4. Materials and Methods

4.1. Plant Material and Essential Oils Distillation

The aerial parts of well-selected individuals from two native *C. nepeta* subspecies, i.e., subsp. *nepeta* and subsp. *glandulosum*, were harvested during the flowering stages from different regions/localities in North and North-Western of Tunisia (Table 1). Botanical identification of uninfected plant materials was authenticated by Dr. Ridha El Mokni, affiliated to the Department of Pharmaceutical Sciences "A", Laboratory of Botany, Cryptogamy and Plant Biology, Faculty of Pharmacy of Monastir, Tunisia, where the voucher specimens have been preserved (Table 1). A quantity of 100 g of each sample was subjected to hydrodistillation in a glass Clevenger-type apparatus for 3 h. The oily fraction obtained on top of the aqueous phase at the end of each extraction was separated, dried over anhydrous sodium sulfate, filtered, and stored in the refrigerator until further analysis. The extraction yields were estimated on a dry weight basis (*w*/*w*).

4.2. Gas Chromatography–Mass Spectrometry (GC–MS)

Analyses of essential oil chemical compositions were performed using an Agilent 7890B gas chromatograph equipped with an auto sampler (PAL RSI 85) and coupled to a 5977B single quadrupole mass analyzer (Santa Clara, Californy, USA). Injection of 1 μL of the diluted sample (1:2000 dilution) in *n*-hexane (Carlo Erba, Milan, Italy) in the front inlet set at 280 °C was performed in split mode (1:100) with a split flow of 120 mL/min using an Agilent 5190-3983 liner (800 μL). Separation was performed using a 30 m × 0.25 mm i.d. × 0.25 μm film thickness, 5% phenylmethylpolysiloxane, HP-5MS capillary column (Agilent, Folsom, CA, USA) and helium (99.99%) was the carrier gas flown at 1.2 mL/min. The following oven temperature program was used: 60 °C for 5 min, then 4 °C/min up to 160 °C, then 11 °C/min up to 280 °C with a hold time of 15 min, and finally 15 °C/min until 300 °C for a total run time of 57.74 min. MSD transfer line temperature was set at 300 °C. Analysis was made in electron impact (EI) mode (internal ionization source; 70 eV) with a scan range from 29 to 400 m/z, after a solvent delay of 2.5 min. The compounds were identified by two approaches: (i) correspondence of retention indices (RIs) reported in libraries [50] (NIST 17, 2017; FFNSC 2, 2012) with the ones calculated using a mixture of n-alkanes (C_8–C_{30}, Supelco, Bellefonte, CA, USA); (ii) comparison of the obtained mass spectra with those stored in libraries (WILEY275, ADAMS, NIST 17 and FFNSC 2); (iii) co-injection with available analytical standards. The chromatograms have been integrated and the relative abundance (%) of each compound was obtained ($\% = 100 \times peakarea/totalpeakarea$). The repeatability is expressed by coefficient of variation (CV)% obtained performing the GC/MS analyses in triplicate of the different samples. The coefficient of variation obtained ranged from 0.3% to 5.8%. The inter-day repeatability of the GC/MS method was determined by 3-day replicate analyses of volatiles, evaluated on the same aliquot of sample stored in the refrigerator, CV% ranged from 0.1% to 3.9%, accounting for very high constant results.

4.3. Antifungal Activity

4.3.1. Fungal Isolates

The antifungal effectiveness of *C. nepeta* subspecies essential oils was tested against the animal source dermatophytes: *Microsporum canis*, *Microsporum gypseum* and *Trichophyton mentagrophytes*, the environmental origin molds: *Aspergillus flavus*, *Aspergillus terreus* and the yeast *Candida albicans*.

4.3.2. Microdilution Test

The antifungal susceptibility tests were carried out using a microdilution assay, according to the Clinical and Laboratory Standards Institute (CLSI) M38A_2 recommendations for molds [51], and those of CLSI M27A_3 for yeasts [52]. Essential oils were assayed at different concentrations (2, 1.8, 1.6, 1.4, 1.2, 1, 0.8, 0.6, 0.4, 0.2, 0.08, 0.06, 0.04, and 0.02 mg/mL). All procedures were performed in triplicate.

4.4. Insecticidal Activity

4.4.1. Tested Insects and Rearing Conditions

Tribolium confusum and *Sitophilus zeamais* were taken from the Laboratory of Entomology, Regional Research Center on Horticulture and Organic Agriculture, Chott-Mariem (CRRHAB), Tunisia. Insect adults were cultured in a growth cabinet set at the following rearing conditions: 28 ± 1 °C, 60% relative humidity (RI), 16 h light and 8 h dark photoperiod, without exposure to any insecticidal contamination. The food media used were wheat flour for *T. confusum* and whole maize grains for *S. zeamais*.

4.4.2. Repellent Activity

The repellency test against *T. confusum* and *S. zeamais* beetle adults was assessed following McDonald et al. [53] method. Briefly, 200 µL of each essential oil solution, adjusted at a concentration of 2.0%, were applied to a half Whatman filter paper (No.1) disc of 9 cm diameter. The other half, used as a control, was steeped with 200 µL of pure acetone. After air-drying for 10 min, treated and untreated halves were attached together. Then, 20 adult insects of both species were released separately at the center of the filter paper disc then placed into Petri dishes. After 15, 30, 60, and 120 min from the beginning of the assay the numbers of insects present on the control (Nc) and on the treated (Nt) areas were registered. Each experiment was performed in four repetitions. The repellency percentage values (PR) were computed as follows:

$$PR = [(Nc - Nt)/(Nc + Nt)] \times 100$$

The resulting values were used for the classification of essential oils in different repellency classes suggested by Jilani and Su [42].

4.4.3. Contact Toxicity: Topical Application Bioassay

C. nepeta subspecies essential oils were tested against *T. confusum* and *S. zeamais* following the method of Liu and Ho [54]. Aliquots of 1 µL from each sample at 5% concentration (10 µL of each EO dissolved in acetone) were topically applied on the thorax of insect adults using a micropipette. Insect controls were treated only with acetone. After evaporating the solvent, groups of 10 adults were introduced on glass Petri dishes (9 cm diameter). Four repetitions were carried out for each experiment. Petri dishes were kept under the same rearing conditions described above and insects mortality was recorded after 24 h of treatment (until the number of dead insects stabilized). Insects that did not record any movements were considered as dead. Abbott's formula [55] was used to correct the mortality rate:

$$\%Mc = [(M0 - Mt)/(100 - Mt)] \times 100$$

where Mc: Corrected mortality rate; and M_0 and Mt: Mortality rate of treated and control insects, respectively.

4.4.4. Data Analysis

Statistical analyses were performed using ANOVA followed by Tukey'HSD test. SPSS 20 software was used to perform all tests.

5. Conclusions

The obtained results showed that *Clinopodium nepeta* subsp. *nepeta* and *C. nepeta* subsp. *glandulosum* essential oils presented chemical variability depending on the subspecies and the geographical location of plant materials. The monoterpenoid-rich essential oils demonstrated high antifungal activities against dermatophytes, molds and yeasts with different efficiencies. Given the pronounced repellent and toxic effects towards *T. confusum* and *S. zeamais*, *C. nepeta* subsp. *glandulosum* oils may be considered as promising candidates to control those insect pests during storage. To better understand the pharmacological effects of the analyzed samples, further investigation on the effective major compounds will be carried out. Added to that, an exploration of the synergistic interactions, the antagonist effects and the environmental safety is imperative before suggesting them as fungal drugs or safe alternatives to grain protectants.

Author Contributions: H.D.; preparation of essential oils and performance of insecticidal tests; R.E.M.; plants selection, collection and identification. I.C.; supervision of insecticidal tests; S.N.; performance of antifungal tests; F.M. and G.C.; composition analysis; paper correction and editing.; S.H.; experimental supervision and revision of original draft. All authors have read and agreed to the published version of the manuscript.

Funding: This research received no external funding.

Conflicts of Interest: The authors declare no conflict of interest.

References

1. Boyer, S.; Zhang, H.; Lempérière, G. A review of control methods and resistance mechanisms in stored-product insects. *Bull. Entomol. Res.* **2012**, *102*, 213–229. [CrossRef]
2. Pavela, R. Essential oils for the development of eco-friendly mosquito larvicides: A review. *Ind. Crops Prod.* **2015**, *76*, 174–187. [CrossRef]
3. Abdel-Kawy, M.A.; Michel, C.G.; Kirollos, F.N.; Hussien, R.A.A.; Al-Mahallawi, A.M.; Sedeek, M.S. Chemical composition and potentiation of insecticidal and fungicidal activities of *Citrus trifoliata* L. fruits essential oil against *Spodoptera littoralis, Fusarium oxysporum* and *Fusarium solani* via nano-cubosomes. *Nat. Prod. Res.* **2019**, 1–6. [CrossRef] [PubMed]
4. Rezgui, M.; Majdoub, N.; Mabrouk, B.; Baldisserotto, A.; Bino, A.; Ben Kaab, L.B.; Manfredini, S. Antioxidant and antifungal activities of *marrubiin*, extracts and essential oil from *Marrubium vulgare* L. against pathogenic dermatophyte strains. *J. Mycol. Med.* **2020**, *30*, 100927. [CrossRef] [PubMed]
5. Benelli, G.; Pavela, R.; Zorzetto, C.; Sánchez-Mateo, C.C.; Santini, G.; Canale, A.; Maggi, F. Insecticidal activity of the essential oil from Schizogyne sericea (Asteraceae) on four insect pests and two non-target species. *Entomol. Gen.* **2019**, *39*, 9–18. [CrossRef]
6. Pavela, R.; Maggi, F.; Iannarelli, R.; Benelli, G. Plant extracts for developing mosquito larvicides: From laboratory to the field, with insights on the modes of action. *Acta Trop.* **2019**, *193*, 236–271. [CrossRef]
7. Benelli, G.; Maggi, F.; Canale, A.; Mehlhorn, H. Lyme disease is on the rise—How about tick repellents? A global view. *Entomol. Gen.* **2019**, *39*, 61–72. [CrossRef]
8. Benelli, G.; Pavela, R. Repellence of essential oils and selected compounds against ticks—A systematic review. *Acta Trop.* **2018**, *179*, 47–54. [CrossRef]
9. Ramezani, S.; Rahmanian, M.; Jahanbin, R.; Mohajeri, F.; Reza Rezaei, M.; Solaimani, B. Diurnal Changes in Essential Oil Content of Coriander (*Coriandrum sativum* L.) Aerial Parts from Iran. *Res. J. Biol. Sci.* **2009**, *4*, 277–281.
10. Mishra, R.; Gupta, A.K.; Kumar, A.; Lal, R.K.; Saikia, D.; Chanotiya, C.S. Genetic diversity, essential oil composition, and in vitro antioxidant and antimicrobial activity of *Curcuma longa* L. germplasm collections. *J. Appl. Res. Med. Aromat. Plants* **2018**, *10*, 1–10. [CrossRef]
11. Santos, V.M.C.S.; Pinto, M.A.S.; Bizzo, H.; Deschamps, C. Seasonal variation of vegetative growth, essential oil yield and composition of menthol mint genotypes at southern brazil. *Biosci. J.* **2012**, *28*, 790–798.
12. Noroozisharaf, A.; Kaviani, M. Effect of soil application of humic acid on nutrients uptake, essential oil and chemical compositions of garden thyme (*Thymus vulgaris* L.) under greenhouse conditions. *Physiol. Mol. Biol. Plants* **2018**, *24*, 423–431. [CrossRef] [PubMed]
13. Dunkić, V.; Kremer, D.; Jurišić Grubešić, R.; Vuković Rodríguez, J.; Ballian, D.; Bogunić, F.; Stešević, D.; Kosalec, I.; Bezić, N.; Stabentheiner, E. Micromorphological and phytochemical traits of four *Clinopodium* L. species (Lamiaceae). *South African J. Bot.* **2017**, *111*, 232–241.
14. Castilho, P.; Liu, K.; Rodrigues, A.I.; Feio, S.; Tomi, F.; Casanova, J. Composition and antimicrobial activity of the essential oil of *Clinopodium ascendens* (Jordan) Sampaio from Madeira. *Flavour Fragr. J.* **2007**, *22*, 139–144. [CrossRef]
15. Moattar, F.S.; Sariri, R.; Giahi, M.; Yaghmaee, P. Essential oil composition and antioxidant activity of *Calamintha officinalis* Moench. *J. Appl. Biotechnol. Rep.* **2018**, *5*, 55–58. [CrossRef]
16. Hidalgo, P.J.; Libera, J.L.; Santos, J.A.; LaFont, F.; Castellanos, C.; Palomino, A.; Román, M. Essential oils in *Calamintha sylvatica* Bromf. ssp. ascendens (Jordan) P.W. Ball: Wild and cultivated productions and antifungal activity. *J. Essent. Oil Res.* **2002**, *14*, 68–71. [CrossRef]

17. Li, H.Y.; Liu, X.C.; Chen, X.B.; Liu, Q.Z.; Liu, Z.L. Chemical composition and insecticidal activities of the essential oil of *Clinopodium chinense* (Benth.) kuntze aerial parts against Liposcelis bostrychophila Badonnel. *J. Food Prot.* **2015**, *78*, 1870–1874. [CrossRef]
18. Motti, R.; Ippolito, F.; Bonanomi, G. Folk Phytotherapy in Paediatric Health Care in Central and Southern Italy: A Review. *Hum. Ecol.* **2018**, *46*, 573–585. [CrossRef]
19. Khodja, N.K.; Boulekbache, L.; Chegdani, F.; Dahmani, K.; Bennis, F.; Madani, K. Chemical composition and antioxidant activity of phenolic compounds and essential oils from *Calamintha nepeta* L. *J. Complement. Integr. Med.* **2018**, *15*, 1–12. [CrossRef]
20. Božović, M.; Ragno, R. *Calamintha nepeta* (L.) Savi and its main essential oil constituent Pulegone: Biological activities and chemistry. *Molecules* **2017**, *22*, 290. [CrossRef]
21. Baldovini, N.; Ristorcelli, D.; Tomi, F.; Casanova, J. Intraspesific variability of the essential oil of *Calamintha nepeta* from Corsica (France). *Flavour Fragr. J.* **2000**, *15*, 50–54. [CrossRef]
22. Araniti, F.; Lupini, A.; Sorgonà, A.; Statti, G.A.; Abenavoli, M.R. Phytotoxic activity of foliar volatiles and essential oils of *Calamintha nepeta* (L.) Savi. *Nat. Prod. Res.* **2013**, *27*, 1651–1656. [CrossRef] [PubMed]
23. Mancini, E.; De Martino, L.; Malova, H.; De Feo, V. Chemical composition and biological activities of the essential oil from *Calamintha nepeta* plants from the wild in southern Italy. *Nat. Prod. Commun.* **2013**, *8*, 139–142. [CrossRef] [PubMed]
24. Kitic, D.; Jovanovic, T.; Ristic, M.; Palic, R.; Stojanovic, G. Chemical composition and antimicrobial activity of the essential oil of *Calamintha nepeta* (L.) Savi ssp. *glandulosa* (Req.) P.W. Ball from Montenegro. *J. Essent. Oil Res.* **2002**, *14*, 150–152. [CrossRef]
25. Sarac, N.; Ugur, A. The In Vitro Antimicrobial Activities of the Essential Oils of Some Lamiaceae Species from Turkey. *J. Med. Food* **2009**, *12*, 902–907. [CrossRef] [PubMed]
26. Božovic, M.; Garzoli, S.; Sabatino, M.; Pepi, F.; Baldisserotto, A.; Andreotti, E.; Romagnoli, C.; Mai, A.; Manfredini, S.; Ragno, R. Essential oil extraction, chemical analysis and anti-Candida activity of *Calamintha nepeta* (L.) Savi subsp. *glandulosa* (Req.) Ball-New approaches. *Molecules* **2017**, *22*, 1–12.
27. Popović, A.; Šućur, J.; Orčić, D.; Štrbac, P. Effects Of Essential Oil Formulations On The Adult Insect *Tribolium Castaneum* (Herbst) (Col., Tenebrionidae). *J. Cent. Eur. Agric.* **2013**, *14*, 181–193. [CrossRef]
28. Conforti, F.; Marrelli, M.; Statti, G.; Menichini, F.; Uzunov, D.; Solimene, U.; Menichini, F. Comparative chemical composition and antioxidant activity of *Calamintha nepeta* (L.) Savi subsp. *glandulosa* (Req.) Nyman and *Calamintha grandiflora* (L.) Moench (Labiatae). *Nat. Prod. Res.* **2012**, *26*, 91–97. [CrossRef]
29. Ambrico, A.; Trupo, M.; Martino, M.; Sharma, N. Essential oil of Calamintha nepeta (L.) Savi subsp. *nepeta* is a potential control agent for some postharvest fruit diseases. *Org. Agric.* **2019**, *10*, 35–48. [CrossRef]
30. Vlachou, G.; Papafotiou, M.; Bertsouklis, K.F. Studies on Seed Germination and Micropropagation of *Clinopodium nepeta*: A medicinal and aromatic plant. *HortScience* **2019**, *54*, 1558–1564. [CrossRef]
31. Marongiu, B.; Piras, A.; Porcedda, S.; Falconieri, D.; Maxia, A.; Goncalves, M.J.; Cavaleiro, C.; Salgueiro, L. Chemical composition and biological assays of essential oils of *Calamintha nepeta* (L.) Savi subsp. *nepeta* (Lamiaceae). *Nat. Prod. Res.* **2010**, *24*, 1734–1742. [CrossRef] [PubMed]
32. Arantes, S.M.; Piçarra, A.; Guerreiro, M.; Salvador, C.; Candeias, F.; Caldeira, A.T.; Martins, M.R. Toxicological and pharmacological properties of essential oils of *Calamintha nepeta*, *Origanum virens* and *Thymus mastichina* of Alentejo (Portugal). *Food Chem. Toxicol.* **2019**, *133*, 110747. [CrossRef] [PubMed]
33. Kitic, D.; Stojanovic, G.; Palic, R.; Randjelovic, V. Chemical Composition and Microbial Activity of the Essential Oil of *Calamintha nepeta* (L.) Savi ssp. *nepeta* var. subisodonda (Borb.) Hayek from Serbia. *J. Essent. Oil Res.* **2005**, *17*, 701–703. [CrossRef]
34. Espina, L.; Somolinos, M.; Lorán, S.; Conchello, P.; García, D.; Pagán, R. Chemical composition of commercial citrus fruit essential oils and evaluation of their antimicrobial activity acting alone or in combined processes. *Food Control* **2011**, *22*, 896–902. [CrossRef]
35. Ćavar, S.; Vidic, D.; Maksimović, M. Volatile constituents, phenolic compounds, and antioxidant activity of *Calamintha glandulosa* (Req.) Bentham. *J. Sci. Food Agric.* **2013**, *93*, 1758–1764. [CrossRef]
36. Ristorcelli, D.; Tomi, F.; Casanova, J. Essential oils of *Calamintha nepeta* subsp. *nepeta* and subsp. *glandulosa* from Corsica (France). *J. Essent. Oil Res.* **1996**, *8*, 363–366. [CrossRef]
37. Şanli, A.; Karadoğan, T. Geographical Impact on Essential Oil Composition of Endemic Kundmannia Anatolica HUB.-MOR. (Apiaceae). *African J. Tradit. Complement. Altern. Med. AJTCAM* **2017**, *14*, 131–137. [CrossRef]

38. De Pooter, L.H.; Goetghebeur, P.; Schamp, N. Variability in composition of the essential oil of *Calamintha nepeta*. *Phytochemistry* **1987**, *26*, 3355–3356. [CrossRef]
39. Fraternale, D.; Giamperi, L.; Ricci, D.; Manunta, A. Composition of essential oil as a taxonomic marker for *Calamintha nepeta* (L.) Savi ssp. *nepeta*. *J. Essent. Oil Res.* **1998**, *10*, 568–570. [CrossRef]
40. Alan, S.; Kürkçüoglu, M.; Can Baser, K.H. Composition of essential oils of *Calamintha nepeta* (L.) Savi subsp. *nepeta* and *Calamintha nepeta* (L.) Savi subsp. *glandulosa* (Req.) P.W. Ball. *Asian J. Chem.* **2011**, *23*, 2357–2360.
41. Kumar, P.; Mishra, S.; Malik, A.; Satya, S. Insecticidal properties of Mentha species: A review. *Ind. Crops Prod.* **2011**, *34*, 802–817. [CrossRef]
42. Jilani, G.; Su, H.C.F. Laboratory Studies on Several Plant Materials as Insect Repellants for Protection of Cereal Grains. *J. Econ. Entomol.* **1983**, *76*, 154–157. [CrossRef]
43. Drapeau, J.; Fröhler, C.; Touraud, D.; Kröckel, U.; Geier, M.; Rose, A.; Kunz, W. Repellent studies with Aedes aegypti mosquitoes and human olfactory tests on 19 essential oils from Corsica, France. *Flavour Fragr. J.* **2009**, *24*, 160–169. [CrossRef]
44. Bakkali, F.; Averbeck, S.; Averbeck, D.; Idaomar, M. Biological effects of essential oils-A review. *Food Chem. Toxicol.* **2008**, *46*, 446–475. [CrossRef] [PubMed]
45. Oumzil, H.; Ghoulami, S.; Rhajaoui, M.; Ilidrissi, A.; Fkih-Tetouani, S.; Faid, M.; Benjouad, A. Antibacterial and antifungal activity of essential oils of *Mentha suaveolens*. *Phytother. Res.* **2002**, *16*, 727–731. [CrossRef] [PubMed]
46. Kavallieratos, N.G.; Boukouvala, M.C.; Ntalli, N.; Skourti, A.; Karagianni, E.S.; Nika, E.P.; Kontodimas, D.C.; Cappellacci, L.; Petrelli, R.; Morshedloo, M.R.; et al. Effectiveness of eight essential oils against two key stored-product beetles, *Prostephanus truncatus* (Horn) and *Trogoderma granarium*. *Food Chem. Toxicol.* **2020**, *139*, 111255. [CrossRef] [PubMed]
47. Božović, M.; Pirolli, A.; Ragno, R. *Mentha suaveolens* Ehrh. (Lamiaceae) essential oil and its main constituent piperitenone oxide: Biological activities and chemistry. *Molecules* **2015**, *20*, 8605–8633. [CrossRef]
48. Wink, M.; Schimmer, O. *Modes of Action of Defensive Secondary Metabolites*; Wink, M., Ed.; CRC: Boca Raton, FL, USA, 1999; pp. 17–133.
49. Tripathi, A.K.; Prajapati, V.; Ahmad, A.; Aggarwal, K.K.; Khanuja, S.P.S. Piperitenone Oxide as Toxic, Repellent, and Reproduction Retardant toward Malarial Vector *Anopheles stephensi* (Diptera: Anophelinae). *J. Med. Entomol.* **2004**, *41*, 691–698. [CrossRef]
50. Adams, R.P. *Identification of Essential Oil Components by Gas Chromatography/Mass Spectrometry*, 4th ed.; Allured Publishing Corporation: Carol Stream, IL, USA, 2007.
51. Wayne, P.A. *Clinical and Laboratory Standards Institute (CLSI): Reference Method for Broth Dilution Antifungal Susceptibility Testing of Filamentous Fungi*, 2nd ed.; Allured Publishing Corporation: Carol Stream, IL, USA, 2008; pp. M38–A2.
52. Wayne, P.A. *Clinical and Laboratory Standards Institute (CLSI): Reference Method for Broth Dilution Antifungal Susceptibility Testing of Yeasts*, 3rd ed.; Allured publishing corporation: Carol Stream, IL, USA, 2008; pp. M27–A3.
53. McDonald, L.L.; Guy, R.H.; Speirs, R.D. *Preliminary Evaluation of New Candidate Materials as Toxicants, Repellents and Attractants Against Stored Product Insects, Marketing Research Report*; Agricultural Research Service, United State Department of Agriculture: Washington, DC, USA, 1970.
54. Liu, Z.L.; Ho, S.H. Bioactivity of the essential oil extracted from *Evodia rutaecarpa* Hook f. et Thomas against the grain storage insects, *Sitophilus zeamais* Motsch. and *Tribolium castaneum* (Herbst). *J. Stored Prod. Res.* **1999**, *35*, 317–328. [CrossRef]
55. Abbott, W.S. A method for computing the effectiveness of an insecticidee. *J. Econ. Entomol.* **1925**, *18*, 265–267. [CrossRef]

molecules

MDPI

Article

Chemical Compositions, Mosquito Larvicidal and Antimicrobial Activities of Essential Oils from Five Species of *Cinnamomum* Growing Wild in North Central Vietnam

Do N. Dai [1,2,*], Nguyen T. Chung [1], Le T. Huong [3], Nguyen H. Hung [4], Dao T.M. Chau [5], Nguyen T. Yen [3] and William N. Setzer [6,7,*]

1 Graduate University of Science and Technology, Vietnam Academy of Science and Technology, 18-Hoang Quoc Viet, Cau Giay, Hanoi 10072, Vietnam; chungpuhoat@gmail.com
2 Faculty of Agriculture, Forestry and Fishery, Nghe An College of Economics, 51-Ly Tu Trong, Vinh City 4300, Nghe An Province, Vietnam
3 School of Natural Science Education, Vinh University, 182 Le Duan, Vinh City 4300, Nghệ An Province, Vietnam; lehuong223@gmail.com (L.T.H.); nguyenthiyenth92@gmail.com (N.T.Y.)
4 Center for Advanced Chemistry, Institute of Research and Development, Duy Tan University, 03 Quang Trung, Da Nang 5000, Vietnam; nguyenhuyhung@duytan.edu.vn
5 Institute of Environmental Biochemistry, Vinh University, 182 Le Duan, Vinh City 4300, Nghệ An Province, Vietnam; daochau27@gmail.com
6 Aromatic Plant Research Center, 230 N 1200 E, Suite 100, Lehi, UT 84043, USA
7 Department of Chemistry, University of Alabama in Huntsville, Huntsville, AL 35899, USA
* Correspondence: daidn23@gmail.com (D.N.D.); wsetzer@chemistry.uah.edu (W.N.S.)

Academic Editor: Giovanni Benelli
Received: 20 January 2020; Accepted: 17 February 2020; Published: 12 March 2020

Abstract: Members of the genus *Cinnamomum* (Lauraceae) have aromatic volatiles in their leaves and bark and some species are commercially important herbs and spices. In this work, the essential oils from five species of *Cinnamomum* (*C. damhaensis*, *C. longipetiolatum*, *C. ovatum*, *C. polyadelphum* and *C. tonkinense*) growing wild in north central Vietnam were obtained by hydrodistillation, analyzed by gas chromatography and screened for antimicrobial and mosquito larvicidal activity. The leaf essential oil of *C. tonkinense*, rich in β-phellandrene (23.1%) and linalool (32.2%), showed excellent antimicrobial activity (MIC of 32 μg/mL against *Enterococcus faecalis* and *Candida albicans*) and larvicidal activity (24 h LC_{50} of 17.4 μg/mL on *Aedes aegypti* and 14.1 μg/mL against *Culex quinquefasciatus*). *Cinnamomum polyadelphum* leaf essential oil also showed notable antimicrobial activity against Gram-positive bacteria and mosquito larvicidal activity, attributable to relatively high concentrations of neral (11.7%) and geranial (16.6%). Thus, members of the genus *Cinnamomum* from Vietnam have shown promise as antimicrobial agents and as potential vector control agents for mosquitoes.

Keywords: Lauraceae; *Aedes aegypti*; *Aedes albopictus*; *Culex quinquefasciatus*; antibacterial; antifungal

1. Introduction

The Lauraceae is a large family of tropical and subtropical trees and shrubs [1]. In this family, the genus *Cinnamomum* is comprised of around 250 species with concentrations in east and southeast Asia [1]. Vietnam is home to 45 species of *Cinnamomum* [2,3], many of which are used in traditional medicine, for essential oils, as well as for timber [4,5].

We are in the midst of a post-antibiotic era. Numerous pathogenic microorganisms have developed resistance to commonly used antibiotic agents [6,7]. For example, *Klebsiella pneumoniae* [8], *Pseudomonas aeruginosa* [9] and *Staphylococcus aureus* [10], three organisms that are major causes of nosocomial infections, have developed extremely drug resistant (XDR) strains. Likewise, antibiotic resistance is increasing in fungi such as *Candida* ssp. and *Aspergillus* spp. [11]. Essential oils have shown promise as complementary or adjuvant therapies for combating antimicrobial resistance [12–19].

Mosquitoes have been and continue to be the deadliest animals on earth. *Aedes aegypti* (L.) (Diptera: Culicidae) and *Ae. albopictus* (Skuse) are vectors for the arboviral diseases dengue, Zika, chikungunya and yellow fever and *Ae. aegypti* is also a vector for the emerging Rift Valley fever virus [20]. *Culex quinquefasciatus* (Say) is a vector of West Nile virus, Saint Louis encephalitis virus and lymphatic filariasis [21]. *Culex quinquefasciatus* may also serve as a vector in emerging viral diseases such as Zika virus [22], Sindbis virus [23] and Usutu virus [24]. Unfortunately, insecticidal resistance of these mosquito species is increasing leading to failure of vector control programs in many locations [25]. Furthermore, populations of *Ae. aegypti* [26], *Ae. albopictus* [27] and *Cx. quinquefasciatus* [28] are showing widespread resistance to commonly used larvicidal agents. It has been suggested that essential oils may serve as alternative and more ecologically benign mosquito larvicidal agents [29–31].

Because of the biological activities and traditional uses of members of the *Cinnamomum* genus, we hypothesize that *Cinnamomum* species from Vietnam may also exhibit potentially useful biological activities. As part of our ongoing investigations into the essential oils of Vietnamese *Cinnamomum* [32–34], we have obtained, analyzed and carried out antimicrobial and larvicidal screening of *Cinnamomum ovatum* C.K. Allen (syn. *Cinnamomum rigidissimum* H.T. Chang), *Cinnamomum tonkinense* (Lecomte) A. Chev. (syn. *Cinnamomum albiflorum* var. *tonkinense* Lecomte), *Cinnamomum damhaensis* Kosterm., *Cinnamomum longipetiolatum* H.W. Li and *Cinnamomum polyadelphum* (Lour.) Kosterm. (syn. *Laurus polyadelpha* Lour., *Cinnamomum litseafolium* Lecomte, *Cinnamomum litseafolium* var. *denticupulatum* Liou, *Cinnamomum saigonicum* Farw, *Camphorina saigonica* Farw).

2. Results

2.1. Essential Oil Collection and Analysis

Plant materials were collected from mature *Cinnamomum* trees from different locations in north central Vietnam. The collection details and essential oil yields of the *Cinnamomum* species are summarized in Table 1. The essential oils were analyzed by gas-chromatography–mass spectrometry (GC-MS) and gas chromatography–flame ionization detector (GC–FID). The chemical compositions of the *Cinnamomum* species are presented in Table 2.

Table 1. Collection details for *Cinnamomum* species from north central Vietnam.

Cinnamomum Species	Vietnamese Name	Voucher Numbers	Part	Yield, % v/w	Collection Month/Year	Collection Location
Cinnamomum ovatum	Re trứng	DND-762	Leaf Stems	0.60 0.21	April/2019	Chau Hoan Commune, Pù Huống Nature Reserve 19°28′12″N, 104°56′45″E, elev. 374 m
Cinnamomum tonkinense	Re xanh, Re bắc, Quế bắc	DND-768	Leaf	0.33	April/2019	Chau Hoan Commune, Pù Huống Nature Reserve 19°28′12″N, 104°56′45″E, elev. 374 m
Cinnamomum damhaensis	Re đầm hà	DND-786	Leaf	0.30	July/2019	Huong Phu Commune, Nam Đông District, Bach Ma National Park 16°12′47″N, 107°43′33″E, elev. 101 m
Cinnamomum longipetiolatum	Re cuống dài	DND-800	Leaf	1.35	August/2019	Nam Nhong Commune, Que Phong District, Pù Hoạt Nature Reserve 19°30′24″N, 104°42′52″E, elev. 667 m
Cinnamomum polyadelphum	Quế bời lời, Miếng sành, Tà Dúi, Ô dược, Đam dao, Hậu phát	DND-813	Leaf	1.20	August/2019	Nam Nhong Commune, Que Phong District, Pù Hoạt Nature Reserve 19°30′24″N, 104°42′52″E, elev. 667 m

Table 2. Chemical compositions (%) of *Cinnamomum* essential oils from north central Vietnam.

Nº	Compounds	RI[a]	RI[b]	*C. ov.*[c]	*C. ov.*[c]	*C. to.*[d]	*C. da.*[e]	*C. lo.*[f]	*C. po.*[g]
				Leaf	**Stem**	**Leaf**	**Leaf**	**Leaf**	**Leaf**
1	α-Thujene	930	924	-	-	0.4	-	-	0.3
2	α-Pinene	939	932	2.1	1.6	4	0.3	2.9	4.3
3	α-Fenchene	953	945	-	-	-	0.1	-	0.2
4	Camphene	955	946	0.7	0.6	0.4	0.2	0.3	1.9
5	Sabinene	978	969	-	-	3.4	0.7	0.5	0.4
6	β-Pinene	984	974	0.9	0.6	2.1	0.2	1.7	2.4
7	Myrcene	992	988	0.2	0.1	3.1	0.1	0.2	2.1
8	Dehydroxy-*trans*-linalool oxide	995	991	-	-	-	-	0.7	-
9	Dehydroxy-*cis*-linalool oxide	1008	1006	-	-	-	-	0.6	-
10	α-Phellandrene	1010	1002	1.3	0.3	4.8	-	-	0.2
11	α-Terpinene	1022	1014	-	-	0.4	-	0.2	
12	*p*-Cymene	1030	1020	0.7	0.4	0.5	0.7	0.2	0.7
13	Limonene	1035	1024	0.9	0.8	3.4	0.2	0.3	5.4
14	β-Phellandrene	1036	1025	-	-	23.1	-	-	-
15	1,8-Cineole	1038	1026	0.2	0.6	9.8	1	2.5	0.8
16	(*E*)-β-Ocimene	1049	1044	0.3	0.2	0.3	-	0.8	-
17	γ-Terpinene	1063	1054	-	-	0.5	-	0.3	-
18	Terpinolene	1094	1086	0.2	0.1	0.2	-	-	-
19	Rosefuran	1098	1091	-	-	-	-	-	0.1
20	Perillene	1104	1102	-	-	-	-	-	0.2
21	Linalool	1105	1095	5.9	8.3	32.2	44.8	75.7	3.2
22	Hotrienol	1107	1104	-	-	-	-	3.2	-
23	Isocitral	1147	1140	-	-	-	-	-	0.2
24	Camphor	1156	1141	-	-	-	-	-	32.2
25	Nerol oxide	1158	1154	-	-	-	-	0.2	-
26	Isoneral	1166	1162	-	-	-	-	-	0.6
27	*cis*-Linalool oxide (pyranoid)	1174	1170	-	-	-	-	3.2	-
28	*trans*-Linalool oxide (pyranoid)	1177	1173	-	-	-	-	2.7	-
29	Borneol	1178	1165	-	-	0.2	-	-	1.6
30	Isogeranial	1184	1180	-	-	-	-	-	0.9
31	Terpinen-4-ol	1187	1174	-	-	0.7	0.4	0.5	0.4
32	α-Terpineol	1200	1186	-	-	1.7	0.2	0.6	0.6
33	Decanal	1208	1201	-	-	0.2	-	-	0.2
34	Citronellol	1228	1223	-	-	-	-	-	0.4
35	Nerol	1231	1227	-	-	-	-	-	0.8
36	Cuminal	1238	1238	-	-	-	0.1	-	-
37	Neral	1245	1235	-	-	-	-	-	11.7
38	Geraniol	1255	1249	-	-	-	-	0.2	1.9
39	Geranial	1274	1264	-	-	-	-	-	16.6
40	(*E*)-Cinnamaldehyde	1278	1267	-	-	-	-	-	0.2
41	Bornyl acetate	1294	1287	0.5	0.5	0.1	-	-	0.1
42	Safrole	1299	1285	-	0.2	-	-	-	-
43	δ-Elemene	1348	1335	-	-	0.2	-	-	-
44	Eugenol	1367	1356	70.5	71.2	0.2	-	-	0.4
45	α-Ylangene	1385	1373	0.4	0.3	-	-	-	-
46	α-Copaene	1389	1374	-	-	0.4	-	-	0.1
47	β-Elemene	1403	1389	-	-	0.3	0.7	-	-
48	Methyl eugenol	1409	1403	-	0.3	-	0.1	-	0.3
49	β-Caryophyllene	1437	1417	1.9	1	1.8	0.2	-	1.3
50	*trans*-α-Bergamotene	1445	1432	0.2	0.2	0.1	-	-	-
51	*allo*-Aromadendrene	1457	1458	0.1	-	-	0.3	-	-
52	α-Humulene	1471	1452	0.3	-	0.6	-	-	0.2
53	α-Amorphene	1483	1483	-	-	-	0.3	-	-
54	β-Selinene	1489	1489	-	-	-	19.1	-	-
55	*trans*-β-Bergamotene	1496	1480	-	0.1	-	-	-	-
56	α-Selinene	1498	1498	-	-	-	0.5	-	-
57	Germacrene D	1498	1484	0.3	0.1	2.7	-	-	-
58	Bicyclogermacrene	1513	1500	0.7	0.6	1	-	0.2	0.2
59	β-Bisabolene	1517	1505	-	0.2	-	-	-	-
60	Eugenyl acetate	1533	1521	9.5	9.3	0.1	-	-	-
61	δ-Cadinene	1537	1522	0.2	0.3	0.2	-	-	0.1
62	(*E*)-α-Bisabolene	1551	1544	-	0.3	-	-	-	-
63	(*E*)-Nerolidol	1570	1561	-	-	0.3	0.9	-	-
64	Germacrene B	1577	1559	-	-	0.2	-	-	-
65	Spathulenol	1598	1577	-	0.6	-	0.4	1.1	0.8

Table 2. *Cont.*

N°	Compounds	RI[a]	RI[b]	*C. ov.*[c] Leaf	*C. ov.*[c] Stem	*C. to.*[d] Leaf	*C. da.*[e] Leaf	*C. lo.*[f] Leaf	*C. po.*[g] Leaf
66	Caryophyllene oxide	1605	1582	-	0.5	0.2	1.1	0.5	0.9
67	Intermedeol isomer	1616	-	-	-	-	5.8	-	-
68	Selin-11-en-4-one	1626	1626	-	-	-	1.5	-	-
69	Selina-3,11-dien-6α-ol	1644	1642	-	-	-	0.6	-	-
70	α-Cadinol	1652	1652	-	-	-	0.5	-	-
71	Selin-11-en-4α-ol	1660	1658	-	-	-	7.3	-	-
72	Germacra-4(15),5,10(14)-trien-1α-ol	1685	1685	-	-	-	1	-	-
73	Aromadendrane-4,10-diol	1717	-	-	-	-	0.5	-	-
74	Oplopanone	1735	1739	-	-	-	0.3	-	-
75	α-Cyperone	1747	-	-	-	-	4	-	-
76	Cyclocolorenone	1763	1759	-	-	-	0.1	-	-
	Monoterpene hydrocarbons			7.3	4.7	46.6	2.5	7.4	17.9
	Oxygenated monoterpenoids			6.6	9.4	44.7	46.5	90.1	72.3
	Sesquiterpene hydrocarbons			4.1	3.1	7.5	21.1	0.2	1.9
	Oxygenated sesquiterpenoids			0	1.1	0.5	24	1.6	1.7
	Phenylpropanoids			80	81	0.3	0.1	0	0.9
	Others			0	0	0.2	0	0	0.2
	Total identified			98	99.3	99.8	94.2	99.3	94.9

[a] RI = Retention Index determined on an HP-5ms column. [b] RI from the databases. [c] *C. ov.* = *Cinnamomum ovatum*. [d] *C. to.* = *Cinnamomum tonkinense*. [e] *C. da.* = *Cinnamomum damhaensis*. [f] *C. lo.* = *Cinnamomum longipetiolatum*. [g] *C. po.* = *Cinnamomum polyadelphum*.

2.2. Antimicrobial Screening

The *Cinnamomum* essential oils were screened for antimicrobial activity against Gram-positive (*Enterococcus faecalis, Staphylococcus aureus, Bacillus cereus*) and Gram-negative (*Escherichia coli, Pseudomonas aeruginosa, Salmonella enterica*) bacteria and a yeast (*Candida albicans*). Minimum inhibitory concentrations and IC_{50} values were determined using the microbroth dilution assay (Table 3).

Table 3. Antimicrobial activities of *Cinnamomum* essential oils from north central Vietnam.

Sample	Gram-Positive *Enterococcus faecalis*	*Staphylococcus aureus*	*Bacillus cereus*	Gram-Negative *Escherichia coli*	*Pseudomonas aeruginosa*	*Salmonella enterica*	Yeast *Candida albicans*
				MIC (μg/mL[a])			
C. ovatum leaf	64	64	128	64	128	64	64
C. ovatum stem	64	64	64	64	16	64	32
C. tonkinense	32	128	128	-	-	-	32
C. damhaensis	-	-	-	-	-	-	-
C. longipetiolatum	64	128	128	256	256	128	256
C. polyadelphum	32	64	64	-	-	128	256
Streptomycin	32	128	64	32	128	64	-
Nistatin	-	-	-	-	-	-	8
Cyclohexamide	-	-	-	-	-	-	32
				IC_{50} (μg/mL[a])			
C. ovatum leaf	32.33	32.33	65.45	32.56	65.44	33.22	33.22
C. ovatum stem	32.44	32.78	33.56	32.56	8.77	31.22	15.67
C. tonkinense	5.67	37.78	56.67	-	-	-	15.67
C. damhaensis	-	-	-	-	-	-	-
C. longipetiolatum	17.88	36.78	56.79	100.34	105.67	56.78	112.45
C. polyadelphum	10.67	24.78	30.24	-	-	57.45	123.45

[a] Micrograms of essential oil per milliliter of test solution.

2.3. Larvicidal Screening

The *Cinnamomum* essential oils were screened for mosquito larvicidal activity against *Aedes aegypti*, *Aedes albopictus* and *Culex quinquefasciatus*. The 24 h and 48 h LC_{50} and LC_{90} values are summarized in Tables 4 and 5.

Table 4. Twenty-four-hour mosquito larvicidal activities (μg/mL) of *Cinnamomum* essential oils from north central Vietnam.

Sample	LC_{50} (95% Confidence Limits)	LC_{90} (95% Confidence Limits)	χ^2	*p*
		Aedes aegypti		
C. ovatum leaf EO	24.12 (20.92–27.45)	50.61 (45.02–58.65)	48.86	0.000
C. ovatum stem EO	52.51 (48.77–57.69)	71.23 (64.50–82.64)	0.4722	0.790
C. tonkinensis leaf EO	17.44 (15.53–19.58)	31.40 (27.93–36.64)	0.1354	0.987
C. damhaensis leaf EO	21.43 (18.66–24.15)	38.98 (34.75–45.58)	0.5494	0.760
C. longepetiolatum leaf EO	64.20 (55.67–73.61)	127.9 (111.0–156.5)	8.805	0.003
C. polyadelphum leaf EO	23.41 (21.37–25.78)	36.69 (33.27–41.52)	8.277	0.041
		Aedes albopictus		
C. ovatum leaf EO	n.t.	n.t.	—	—
C. ovatum stem EO	61.45 (55.66–68.20)	103.3 (93.3–117.1)	34.38	0.000
C. tonkinensis leaf EO	42.89 (39.73–46.59)	61.65 (56.52–69.09)	2.595	0.273
C. damhaensis leaf EO	43.91 (41.25–46.46)	56.16 (52.95–60.79)	0.04480	0.978
C. longepetiolatum leaf EO	n.t.	n.t.	—	—
C. polyadelphum leaf EO	20.66 (18.02–23.28)	37.21 (33.04–43.97)	2.577	0.276
		Culex quinquefasciatus		
C. ovatum leaf EO	34.19 (31.18–37.65)	56.01 (50.85–63.12)	10.73	0.013
C. ovatum stem EO	28.79 (22.07–34.79)	78.3 (67.72–94.57	8.295	0.016
C. tonkinensis leaf EO	14.05 (12.28–15.75)	25.70 (23.06–29.59)	16.31	0.001
C. damhaensis leaf EO	46.74 (41.58–52.63)	86.80 (77.37–100.39)	13.53	0.001
C. longepetiolatum leaf EO	126.8 (108.3–151.4)	293.9 (248.2–368.9)	21.47	0.000
C. polyadelphum leaf EO	18.33 (13.26–22.87)	58.95 (50.62–72.05)	5.639	0.131

Table 5. Forty-eight-hour mosquito larvicidal activities (μg/mL) of *Cinnamomum* essential oils from north central Vietnam.

Sample	LC_{50} (95% Confidence Limits)	LC_{90} (95% Confidence Limits)	χ^2	*p*
		Aedes aegypti		
C. ovatum leaf EO	13.76 (11.42–15.95)	30.17 (26.76–35.17)	46.16	0.000
C. ovatum stem EO	46.74 (43.17–51.21)	67.53 (61.30–77.07)	2.744	0.254
C. tonkinensis leaf EO	15.83 (13.76–17.99)	31.17 (27.53–36.80)	2.196	0.533
C. damhaensis leaf EO	17.36 (13.67–20.36)	37.53 (32.94–45.22)	5.494	0.064
C. longepetiolatum leaf EO	39.50 (29.92–47.24)	95.24 (83.27–114.77)	2.513	0.113
C. polyadelphum leaf EO	17.30 (15.44–19.41)	30.80 (27.43–35.89)	3.650	0.302
		Aedes albopictus		
C. ovatum leaf EO	n.t.	n.t.	—	—
C. ovatum stem EO	50.18 (45.07–56.12)	87.98 (78.88–100.81)	35.66	0.000
C. tonkinensis leaf EO	42.74 (39.48–46.59)	62.40 (57.00–70.25)	4.098	0.129
C. damhaensis leaf EO	39.85 (37.05–42.91)	56.02 (51.91–61.70)	0.06006	0.970
C. longepetiolatum leaf EO	n.t.	n.t.	—	—
C. polyadelphum leaf EO	20.79 (17.84–23.61)	39.45 (34.97–46.59)	6.980	0.031
		Culex quinquefasciatus		
C. ovatum leaf EO	30.48 (27.00–34.48)	59.19 (52.54–68.81)	1.181	0.757
C. ovatum stem EO	20.54 (11.92–27.10)	72.40 (62.01–89.54)	5.799	0.055
C. tonkinensis leaf EO	8.721 (6.874–10.253)	18.81 (16.70–22.01)	26.83	0.000
C. damhaensis leaf EO	18.63 (9.90–25.06)	67.93 (58.16–84.15)	6.243	0.001
C. longepetiolatum leaf EO	76.88 (52.08–101.93)	314.5 (249.3–447.0)	47.36	0.000
C. polyadelphum leaf EO	11.03 (4.50–15.93)	52.40 (44.35–65.76)	10.30	0.016

3. Discussion

3.1. *Cinnamomum ovatum*

The leaf and stem bark essential oils of *C. ovatum* demonstrated broad antimicrobial activity against the organisms tested with MIC values ranging from 16 to 128 μg/mL (Table 3). The major components of the leaf and stem essential oils were eugenol (70.5% and 71.2%, respectively), eugenyl acetate (9.5% and 9.3%, respectively) and linalool (5.9% and 8.3%, respectively) (Table 2). The high concentration of eugenol in these two essential oils is likely responsible for the observed antimicrobial effects. Eugenol has shown broad spectrum antibacterial [35,36] and antifungal [37–39] activities. Likewise, the mosquito larvicidal activity of *C. ovatum* leaf essential oil is likely due to eugenol; that compound has shown larvicidal activity against *Ae. aegypti* [40], *Ae. albopictus* [41] and *Cx. quinquefasciatus* [42].

Cinnamomum cambodianum leaf [34] and stem bark [33] essential oils from Vietnam have also shown high concentrations of linalool (27.0% and 33.1%, respectively).

3.2. *Cinnamomum tonkinense*

Cinnamomum tonkinense leaf essential oil showed excellent antimicrobial activity against *E. faecalis* and *C. albicans* with MIC of 32 μg/mL and good activity against *B. cereus* and *S. aureus* (Table 3). The essential oil is rich in monoterpenes, α-pinene (4.0%), sabinene (3.4%), α-phellandrene (4.8%), β-phellandrene (23.1%), 1,8-cineole (9.8%), linalool (32.2%) (Table 2). Both α-pinene and linalool have shown antibacterial activity against *E. faecalis* [35] and *S. aureus* [43]; α-pinene and 1,8-cineole have shown antifungal activity against *C. albicans* [43]. Sabinene, on the other hand, has shown little [44] or no [45] antimicrobial activity. Likewise, α-phellandrene has shown no activity against *C. albicans* [46]. The leaf essential oils of *C. cordatum* and *C. scortechini* from Pahang, Malaysia, both rich in β-phellandrene (9.0% and 17.3%, respectively) and linalool (17.3% and 16.4%, respectively), have shown antifungal activities against several fungal strains [47].

The leaf essential oil of *C. tonkinense* is one of the most larvicidal in this study (Tables 4 and 5). The major components in the essential oil likely account for the observed larvicidal activity. α-Pinene, has been shown to be larvicidal against *Ae. aegypti*, *Ae. albopictus* and *Cx. quinquefasciatus* [48]; sabinene and linalool have both demonstrated larvicidal against *Ae. aegypti* and *Cx. quinquefasciatus* [49]; and α-phellandrene has shown activity against *Ae. aegypti* and *Ae. albopictus* [50] as well as *Culex pipiens molestus* [51]. The leaf essential oil of *C. scortechinii*, rich in β-phellandrene (17.3%) and linalool (16.4%), had shown excellent larvicidal activity against *Ae. aegypti* and *Ae. albopictus* (LC_{50} = 21.5 and 16.7 μg/mL, respectively) [52].

3.3. *Cinnamomum damhaensis*

The major components of *C. damhaensis* leaf essential oil were linalool (44.8%) and β-selinene (19.1%) (Table 2). The essential oil also showed pronounced larvicidal activity against *Ae. aegypti* and *Cx. quinquefasciatus* with 48 h LC_{50} values of 17.4 and 18.6 μg/mL, respectively (Table 5), which can be attributed to the high concentration of linalool (see above). Note that *Piper gaudichaudianum* and *Piper humaytanum* leaf essential oils, rich in β-selinene (10.5% and 15.8%, respectively), but devoid of linalool, showed only marginal larvicidal activity against *Ae. aegypti* [53].

3.4. *Cinnamomum longipetiolatum*

The leaf essential oil of *C. longipetiolatum* was dominated by linalool (75.7%, Table 2), which likely accounts for the observed antimicrobial (Table 3) activity; linalool has shown broad antibacterial and antifungal activity [35,54]. Although linalool has shown larvicidal activity against *Ae. aegypti* (LC_{50} = 38.6 μg/mL) and *Cx. quinquefasciatus* (LC_{50} = 42.3 μg/mL) [49], the larvicidal activity of *C. longipetiolatum* leaf oil was less (24 h LC_{50} = 64.2 and 126.8 μg/mL against *Ae. aegypti* and *Cx. quinquefasciatus*, respectively, Table 4).

3.5. *Cinnamomum polyadelphum*

The leaf essential oil of *C. polyadelphum* showed good activity against the Gram-positive organisms tested with MIC values of 32, 64 and 64 μg/mL on *E. faecalis*, *S. aureus* and *B. cereus*, respectively (Table 3). The essential oil also showed notable larvicidal activity against all three mosquito species with 48-h LC_{50} values of 17.3, 20.8 and 11.0 μg/mL against *Ae. aegypti*, *Ae. albopictus* and *Cx. quinquefasciatus*, respectively (Table 5). The major components in *C. polyadelphum* leaf essential oil were camphor (32.2%), neral (11.7%) and geranial (16.6%) (Table 2). The antimicrobial properties of camphor are relatively marginal [55,56]. Citral (mixture of neral and geranial), on the other hand, has shown greater antimicrobial activity on Gram-positive bacteria [57–59] and fungi [60,61]. Likewise, citral has exhibited mosquito larvicidal activity against *Ae. albopictus* [62] but camphor is inactive against larvae of *Ae. aegypti*, *Ae. albopictus* [52,62] or *Cx. pipiens* [63].

4. Materials and Methods

4.1. Plant Material

Leaves or stem bark of the *Cinnamomum* species were collected from locations in north central Vietnam (see Table 1). Plants were identified by Do N. Dai and voucher specimens (Table 1) have been deposited in the plant specimen room, Faculty Agriculture, Forestry and Fishery, Nghe An, College of Economics. The fresh plant materials (2.0 kg each) were shredded and hydrodistilled using a Clevenger apparatus for 4 h to give the essential oils. The essential oil yields are summarized in Table 1.

4.2. Gas Chromatographic Analysis

Gas chromatography (GC) analysis was performed on an Agilent Technologies (Santa Clara, CA, USA) HP 7890A Plus Gas chromatograph equipped with a flame ionization detector (FID) and fitted with HP-5ms column (30 m × 0.25 mm, film thickness 0.25 μm, Agilent Technologies). The analytical conditions were—carrier gas H_2 (1 mL/min), injector temperature (PTV) 250 °C, detector temperature 260 °C, column temperature programmed from 60 °C (2 min hold) to 220 °C (10 min hold) at 4 °C/min. Samples were injected by splitting and the split ratio was 10:1. The volume injected was 1.0 μL. Inlet pressure was 6.1 kPa.

An Agilent Technologies (Santa Clara, California, USA) HP 7890A Plus Chromatograph fitted with a fused silica capillary HP-5ms (30 m × 0.25 mm, film thickness 0.25 μm) and interfaced with a mass spectrometer (HP 5973 MSD) was used for the GC-MS analysis, under the same conditions as those used for GC-FID analysis. The conditions were the same as described above with He (1 mL/min) as carrier gas. The MS conditions were as follows—ionization voltage 70 eV; emission current 40 mA; acquisitions scan mass range of 35–350 amu at a sampling rate of 1.0 scan/s.

The identification of constituents was performed on the basis of retention indices (RI) determined with reference to a homologous series of *n*-alkanes, under identical experimental conditions, co-injection with standards (Sigma-Aldrich, St. Louis, MO, USA) or known essential oil constituents, MS library search (NIST 08 and Wiley 9th Version) and by comparing with MS literature data [64]. The relative amounts of individual components were calculated based on the GC peak area (FID response) without using correction factors.

4.3. Antimicrobial Screening

The antimicrobial activity of the essential oils was evaluated using three strains of Gram-positive test bacteria, *Enterococcus faecalis* (ATCC299212), *Staphylococcus aureus* (ATCC25923), *Bacillus cereus* (ATCC14579), three strains of Gram-negative test bacteria, *Escherichia coli* (ATCC 25922), *Pseudomonas aeruginosa* (ATCC27853), *Salmonella enterica* (ATCC13076) and one strain of yeast, *Candida albicans* (ATCC 10231).

Minimum inhibitory concentration (MIC) and median inhibitory concentration (IC_{50}) values were measured by the microdilution broth susceptibility assay [65]. Stock solutions of the oil were prepared in dimethylsulfoxide. Dilution series were prepared from 16,384 to 2 μg/mL (2^{14}, 2^{13}, 2^{12}, 2^{11}, 2^{10}, 2^9, 2^7, 2^5, 2^3 and 2^1 μg/mL) in sterile distilled water in micro-test tubes from where they were transferred to 96-well microtiter plates. Bacteria grown in double-strength Mueller-Hinton broth or double-strength tryptic soy broth and fungi grown in double-strength Sabouraud dextrose broth were standardized to 5×10^5 and 1×10^3 CFU/mL, respectively. The last row, containing only the serial dilutions of sample without microorganisms, was used as a positive (no growth) control. Sterile distilled water and medium served as a negative (no antimicrobial agent) control. Streptomycin was used as the antibacterial standard, nystatin and cycloheximide were used as antifungal standards. After incubation at 37 °C for 24 h, the MIC values were determined to be the well with the lowest concentration of agents completely inhibiting the growth of microorganisms. The IC_{50} values were determined by the percentage of microorganisms that inhibited growth based on the turbidity measurement data of EPOCH2C spectrophotometer (BioTeK Instruments, Inc Highland Park Winooski, VT, USA) and Rawdata computer software (Brussels, Belgium) according to the following equations:

$$\%_{inhibition} = \frac{OD_{control(-)} - OD_{test\ agent}}{OD_{control\ (-)} - OD_{control(+)}} \quad (1)$$

$$IC_{50} = High_{conc} - \frac{(High_{inh\%} - 50\%) \times (High_{conc} - Low_{conc})}{(High_{inh\%} - Low_{inh\%})} \quad (2)$$

where OD is the optical density, control(–) are the cells with medium but without antimicrobial agent, test agent corresponds to a known concentration of antimicrobial agent, control(+) is the culture medium without cells, $High_{conc}/Low_{conc}$ is the concentration of test agent at high concentration/low concentration and $High_{inh\%}/Low_{inh\%}$ is the % inhibition at high concentration/% inhibition at low concentration). Each of the antimicrobial screens were carried out in triplicate.

4.4. Larvicidal Screening

Eggs of *Aedes aegypti* were purchased from Institute of Biotechnology, Vietnam Academy of Science and Technology and maintained at the Laboratory of Department of Pharmacy of Duy Tan University, Da Nang, Vietnam. Adults of *Culex quinquefasciatus* and *Aedes albopictus* collected in Hoa Khanh Nam ward, Lien Chieu district, Da Nang city (16°03′14.9″N, 108°09′31.2″E) and were identified by National institute of Malariology, Parasitology and Entomology, Ho Chi Minh City. Adult mosquitoes were maintained in entomological cages (40 × 40 × 40 cm) and fed a 10% sucrose solution and were allowed to blood feed on 1-week-old chicks and mice, respectively. Egg hatchings were induced with tap water. Larvae were reared in plastic trays (24 × 35 × 5 cm). The larvae were fed on Koi fish food. All developmental stages were maintained at 25 ± 2 °C, 65–75% relative humidity and a 12:12 h light:dark cycle at the Laboratory of the Faculty of Environmental and Chemical Engineering of Duy Tan University, Da Nang, Vietnam.

Larvicidal activities of the *Cinnamomum* essential oils were evaluated according to the protocol Liu and co-workers [66] with slight modifications. For the assay, 150 mL of water that contained 20 larvae (fourth instar) was placed in 250-mL beakers and aliquots of the *Cinnamomum* essential oils dissolved in EtOH (1% stock solution) were then added. With each experiment, a set of controls using EtOH only (negative control) and permethrin (positive control) were also run for comparison. Mortality was recorded after 24 h and again after 48 h of exposure during which no nutritional supplement was added. The experiments were carried out at 25 ± 2 °C. Each test was conducted with four replicates with five concentrations (100, 50, 25, 12.5 and 6 μg/mL). The data obtained were subjected to log-probit analysis [67] to obtain LC_{50} values, LC_{90} values and 95% confidence limits using Minitab® 19 (Minitab, LLC, State College, PA, USA).

5. Conclusions

The essential oils of five species of *Cinnamomum* were collected from north central Vietnam and screened for antimicrobial and mosquito larvicidal activities. According to Duarte and co-workers [68], essential oils with MIC values between 50 and 500 μg/mL can be considered to have strong antimicrobial activity. Similarly, Dias and Moraes have concluded that essential oils with $LC_{50} < 100$ μg/mL are considered to be active [69]. Therefore, all of the *Cinnamomum* essential oils in this study can be considered to be active and show promise as antimicrobial agents and as alternative insecticidal agents against mosquito larvae.

Author Contributions: Conceptualization, D.N.D. and W.N.S.; methodology, D.N.D., L.T.H., N.H.H., W.N.S.; validation, D.N.D. and W.N.S.; formal analysis, L.T.H., W.N.S.; investigation, N.T.C., L.T.H., N.T.Y., D.T.M.C.; resources, D.N.D.; data curation, W.N.S.; writing—original draft preparation, W.N.S., D.N.D.; writing—review and editing, D.N.D., L.T.H., W.N.S.; supervision, D.N.D.; project administration, D.N.D.; funding acquisition, D.N.D. All authors have read and agreed to the published version of the manuscript.

Funding: This research was funded by Vietnam National Foundation for Science and Technology Development (NAFOSTED) under grant number: 106.03-2018.02.

Acknowledgments: W.N.S. participated in this work as part of the activities of the Aromatic Plant Research Center (APRC, https://aromaticplant.org/).

Conflicts of Interest: The authors declare no conflict of interest.

References

1. Mabberley, D.J. *Mabberley's Plant-Book*, 3rd ed.; Cambridge University Press: Cambridge, UK, 2008.
2. Ho, P.-H. *An Illustrated Flora of Vietnam, Vol. 3*; Youth Publishing House: Ho Chi Minh City, Vietnam, 2000.
3. Dao, N.K. *Flora of Vietnam, Lauraceae Juss, Vol. 17*; Publishing House for Science & Technology: Hanoi, Vietnam, 2017.
4. Chi, V.V. *Dictionary of Vietnamese Medicinal Plants, Vol. I–II*; Medical Publishing House: Hanoi, Vietnam, 2012.
5. Moi, L.D.; Cu, L.D.; Hoi, T.M.; Thuy, N.T.; Thao, N.T.; Thai, T.H.; Ban, N.K. *The Essential Oil Resources in Vietnam, Vol. 1*; Agriculture Publishing House: Hanoi, Vietnam, 2001.
6. Zaman, S.B.; Hussain, M.A.; Nye, R.; Mehta, V.; Mamun, K.T.; Hossain, N. A review on antibiotic resistance: Alarm bells are ringing. *Cureus* **2017**, *9*, e1403. [CrossRef] [PubMed]
7. Banin, E.; Hughes, D.; Kuipers, O.P. Bacterial pathogens, antibiotics and antibiotic resistance. *FEMS Microbiol. Rev.* **2017**, *41*, 450–452. [CrossRef] [PubMed]
8. Navon-Venezia, S.; Kondratyeva, K.; Carattoli, A. *Klebsiella pneumoniae*: A major worldwide source and shuttle for antibiotic resistance. *FEMS Microbiol. Rev.* **2017**, *41*, 252–275. [CrossRef] [PubMed]
9. Pierce, G.E. *Pseudomonas aeruginosa, Candida albicans* and device-related nosocomial infections: Implications, trends and potential approaches for control. *J. Ind. Microbiol. Biotechnol.* **2005**, *32*, 309–318. [CrossRef] [PubMed]
10. Sakoulas, G.; Moellering, R.C., Jr. Increasing antibiotic resistance among methicillin-resistant *Staphylococcus aureus* strains. *Clin. Infect. Dis.* **2008**, *46*, S360–S367. [CrossRef] [PubMed]
11. Sweileh, W.M.; Sawalha, A.F.; Al-Jabi, S.; Zyoud, S.H. Bibliometric analysis of literature on antifungal triazole resistance: 1980–2015. *Germs* **2017**, *7*, 19–27. [CrossRef]
12. Rajkowska, K.; Kunicka-Styczyńska, A.; Maroszyńska, M. Selected essential oils as antifungal agents against antibiotic-resistant *Candida* spp.: In vitro study on clinical and food-borne isolates. *Microb. Drug Resist.* **2017**, *23*, 18–24. [CrossRef]
13. Orchard, A.; van Vuuren, S. Commercial essential oils as potential antimicrobials to treat skin diseases. *Evid. Based Complement. Altern. Med.* **2017**, *2017*, 4517971. [CrossRef]
14. Orchard, A.; Viljoen, A.; van Vuuren, S. Wound pathogens: Investigating antimicrobial activity of commercial essential oil combinations against reference strains. *Chem. Biodivers.* **2018**, *15*, e1800405. [CrossRef]
15. Serra, E.; Hidalgo-Bastida, L.A.; Verran, J.; Williams, D.; Malic, S. Antifungal activity of commercial essential oils and biocides against *Candida albicans*. *Pathogens* **2018**, *7*, 15. [CrossRef]
16. Powers, C.N.; Osier, J.L.; McFeeters, R.L.; Brazell, C.B.; Olsen, E.L.; Moriarity, D.M.; Satyal, P.; Setzer, W.N. Antifungal and cytotoxic activities of sixty commercially-available essential oils. *Molecules* **2018**, *23*, 1549. [CrossRef] [PubMed]
17. Wińska, K.; Mączka, W.; Łyczko, J.; Grabarczyk, M.; Czubaszek, A.; Szumny, A. Essential oils as antimicrobial agents—Myth or real alternative? *Molecules* **2019**, *24*, 2130. [CrossRef] [PubMed]
18. D'agostino, M.; Tesse, N.; Frippiat, J.P.; Machouart, M.; Debourgogne, A. Essential oils and their natural active compounds presenting antifungal properties. *Molecules* **2019**, *24*, 3713. [CrossRef] [PubMed]
19. Kozics, K.; Bučková, M.; Puškárová, A.; Kalászová, V.; Cabicarová, T.; Pangallo, D. The Effect of ten essential oils on several cutaneous drug-resistant microorganisms and their cyto/genotoxic and antioxidant properties. *Molecules* **2019**, *24*, 4570. [CrossRef] [PubMed]
20. Leta, S.; Beyene, T.J.; De Clercq, E.M.; Amenu, K.; Kraemer, M.U.G.; Revie, C.W. Global risk mapping for major diseases transmitted by *Aedes aegypti* and *Aedes albopictus*. *Int. J. Infect. Dis.* **2018**, *67*, 25–35. [CrossRef] [PubMed]
21. Samy, A.M.; Elaagip, A.H.; Kenawy, M.A.; Ayres, C.F.J.; Peterson, A.T.; Soliman, D.E. Climate change influences on the global potential distribution of the mosquito *Culex quinquefasciatus*, vector of West Nile virus and lymphatic filariasis. *PLoS ONE* **2016**, *11*, e0163863. [CrossRef]

22. Song, S.; Li, Y.; Fu, S.; Liu, H.; Li, X.; Gao, X.; Xu, Z.; Liu, G.; Wang, D.; Tian, Z.; et al. Could Zika virus emerge in Mainland China? Virus isolation from nature in *Culex quinquefasciatus*, 2016. *Emerg. Microbes Infect.* **2017**, *6*, e93. [CrossRef]
23. Bhattacharya, S.; Basu, P. The southern house mosquito, *Culex quinquefasciatus*: Profile of a smart vector. *J. Entomol. Zool. Stud.* **2016**, *4*, 73–81.
24. Cook, C.L.; Huang, Y.-J.S.; Lyons, A.C.; Alto, B.W.; Unlu, I.; Higgs, S.; Vanlandingham, D.L. North American *Culex pipiens* and *Culex quinquefasciatus* are competent vectors for Usutu virus. *PLoS Negl. Trop. Dis.* **2018**, *12*, e0006732. [CrossRef]
25. Naqqash, M.N.; Gökçe, A.; Bakhsh, A.; Salim, M. Insecticide resistance and its molecular basis in urban insect pests. *Parasitol. Res.* **2016**, *115*, 1363–1373. [CrossRef]
26. Grisales, N.; Poupardin, R.; Gomez, S.; Fonseca-Gonzalez, I.; Ranson, H.; Lenhart, A. Temephos resistance in *Aedes aegypti* in Colombia compromises dengue vector control. *PLoS Negl. Trop. Dis.* **2013**, *7*, e2438. [CrossRef]
27. Elia-Amira, N.M.R.; Chen, C.D.; Lau, K.W.; Lee, H.L.; Low, V.L.; Norma-Rashid, Y.; Sofian-Azirun, M. Organophosphate and organochlorine resistance in larval stage of *Aedes albopictus* (Diptera: Culicidae) in Sabah, Malaysia. *J. Econ. Entomol.* **2018**, *111*, 2488–2492. [CrossRef]
28. Delisi, N.; Ottea, J.; Healy, K. Susceptibility of *Culex quinquefasciatus* (Diptera: Culicidae) in southern Louisiana to larval insecticides. *J. Econ. Entomol.* **2017**, *110*, 2562–2567. [CrossRef]
29. Pitasawat, B.; Champakaew, D.; Choochote, W.; Jitpakdi, A.; Chaithong, U.; Kanjanapothi, D.; Rattanachanpichai, E.; Tippawangkosol, P.; Riyong, D.; Tuetun, B.; et al. Aromatic plant-derived essential oil: An alternative larvicide for mosquito control. *Fitoterapia* **2007**, *78*, 205–210. [CrossRef]
30. Silva, W.J.; Dória, G.A.A.; Maia, R.T.; Nunes, R.S.; Carvalho, G.A.; Blank, A.F.; Alves, P.B.; Marçal, R.M.; Cavalcanti, S.C.H. Effects of essential oils on *Aedes aegypti* larvae: Alternatives to environmentally safe insecticides. *Bioresour. Technol.* **2008**, *99*, 3251–3255. [CrossRef]
31. Pavela, R. Essential oils for the development of eco-friendly mosquito larvicides: A review. *Ind. Crop. Prod.* **2015**, *76*, 174–187. [CrossRef]
32. Son, L.C.; Dai, D.N.; Thai, T.H.; Huyen, D.D.; Thang, T.D.; Ogunwande, I.A. The leaf essential oils of four Vietnamese species of *Cinnamomum* (Lauraceae). *J. Essent. Oil Res.* **2013**, *25*, 267–271. [CrossRef]
33. Son, L.C.; Dai, D.N.; Thang, T.D.; Huyen, D.D.; Olayiwola, T.O.; Ogunmoye, A.R.; Ogunwande, I.A. Chemical composition of essential oils from the stem barks of three *Cinnamomum* species. *Br. J. Appl. Sci. Technol.* **2015**, *11*, 20442. [CrossRef]
34. Son, L.C.; Dai, D.N.; Thang, T.D.; Huyen, D.D.; Ogunwande, I.A. Study on *Cinnamomum* oils: Compositional pattern of seven species grown in Vietnam. *J. Oleo Sci.* **2014**, *63*, 1035–1043. [CrossRef]
35. Dorman, H.J.D.; Deans, S.G. Antimicrobial agents from plants: Antibacterial activity of plant volatile oils. *J. Appl. Microbiol.* **2000**, *88*, 308–316. [CrossRef]
36. Gallucci, M.N.; Oliva, M.; Casero, C.; Dambolena, J.; Luna, A.; Zygadlo, J.; Demo, M. Antimicrobial combined action of terpenes against the food-borne microorganisms *Escherichia coli*, *Staphylococcus aureus* and *Bacillus cereus*. *Flavour Fragr. J.* **2009**, *24*, 348–354. [CrossRef]
37. Boonchird, C.; Flegel, T.W. In vitro antifungal activity of eugenol and vanillin against *Candida albicans* and *Cryptococcus neoformans*. *Can. J. Microbiol.* **1982**, *28*, 1235–1241. [CrossRef]
38. Cheng, S.-S.; Liu, J.-Y.; Chang, E.-H.; Chang, S.-T. Antifungal activity of cinnamaldehyde and eugenol congeners against wood-rot fungi. *Bioresour. Technol.* **2008**, *99*, 5145–5149. [CrossRef] [PubMed]
39. Abbaszadeh, S.; Sharifzadeh, A.; Shokri, H.; Khosravi, A.R.; Abbaszadeh, A. Antifungal efficacy of thymol, carvacrol, eugenol and menthol as alternative agents to control the growth of food-relevant fungi. *J. Mycol. Med.* **2014**, *24*, 51–56. [CrossRef] [PubMed]
40. Cheng, S.-S.; Liu, J.-Y.; Tsai, K.-H.; Chen, W.-J.; Chang, S.-T. Chemical composition and mosquito larvicidal activity of essential oils from leaves of different *Cinnamomum osmophloeum* provenances. *J. Agric. Food Chem.* **2004**, *52*, 4395–4400. [CrossRef]
41. Govindarajan, M.; Rajeswary, M.; Hoti, S.L.; Bhattacharyya, A.; Benelli, G. Eugenol, α-pinene and β-caryophyllene from *Plectranthus barbatus* essential oil as eco-friendly larvicides against malaria, dengue and Japanese encephalitis mosquito vectors. *Parasitol. Res.* **2016**, *115*, 807–815. [CrossRef]
42. Pavela, R. Insecticidal properties of phenols on *Culex quinquefasciatus* Say and *Musca domestica* L. *Parasitol. Res.* **2011**, *109*, 1547–1553. [CrossRef]

43. Schmidt, J.M.; Noletto, J.A.; Vogler, B.; Setzer, W.N. Abaco bush medicine: Chemical composition of the essential oils of four aromatic medicinal plants from Abaco Island, Bahamas. *J. Herbs. Spices Med. Plants* **2006**, *12*, 43–65. [CrossRef]
44. Wanner, J.; Schmidt, E.; Bail, S.; Jirovetz, L.; Buchbauer, G.; Gochev, V.; Girova, T.; Atanasova, T.; Stoyanova, A. Chemical composition and antibacterial activity of selected essential oils and some of their main compounds. *Nat. Prod. Commun.* **2010**, *5*, 1359–1364. [CrossRef]
45. Filipowicz, N.; Kaminski, M.; Kurlenda, J.; Asztemborska, M.; Ochocka, J.R. Antibacterial and antifungal activity of juniper berry oil and its selected components. *Phyther. Res.* **2003**, *17*, 227–231. [CrossRef]
46. Hammer, K.A.; Carson, C.F.; Riley, T. V Antifungal activity of the components of *Melaleuca alternifolia* (tea tree) oil. *J. Appl. Microbiol.* **2003**, *95*, 853–860. [CrossRef] [PubMed]
47. Jantan, I.; Moharam, B.A.K.; Santhanam, J.; Jamal, J.A. Correlation between chemical composition and antifungal activity of the essential oils of eight *Cinnamomum* species. *Pharm. Biol.* **2008**, *46*, 406–412. [CrossRef]
48. Hung, N.H.; Satyal, P.; Hieu, H.V.; Chuong, N.T.H.; Dai, D.N.; Huong, L.T.; Tai, T.A.; Setzer, W.N. Mosquito larvicidal activity of the essential oils of *Erechtites* species growing wild in Vietnam. *Insects* **2019**, *10*, 47. [CrossRef] [PubMed]
49. Govindarajan, M. Chemical composition and larvicidal activity of leaf essential oil from *Clausena anisata* (Willd.) Hook. f. ex Benth (Rutaceae) against three mosquito species. *Asian Pac. J. Trop. Med.* **2010**, *3*, 874–877. [CrossRef]
50. Cheng, S.-S.; Huang, C.-G.; Chen, Y.-J.; Yu, J.-J.; Chen, W.-J.; Chang, S.-T. Chemical compositions and larvicidal activities of leaf essential oils from two eucalyptus species. *Bioresour. Technol.* **2009**, *100*, 452–456. [CrossRef]
51. Evergetis, E.; Michaelakis, A.; Haroutounian, S.A. Exploitation of Apiaceae family essential oils as potent biopesticides and rich source of phellandrenes. *Ind. Crops Prod.* **2013**, *41*, 365–370. [CrossRef]
52. bin Jantan, I.; Yalvema, M.F.; Ahmad, N.W.; Jamal, J.A. Insecticidal activities of the leaf oils of eight *Cinnamomum* species against *Aedes aegypti* and *Aedes albopictus*. *Pharm. Biol.* **2005**, *43*, 526–532. [CrossRef]
53. de Morais, S.M.; Facundo, V.A.; Bertini, L.M.; Cavalcanti, E.S.B.; dos Anjos Júnior, J.F.; Ferreira, S.A.; de Brito, E.S.; de S Neto, M.A. Chemical composition and larvicidal activity of essential oils from *Piper* species. *Biochem. Syst. Ecol.* **2007**, *35*, 670–675. [CrossRef]
54. Pattnaik, S.; Subramanyam, V.R.; Bapaji, M.; Kole, C.R. Antibacterial and antifungal activity of aromatic constituents of essential oils. *Microbios* **1997**, *89*, 39–46.
55. Setzer, W.N.; Vogler, B.; Schmidt, J.M.; Leahy, J.G.; Rives, R. Antimicrobial activity of *Artemisia douglasiana* leaf essential oil. *Fitoterapia* **2004**, *75*. [CrossRef]
56. Satyal, P.; Paudel, P.; Poudel, A.; Dosoky, N.S.; Pokharel, K.K.; Setzer, W.N. Bioactivities and compositional analyses of *Cinnamomum* essential oils from Nepal: *C. camphora*, *C. tamala* and *C. glaucescens*. *Nat. Prod. Commun.* **2013**, *8*, 1777–1784. [CrossRef] [PubMed]
57. Setzer, W.N.; Schmidt, J.M.; Eiter, L.C.; Haber, W.A. The leaf oil composition of *Zanthoxylum fagara* (L.) Sarg. from Monteverde, Costa Rica and its biological activities. *J. Essent. Oil Res.* **2005**, *17*, 333–335. [CrossRef]
58. Reichling, J.; Suschke, U.; Schneele, J.; Geiss, H.K. Antibacterial activity and irritation potential of selected essential oil components—Structure-activity relationship. *Nat. Prod. Commun.* **2006**, *1*, 1003–1012. [CrossRef]
59. Fisher, K.; Phillips, C.A. The effect of lemon, orange and bergamot essential oils and their components on the survival of *Campylobacter jejuni*, *Escherichia coli* O157, *Listeria monocytogenes*, *Bacillus cereus* and *Staphylococcus aureus* in vitro and in food systems. *J. Appl. Microbiol.* **2006**, *101*, 1232–1240. [CrossRef]
60. Mesa-Arango, A.C.; Montiel-Ramos, J.; Zapata, B.; Durán, C.; Betancur-Galvis, L.; Stashenko, E. Citral and carvone chemotypes from the essential oils of Colombian *Lippia alba* (Mill.) N.E. Brown: Composition, cytotoxicity and antifungal activity. *Mem. Inst. Oswaldo Cruz* **2009**, *104*, 878–884. [CrossRef]
61. Silva, C.D.B.D.; Guterres, S.S.; Weisheimer, V.; Schapoval, E.E. Antifungal activity of the lemongrass oil and citral against *Candida* spp. *Braz. J. Infect. Dis.* **2008**, *12*, 63–66. [CrossRef]
62. Cheng, S.-S.; Liu, J.-Y.; Huang, C.-G.; Hsui, Y.-R.; Chen, W.-J.; Chang, S.-T. Insecticidal activities of leaf essential oils from *Cinnamomum osmophloeum* against three mosquito species. *Bioresour. Technol.* **2009**, *100*, 457–464. [CrossRef]

63. Radwan, M.A.; El-Zemity, S.R.; Mohamed, S.A.; Sherby, S.M. Larvicidal activity of some essential oils, monoterpenoids and their corresponding *N*-methyl carbamate derivatives against *Culex pipiens* (Diptera: Culicidae). *Int. J. Trop. Insect Sci.* **2008**, *28*, 61–68. [CrossRef]
64. Adams, R.P. *Identification of Essential Oil Components by Gas Chromatography/Mass Spectrometry*, 4th ed.; Allured Publishing: Carol Stream, IL, USA, 2007.
65. Sahm, D.H.; Washington, J.A. Antibacterial susceptibility tests: Dilution methods. In *Manual of Clinical Microbiology*; Balows, A., Hausler, W.J., Herrmann, K.L., Isenberg, H.D., Shamody, H.J., Eds.; American Society for Microbiology: Washington, DC, USA, 1991.
66. Liu, Z.L.; He, Q.; Chu, S.S.; Wang, C.F.; Du, S.S.; Deng, Z.W. Essential oil composition and larvicidal activity of *Saussurea lappa* roots against the mosquito *Aedes albopictus* (Diptera: Culicidae). *Parasitol. Res.* **2012**, *110*, 2125–2130. [CrossRef]
67. Finney, D. *Probit Analysis*; Vincent, K., Ed.; Cambridge University Press: Cambridge, UK, 2009; ISBN 978-0521135900.
68. Duarte, M.C.T.; Leme, E.E.; Delarmelina, C.; Soares, A.A.; Figueira, G.M.; Sartoratto, A. Activity of essential oils from Brazilian medicinal plants on *Escherichia coli*. *J. Ethnopharmacol.* **2007**, *111*, 197–201. [CrossRef]
69. Dias, C.N.; Moraes, D.F.C. Essential oils and their compounds as *Aedes aegypti* L. (Diptera: Culicidae) larvicide: Review. *Parasitol. Res.* **2014**, *113*, 565–592. [CrossRef] [PubMed]

Sample Availability: The Cinnamomum essential oils are no longer available.

molecules

MDPI

Review

Plant-Derived Pesticides as an Alternative to Pest Management and Sustainable Agricultural Production: Prospects, Applications and Challenges

Augusto Lopes Souto [1], Muriel Sylvestre [2], Elisabeth Dantas Tölke [3], Josean Fechine Tavares [1], José Maria Barbosa-Filho [1] and Gerardo Cebrián-Torrejón [2,*]

[1] Programa de Pós-Graduação em Produtos Naturais e Sintéticos Bioativos, Universidade Federal da Paraíba, João Pessoa 58051-900, Brazil; augustosouto@gmail.com (A.L.S.); josean@ltf.ufpb.br (J.F.T.); jbarbosa@ltf.ufpb.br (J.M.B.-F.)

[2] COVACHIM-M2E Laboratory EA 3592, Department of Chemistry, Fouillole Campus, University of the French West Indies, UFR Sciences Exactes et Naturelles, CEDEX, 97157 Pointe-à-Pitre, France; muriel.sylvestre@univ-antilles.fr

[3] Departamento de Botânica, Instituto de Biociências, Universidade de São Paulo, São Paulo 05508-090, Brazil; elisabeth.tolke@gmail.com

* Correspondence: gerardo.cebrian-torrejon@univ-antilles.fr

Citation: Souto, A.L.; Sylvestre, M.; Tölke, E.D.; Tavares, J.F.; Barbosa-Filho, J.M.; Cebrián-Torrejón, G. Plant-Derived Pesticides as an Alternative to Pest Management and Sustainable Agricultural Production: Prospects, Applications and Challenges. *Molecules* **2021**, *26*, 4835. https://doi.org/10.3390/molecules26164835

Academic Editor: Francesca Mancianti

Received: 29 June 2021
Accepted: 4 August 2021
Published: 10 August 2021

Abstract: Pests and diseases are responsible for most of the losses related to agricultural crops, either in the field or in storage. Moreover, due to indiscriminate use of synthetic pesticides over the years, several issues have come along, such as pest resistance and contamination of important planet sources, such as water, air and soil. Therefore, in order to improve efficiency of crop production and reduce food crisis in a sustainable manner, while preserving consumer's health, plant-derived pesticides may be a green alternative to synthetic ones. They are cheap, biodegradable, ecofriendly and act by several mechanisms of action in a more specific way, suggesting that they are less of a hazard to humans and the environment. Natural plant products with bioactivity toward insects include several classes of molecules, for example: terpenes, flavonoids, alkaloids, polyphenols, cyanogenic glucosides, quinones, amides, aldehydes, thiophenes, amino acids, saccharides and polyketides (which is not an exhaustive list of insecticidal substances). In general, those compounds have important ecological activities in nature, such as: antifeedant, attractant, nematicide, fungicide, repellent, insecticide, insect growth regulator and allelopathic agents, acting as a promising source for novel pest control agents or biopesticides. However, several factors appear to limit their commercialization. In this critical review, a compilation of plant-derived metabolites, along with their corresponding toxicology and mechanisms of action, will be approached, as well as the different strategies developed in order to meet the required commercial standards through more efficient methods.

Keywords: biopesticides; bio-based pesticides; chemical ecology; pest control; natural products

1. Introduction

Pesticides may be defined as any compound or mixture of components intended for preventing, destroying, repelling or mitigating any pest [1]. Additionally, herbicides or weed-killers may also be considered as pesticides, and are used to kill unwanted plants in order to leave the desired crop relatively unharmed and well provided with nutrients, leading to a more profitable harvest [2].

Nevertheless, the world food production is constantly affected by insects and pests during crop growth, harvest and storage. As a matter of fact, there is an estimated loss of 18–20% regarding the annual crop production worldwide, reaching a value of more than USD 470 billion [3]. Furthermore, insects and pests not only represent a menace to our homes, gardens and reservoirs of water, but also, they transmit a number of diseases by acting as hosts to some disease-causing parasites. Therefore, the mitigation or control of pests' activities may lead to a substantial reduction of the world food crisis as well as the improvement of human and animal health [2].

The great demand for food has led to the intensification of agricultural technology in order to achieve maximum productivity per hectare, through expansion of irrigation facilities, introduction of high-yielding varieties and application of increased amounts of agrochemicals, mostly synthetic [1] (for example, the 1,1,1-trichloro-2,2-bis(4-chlorophenyl)-ethane (DDT) used from 1939 to 1962 [4] and the polychlorinatedbiphenyls (PCBs) used from 1926 to 1970 [5]). In spite of various technological achievements over the years, serious problems have come along, especially due to the indiscriminate use of synthetic pesticides. As they remain in our planet for an extremely long period, its long persistence in our biosphere allows insects to develop resistance against them; they are also known to contaminate indispensable resources, such as water, air and soil. This is illustrated by the problem of chlordecone (CLD), a pesticide widely used from 1972 to 1993 in the French West Indies (FWI) [6] which is often called as the "monster of the Antilles" because of the extent of soil and biomass contamination. Chemically, CLD is an organochlorine ketone, with high steric hindrance and high hydrophobicity, which allows it to adsorb strongly into soils rich in organic matter. At the same time, it is non-volatile and has a low biodegradability. It has also been demonstrated that CLD is a particularly persistent molecule, which remains almost perennial in soils, and the phenomenon of bioaccumulation in living organisms has been observed. Despite the ban on its use in the early 1990s, this molecule is still present today in the waters and soils of the French West Indies [7]. Extremely few pesticides used in the post-DDT era have long half-lives in the environment and none bioaccumulate in the way that the organochlorines did. Moreover, most of the synthetic pesticides act against non-target organisms (mammals, fish and plant species), becoming a potential health hazard to consumers. Furthermore, they are too expensive for the farmers in developing countries [8–11].

This problematic is illustrated as well by the neonicotinoid insecticides (such as imidacloprid, clothianidin, thiamethoxam, acetamiprid and thiacloprid). This more recent class of molecules is active against several pests (targeting the acetylcholine receptors) and can be applied to different cultures (such as tobacco, cotton, peach and tomato) [12].

The use of neonicotinoid insecticides was allowed in Europe in 2005 until 2013 (see Regulation (EU) No 485/2013) when the employ was restricted to protect wildlife, such as pollinators (honeybees), mammals, birds, fish, amphibians and reptiles, and the effects on vertebrates—mammals, birds, fish, amphibians and reptiles [13].

This awareness regarding pest problems and the environment has led to the search for powerful and eco-friendly pesticides that degrade after some time, avoiding pest resistance, which is also pest-specific, non-phytotoxic, nontoxic to mammals and relatively less expensive in order to obtain a sustainable crop production [14,15]. In addition to the awareness achieved in the various countries (such as China, United States, Brazil or Turkey) [16,17] and also in European Union (EU), the legislation has become increasingly stringent and binding with the consequences of a green choice in the use of pesticides and approbation of plant derivatives allowed in the biological control regulation (for example: clove (*Syzygium aromaticum* [L.] Merr. and L.M. Perry [Myrtaceae]) essential oil in the EU, the derived terpenes from *Chenopodium ambrosioides* L. (Amaranthaceae) in US, Ginkgo (*Ginko biloba* L. [Ginkgoaceae]) fruit extract and *Psoralea corylifolia* L. (Leguminosae) seed extract in China or the extract of *Tephrosia candida* DC. (Fabaceae) in Brazil).

In this context, biopesticides may meet those required standards and become the key to solve pest problems and promote sustainable production, once they are cheap, target-specific, less hazardous to human health, bio-degradable and therefore environmentally friendly. Biopesticides are pest management agents based on biochemicals derived from living microorganisms, insects and plants. [2,18,19]. In this review, we will focus on biopesticides from plant origin.

Among the plant-derived pesticides, there are the insecticides (constituents that kill insects in any stage of development: adults, ova and larvae), which act by several different mechanisms affecting one or more biological systems, including nervous, respiratory and endocrine systems, as well as water balance. Additionally, insecticides can also be classified depending on the mode of its entry into the insect, namely: stomach poisons, contact poisons and fumigants [11]. A compilation of natural insecticides with their corresponding toxicity and mechanisms of action may be found in Table 1.

2. Plant Derived Insecticides That Affect the Nervous System

The majority of insecticides, whether biological or synthetic, fit in this category, acting on several targets, such as: voltage-gated sodium channels, voltage-gated calcium channels, acetylcholinesterase enzyme (AChE), nicotinic acetylcholine receptors, GABA receptors and octopamine receptors.

2.1. Voltage-Gated Sodium Channels

Pyrethrum, an oleoresin extracted from the dried flowers of the pyrethrum daisy, *Tanacetum cinerariifolium* (Trevir.) Sch. Bip. (Asteraceae), contains two major compounds, namely Pyrethryns I and II (cyclopropylmonoterpene esters) [20] that act as a modulator on voltage-gated sodium channels, which are essential for proper electrical signaling in the nervous system, causing a delay in sodium channel closing, resulting in over neuroexcitation, leading to loss of control of the coordinated movement, paralysis and death. This mode of actions is highly similar to other synthetic insecticides (such as synthetic pyrethryns). However, natural pyrethrins used to be more target-specific than synthetic ones [21]. It is defined as a contact and stomach poison that provides an immediate knockdown when applied, which has also demonstrated low toxicity to mammals and a particularly short residual activity, once it is rapidly degraded by sunlight, air and moisture; therefore, frequent application may be required [22]. Pyrethrins may be used against a wide range of insects and mites, including spider mites, flies, fleas and beetles [23,24]. Its activity may be enhanced by incorporating piperonylbutoxide (PBO) as a synergist [2].

Other natural products with a similar mechanism of action have already been reported, such as decalesides I and II, firstly isolated from the roots of *Decalepis hamiltonii* Wight and Arn. (Apocynaceae). They are classified as trisaccharides, and are toxic to a variety of insects by contact exposure, not orally, but through contact to the gustatory receptors located in the tarsi of the insect [25].

Sabadilla, an insecticidal preparation from pulverized seeds of *Schoenocaulon officinale* (Schltdl. and Cham.) A. Gray ex Benth. (Melanthiaceae), has been used by native people from south and central America for many years. Its alkaloid preparation contains mainly two major alkaloids: cevadine and veratridine at a proportion of 2:1, which have a mode of action similar to pyrethrins, although the binding site seems to be different [26]. Sabadilla is one of the least toxic plant extracts with pesticide activity, considered a contact and stomach poison with minimal residual activity. However, its major alkaloids, when isolated, are much more toxic to humans, and affect mostly stinks, squash bugs, thrips, leafhoppers and caterpillars [11,27].

2.2. Voltage-Gated Calcium Channels

Ryanodine, an active component found in the roots and woody stems of *Ryania speciosa* Vahl (Salicaceae), native to Trinidad, activates the calcium channels from the sarcoplasmic reticulum of skeletal muscle cells. Once activated, the calcium channels release an excess of calcium ions into actin and myosin protein filaments, leading to skeletal muscle contraction and paralysis [28]. Ryanodine is a "fast act" poison, promoting its insecticidal activity either by contact or stomach, with a low mammalian toxicity and a long residual activity, providing up to two weeks of control after the first application. Ryania crude extracts insecticidal activity is synergized by piperonylbutoxide (PBO), and is reported to be most effective in hot waters, working efficiently against caterpillars, worms, potato beetles, lace bugs, aphids and squash bugs. [11,29]. Ryania is almost no longer used in the United States.

2.3. *Acetylcholinesterase Enzyme (AChE)*

Acetylcholinesterase (AChE), an enzyme that hydrolyzes the neurotransmitter acetylcholine, plays an important role regulating the transmission of the cholinergic nervous impulse, and may also be a target for biopesticides. Coumaran (2,3-dihydrobenzofuran), an active ingredient found in *Lantana camara* L. (Verbenaceae), inhibits this enzyme, building up the concentration levels of acetylcholine in the synapse cleft, causing an excessive neuroexcitation due to the prolonged biding of the neurotransmitter to its postsynaptic receptor, leading to restlessness, hyperexcitability, tremors, convulsion, paralysis and death [30]. It presents low toxicity to mammals and works rapidly against houseflies and grain storage pests, in spite of its short residual activity. Regarding its mechanism of action, coumaran may also be compared to the monoterpene 1,8-cineole [31] or other synthetic pesticides such as organophosphates and carbamates [32]. Moreover, Khorshid, et al. [33] have presented the inhibitory activity of methanolic extract from *Cassia fistula* L. (Fabaceae) roots and proposed the indole alkaloids as new potential active agents. However, it should be noted that many essential oils and terpenes therefrom have demonstrated anti-AChE activity in vitro, but their contribution to insect mortality is questionable [34].

2.4. *Nicotinic Acetylcholine Receptors*

In relation to nicotinic acetylcholine receptors, they are present in the insect nervous system, either on pre or postsynaptic nerve terminal, as well as the cell bodies of the inter neurons, motor neurons and sensory neurons [35]. Nicotine, an alkaloid firstly isolated from *Nicotiana tabacum* L. (Solanaceae), can mimic acetylcholine by acting as an agonist of the acetylcholine receptor, leading to an influx of sodium ion and generation of action potential. Under normal conditions, the synaptic action of acetylcholine is terminated by AChE. However, since nicotine cannot be hydrolyzed by AChE, the persistent activation caused by the nicotine leads to an overstimulation of the cholinergic transmission, resulting in convulsion, paralysis and finally death [35]. Nicotine is an extremely fast nerve toxin, most effective towards soft-bodied insects and mites. However, is the most toxic of all botanicals and extremely harmful to humans [2]. Alternatively, there is another class of insecticides inspired on nicotine chemical structure, called neonicotinoids, which may be represented by imidacloprid, acetamiprid and thiamethoxam. Similarly to nicotine, neonicotinoids interact with nicotinic acetylcholine receptors. However, they are more specific, being much more toxic to invertebrates such as insects than to mammals. Additionally, they have higher water solubility, which permits its application to soils and therefore, its absorption by plants, promoting a more efficient defense [36]. They may act as a contact or ingestion poison, leading to the cessation of feeding within several hours of contact followed by death shortly after [37].

2.5. *GABA-Gated Chloride Channels*

GABA-gated chloride channels are potential targets for insecticides. Once they are blocked by its antagonists (such as the α-Thujone (isolated from *Artemesia absinthium* L. (Asteraceae)) or the picrotoxine (*Anamirta cocculus* (L.) Wight and Arn (Menispermaceae)), neuronal inhibition is reduced, leading to hyper-excitation of the central nervous system (CNS), convulsion and death. Previous research have reported the monoterpene thujone as a neurotoxin, since it acts on GABAA receptors as an allosteric reversible modulator, and as a competitive inhibitor of [3H]Ethynylbicycloorthobenzoate ([3H]EBOB binding) [38]. Additionally, the GABA receptor may be inhibited by the monoterpenoids carvacrol, pulegone and thymol through [3H]TBOB binding [39]. Similarly, the silphinene-type sesquiterpenes, plant-derived natural compounds, antagonize the action of aminobutyricacid (GABA), by stabilizing non-conducting conformations of the chloride channel [24,38]. As GABA is an endogenous ligand related to stimulate feeding and evoke taste cell responses on most herbivorous insects, the chemicals that antagonize GABA receptors may also be considered as antifeedant or deterrent compounds, affecting mostly aphids, lepidopterans and beetles [40,41].

2.6. Octopamine Receptors

Octopamine is a multi-functional endogenous amine that acts as a neurotransmitter, neurohormone and neuromodulator on invertebrates [42]. Its receptors are widely distributed in the central and peripheral nervous systems of insects, comprising the octopaminergic system, constituting of several subtypes of octopamine receptors, which are coupled to different second messenger systems, therefore playing a key role in mediating physiological functions and behavioral aspects [43–45]. For instance, octopamine1 receptor modulates myogenic rhythm of contraction in locust extensor-tibiae through changes in intracellular calcium concentrations, whereas octopamine2A and octopamine2B receptors mediate their effects through the activation of adenylatecyclase. Moreover, octopamine3 receptors mediate changes in cyclic adenosine monophosphate (CAMP) levels in the locust central nervous system [46].

The rapid action of monoterpenes against some pests suggests a neurotoxic mode of action. This hypothesis was confirmed by Reynoso, et al. [47], who have demonstrated repellent and insecticidal activity of eugenol against the blood-sucking bug *Triatoma infestans* (Klug; Reduviidae) through activation of the octopamine receptor.

Previous studies have reported the presence of octopamine receptors in a large variety of insects, including, firefly, flies, nymphs, cockroaches and lepidopterans [46–48]. As these receptors do not conform to the receptor categories that have been recognized in vertebrates, agonists of octopamine receptors may be a valuable candidate for a commercial pesticide, once they are target-specific, less toxic to mammals and have a different mechanism of action when compared to the majority of pesticides currently in the market [47].

3. Plant Derived Insecticides That Affect Respiratory or Energy System

Cellular respiration is a process that converts nutrient compounds into energy or adenosine triphosphate (ATP) at a molecular level. More specifically, this process is performed by the electron transport chain of the mitochondria, which comprises several important enzymes that are potential targets for insecticides. Rotenone is the most common natural product among rotenoids, a type of isoflavonoid and is usually found in species from Derris and Lonchocarpus (in Fabaceae) and Rhododendron (in Ericaceae), spread throughout East Indies, Malaya and South America [20].

Rotenone is defined as a complex I inhibitor of the mitochondrial respiratory chain, which works both as contact and stomach poison. It blocks the nicotinamide adenine dinucleotide (NADH) dehydrogenase, stopping the flow of electrons from NADH to coenzyme Q, therefore, preventing ATP formation from NADH, but maintaining ATP formation through flavine adenine dinucleotide ($FADH_2$); therefore, it is one of the slowest acting botanical insecticide, and yet readily degradable by air and sunlight, taking several days to kill insects, affecting primarily nerve and muscle cells, leading to cessation of feeding, followed by death, from several hours to a few days after exposure. Moreover, this bio-based pesticide is constantly applied to protect lettuce and tomato crops as it has a broad spectrum of activity against mite pests, including leaf-feeding beetles, lice, caterpillars, mosquitoes, ticks, fire ants and fleas. Furthermore, its effects are substantially synergized by PBO or pyrodone (MGK 264).

Rotenone is highly toxic to mammals and fish [24,49]. Its activity and persistence are comparable to dichlorodiphenyltrichloroethane (DDT) [2]; moreover, previous studies have correlated a possible link between its exposure and Parkinson's Disease (PD) [50]. However, in spite of its high toxicity, rotenoids may be a potential source of novel complex I inhibitors, acting as a prototype for the development of safer and more efficient pesticide derivatives [51].

Acetogenins (annonins, asimicin, squamocin, annonacins) obtained from *Annona squamosa* L. (Annonaceae) are well known for their pest control properties. A botanical formulation based on annonins wherein asimicin is the major pesticidal compound has been patented [52].

4. Plant Derived Insecticides That Affect the Endocrine System

Chemical constituents that interfere with the endocrine system of insects are classified as insect growth regulators (IGR). They may act either as juvenile insect hormone mimics or inhibitors, as well as chitin synthesis inhibitors (CSI). Normally, the juvenile hormones are produced by insects in order to keep its immature state. When a sufficient growth has been reached, the production of the hormone stops, triggering the molt to the adult stage [53]. Triterpenes from *Catharanthus roseus* (L.) G. Don (Apocynaceae), such as α-amyrin acetate and oleanolic acid, have demonstrated interesting growth regulator activity [54]. Acyclic sesquiterpenes such as davanone, ipomearone and the juvenile hormone from silkworm are perfect examples of natural products with IGR activity as well. Therefore, the constant application of IGR towards the crops will maintain the insects in its larvae state, preventing a successful molting and resulting in an efficient pest control [55]. On the other hand, it has been reported the antijuvenile hormone activity of two chromenes found in *Ageratum conyzoides* L. (Asteraceae), precocene I and II promotes a precocious metarmophosis of the larvae and production of sterile, moribund and dwarfish adults after exposure [56]. Although, resistance to azadirachtin has been demonstrated [57], indicating that insects can develop resistance to natural hormones or hormone-related compounds; however, this class of compounds remains a natural potential for commercial bio-based pesticides [55]. Additionally, complex polyphenolic fractions also present a wide range of insecticidal activities, interfering with the fecundity and inducing the disruption of the oogenesis [58,59] (WO 94/13141).

Moreover, previous researches have reported a natural insecticide of broad-spectrum activity, which has low mammalian toxicity and is the least toxic among botanical insecticides. It is called azadirachtin, a complex tetranortriterpenoid limonoid, majorly found in the seeds of *Azadirachta indica* A. Juss. (Meliaceae), a plant species commonly known as the Neem tree which originated from Burma, but is currently grown in more arid, tropical and subtropical zones of Southeast Asia, Africa, Americas and Australia [24,26,60]. Azadirachtin is considered a contact poison of systemic activity, which may be categorized in two ways: direct effects towards cells and tissues, or indirect effects, represented by endocrine system interference. It is a powerful compound that acts mainly as a feeding deterrent and insect growth regulator, comprising a wide variety of insect taxa including Lepidoptera, Diptera, Hemiptera, Orthoptera, Hymenoptera [60]. As for its growth regulatory effects, azadirachtin affects the neurosecretory system of the brain insect, blocking the release of morphogenetic peptide hormones (e.g., prothoracicotropic hormone (PTTH) and allatostatins). These hormones control the function of the prothoracic glands and the corpora allata, respectively. Therefore, as the moulting hormone (which controls new cuticle formation and ecdyses) and the juvenile hormone (JH) (which controls the juvenile stage at each moult) are regulated by prothoracic glands and the corpora allata, any disruption on this biochemical cascade may lead to moult disruption, moulting defects or sterility. The effects on feeding, developmental and reproductive disruption are caused by effects of the molecule directly on somatic and reproductive tissues and indirectly through the disruption of endocrine processes [60].

Neem-based non-commercial products are normally found as neem oil, obtained from the cold pressing of its seeds, in order to control phytopathogens (including insects). The other product is a medium-polarity extract containing azadirachtin (0.2–0.6% of seed/weight) [2], whereas the actual commercial product is a 1 to 4.5% azadirachtin solution [61]. Despite its 20 h half-life, it ensures a reasonable persistence in field applications due to its systemic action [2].

In relation to CSIs, they inhibit the production of chitin, a β-(1,4)-linked homopolymer of N-acetyl-D-glucosamine, one of the most important structural components of nearly all fungi cell walls, and also a major component of the insect exoskeleton, which provides physical protection and osmoregulation. As chitin is absent on plant and mammalian species, while it is abundant in arthropods and most fungi, chitin biosynthesis has become an important target for developing more specific insecticides and antifungal agents. Previous research has reported chitin synthase inhibition activity of 2-benzoyloxycinnamaldehyde (2-BCA), a natural product isolated from the roots of *Pleuropterus ciliinervis* Nakai (Polygonaceae), which is a plant species traditionally used in Chinese folk medicine to treat inflammation and several types of infection [62].

5. Plant Derived Insecticides That Affect the Water Balance

Insects have a thin layer of wax covering their body, which provides the ecological function of preventing water loss from the cuticular surface. For instance, vegetable crude oils of rice bran, cotton seed and palm kernel, as well as saponins (natural soaps) may act by disrupting this protective waxy covering, affecting the water balance of insects through a rapid water loss from the cuticle, therefore leading to death by desiccation. Interestingly, the action of soaps affects the wax covering of insects [63]. The action of soaps on the wax covering of insects is influenced by the temperature [64]. Additionally, the crude oils may also act by interfering with insect respiration by plugging the orifices called spiracles, resulting in death by asphyxiation, controlling several types of insects such as whiteflies, mites, caterpillars, leafhoppers and beetles [1].

6. Other Classes of Pesticides

The botanical pesticide agents may also be categorized into repellents, attractants, antifeedants or deterrents, molluscisides, fungicides, phytotoxins (herbicides) and photoxins [15]. These classes are less common in plant sources than the insecticides [65]. Sometimes, a given compound may act as an insecticide and/or as a repellent. The major difference between those two is that the repellent does not kill insects, but only keeps them away by releasing pungent vapors or exhibiting a slight toxic effect [66].

7. Repellents

There are several essential oils which are majorly constituted of monoterpenes and are considered extremely effective repellents, including lemongrass (*Cymbopogon flexuosus* (Nees) Will. Watson (Poaceae)), eucalyptus (*Eucalyptus globulus* Labill. (Myrtaceae)), rosemary (*Rosmarinus officinalis* L. (Lamiaceae)), vetiver (*Vetiveria zizanioides* (L.) Nash (Poaceae)), clove (*Eugenia caryophyllus* (Spreng.) Bullock and S.G. Harrison (Myrtaceae)) and thyme (*Thymus vulgaris* L. (Lamiaceae)) [67]. Catnip oil, for example, extracted from *Nepeta cataria* L. (Lamiaceae), is considered a highly effective repellent of mosquitoes, bees and other flying insects. As a matter of fact, this oil repels *Aedes aegypti* L. (Culicidae) ten times more than DEET, which is probably related to its most effective constituent, nepetalactone, a monoterpene lactone [68], which is also reported as a repellent for lady beetles, cockroaches, flies and termites [69,70]. The anthraquinone tectoquinone was also described as a repellent against termites [71,72], and alstonine alkaloid has a repellent and larvicidal activity against *Anopheles gambiae* Giles (Culicidae) [73].

8. Attractants

In relation to attractants, they are considered semio-chemicals or communication compounds, released by plants in order to attract insects or to attract natural predators of the insects that feed on the plant [74]. Miller [75] have related the release of (−) and (+) limonene from white pine (*Pinus strobus* L. (Pinaceae)) to the attraction of the white pine cone beetle, *Conophthorus coniperda* Schwarz (Curculionidae), as well as the attraction of the predator beetle, *Enoclerus nigripes* Say (Cleridae), through the release of (−)-α-pinene, as well as the sesquiterpene caryophyllene [76].

9. Antifeedants or Deterrents

Previous studies have correlated antifeedant activity to a chemoreception mechanism, consisting in the blockage of receptors that normally respond to phagostimulants or through stimulation of deterrent cells (primary antifeedancy). According to Qiao et al., [77] azadirachtin reduces the cholinergic transmission of neurons related to the suboesophageal ganglion (SOG) of *Drosophyla melanogaster* Meigen (Drosophylinae), which are strongly related to feeding behavior. Additionally, food consumption may also be reduced due to its toxic effects after the first intake (secondary antifeedancy), promoting astringency, bitter taste or anti-digestive activity to certain herbivores [78,79]. For instance, Okwute and Nduji [80] have reported that schimperii, a gallotannin isolated from *Anogeissus schimperi* (Hochst. ex Hutch and Dalziel) (Combretaceae) was responsible for conferring this unattractive taste to herbivores. Similar effects were reported to, isoflavonoids [81], acetogenines [82,83] or cyanogenic glycosides, such as linamarin [84].

Moreover, Lajide, Escoubas and Mizutani [66] have reported feed deterrent activity of ent-kaurane diterpenoids isolated from *Xylopia aethiopica* (Dunal) A. Rich. (Annonaceae), among which, (−)-kau-16-en-19-oic acid has demonstrated the strongest antifeedant activity. According to Okwute [2], 15-epi-4E-jatrogrossidentadione, a diterpene from *Jatropha podagrica* L. (Euphorbiaceae) have also demonstrated its antifeedant activity towards *Chilo partellus* Swinhoe (Crambidae). Moreover, silphinene sesquiterpenes (*Senecio palmensis* C. Sm. (Asteraceae)) and thymol (*Thymus vulgaris* L. (Lamiaceae)) have been described as model of insect antifeedants [40].

However, as demonstrated by Huang et al., in spite of numerous natural plant natural products acting as antifeedants, no commercial product based on this mode of action have been produced. Insect habituation to feeding deterrents considerably limits their utility in crop protection [85].

10. Phytotoxines or Herbicides

Regarding phytotoxins, they may be defined as natural herbicides that are naturally released by plant species in order to interfere with the growth or germination of specific targets around them, such as weeds, leaving the emitting plant with more chances to survive. In nature, such action is called allelopathy, and the compounds that promote this action are defined as allelopathic agents [86–88]. Clay, et al. [89] have reported a study regarding herbicidal activity of citronella oil against different weed species: the oil at a dose of 504 kg a.i. ha-1 largely killed the foliage of the weed species within one application. However, most species have regrown substantially after two months, except for *Senecio jacobaea* L. (Asteraceae), which was the most susceptible one. According to Ismail, et al. [76], its herbicidal activity occurs through inhibition of photosynthesis. Besides essential oils herbicidal activity, Ismail, Hamrouni, Hanana and Jamoussi [90] have also reported plant-derived isolated compounds, such as eugenol and 1,8-cineole, with herbicidal activity promoted through inhibition of DNA synthesis and mitosis. Furthermore, several classes of secondary metabolites have been already described as phytotoxins, including naphtoquinones, such as juglone [91,92], amino acids such as m-tyrosin e [93] and L-tryptophane [94], terpenoids as 5,6-dihydroxycadinan-3-ene-2,7-dione [2,95] and citronnellol [90], catechins [2,96], polyphenols [97] and alkylamides [98].

11. Phototoxins

There is a class of phytochemicals called phototoxins or light-activated compounds that instead of losing their efficiency due to sunlight degradation, they are actually increased or activated by two different mechanisms. In the first mechanism (less common), molecular oxygen from the phototoxin absorbs the energy from the light, generating activated species of oxygen which ultimately damage important biomolecules [99]. The other mechanism of action is photogenotoxic, where phytochemicals cause damage to DNA, triggered by sunlight activation, regardless of the presence of oxygen in the phototoxin. In actuality, thephototoxin on its ground state, absorbs the photon, reaching its excited state, which interacts with ground state O_2 located in the tissue of its target, generating singlet oxygen and enabling insecticidal activity. This peculiar mode of action of photoxins is so different from conventional synthetic pesticides that cross resistence among them is unlikely [100,101].

Light-activated phototoxins may be exemplified by several classes such as quinones, furanocoumarins, substituted acetylenes and thiophenes. For instance, Marchant and Cooper [102] have reported several phototoxins, such as 3-methyl-3-phenyl-1,4-pentadiyne, an oil constituent from *Artemisia monosperma* Delile (Asteraceae), which under sunlight-induced conditions exerts an activity similar to DDT against the housefly *Musca domestica* L. (Muscidae) and cotton leaf worm *Spodoptera littoralis* Boisduval (Noctuidae) larvae. They have also discovered that an acetylenic epoxide from *Artemisia pontica* L. (Asteraceae), called ponticaepoxide, exhibits an LC_{50} of 1.47 ppm against mosquito larvae when submitted to UV light. Additionally, Nivsarkar, et al. [103] have also found that the major compound from the roots of *Tagetes minuta* L. (Asteraceae), a thiophene called terthiophene or α-terthienyl (αT), is highly toxic against several organisms when co-submitted to near UV light radiation, such as nematodes, red flower beetles (*Tribolium casteneum* Herbst (Neoptera)), blood-feeding insects such as *Manduca sexta* L. (Sphingidae) and mosquito larvae (dipteres): *Aedes aegypti* L. (Culicidae), *Aedes atropalpus* Coquillett (Culicidae), *Aedes intrudens* Dyar (Culicidae) and yet, to our knowledge no commercial product has been generated. Plant-based natural product chemical structures with their corresponding pesticide activity and targets may be found in Table 2.

12. Discussion

Since ancient times, efforts to protect the agricultural harvest against pests have been reported. The use of inorganic compounds to control pests was reported between 500 B.C and the 19th century. They included products based on sulphur, lead, arsenic and mercury [2]. On the other hand, plant biodiversity has proved to be an endless source of biologically active ingredients, used for traditional crop and storage protection. Egyptian and Indian farmers used to mix the stored grain with fire ash [104]. The ancient Romans used false hellebore (*Veratrum viride* Aiton (Melanthiaceae)) as a rodenticide. Moreover, pyrethrum (extract from *Tanacetum cinerariifolium* (Trevir.) Sch.Bip (Asteraceae)) was used as an insecticide in Persia and Dalmatia, whereas the Chinese have discovered the insecticidal properties of *Derris* spp. (Fabaceae) [105].

Previous studies have already reported more than 2500 plant species belonging to 235 families, which have demonstrated their biological activity against several types of pests [1]. However, in spite of the remarkable potential as natural sources for commercial botanical pesticide development, not many have been found on the market, remaining in use only for small organic crops and commonly classified as so-called farming products [106].

Plant-derived pesticides can be processed in various ways: as crude plant material in the form of dust or powder; as extracts or as pure plant natural products, formulated into solutions or suspensions [2]. Several different classes of natural compounds that promote pesticide activity have already been reported, namely: fatty acids, glycolipids, aromatic phenols, aldehydes, ketones, alcohols, terpenoids, flavonoids, alkaloids, limonoids, naphtoquinones, saccharides, polyolesthers, saponins and sapogenins [20,107–110]. However, several factors appear to limit the commercialization of botanical pesticides, such as: problems in large scale production, non-availability of raw materials, poor shelf life, diminished residual toxicity under field conditions, limitations regarding standardization and refinement of the final product. Additionally, as the phytochemical profile of plant species may vary according to its genome/transcriptome/proteome/metabolome, and this variation depends on several edaphic-climatic factors (i.e., temperature, relative humidity, level of sunlight radiation, altitude, photoperiod and type of soil) as well as ecologic interactions, (i.e., herbivory or mutualism), manufacturers must take additional care in order to maintain efficiency and ensure that their products will perform consistently (standardization). Finally, even if all these issues are addressed, regulatory approval remains as the major barrier. A serious drawback to commercialization of botanicals is the high cost of processing plant materials to meet the standards of pesticide regulatory authorities [111]. Marrone [112] provided an overview of the current state of biopesticides and offered some ideas for improving their adoption, including conducting on-farm demonstrations and more education and training on how the products work and how to incorporate them into integrated pest management. In many jurisdictions, no distinction is made between synthetic pesticides and biopesticides. Simply because a compound is a natural product, it does not mean that it is safe, since most of the toxic poisons are natural products or inspired by them. Furthermore, if biopesticides are used indiscriminately, as wells as the synthetics, they may also lead to the development of pest resistance [113].

In this context, only a few new sources of botanicals have reached commercial status in the past twenty years. Thus, the major commercial botanic pesticides currently in use include: pyrethrin, rotenone, azadirachtin and essential oils in general [111], along with three other products, commercialized in a more limited way: ryania, nicotine and sabadilla or veratrine alkaloids [20].

Therefore, the best strategy for a botanical pesticide to meet all the standards required and reach commercial status in a more efficient and pragmatic way is by performing bioassay-guided fractionation in a high scale [82,109,114]. In other words, the bioassays assessment of several plant extracts and its fractions, obtained whether by sophisticated or unsophisticated purification procedures, may lead to the discovery of the most effective compound or mixture of compounds correlated to the pesticide activity of each corresponding species. The isolated compound may act as a lead compound or prototype for the synthesis or semi synthesis of pesticide derivatives, which, by structure–activity relationship (SAR) techniques may result in more effective and safer products. However, sometimes, when the compound is presented on its isolated form, it may promote no activity at all, proving that extracts or fractions from a certain plant species are more effective than its isolated compounds, due to synergic effect of compound mixture, which may also lead to the manufacture of potential raw material for commercial biopesticides [114,115].

Table 1. Toxicity and mechanism of action of bio-based natural insecticides.

Product Name	Toxicity	Mammalian Toxicity (LD_{50} (mg/kg bw))	Reference	Mechanism of Action	Reference
Azadirachtin	IGR, R	13000 (Orally)	[116]	Prothoracicotropic hormone (PTTH) inhibitor; phagostimulant disruptor by cholinergic transmission reduction	[60]
Nicotine	C	50 (Orally)	[116]	Acetylcholine mimic; agonist of nicotinic acetylcholine receptor	[35]
Rotenone	S	350 (Orally)	[116]	Complex I inhibitor of the mitochondrial respiratory chain	[24,49]
Pyrethrins I and II	C, S	1200 (Orally)	[116]	Voltage-gated Sodium channels modulator	[21]
Ryania	C, S	750 (Orally)	[116]	Activation of sarcoplasmic reticulum calcium channels (ryanodine channels)	[11,29]
Sabadilla	C, S	5000 (Orally)	[116]	Voltage-gated Sodium channels modulator	[27]
trans-Cinnamaldeyde	C	1160 (Orally)	[116]	Inhibitor of β-(1,3)-glucan synthase and chitin synthase	[117]
1,8 cineole	C, F	2480 (Orally)	[116]	AChE inhibitor	[31]
Eugenol	C, F	500 (Orally)	[116]	Agonist of octopamine receptor	[47]
Citronella oil	C, F	7200 (Orally)	[89]	Inhibition of (AChE) and glutathione-S-transferase	[118]
Thujone	C, F	230 (Orally)	[119]	Allosteric reversible modulator of $GABA_A$ receptors	[24,38,120]
Terthiophene	C	110 (intraperitoneally)	[121]	Light activated phototoxin. Activated species Oxygen generator	[102]
Palm kernel crude oil	C	>5000 (orally)	[122]	Disturbance of water balance caused by disruption of the protective waxy covering of insects	[1]

C—Contact. S—Stomach. F—Fumigant. IGR—Insect Growth Regulator. R—Repellent.

Table 2. Plant-based natural products with pesticide activity.

Compound/Trade Name	Plant Species	Class	Mode of Action	Targets	References
Nervous system: Voltage-gated sodium channels					
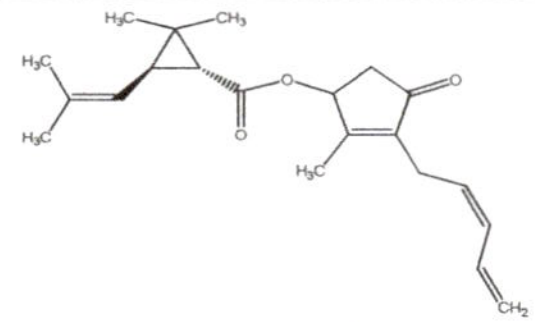 Pyrethrin I	*Tanacetum cinerariifolium* (Trevir.) Sch. Bip. (Asteraceae)	Cyclopropylmonoterpene esters	Insecticidal	Spider mites; flies; fleas; beetles	[23,24]
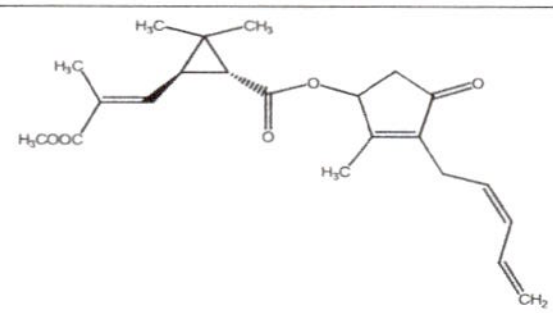 Pyrethrin II	*T. cinerariifolium*	Cyclopropylmonoterpene esters	Insecticidal	Spider mites; flies; fleas; beetles	[23,24]
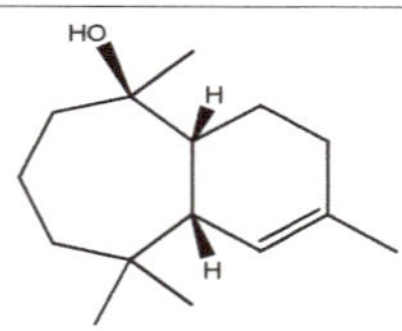 Himachalol	*Cedrus deodara* (Roxb. ex D. Don) G. Don (Pinaceae)	Sesquiterpene	Insecticidal	Pulse beetle (*Callosobruchus analis* Fabriciuss) and Housefly (*Musca domestica* L.)	[123]
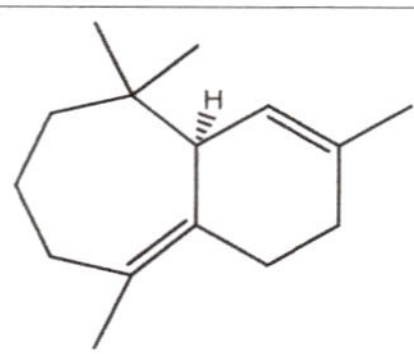 β-Himachalene	*C. deodara*	Sesquiterpene	Insecticidal	Pulse beetle (*Callosobruchus analis.*) and Housefly (*Musca domestica*)	[123]

Table 2. *Cont.*

Compound/Trade Name	Plant Species	Class	Mode of Action	Targets	References
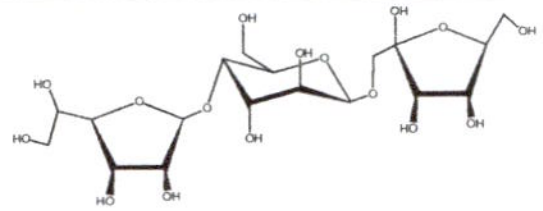 Decaleside II	*Decalepis hamiltonii* Wight and Arn. (Apocynaceae)	Trisaccharide	Insecticidal	Houseflies; cockroaches; stored grain pests	[25]
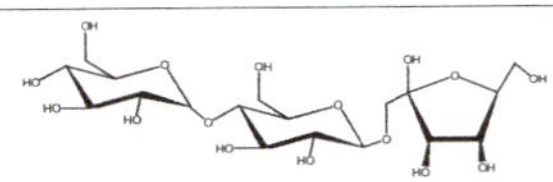 Decaleside I	*D. hamiltonii*	Trisaccharide	Insecticidal	Houseflies; cockroaches; stored grain pests	[25]
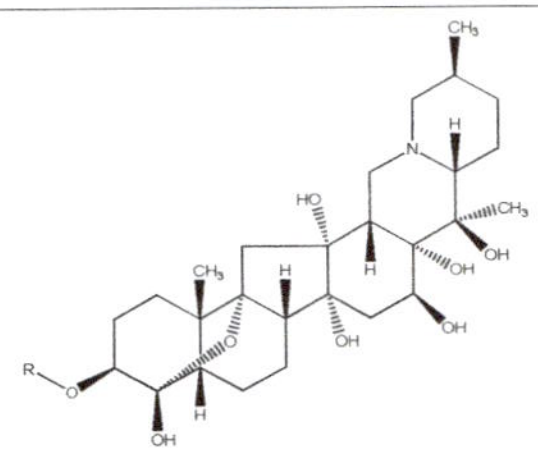 Cevadine R = (z)-$CH_3CH{=}C(CH_3)CO$ Veratridine R = 3,4-$(CH_3O)_2PhCO$	*Schoenocaulon officinale* (Schltdl. and Cham.) A. Gray ex Benth. (Melanthiaceae)	Sabadilla alkaloid or *Veratrum* alkaloid	Insecticidal	Stinks, leafhoppers, caterpillars; leafhoppers; housefly (*Musca domestica*) and thrips (*Scirtothrips* spp.)	[11,26,27,124]
Nervous system: Voltage-gated calcium channels					
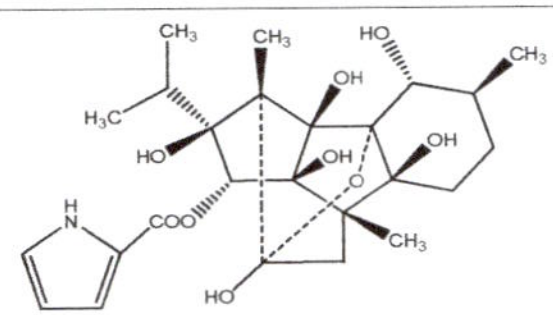 Ryanodine	*Ryania speciosa* Vahl (Salicaceae)	Diterpene	Insecticidal	Caterpillars; worms; potato beetle; lace bugs; aphids and squash bugs	[11,29]

Table 2. *Cont.*

Compound/Trade Name	Plant Species	Class	Mode of Action	Targets	References
		Acetylcholinesterase			
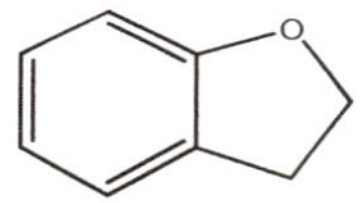 (Coumaran)2,3-dihydrobenzofuran	*Lantana camara* L. (Verbenaceae)	Benzofuran	Insecticidal	Stored grain pests (*Sitophilus oryzae* L.; *Tribolium castaneum herbst*); Housefly pests (*Musca domestica*)	[30]
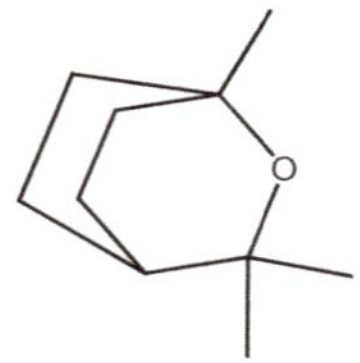 1,8-Cineole (Eucalyptol)	*Eucalyptus globulus* Labill. (Myrtaceae)	Monoterpene	Insecticidal	Head lice (*Pediculus humanus capitis* De geer)	[31]
		Nicotinic acetylcholine receptors			
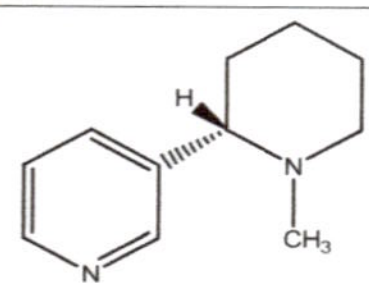 Nicotine	*Nicotiana tabacum* Velloso (Solanaceae)	Pyridine Alkaloid	Insecticidal Antifungal	Aphids; thrips; mites; leaf hoppers; spider mites; fungus	[24]

Table 2. *Cont.*

Compound/Trade Name	Plant Species	Class	Mode of Action	Targets	References
GABA-gated chloride channels					
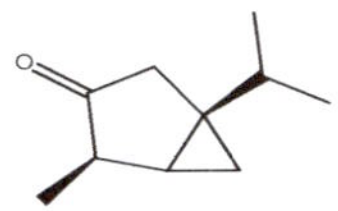 Thujone	*Artemisia absinthium* L. (Asteraceae); *Juniperus* sp. (Cupressaceae); *Cedrus* sp. (Pinaceae)	Monoterpene	Larvicidal Insecticidal	Western corn rootworm larvae (*Diabrotica virgifera*) Fruit fly (*Drosophila melanogaster* Meigen)	[38]
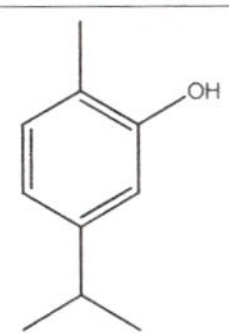 Carvacrol	**	Monoterpene	Insecticidal	*Periplaneta americana* L.	[39]
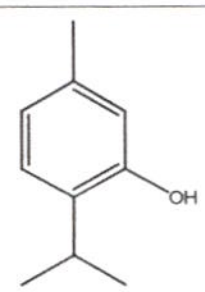 Thymol	**	Monoterpene	Insecticidal	*Periplaneta americana*	[39]
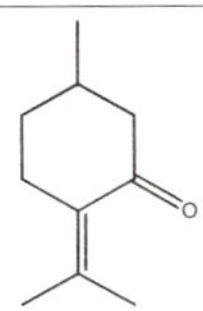 Pulegone	**	Monoterpene	Insecticidal	*Periplaneta americana*	[39]

Table 2. *Cont.*

Compound/Trade Name	Plant Species	Class	Mode of Action	Targets	References
Octopamine receptors					
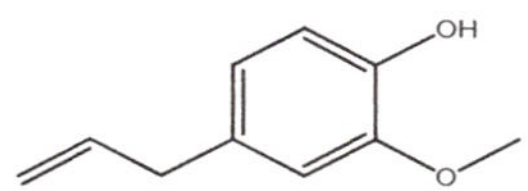 Eugenol	*Syzygium aromaticum* (L.) Merr. and L.M. Perry (Myrtaceae)	Monoterpene	Herbicidal	*Cassia occidentalis* and *Biden spilosa*	[90]
			Insecticidal Repellent	Blood-sucking bug *Triatoma infestans* (Klug); fruit fly (*Drosophila melanogaster* (*Meigen*); American cockroach (*Periplaneta americana*)	[46–48]
Respiratory or energy system					
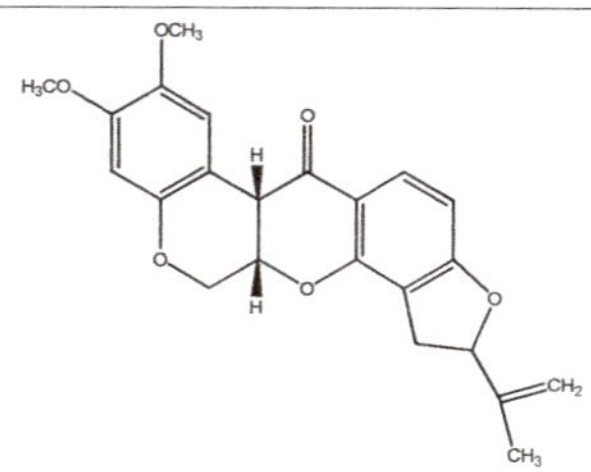 Rotenone	*Lonchocarpus* Kunth (Fabaceae); *Derris* Lour (Fabaceae); *Rhododendron* L. (Ericaceae)	Isoflavonoid/Rotenoid	Insecticidal; Piscicidal	Beetles; caterpillars; lice; mosquitoes; ticks; fleas; fire ants	[24,125]
Endocrine system					

Table 2. *Cont.*

Compound/Trade Name	Plant Species	Class	Mode of Action	Targets	References
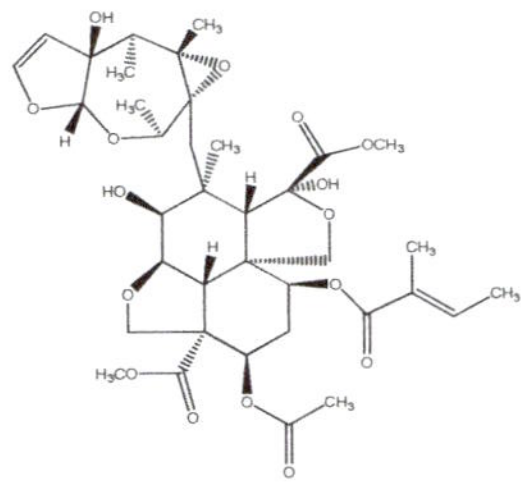 Azadirachtin	*Azadirachta indica* A. Juss. (Meliaceae)	Tetranortriterpenoid limonoid	Repellent; Antifeedant; Nematicidal; Sterilant; Antifungal; Insect Growth Regulator	Dandruffs eczema; stored grain pests; aphids; caterpillars; thrips; mealy bugs	[2,26,60,77,78,126]
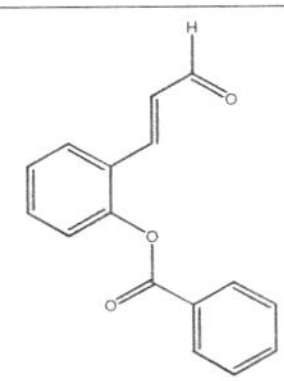 2′-Benzoyloxycinnamaldehyde (2′-BCA)	*Pleuropterus ciliinervis* Nakai (Polygonaceae)	Aromatic aldehyde	Antifungal	*Saccharomyces cerevisiae* Meyen ex Hansen.	[62]
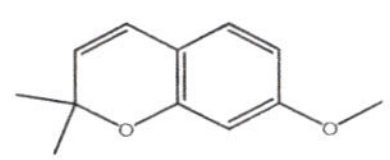 Precocene I (7-methoxy-2,2-dimethylchromene)	*Ageratum conyzoides* L. (Asteraceae)	Chromene	IGR	Sawtoothed grain beetle (*Oryzaephilus surinamensis* L.); Milkweed bug (*Oncopeltus fasciatu* Dallas); Noctuid moth (*Spodoptera litura* Fabricius); Parasitic wasp (*Microplitis rufiventris* Nees)	[127–130]

Table 2. *Cont.*

Compound/Trade Name	Plant Species	Class	Mode of Action	Targets	References
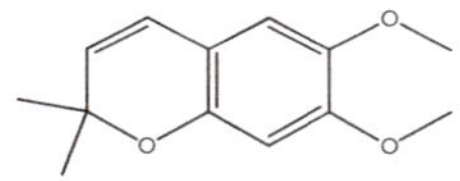 Precocene II (6,7-dimethoxy-2,2-dimethylchromene)	*A. conyzoides*	Chromene	IGR	Desert locust (*Schistocerca gregaria* Forskål); Milkweed bug (*Oncopeltus fasciatus* Dallas); Noctuid moth (*Spodoptera litura* Fabricius); Parasitic wasp (*Microplitis rufiventris* Nees)	[128–131]
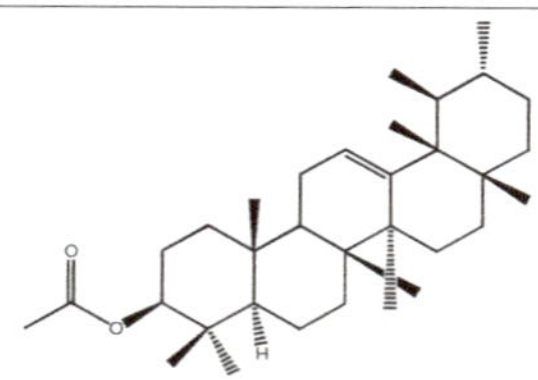 α-amyrin acetate	*Catharanthus roseus* (L.) G. Don (Apocynaceae)	Steroid	IGR	*Helicoverpa armigera* Hübner	[54]
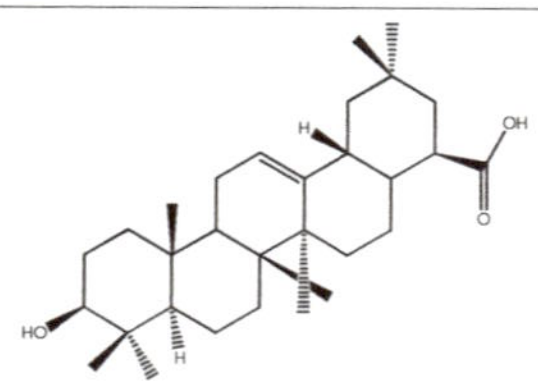 Oleanolic acid	*Catharanthus roseus* (L.) G.Don (Apocynaceae)	Steroid	IGR	*Helicoverpa armigera*	[54]
Antifeedants					

Table 2. *Cont.*

Compound/Trade Name	Plant Species	Class	Mode of Action	Targets	References
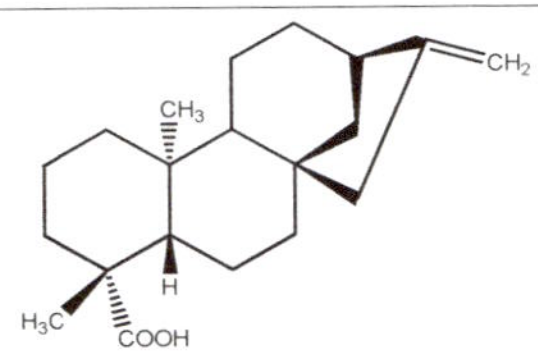 (−)Kau-16-en-19-oic acid	*Xylopia aethiopica* (Dunal) A. Rich. (Annonaceae)	Kaurane diterpene	Antifeedant	Termites (*Reticulitermes speratus* Kolbe)	[79]
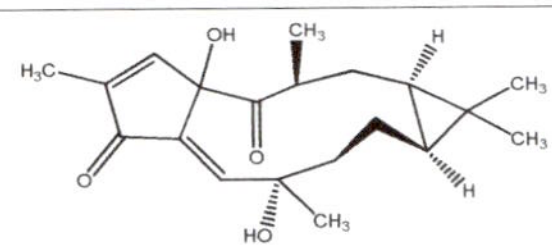 15-epi-4E-jatrogrossidentadione	*Jatropha podagrica* Hook. (Euphorbiaceae)	Diterpene	Antifeedant	Moth (*Chilo partellu* Swinhoes)	[2]
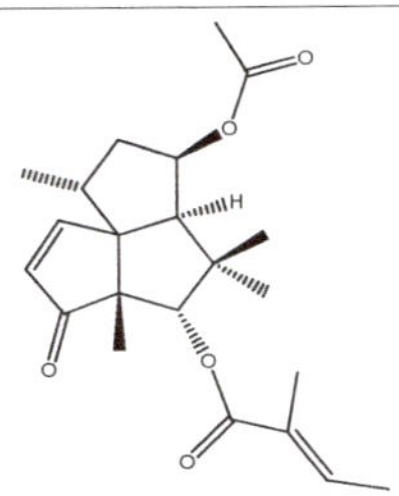 11-Acetoxy-5-isobutyryloxysilphinen-3-one	*Senecio palmensis* C. Sm. (Asteraceae)	Silphinene sesquiterpene	Antifeedant	Colorado potatobeetle (*Leptinotarsa decemlineata Say*);Aphids (*Myzus persicae* Sulzer, *Diuraphis noxia* Kurdjumov, *Rhopalosiphum padi L.*, *Metopolophium dirhodum Walker*, *Sitobiona venae* Fabricius)	[40]

Table 2. *Cont.*

Compound/Trade Name	Plant Species	Class	Mode of Action	Targets	References
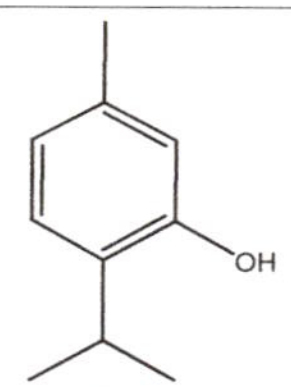 Thymol	*Thymus vulgaris* L. (Lamiaceae)	Monoterpene	Antifeedant	Colorado potato beetle (*Leptinotarsa decemlineata*);Aphids (*Myzus persicae*, *Diuraphis noxia*, *Rhopalosiphum padi*, *Metopolophium dirhodum*, *Sitobiona venae*)	[40]
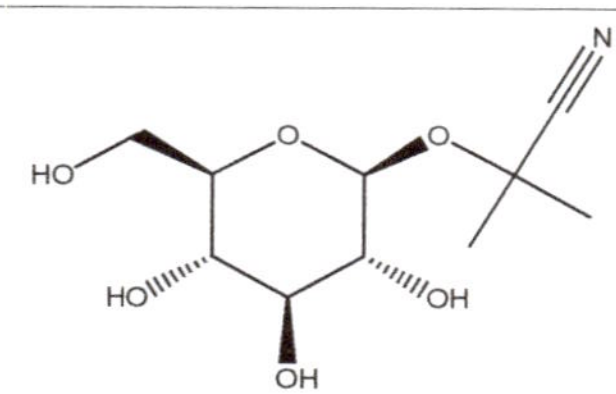 Linamarin	*Lotus corniculatus* L. (Fabaceae); *Trifolium repens* L. (Fabaceae)	Cyanogenic glycoside	Antifeedant	Snails (*Arianta arbustorum* L. and *Helix* aspersa O.F. Müller; slugs (*Agriolimax reticulates* O.F. Müller); lemmings (*Lemmus lemmus* L.); aphids (*Aphis craecivora* Koch, *Nearctaphis bakeri* Cowen ex Gillette and Baker)	[84]
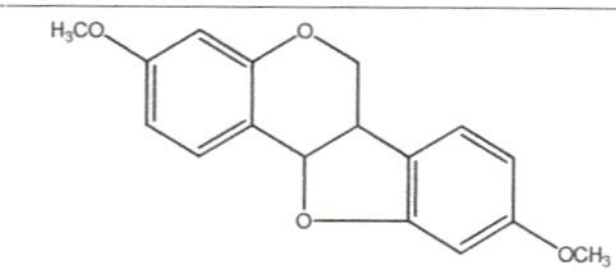 (−)-homopterocarpin	*Pterocarpus macrocarpus* Kurz (Fabaceae)	Isoflavonoid/Pterocarpan	Antifeedant	Common cutworm (*Spodoptera litura* F.) and the subterranean termite (*Reticulitermes speratus*)	[81]

Table 2. *Cont.*

Compound/Trade Name	Plant Species	Class	Mode of Action	Targets	References
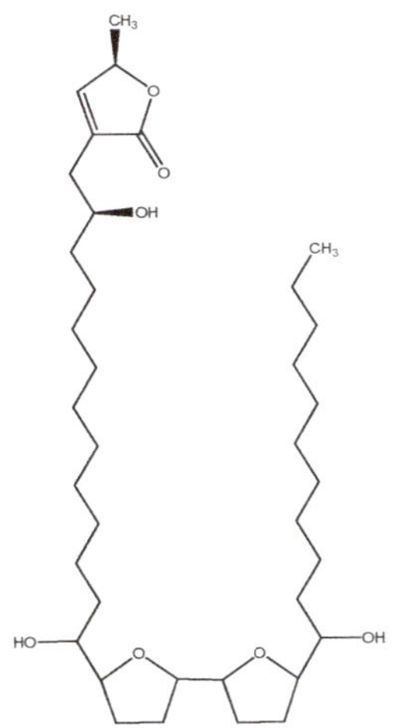 Asimicin	*Asimina triloba* (L.) Dunal (Annonaceae)	Polyketide/Acetogenin	Insecticidal; Antifeedant Nematicidal Larvicidal	Mexican bean beetle (*Epilachna varivestis* Mulsant); striped cucumber beetle (*Acalymma bivittatum Fabricius*); two-spotted spider mite (*Tetranychus urticae* Koch); melon aphid (*Aphis gossyphii Glover*) *Caenorhabditis elegans* (Maupas) Blowfly larvae (*Calliphora vicina* Robineau-Desvoidy); mosquito larvae (*Aedes aegypti*)	[82,83]
Repellents					
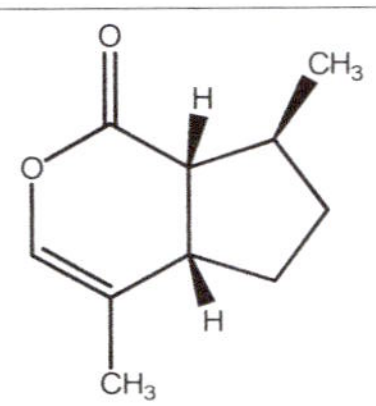 Nepetalactone	*Nepeta cataria* L. (Lamiaceae)	Monoterpene lactone	Repellent	Mosquitoes (*Aedes aegypti*); bees;lady beetle; cockroaches; flies; termites	[68–70]

Table 2. *Cont.*

Compound/Trade Name	Plant Species	Class	Mode of Action	Targets	References
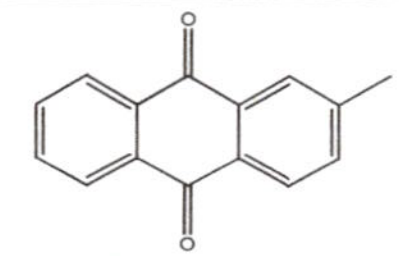 Tectoquinone	*Tectona grandis* L. f. (Lamiaceae)	Anthraquinone	Repellent	Termites (*Cryptotermes brevis* Walker and *Reticulitermes flavipes* Kollar)	[71,72,132]
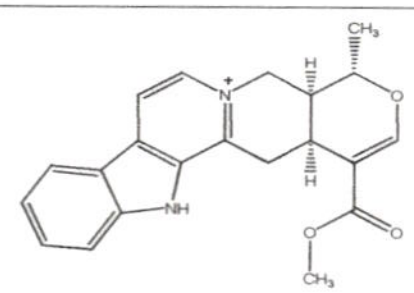 Alstonine	*Alstonia boonei De* Wild. (Apocynaceae)	Indoloquinolizidine alkaloid	Repellent; Larvicidal	Mosquito (*Anopheles gambiae Giles*)	[73]
Attractants					
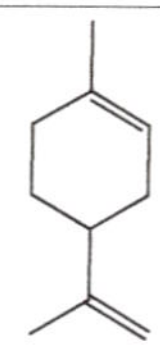 Limonene	*Pinus strobes* D. Don (Pinaceae)	Monoterpene	Attractant	White pine cone beetle (*Conophthorus coniperda* Schwarz)	[75]
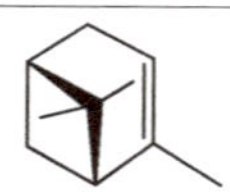 (−)-α–Pinene	*P. strobus*	Monoterpene	Attractant	*Enoclerus nigripes Say*	[75]

Table 2. *Cont.*

Compound/Trade Name	Plant Species	Class	Mode of Action	Targets	References
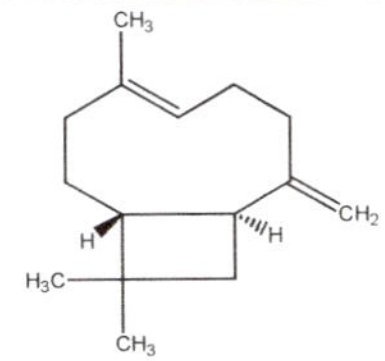 (*E*)-β-caryophyllene	*Zea mays* L. (Poaceae)	Sesquiterpene	Attractant	Nematodes (*Heterorhabditis megidis* Poinar, Jackson and Klein), natural enemy/parasite of corn root worm (*Diabrotica virgifera* Leconte)	[76]
		Phytotoxins			
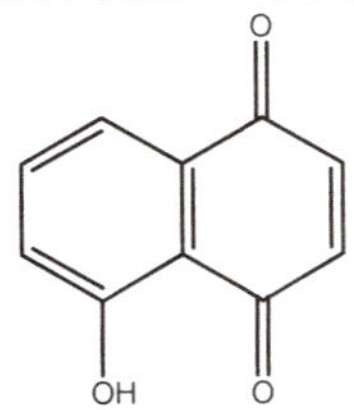 Juglone	*Juglans nigra* L. (Juglandaceae)	Naphtoquinone	Herbicidal	*Echinochloa crus-galli* L.; *Amaranthus retroflexus* L.; *Abutilon theopharasti* Medik	[91,92]
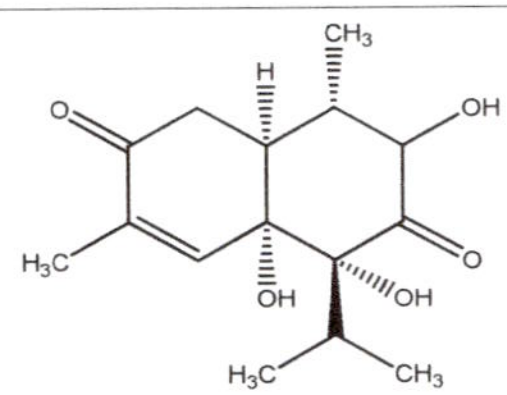 5,6-dihydroxycadinan-3-ene-2,7-dione	*Eupatorium adenophorum* Spreng. (Asteraceae)	Cadinane sesquiterpene	Herbicidal	*Arabidopsis thaliana* (L.) Heynh	[95]

Table 2. *Cont.*

Compound/Trade Name	Plant Species	Class	Mode of Action	Targets	References
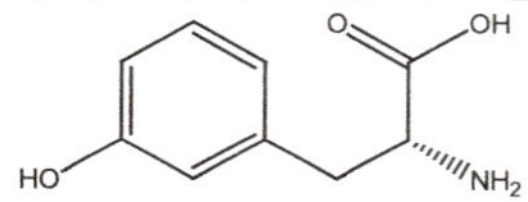 m-Tyrosine	Poaceae spp.	Amino acid	Herbicidal	Weeds	[93]
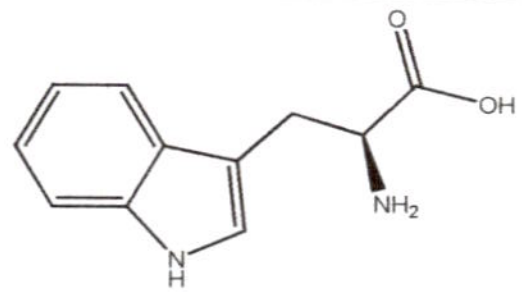 Tryptophan	*Prosopis juliflora* (Sw.) (Fabaceae)	Amino acid	Herbicidal	Barnyard grass (*Echinochloa crus-galli* L.)	[133]
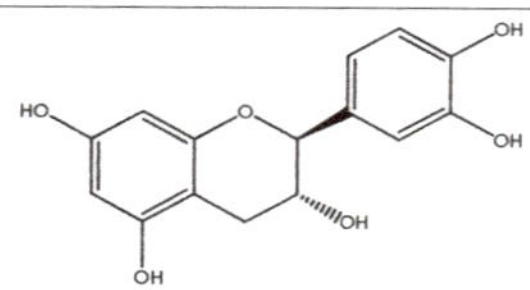 (−)-Catechin	*Centaurea stoebe* L. (Asteraceae)	Flavanol	Herbicidal	*Koeleria macrantha* (Ledeb.) Schult., and *Festuca idahoensis* Elmer	[96]
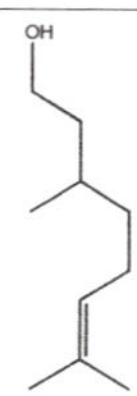 Citronellol Citronellol	*Cymbopogon citratus* (DC.) Stapf (Poaceae)	Monoterpene	Herbicidal	*Cassia occidentalis* L.	[90]

Table 2. *Cont.*

Compound/Trade Name	Plant Species	Class	Mode of Action	Targets	References
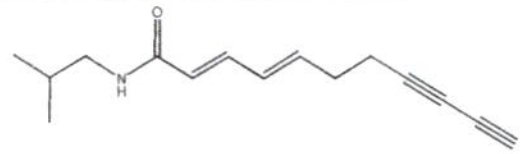 (*E*,*E*)-2,4-undecadien-8,10-diynoic acid isobutylamide	*Acmella oleracea* (L.) R. K. Jansen (Asteraceae)	Isobutylamide	Herbicidal	Cress (*Lepidum sativum* L.) and barnyard grass (*Echinochloa crus-galli* (L.) P. Beauv)	[98]
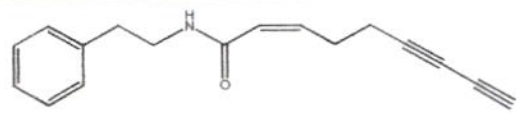 Nona-(2Z)-en-6,8-diynoic acid 2-phenylethylamide	*A. oleracea*	Phenylethylamide	Herbicidal	Cress (*Lepidum sativum*) and barnyard grass (*Echinochloa crus-galli*)	[98]
		Antifungals			
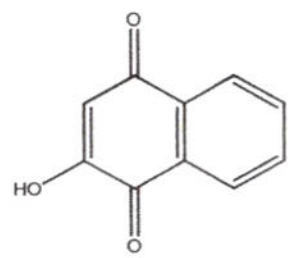 Lawsone	*Lawsonia inermis* L. (Lythraceae)	Naphtoquinone	Antifungal	*Alternaria solani* (Fr.) Keissl.; *Alternaria tenuis* Nees; *Aspergilus niger* Tieghem; *Aspergilus wentii* Whemer; *Absidia ramosa* (Zopf) Vuil; *Absidia corymbifera* Cohn; *Acrophialophora fusispora* (S.B. Saksena) Samson; *Circinella umbellate* Tiegh. and G. Le Monn	[134]
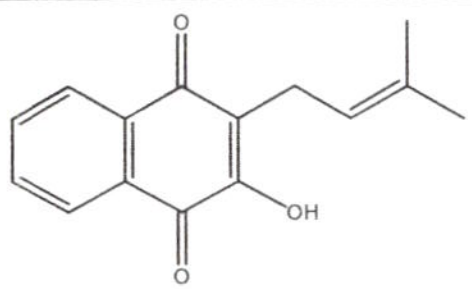 Lapachol	*Tabebuia serratifolia* (Vahl) G. Nicholson (Bignoniaceae)	Naphtoquinone	Antifungal; Larvicidal	*Aedes aegypti*; *Gloeophyllum trabeum* ATCC 11539; *Trametes versicolor* (L.) Lloyd	[135,136]

Table 2. *Cont.*

Compound/Trade Name	Plant Species	Class	Mode of Action	Targets	References
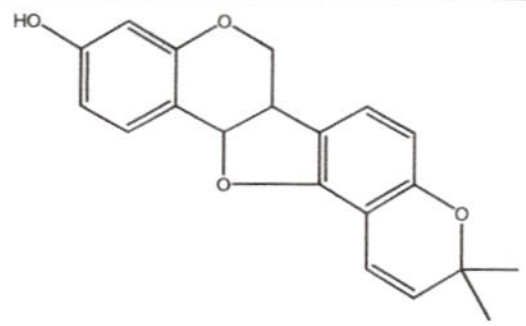 Phaseollin	*Z. mays*	Isoflavonoid/Pterocarpan	Antifungal	*Botrytis cinerea* Pers.; *Colletotrichum lindemuthianum* (Sacc. and Magnus) Briosi and Cavara,; *Fusarium solani* Mart.; *Rhizoctonia solani* J. G. Kühn and *Thielaviopsis basicola* (Berk. and Broome) Ferraris	[137,138]
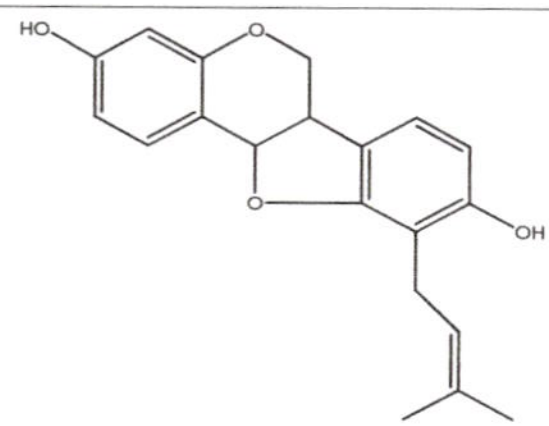 Phaseollidin	*Z. mays*	Isoflavonoid/Pterocarpan	Antifungal	*Botrytis cinerea*; *Colletotrichum lindemuthianum*; *Fusarium solani*; *Rhizoctonia solani* and *Thielaviopsis basicola*	[137,138]
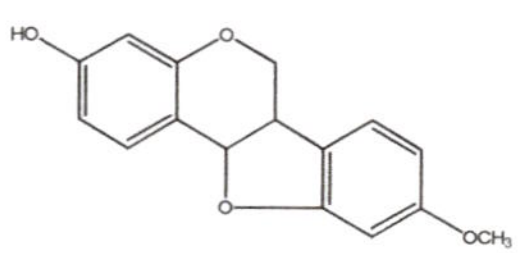 Medicarpin	*Medicago sativa* L. (Fabaceae)	Isoflavonoid/Pterocarpan	Antifungal	*Colletotrichum phomoide* (Sacc.) Chester.; *Stemphylium loti* J.H. Graham,; *Stemphylium botryosum* Walroth; *Phoma herbarum* Westendorp; *Leptosphaeria briossiana* (Higgings) and *Cladosporium cladosporoides* (Fresen.) G.A. de Vries	[137,138]

Table 2. *Cont.*

Compound/Trade Name	Plant Species	Class	Mode of Action	Targets	References
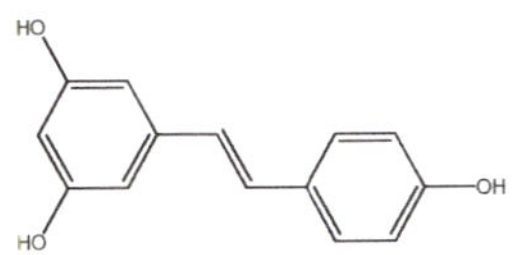 Resveratrol	*Polygonum cuspidatum* Siebold and Zucc. (Polygonaceae)	Polyphenol/Stilbene	Anti-viral	Tobacco Mosaic Virus (TMV)	[97]
			Antifungal	*Alternaria solani; Botrytis cinerea; Fusarium graminearum* Schwabe; *Phytophthora capsici Leonian; Phytophthora infestans (Mont.) de Bary; Rhizoctonia* solani J.G. Kühn; *Sclerotinia sclerotiorum* (Lib.) (Y. Nisik. and C. Miyake) Shoemaker, *Rhizoctonia cerealis* D. I. Murray and Burpee; *Watermelon anthracnose*	
			Herbicidal	*Digitaria sanguinalis* (L.) Scop.; *Echinochloa crus-galli*	
			Insecticidal	Oriental armyworm (*Mythimna separata* Walker); Cotton bollworm (*Helicoverpa armigera*); Corn borer (*Ostrinia nubilalis* Hubner)	
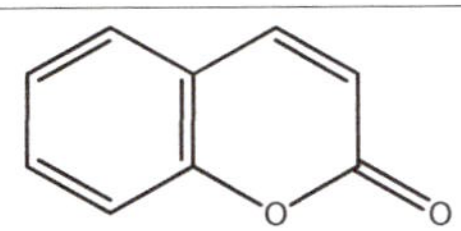 2H-chromen-2-one	*Lavandula angustifolia Mill. (Lamiaceae)*	Polyphenol/Coumarin	Antibacterial	*Ralstonia solanacearum* Smith	[139]
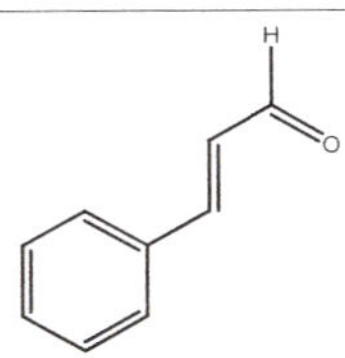 *Trans*-Cinnamaldehyde	*Cinnamomi cortex* J. Presl (Lauraceae)	Aromatic aldehyde	Antifungal	*Saccharomyces cerevisiae*	[117]

Table 2. *Cont.*

Compound/Trade Name	Plant Species	Class	Mode of Action	Targets	References
		Phototoxins			
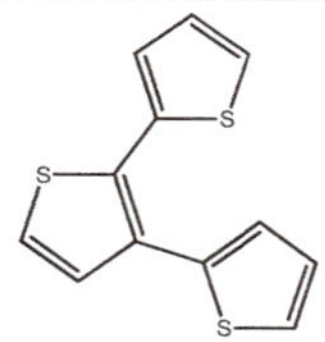 Terthiophene	*Tagetes minuta* L. (Asteraceae)	Thiophene	Nematicidal	Nematodes	[103]
			Insecticidal	Tobacco hornworm (*Manduca sexta*); Lepidopteran (*Pieris rapae* L.); housefly (*Musca domestica*); Red flower beetle (*Tribolium casteneum Herbst*); mosquito larvae (*Aedes atropalpus*, *Aedes aegypti* and *Aedes intrudens*)	
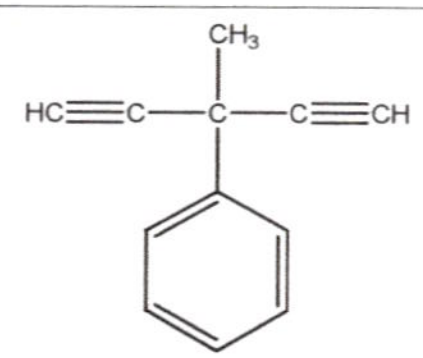 3-methyl-3-phenyl-1,4-pentadiyne	*Artemisia monosperma* Delile (Asteraceae)	Substituted Acetylene	Insecticidal	Housefly (*Musca domestica*) and Cotton Leaf worm (*Spodoptera littoralis* Boisduval)	[102]

** Purchased from Sigma–Aldrich Chemical Co.

13. Conclusions

Despite several factors that appear to limit botanical pesticide commercialization, such as problems in large scale production, non-availability of raw materials, poor shelf life, diminished residual toxicity under field conditions and lack of extract standardization, a multidisciplinary approach, comprising bioassay-guided fractionation, combined with structure-activity relationship (SAR) and analytical techniques, has revealed to be an extremely efficient strategy in order to develop bio-based pesticides that meet all the commercial standards required. In summary, plant-derived pesticides have indicated their potential as a great alternative for pest management, once they become cheap, target-specific, less hazardous to human health, biodegradable and ecofriendly; therefore, they may improve crop efficiency and reduce food crisis while maintaining sustainability.

Funding: This research was funded by "Plan Chlordecone III en Guadeloupe et Martinique" in the framework of DICHOAL-MAREC (DIREC) project.

Institutional Review Board Statement: Not applicable.

Informed Consent Statement: Not Applicable.

Acknowledgments: Special thanks are due to Tara Massad (expert in insect chemical ecology from Gorongosa National Park (Mozambique)) for her attentive lecture and fruitful discussion.

Conflicts of Interest: The authors declare no conflict of interest.

References

1. Walia, S.; Saha, S.; Rana, V. Phytochemical Pesticides. In *Advances in Plant Biopesticides*; Singh, D., Ed.; Springer: New Delhi, India, 2014; pp. 295–322.
2. Okwute, S.K. Plants as Potential Sources of Pesticidal Agents: A Review. In *Pesticides—Advances in Chemical and Botanical Pesticides*, 1st ed.; eBook; Soundararajan, R.P., Ed.; IntechOpen: London, UK, 2012; pp. 207–232.
3. Sharma, S.; Kooner, R.; Arora, R. Insect Pests and Crop Losses. In *Breeding Insect Resistant Crops for Sustainable Agriculture*; Arora, R., Sandhu, S., Eds.; Springer: Singapore, 2017; pp. 45–66.
4. Jarman, W.K.; Ballschmiter, K. From coal to DDT: The history of the development of the pesticide DDT from synthetic dyes till Silent Spring. *Endeavour* **2012**, *36*, 131–142. [CrossRef] [PubMed]
5. Rattner, B.A. History of wildlife toxicology. *Ecotoxicology* **2009**, *18*, 773–783. [CrossRef]
6. Multigner, L.; Ndong, J.R.; Giusti, A.; Romana, M.; Delacroix-Maillard, H.; Cordier, S.; Jégou, B.; Thome, J.P.; Blanchet, P. Chlordecone Exposure and Risk of Prostate Cancer. *J. Clin. Oncol.* **2010**, *28*, 3457–3462. [CrossRef] [PubMed]
7. de Boer, J.; Leslie, H.; van Leeuwen, S.P.J.; Wegener, J.W.; van Bavel, B.; Lindström, G.; Lahoutifard, N.; Fiedler, H. United Nations Environment Programme Capacity Building Pilot Project—Training and interlaboratory study on persistent organic pollutant analysis under the Stockholm Convention. *Anal. Chim. Acta* **2008**, *617*, 208–215. [CrossRef] [PubMed]
8. Multigner, L.; Kadhel, P.; Rouget, F.; Blanchet, P.; Cordier, S. Chlordecone exposure and adverse effects in French West Indies populations. *Environ. Sci. Pollut. Res.* **2016**, *23*, 3–8. [CrossRef] [PubMed]
9. Boucher, O.; Simard, M.N.; Muckle, G.; Rouget, F.; Kadhel, P.; Bataille, H.; Cordier, S. Exposure to an organochlorine pesticide (chlordecone) and development of 18-month-old infants. *Neurotoxicology* **2013**, *35*, 162–168. [CrossRef] [PubMed]
10. Cabidoche, Y.-M.; Achard, R.; Cattan, P.; Clermont-Dauphin, C.; Massat, F.; Sansoulet, J. Long-term pollution by chlordecone of tropical volcanic soils in the French West Indies: A simple leaching model accounts for current residue. *Environ. Pollut.* **2009**, *157*, 1697–1705. [CrossRef]
11. Shivanandappa, T.; Rajashekar, Y. Mode of Action of Plant-Derived Natural Insecticides. In *Advances in Plant Biopesticides*, 1st ed.; Singh, D., Ed.; Springer: New Delhi, India, 2014; pp. 323–345.
12. Panagiotis, J.S.; Brokaki, M.; Stathas, G.J.; Demopoulos, V.; Giannis, L.; Margaritopoulos, J.T. Lethal and sub-lethal effects of imidacloprid on the aphidophagous coccinellid *hippodamia variegate*. *Chemosphere* **2019**, *229*, 392–400.
13. Gibbons, D.; Morrissey, C.; Mineau, P. A review of the direct and indirect effects of neonicotinoids and fipronil on vertebrate wildlife. *Environ. Sci. Pollut. Res.* **2015**, *22*, 103–118. [CrossRef]
14. Hermawan, W.; Nakajima, S.; Tsukuda, R.; Fujisaki, K.; Nakasuji, F. Isolation of an Antifeedant Compound from *Andrographis paniculata* (Acanthaceae) against the Diamondback Moth, *Plutella xylostella* (Lepidoptera: Yponomeutidae). *Appl. Entomol. Zool.* **1997**, *32*, 551–559. [CrossRef]
15. Miresmailli, S.; Isman, M.B. Botanical insecticides inspired by plant–herbivore chemical interactions. *Trends Plant Sci.* **2014**, *19*, 29–35. [CrossRef]
16. Arora, N.K.; Verma, M.; Jai rakash, J.; Mishra, J. *Regulation of Biopesticides: Global Concerns and Policies*; Springer: New Delhi, India, 2016.

17. Friedich, K.; Rodriguez da Silveira, G.; Costa Amazonas, J.; Monte Gurgel, A.; Soares de Almeida, V.; Sarpa, M. International regulatory situation of pesticides authorized for use in Brazil: Potential for damage to health and environmental impacts. *Cad. Saúde Pública* **2021**, *37*, e00061820.
18. Rajashekar, Y.; Gunasekaran, N.; Shivanandappa, T. Insecticidal activity of the root extract of Decalepishamiltonii against stored-product insect pests and its application in grain protection. *J. Food Sci. Technol.* **2010**, *47*, 310–314. [CrossRef]
19. Yao, Y.; Cai, W.; Yang, C.; Xue, D.; Huang, Y. Isolation and characterization of insecticidal activity of (Z)-asarone from *Acorus calamus* L. *Insect Sci.* **2008**, *15*, 229–236. [CrossRef]
20. Isman, M. Botanical Insecticides, Deterrents, and Repellents in Modern Agriculture and an Increasingly Regulated World. *Annu. Rev. Entomol.* **2006**, *51*, 45–66. [CrossRef]
21. Du, Y.; Nomura, Y.; Satar, G.; Hu, Z.; Nauen, R.; He, S.; Zhorov, B.; Dong, K. Molecular evidence for dual pyrethroid-receptor sites on a mosquito sodium channel. *Proc. Natl. Acad. Sci. USA* **2013**, *110*, 11785–11790. [CrossRef] [PubMed]
22. Casida, J.E.; Quistad, G.B. *Pyrethrum Flowers: Production, Chemistry, Toxicology, and Uses*, 1st ed.; Oxford University Press: Oxford, UK, 1995.
23. Casida, J.; Quistad, G. Golden age of insecticide research: Past, present, or future? *Annu. Rev. Entomol.* **1998**, *43*, 1–16. [CrossRef] [PubMed]
24. Rattan, R.S. Mechanism of action of insecticidal secondary metabolites of plant origin. *Crop. Prot.* **2010**, *29*, 913–920. [CrossRef]
25. Rajashekar, Y.; Rao, L.; Shivanandappa, T. Decaleside: A new class of natural insecticide targeting tarsal gustatory sites. *Die Nat.* **2012**, *99*, 843–852. [CrossRef] [PubMed]
26. Copping, L.G.; Duke, S.O. Natural products that have been used commercially as crop protection agents. *Pest Manag. Sci.* **2007**, *63*, 524–554. [CrossRef] [PubMed]
27. Hare, J.; Morse, J. Toxicity, Persistence, and Potency of Sabadilla Alkaloid Formulations to Citrus Thrips (Thysanoptera: Thripidae). *J. Econ. Entomol.* **1997**, *90*, 326–332. [CrossRef]
28. Nauen, R. Insecticide mode of action: Return of the ryanodine receptor. *Pest Manag. Sci.* **2006**, *62*, 690–692. [CrossRef]
29. Bloomquist, J.R. Insecticides: Chemistries and Characteristics. In *Radcliffe's IPM World Textbook*, 2nd ed.; University of Minesota: St. Paul, MI, USA, 1999.
30. Rajashekar, Y.; Raghavendra, A.; Bakthavatsalam, N. Acetylcholinesterase Inhibition by Biofumigant (Coumaran) from Leaves of *Lantana camara* in Stored Grain and Household Insect Pests. *BioMed Res.* **2014**, *2014*, 187019.
31. Picollo, M.I.; Toloza, A.C.; Mougabure Cueto, G.; Zygadlo, J.; Zerba, E. Anticholinesterase and pediculicidal activities of monoterpenoids. *Fitoterapia* **2008**, *79*, 271–278. [CrossRef] [PubMed]
32. Hansen, M.R.H.; Jørs, E.; Sandbæk, A.; Sekabojja, D.; Ssempebwa, J.C.; Mubeezi, R.; Staudacher, P.; Fuhrimann, S.; Sigsgaard, T.; Burdorf, A.; et al. Organophosphate and carbamate insecticide exposure is related to lung function change among smallholder farmers: A prospective study. *Thorax* **2021**, *76*, 780–789. [CrossRef]
33. Khorshid, M.A.; Hassan, A.F.; Abd El- Gawad, M.; Enab, A.K. Effect of some Plants and Pesticides on Acetylcholinesterase. *Am. J. Food Technol.* **2015**, *10*, 93–99. [CrossRef]
34. Isman, M.; Tak, J.H. Inhibition of acetylcholinesterase by essential oils and monoterpenoids: A Relevant mode of action for insecticidal essential oils? *Biopestic. Int.* **2017**, *13*, 71–78.
35. Green, B.; Welch, K.; Panter, K.; Lee, S. Plant Toxins That Affect Nicotinic Acetylcholine Receptors: A Review. *Chem. Res. Toxicol.* **2013**, *26*, 1129–1138. [CrossRef] [PubMed]
36. Elbert, A.; Nauen, R.; Cahill, M.; Devonshire, A.L.; Scarr, A.; Sone, S.; Steffens, R. Resistance management for chloronicotinyl insecticides using imidacloprid as an example. *Pflanzenschutz Nachr. Bayer* **1996**, *49*, 5–54.
37. Tomizawa, M.; Casida, J.E. Neonicotinoid insecticide toxicology: Mechanisms of selective action. *Annu. Rev. Pharmacol. Toxicol.* **2005**, *45*, 247–268. [CrossRef] [PubMed]
38. Höld, K.; Sirisoma, N.; Ikeda, T.; Narahashi, T.; Casida, J. α-Thujone (the active component of absinthe): γ—Aminobutyric acid type A receptor modulation and metabolic detoxification. *Proc. Natl. Acad. Sci. USA* **2000**, *97*, 3826–3831. [CrossRef] [PubMed]
39. Tong, F.; Coats, J.R. Effects of monoterpenoid insecticides on [3H]-TBOB binding in house fly GABA receptor and 36Cl− uptake in American cockroach ventral nerve cord. *Pestic. Biochem. Physiol.* **2010**, *98*, 317–324. [CrossRef]
40. González-Coloma, A.; Valencia, F.; Martín, N.; Hoffmann, J.J.; Hutter, L.; Marco, J.A.; Reina, M. Silphinene Sesquiterpenes as Model Insect Antifeedants. *J. Chem. Ecol.* **2002**, *28*, 117–129. [CrossRef] [PubMed]
41. Mullin, C.A.; González-Coloma, A.; Gutiérrez, C.; Reina, M.; Eichenseer, H.; Hollister, B.; Chyb, S. Antifeedant Effects of Some Novel Terpenoids on Chrysomelidae Beetles: Comparisons with Alkaloids on an Alkaloid-Adapted and Nonadapted Species. *J. Chem. Ecol.* **1997**, *23*, 1851–1866. [CrossRef]
42. Roeder, T. Tyramine and octopamine: Ruling Behavior and Metabolism. *Annu. Rev. Entomol.* **2005**, *50*, 447–477. [CrossRef] [PubMed]
43. Farooqui, T. Review of octopamine in insect nervous system. *Open Access Insect Physiol.* **2012**, *4*, 1. [CrossRef]
44. Evans, P.D. Multiple receptor types for octopamine in the locust. *J. Physiol.* **1981**, *318*, 99–122. [CrossRef] [PubMed]
45. Evans, P.D. Molecular studies on insect octopamine receptors. *EXS* **1993**, *63*, 286–296. [PubMed]
46. Kostyukovsky, M.; Rafaeli, A.; Gileadi, C.; Demchenko, N.; Shaaya, E. Activation of octopaminergic receptors by essential oil constituents isolated from aromatic plants: Possible mode of action against insect pests. *Pest Manag. Sci.* **2002**, *58*, 1101–1106. [CrossRef] [PubMed]

47. Reynoso, M.M.N.; Lucia, A.; Zerba, E.N.; Alzogaray, R.A. The Octopamine Receptor Is a Possible Target for Eugenol-Induced Hyperactivity in the Blood-Sucking Bug *Triatoma infestans* (Hemiptera: Reduviidae). *J. Med. Entomol.* **2019**, *57*, 627–630. [CrossRef] [PubMed]
48. Enan, E. Molecular and pharmacological analysis of an octopamine receptor from American cockroach and fruit fly in response to plant essential oils. *Arch. Insect Biochem. Physiol.* **2005**, *59*, 161–171. [CrossRef]
49. Casida, J.; Durkin, K. Neuroactive Insecticides: Targets, Selectivity, Resistance, and Secondary Effects. *Annu. Rev. Entomol.* **2013**, *58*, 99–117. [CrossRef] [PubMed]
50. Sherer, T.; Richardson, J.; Testa, C.; Seo, B.; Panov, A.; Yagi, T.; Matsuno-Yagi, A.; Miller, G.; Greenamyre, J. Mechanism of toxicity of pesticides acting at complex I: Relevance to environmental etiologies of Parkinson's disease. *J. Neurochem.* **2007**, *100*, 1469–1479. [CrossRef]
51. Hollingworth, R.; Ahammadsahib, K.; Gadelhak, G.; McLaughlin, J.L. New inhibitors of Complex I of the mitochondrial electron transport chain with activity as pesticides. *Biochem. Soc. Trans.* **1994**, *22*, 230–233. [CrossRef] [PubMed]
52. Walia, S.; Saha, S.; Tripathy, V.; Sharma, K. Phytochemical biopesticides: Some recent developments. *Phytochem. Rev.* **2017**, *16*, 989–1007. [CrossRef]
53. Birnbaum, L. When environmental chemicals act like uncontrolled medicine. *Trends Endocrinol. Metab.* **2013**, *24*, 321–323. [CrossRef]
54. Singh, D.; Mehta, S.S.; Neoliya, N.K.; Shukla, Y.N.; Mishra, M. New possible insect growth regulators from *Catharanthus roseus*. *Curr. Sci.* **2003**, *84*, 1184–1186.
55. Abreu, P.M.; Heggie, W. Terpenoides e Esteroides. In *Biossíntese de Produtos Naturais*, 1st ed.; Lobo, A.M., Lourenço, A.M., Eds.; IST Press: Lisboa, Portugal, 2008; pp. 119–149.
56. Kamboj, A.; Saluja, A. Ageratum conyzoides L.: A review on its phytochemical and pharmacological profile. *Int. J. Green Pharm.* **2008**, *2*, 59–68. [CrossRef]
57. Feng, R.; Isman, M.B. Selection for resistance to azadirachtin in the green peach aphid, Myzuspersicae. *Experientia* **1995**, *51*, 831–833. [CrossRef]
58. Di Ilio, V.; Cristofaro, M. Polyphenolic extracts from the olive mill wastewater as a source of biopesticides and their effects on the life cycle of the Mediterranean fruit fly Ceratitiscapitata (Diptera, Tephriditae). *Int. J. Trop. Insect Sci.* **2021**, *41*, 359–366. [CrossRef]
59. Larif, M.; Zarrouk, A.; Soulaymani, A.; Elmidaoui, A. New innovation in order to recover the polyphenols of olive mill wastewater extracts for use as a biopesticide against the Euphylluraolivina and Aphis citricola. *Res. Chem. Intermed.* **2013**, *39*, 4303–4313. [CrossRef]
60. Mordue, A.J.; Nisbet, A.J. Azadirachtin from the neem tree Azadirachtaindica: Its action against insects. *An. Soc. Do Bras.* **2000**, *29*, 615–632.
61. Bernardi, D.; Botton, M.; Cunha, U.; Bernardi, O.; Malausa, T.; Garcia, M.; Nava, D. Effects of azadirachtin on *Tetranych usurticae* (Acari: Tetranychidae) and its compatibility with predatory mites (Acari: Phytoseiidae) on strawberry. *Pest Manag. Sci.* **2013**, *69*, 75–80. [CrossRef]
62. Kang, T.H.; Hwang, E.I.; Yun, B.S.; Park, K.D.; Kwon, B.; Shin, C.; Kim, S. Inhibition of chitin synthases and antifungal activities by 2′-benzoyloxycinnamaldehyde from Pleuropterusciliinervis and its derivatives. *Biol. Pharm. Bull.* **2007**, *30*, 598–602. [CrossRef] [PubMed]
63. Cui, C.; Yang, Y.; Zhao, T.; Zou, K.; Peng, C.; Cai, H.; Wan, X.; Hou, R. Insecticidal Activity and Insecticidal Mechanism of Total Saponins from *Camellia oleifera*. *Molecules* **2019**, *24*, 4518. [CrossRef] [PubMed]
64. Wigglesworth, V.B. Transpiration Through the Cuticle of Insects. *J. Exp. Biol.* **1945**, *21*, 97–114. [CrossRef]
65. Mwine, J.; Van Damme, P.; Kamoga, G.; Kudamba, K.; Nasuuna, M.; Jumba, F. Ethnobotanical survey of pesticidal plants used in South Uganda: Case study of Masaka district. *J. Med. Plants Res.* **2011**, *5*, 1155–1163.
66. Rajapakse, R.; Ratnasekera, D. Pesticidal potential of some selected tropical plant extracts againsCallosobruchus maculates (F) and *Callosobruchus chinensis* (L) (Coleoptera: Bruchidae). *Trop. Agric. Res. Ext.* **2010**, *11*, 69–71. [CrossRef]
67. Tripathi, A.K.; Prajapati, V.; Aggarwal, K.K.; Kumar, S.; Prajapti, V.; Kumar, S.; Kukreja, A.K.; Dwivedi, S.; Singh, A.K. Effect of volatile oil constituents of Mentha species against stored grain pests, *Callosobruchus maculatus* and *Tribolium castaneum*. *J. Med. Arom. Plant Sci.* **2000**, *22*, 549–556.
68. Agarwal, M.; Walia, S.; Dhingra, S. Pest control properties of turmeric leaf oil against Spilosomaobliqua, Dysdercuskoenigii and Triboliumcastaneum. In Proceedings of the 2nd All India People's Technology Congress, Calcultta, India; 1999; pp. 1–7.
69. Suszkiw, J. *Catnip Compounds Curb Asian Lady Beetles*; Agricultural Research Services (ARS) Report; US Department of Agriculture: Washington, DC, USA, 2009.
70. McElvain, S.M.; Bright, R.D.; Johnson, P.R. The Constituents of the Volatile Oil of Catnip. I. Nepetalic Acid, Nepetalactone and Related Compounds. *J. Am. Chem. Soc.* **1941**, *63*, 1558–1563. [CrossRef]
71. Rudman, P.; Da Costa, E.W.B.; Gay, F.J.; Wetherly, A.H. Relationship of Tectoquinone to Durability in Tectonagrandis. *Nature* **1958**, *181*, 721–722. [CrossRef]
72. Wolcott, G.N. Organic Termite Repellents Tested against Cryptotermes Brevis Walker. *J. Agric. Univ. Puerto Rico* **1955**, *39*, 115–149. [CrossRef]
73. Ileke, K.; Ogungbite, O. Alstoniaboonei De Wild oil extract in the management of mosquito (*Anopheles gambiae*), a vector of malaria disease. *J. Coast. Life Med.* **2015**, *3*, 557–563.

74. Weinzierl, R.; Henn, T.; Koehler, P.G.; Tucker, C.L. Insect Attractants and Traps. In *IFAS Extension, Institute of Food and Agricultural Sciences*; University of Florida: Gainesville, FL, USA, 1995; pp. 1–9.
75. Miller, D.R. Limonene: Attractant kairomone for white pine cone beetles (Coleoptera: Scolytidae) in an eastern white pine seed orchard in western North Carolina. *J. Econ. Entomol.* **2007**, *100*, 815–822. [CrossRef]
76. Degenhardt, J.; Hiltpold, I.; Köllner, T.G.; Frey, M.; Gierl, A.; Gershenzon, J.; Hibbard, B.E.; Ellersieck, M.R.; Turlings, T.C.J. Restoring a maize root signal that attracts insect-killing nematodes to control a major pest. *Proc. Natl. Acad. Sci. USA* **2009**, *106*, 13213–13218. [CrossRef] [PubMed]
77. Qiao, J.; Zou, X.; Lai, D.; Yan, Y.; Wang, Q.; Li, W.; Deng, S.; Xu, H.-H.; Gu, H. Azadirachtin blocks the calcium channel and modulates the cholinergic miniature synaptic current in the Central Nervous System of Drosophila. *Pest Manag. Sci.* **2014**, *70*, 1041–1047. [CrossRef] [PubMed]
78. Mordue, A.J.; Blackwell, A. Azadirachtin: An update. *J. Insect Physiol.* **1993**, *39*, 903–924. [CrossRef]
79. Lajide, L.; Escoubas, P.; Mizutani, J. Termite antifeedant activity in Xylopia aethiopica. *Phytochemistry* **1995**, *40*, 1105–1112. [CrossRef]
80. Okwute, S.K.; Nduji, A.A. Isolation of Schimperiin: A new gallotannin from the leaves of *Anogeissuss chimperii* (Combretaceae). *Proc. Natl. Acad. Sci. USA* **1992**, *4*, 36–41.
81. Morimoto, M.; Fukumoto, H.; Hiratani, M.; Chavasiri, W.; Komai, K. Insect Antifeedants, Pterocarpans and Pterocarpol, in Heartwood of *Pterocarpus macrocarpus* Kruz. *Biosci. Biotechnol. Biochem.* **2006**, *70*, 1864–1868. [CrossRef]
82. Ratnayake, S.; Rupprecht, K.; Potter, W.; McLaughlin, J. Evaluation of the Pawpaw Tree, *Asimina triloba* (Annonaceae), as a Commercial Source of the Pesticidal Annonaceous Acetogenins*. In *New Crops*; Janick, J., Simon, J.E., Eds.; Wiley: New York, NY, USA, 1993; pp. 644–648.
83. Alkofahi, A.; Rupprecht, J.K.; Anderson, J.E.; McLaughlin, J.L.; Mikolajczak, K.L.; Scott, B.A. Search for New Pesticides from Higher Plants. In *Insecticides of Plant Origin*; Arnason, J.T., Philogene, B.J., Morand, P., Eds.; Books on Demand: Norderstedt, Germany, 1989; Volume 387, pp. 25–43.
84. Nahrstedt, A. Cyanogenic compounds as protecting agent for organisms. *Plant Syst. Evol.* **1985**, *150*, 35–47. [CrossRef]
85. Huang, X.P.; Renwick, J.A.A. Cross habituation to feeding deterrents and acceptance of a marginal host plant by *Pieris rapae larvae*. *Entomol. Exp. Appl.* **1995**, *76*, 295–302. [CrossRef]
86. Bais, H.; Walker, T.; Kennan, A.; Stermitz, F.; Vivanco, J. Structure-Dependent Phytotoxicity of Catechins and Other Flavonoids: Flavonoid Conversions by Cell-free Protein Extracts of *Centaurea maculosa* (Spotted Knapweed) Roots. *J. Agric. Food Chem.* **2003**, *51*, 897–901. [CrossRef]
87. Kalinova, J.; Vrchotova, N. Level of Catechin, Myricetin, Quercetin and Isoquercitrin in Buckwheat (*Fagopyrum esculentum* Moench), Changes of Their Levels during Vegetation and Their Effect on The Growth of Selected Weeds. *J. Agric. Food Chem.* **2009**, *57*, 2719–2725. [CrossRef]
88. Bachheti, A.; Sharma, A.; Bachheti, R.K.; Husen, A.; Pandey, D.P. Plant Allelochemicals and Their Various Applications. In *Co-Evolution of Secondary Metabolites*; Mérillon, J.-M., Ramawat, K.G., Eds.; Springer: Cham, Switzerland, 2019; pp. 1–25.
89. Clay, D.V.; Dixon, F.L.; Willoughby, I. Natural products as herbicides for tree establishment. *For. Int. J. For. Res.* **2005**, *78*, 1–9. [CrossRef]
90. Ismail, A.; Hamrouni, L.; Hanana, M.; Jamoussi, B. Review on the phytotoxic effects of essential oils and their individual components: News approach for weed management. *Int. J. Appl. Biol. Pharm. Technol.* **2013**, *4*, 96–114.
91. Curto, M.J.M. Ecologiaquímica: A comunicaçãoquímicananatureza. In *Biossíntese de Produtos Naturais*, 1st ed.; Lobo, A.M., Lourenço, A.M., Eds.; IST Press: Lisboa, Portugal, 2008; pp. 227–240.
92. Narwal, S. Allelopathy in ecological agriculture. In *Allelopathy in Ecological Agriculture and Forestry*; Narwal, S.S., Hoagland, R.E., Dilday, R.H., Eds.; Kluwer Academic: Dharwad, India, 2000; pp. 11–32.
93. Bertin, C.; Weston, L.A.; Huang, T.; Jander, G.; Owens, T.; Meinwald, J.; Schroeder, F.C. Grass roots chemistry: Meta-Tyrosine, an herbicidal nonprotein amino acid. *Proc. Natl. Acad. Sci. USA* **2007**, *104*, 16964. [CrossRef]
94. Kato-Noguchi, H.; Kosemura, S.; Yamamura, S.; Mizutani, J.; Hasegawa, K. Allelopathy of oats. I. Assessment of allelopathic potential of extract of oat shoots and identification of an allelochemical. *J. Chem. Ecol.* **1994**, *20*, 309–314. [CrossRef] [PubMed]
95. Zhao, X.; Zheng, G.W.; Niu, X.M.; Li, W.Q.; Wang, F.S.; Li, S.H. Terpenes from Eupatorium adenophorum and their allelopathic effects on Arabidopsis seeds germination (dagger). *J. Agric. Food Chem.* **2009**, *57*, 478–482. [CrossRef] [PubMed]
96. Duke, S.; Blair, A.; Dayan, F.; Johnson, R.; Cook, D.; Bajsa-Hirschel, J. Is (−)-Catechin a Novel Weapon of Spotted Knapweed (*Centaurea stoebe*)? *J. Chem. Ecol.* **2009**, *35*, 141–153. [CrossRef] [PubMed]
97. Yang, W.; Li, F.; Xing, X.; Wang, Z.; Yu, X. Study in Pesticide Activities of *Polygonum cuspidatum* Extracts and its Active Ingredient Resveratrol. *Nat. Prod. Commun.* **2019**, *14*, 1934578X19861022. [CrossRef]
98. Kato-Noguchi, H.; Suwitchayanon, P.; Boonmee, S.; Iwasaki, A.; Suenaga, K. Plant Growth Inhibitory Activity of the Extracts of *Acmella oleracea* and its Growth Inhibitory Substances. *Nat. Prod. Commun.* **2019**, *14*, 1934578X19858805. [CrossRef]
99. Arnason, T.; Towers, G.H.N.; PhilogÈNe, B.J.R.; Lambert, J.D.H. The Role of Natural Photosensitizers in Plant Resistance to Insects. In *Plant Resistance to Insects*; American Chemical Society: Washington, DC, USA, 1983; Volume 208, pp. 139–151.
100. Berenbaum, M. Charge of the light brigade: Phototoxicity as a defense against insects. In *Light-Activated Pesticides*; American Chemical Society: Washington, DC, USA, 1987.

101. Towers, G. Interactions of light with phytochemicals in some natural and novel systems. *Can. J. Bot.* **2011**, *62*, 2900–2911. [CrossRef]
102. Marchant, Y.Y.; Cooper, G.K. Structure and Function Relationships in Polyacetylene Photoactivity. In *Light-Activated Pesticides*; American Chemical Society: Washington, DC, USA, 1987; Volume 339, pp. 241–254.
103. Nivsarkar, M.; Cherian, B.; Padh, H. Alpha-terthienyl: A plant-derived new generation insecticide. *Curr. Sci.* **2001**, *81*, 667–672.
104. Varma, J.; Dubey, N.K. Perspective of botanical and microbial products as pesticides of tomorrow. *Curr. Sci.* **1999**, *76*, 172–179.
105. Ahmed, S.; Grainge, M. Potential of the neem tree (*Azadirachta indica*) for pest control and rural development. *Econ. Bot.* **1986**, *40*, 201–209. [CrossRef]
106. Pavela, R. History, presence and perspective of using plant extracts as commercial botanical insecticides and farm products for protection against insects@ a review. *Plant Prot. Sci.* **2016**, *52*, 229–241.
107. Diaz, M.; Rossini, C. Bioactive Natural Products from Sapindaceae Deterrent and Toxic Metabolites against Insects. In *Insecticides—Pest Engineering*, 1st ed.; Perveen, F.K., Ed.; IntechOpen: London, UK, 2012; pp. 277–308.
108. Isman, M.; Machial, C.M. Pesticides based on plant essential oils: From traditional practice to commercialization. *Nat. Occur. Bioact. Compd.* **2006**, *3*, 29–44.
109. Marquina, S.; Maldonado, N.; Garduno-Ramirez, M.L.; Aranda, E.; Villarreal, M.; Navarro, V.C.; Bye, R.; Delgado, G.; Alvarez, L. Bioactive oleanolic acid saponins and other constituents from the roots of Viguieradecurrens. *Phytochemistry* **2001**, *56*, 93–97. [CrossRef]
110. Matsuzaki, T.; Koseki, K.; Koiwai, A. Germination and Growth Inhibition of Surface Lipids from Nicotiana Species and Identification of Sucrose Esters. *Agric. Biol. Chem.* **1988**, *52*, 1889–1897. [CrossRef]
111. Isman, M.B. Botanical Insecticides in the Twenty-First Century-Fulfilling Their Promise? *Annu. Rev. Entomol.* **2020**, *65*, 233–249. [CrossRef] [PubMed]
112. Marrone, P. Pesticidal natural products—Status and future potential. *Pest Manag. Sci.* **2019**, *75*, 2325–2340. [CrossRef]
113. Isman, M. Problems and opportunities for the commercialization of botanical insecticides. In *Biopesticides of Plant Origin*; Régnault-Roger, C., Philogéne, B.J., Vincent, C., Eds.; Lavoisier: Paris, France, 2005; pp. 283–291.
114. Mitscher, L.A.; Drake, S.; Gollapudi, S.R.; Okwute, S.K. A modern look at folkloric use of anti-infective agents. *J. Nat. Prod.* **1987**, *50*, 1025–1040. [CrossRef] [PubMed]
115. MacKinnon, S.; Durst, T.; Arnason, J.T.; Angerhofer, C.; Pezzuto, J.; Sanchez-Vindas, P.E.; Poveda, L.J.; Gbeassor, M. Antimalarial Activity of Tropical Meliaceae Extracts and Gedunin Derivatives. *J. Nat. Prod.* **1997**, *60*, 336–341. [CrossRef] [PubMed]
116. Golob, P. *The Use of Spices and Medicinals as Bioactive Protectants for Grains*; Food and Agriculture Organization: Rome, Italy, 1999.
117. Bang, K.H.; Lee, D.W.; Park, H.M.; Rhee, Y.H. Inhibition of fungal cell wall synthesizing enzymes by trans-cinnamaldehyde. *Biosci. Biotechnol. Biochem.* **2000**, *64*, 1061–1063. [CrossRef]
118. Upatoom, P.; Visetson, S. Mechanisms of Essential Oils from Citronella (*Cymbopogon winterianus* Jowitt) against Siamensis Subterranean Termite Workers (*Coptotermes gestroi* Wasmann) and Mice (*Mus musculus* L.). *Thai Agric. Res. J.* **2018**, *35*, 270–287.
119. Pelkonen, O.; Abass, K.; Wiesner, J. Thujone and thujone-containing herbal medicinal and botanical products: Toxicological assessment. *Regul. Toxicol. Pharmacol.* **2013**, *65*, 100–107. [CrossRef]
120. Bloomquist, J.R. Chloride channels as tools for developing selective insecticides. *Arch. Insect Biochem. Physiol.* **2003**, *54*, 145–156. [CrossRef]
121. Marles, R.; Durst, T.; Kobaisy, M.; Soucy-Breau, C.; Abou-Zaid, M.; Arnason, J.; Kacew, S.; Kanjanapothi, D.; Rujjanawate, C.; Meckes, M.; et al. Pharmacokinetics, Metabolism and Toxicity of the Plant-Derived Photoxin α-Terthienyl. *Pharmacol. Toxicol.* **2009**, *77*, 164–168. [CrossRef]
122. Johnson, W. Final report on the safety assessment of *Elaeis guineensis* (Palm) Oil, *Elaeis guineensis* (Palm) Kernel Oil, Hydrogenated Palm Oil and Hydrogenated Palm Kernel Oil. *Int. J. Toxicol.* **2000**, *19*, 7–28.
123. Singh, D.; Agarwal, S.K. Himachalol and β-himachalene: Insecticidal principles of himalayan cedarwood oil. *J. Chem. Ecol.* **1988**, *14*, 1145–1151. [CrossRef]
124. Ujváry, I. Pest Control Agents from Natural Products. In *Hayes' Handbook of Pesticide Toxicology*, 3rd ed.; Krieger, R., Ed.; Academic Press: New York, NY, USA, 2010; pp. 119–229.
125. Tooley, P. Crop Protection. In *Food and Drugs*; John Murray Albermark Street: London, UK, 1971.
126. Inwood, M. Azadirachtin, a scientific gold mine. *Bioorg. Med. Chem.* **2009**, *17*, 4096–4105.
127. Saleem, M.A.; Wilkins, R.M. Precocene-1: An anti-juvenile hormone, a potential 4th generation insecticide against a malathion-resistant strain of *Oryzaephilus surinamensis* (L.). *Pak. J. Zool.* **1984**, *16*, 195–201.
128. Singh, S.; Kumar, K. Anti-JH compounds and Insect Pest Management. In *Emerging Trends in Zoology*; eBook; Srivastava, U.C., Ed.; Narendra Publishing House: Delhi, India, 2011; pp. 335–350.
129. Srivastva, S.; Kumar, K. Precocene I and II induced metamorphosis in a noctuid moth, *Spodoptera litura* Fabr. *Proc. Natl. Acad. Sci. India* **1997**, *67*, 213–226.
130. Khafagi, W. Effects of juvenile hormone I, precocene I and precocene II on the progeny of *Microplitis rufiventris* Kok. female when administered via its host, *Spodoptera littoralis* (Boisd.). *J. Appl. Entomol.* **2004**, *128*, 6–10. [CrossRef]
131. Eid, M.A.A.; Salem, M.S.; Taha, G.Z. Effects of precocene II on morphogenesis of the desert locust *Schistocerca gregaria*. *Biochem. Syst. Ecol.* **1988**, *16*, 515–520. [CrossRef]

132. Sandermann, V.W.; Dietrichs, H.H. Investigations on Termiteproof Wood Species. *Holzals Roh-Und Werkst.* **1957**, *15*, 281–297. [CrossRef]
133. Nakano, H.; Nakajima, E.; Fujii, Y.; Yamada, K.; Shigemori, H.; Hasegawa, K. Leaching of the allelopathic substance, -tryptophan from the foliage of mesquite (*Prosopis juliflora* (Sw.) DC.) plants by water spraying. *Plant Growth Regul.* **2003**, *40*, 49–52. [CrossRef]
134. Tripathi, R.D.; Srivastava, H.S.; Dixit, S.N. A fungitoxic principle from the leaves of lawsoniainermis lam. *Experientia* **1978**, *34*, 51–52. [CrossRef]
135. Velásquez, J.; Rojas, L.; Alfredo, U. Antifungal activity of naphtoquinone from *Tabebuia serratifolia* (Vahl. Nicholson). *Ciencia* **2004**, *12*, 64–69.
136. Oliveira, M.F.; Lemos, T.L.G.; Mattos, M.C.D.; Segundo, T.A.; Santiago, G.M.P.; Braz-Filho, R. New enamine derivatives of lapachol and biological activity. *An. Acad. Bras. Ciências* **2002**, *74*, 211–221. [CrossRef] [PubMed]
137. Jiménez-González, L.; Álvarez-Corral, M.; Muñoz-Dorado, M.; Rodríguez-García, I. Pterocarpans: Interesting natural products with antifungal activity and other biological properties. *Phytochem. Rev.* **2008**, *7*, 125–154. [CrossRef]
138. Soby, S.; Caldera, S.; Bates, R.; VanEtten, H. Detoxification of the phytoalexins maackiain and medicarpin by fungal pathogens of alfalfa. *Phytochemistry* **1996**, *41*, 759–765. [CrossRef]
139. Chen, J.; Yu, Y.; Li, S.; Ding, W. Resveratrol and Coumarin: Novel Agricultural Antibacterial Agent against *Ralstonia solanacearum* In Vitro and In Vivo. *Molecules* **2016**, *21*, 1501. [CrossRef] [PubMed]

molecules

MDPI

Review

Diatomaceous Earth for Arthropod Pest Control: Back to the Future

Valeria Zeni [1,†], Georgia V. Baliota [2,†], Giovanni Benelli [1,*], Angelo Canale [1] and Christos G. Athanassiou [2]

1 Department of Agriculture, Food and Environment, University of Pisa, via del Borghetto 80, 56124 Pisa, Italy; valeriazeni93@gmail.com (V.Z.); angelo.canale@unipi.it (A.C.)
2 Laboratory of Entomology and Agricultural Zoology, Department of Agriculture, Crop Production and Rural Environment, University of Thessaly, Phytokou Str., 38446 Volos, Greece; mpaliota@agr.uth.gr (G.V.B.); athanassiou@agr.uth.gr (C.G.A.)
* Correspondence: giovanni.benelli@unipi.it; Tel.: +39-050-2216141
† These authors contributed equally.

Abstract: Nowadays, we are tackling various issues related to the overuse of synthetic insecticides. Growing concerns about biodiversity, animal and human welfare, and food security are pushing agriculture toward a more sustainable approach, and research is moving in this direction, looking for environmentally friendly alternatives to be adopted in Integrated Pest Management (IPM) protocols. In this regard, inert dusts, especially diatomaceous earths (DEs), hold a significant promise to prevent and control a wide range of arthropod pests. DEs are a type of naturally occurring soft siliceous sedimentary rock, consisting of the fossilized exoskeleton of unicellular algae, which are called diatoms. Mainly adopted for the control of stored product pests, DEs have found also their use against some household insects living in a dry environment, such as bed bugs, or insects of agricultural interest. In this article, we reported a comprehensive review of the use of DEs against different arthropod pest taxa, such as Acarina, Blattodea, Coleoptera, Diptera, Hemiptera, Hymenoptera, Ixodida, Lepidoptera, when applied either alone or in combination with other techniques. The mechanisms of action of DEs, their real-world applications, and challenges related to their adoption in IPM programs are critically reported.

Keywords: urban pests; agricultural pests; aphids; cockroaches; kissing bugs; insect vectors; green insecticides; mosquitoes; moth pests; non-target toxicity; stored product pests; termites

Citation: Zeni, V.; Baliota, G.V.; Benelli, G.; Canale, A.; Athanassiou, C.G. Diatomaceous Earth for Arthropod Pest Control: Back to the Future. *Molecules* **2021**, *26*, 7487. https://doi.org/10.3390/molecules26247487

Academic Editor: Baoan Song

Received: 16 October 2021
Accepted: 3 December 2021
Published: 10 December 2021

1. Introduction

Among different types of inert materials currently adopted in pest control, diatomaceous earths (DEs) hold a prominent position, as they are apparently the most often tested material for this purpose. A search in *Journal of Stored Products Research* for published papers between January 2019 and January 2021 revealed the publication of 13 papers with "diatomaceous earth" on their title, emphasizing the utilization of DEs in stored product protection. DEs are not only used for the management of insects and other arthropods, but they also have multiple uses including the control of different pathogens, such as fungi and bacteria [1–4]. Other types of inert dusts, such as zeolites [5] or kaolin [6], have been also investigated for pest control. This work will be focused solely on the use of DEs in crop protection but also in post-harvest and urban pest control, highlighting their wide applicability.

In a recent review paper, Athanassiou et al. [7] categorized the materials that can be used in pest control and fall into the category of "nano" under the general term of nanoparticles. Although there are cases where DE particles can touch the "nano" scale, DEs are generally classified in the "micro" category and can be considered as "microparticles" in contrast with nanoparticles.

DEs are the fossilized remains of phytoplankton, which are diatoms that occurred mostly during the Miocene and Eocene periods [1]. Diatoms are unicellular eukaryotic algae that are characterized by an external skeleton (frustule) rich in silicon dioxide whose fossilized remains constitute DEs [1,2,8]. These diatoms are abundant either in fresh-water or marine environments, but they are also present in terrestrial ecosystems.

The present review provides a focus on the utilization of DEs to manage different arthropod pest categories when applied either alone or in combination with other techniques.

2. Which Is the Mode of Action of DEs?

There are different theories about the insecticidal effect of DEs [2]. It is generally considered that DE particles attach to the insects' cuticle, causing death through desiccation [1,2,9], although the abrasion is also a complementary action, i.e., through cuticular micro-wounds [2]. The shape of DEs may be a critical factor in this sense, as round-shaped diatom may lead to more rapid water absorption, while sharp-shaped DE acts more as an abrasive factor [1,2,9,10]. Nevertheless, the shape of the diatom, and probably its action (i.e., sorption vs. abrasion) can be chanced through different processing techniques [11].

3. Why Use DEs for Arthropod Pest Control?

Thanks to their characteristics, the use of DEs is advantageous for several types of applications [2,8]. First, DEs are natural substances, and given their low toxicity to mammals and the environment, the registration process is greatly simplified. In addition, being inert materials, DEs have no interaction with the commodity and can be easily removed through standard processing, such as sieving [1,10,12–14], while their presence in the final product, such as flour or semolina, does not alter baking or pasta-making properties [1,12]. For more than two decades, DEs have been used as feed additives and in veterinary pest control [1]. Moreover, DEs are easily accessible [8,15]. The natural deposits from where DEs are extracted are found almost everywhere. Following their extraction, these powders are sieved to obtain a homogeneous mixture of particle sizes and dried at approximately 2–6% moisture content [1,11,15]. Finally, due to their mechanism of action, no physiological pest resistance is expected to occur, while tolerance may be exhibited through reduced contact with the DE particles [16–18].

4. Any Dark Facets for DEs Use in Pest Control?

In general, to be effective, DEs must be applied at elevated concentrations, which are much higher than those of conventional insecticides and often exceed 1000 ppm [2,8,19,20]. In this way, they create a "dusty" appearance on the products and might cause health problems to workers, such as respiratory disorders [1,2,8,10]. In addition, their application on stored products results in the reduction of the test weight (weight to volume ratio), which is a critical characteristic in the international grain market [1,12].

In this scenario, we focused on current knowledge and challenges on the use of DEs in stored products as well as for managing arthropod pests of agricultural importance, urban pests, and vectors of public health relevance. The potential impact of DEs on non-target species is also discussed.

5. DEs to Control Stored Product Pests

Currently, most studies assessing the toxicity of DEs on arthropods of economic importance are focused on stored product pests. Storing durable commodities is significant since it ensures stable food and feed production all year long and on a global scale. However, the storage environment, which may range from warehouses to retail shelves, is also a prosperous place for a range of insects to thrive [21]. Insect infestations have multiple effects on stored food, feed commodities, and seeds. Beyond the direct damage caused by food consumption, insects also pose a quarantine threat. Insect fragments within durable edible products provoke allergic reactions, alter the organoleptic characteristics, and potentially carry disease-causing pathogens [22]. Therefore, even a small percentage of damage may result in profound monetary losses. Despite the technological advantages over the years, most segments of the food industry are very susceptible to insect infestations, especially when it comes to stored grains [23]. On the other hand, pest management currently depends mostly on chemical methods, but such approaches must be at least improved by adopting more sustainable and eco-friendly treatments for raw and processed commodities [24]. Herein, we analyze the various factors routing the efficacy of DEs against stored product pests and their real-world use, even in combination with fungal and plant-borne pesticides.

5.1. Biotic and Abiotic Factors That Influence the Efficacy of DEs

Given their high absorptive power, the efficacy of DEs is highly determined by the levels of relative humidity (R.H.)/moisture content (m.c.). Hence, in humid conditions, some types of DEs may not be as effective as in dry conditions. For instance, Vayias and Athanassiou [25] tested larvae of the confused flour beetle, *Tribolium confusum* Jacquelin du Val (Coleoptera: Tenebrionidae) for their susceptibility to DEs, and they found that the efficacy of commercial DEs was reduced as the R.H. level rose from 55% to 65%. This is particularly important for grain protection, as R.H. levels between 55% and 75% correspond with an equivalent of 10.5% to 14% m.c., which are realistic ranges for long-term storage [26,27]. However, there are studies where the efficacy of DEs was not much affected by the increased R.H., suggesting that certain DE types do not interact much with moisture. A slurry formulation of DE, i.e., a mixture of DEs and water, may not be as effective as dust (powder) formulations [28]. However, a slurry formulation can be more practical in terms of direct application in the commodity with the same technology as traditional grain protectants [28,29].

The temperature might act indirectly on the efficacy of DEs, since at a higher temperature, the water loss occurs faster. In addition, insect mobility is increased at elevated temperatures, causing an increase in the contact with the DE particles. Athanassiou et al. [20] tested a commercially available DE on wheat for the control of adults of *T. confusum* and the rice weevil, *Sitophilus oryzae* (L.) (Coleoptera: Curculionidae), and noticed that there was a positive correlation between the mortality rates and the temperature. Indeed, by increasing the temperature by 10 °C, Athanassiou et al. [20] reported that the mortality rates were raised from approximately 45% at 22 °C to 100% at 32 °C. Other studies show similar results for a wide range of species [2,25,30,31], but some reports show that the increase in temperature decreases mortality [30,32]. The type of commodity on which DEs are applied is another critical aspect that should be considered. In the case of stored grain protection, not all grains are equal in terms of their response to DEs, suggesting that there are specific interactions with the external parts of the grains mass that may partially inactivate the DE particles. In a series of studies [13,14,19] it was shown that DEs are less effective on maize than on small grains, such as wheat, rice, and barley. Kavallieratos et al. [13] used sieves to remove two different DEs from eight grains, and the percentage of DEs removed was always higher on maize and minimal on wheat or barley. In addition, DE adherence was much lower in peeled barley than in non-peeled barley, which is a clear indication that the shape of the external kernel part is critical in maintaining the DE particles [13]. Still, these adherence differences among the different grains did not correlate with adult mortality in the lesser grain borer, *Rhyzopertha dominica* (F.) (Coleoptera: Bostrychidae) [13,14].

Different target species have different levels of susceptibility to DEs. It is generally expected that soft-bodied insects are more vulnerable to DEs, as their cuticula can be easily damaged, causing rapid desiccation [29]. However, this is not always true. For instance, stored product mites, such as Astigmata, are extremely vulnerable to DEs, which is considered as a direct consequence of their sensitivity to water loss and their thin cuticles [29,33]. Nevertheless, another category of soft-bodied stored product pests, psocids (Psocoptera), are extremely tolerant to DEs [34]. Psocids have a certain mechanism that can moderate water loss and absorb moisture from the air to compensate losses [35,36]. Larvae are considered more susceptible to DEs than adults [29]. For instance, Vayias and Athanassiou [25] found that *T. confusum* larvae were more susceptible to DE than adults, with early-stage larvae being the most vulnerable larval instar. However, this is not true for the yellow mealworm, *Tenebrio molitor* L. (Coleoptera: Tenebrionidae), where adults are susceptible to DEs, but larvae remain unaffected due to the occurrence of a mechanism that moderates water loss [37]. The adults of the red flour beetle, *Tribolium castaneum* (Herbst) (Coleoptera: Tenebrionidae) and *T. confusum*, are being considered as the least susceptible beetle species to DEs, with the latter slightly more tolerant [19,25,29,38,39]. On the other hand, adults of the rusty grain beetle, *Cryptolestes ferrugineus* (Stephens) (Coleoptera: Laemophloeidae), are very susceptible to DEs, as they are flat-bodied, and water loss can rapidly occur [1,9,30]. Still, there are dissimilar and not directly comparable results for different species of stored products [15,29,30], but some general conclusions can be drawn based on the above observations. Apart from body size, shape, and characteristics, insect mobility is a critical parameter, as slow-moving insects may have a lower DE particle uptake. This is considered a key feature for the reduced susceptibility of *R. dominica* to DEs [1,30], although some reports show that this species is particularly susceptible to different DEs [13,14].

Some additional parameters that influence the efficacy of DEs have to do with their physicochemical characteristics. For instance, it has been shown that particle size is an important parameter, and the smaller the particles, the highest the DE efficacy against insects [1,8,9,11]. Vayias et al. [9] have shown that DEs with particles that were smaller than 45 μm were more effective than DEs with larger particles against *R. dominica*, *S. oryzae*, and *C. ferrugineus*. Nonetheless, Baliota and Athanassiou [11] have shown that it is the particle shape, rather than the size, that had a certain effect on the insecticidal value of DEs, and that smaller particles do not necessarily mean higher efficacy. Moreover, very small particles may not be desirable for safety issues [2].

Several physicochemical characteristics can be further utilized toward the prediction of the expected insecticidal value of DEs. Korunić [10] summarized these characteristics in standardized testing, which can be carried out for rapid screening of DE samples, without the need to conduct bioassays with insects, which is a time-consuming procedure. The silicon dioxide content and pH are important factors, while clay and other impurities are not desirable [10]. Even more important parameters are the tapped density, the bulk density reduction, and the adherence to grain kernels [1,10]. Diatom species, origin, and other characteristics may be less important [1,2,9,11,15,40].

5.2. Combinations with Contact Synthetic Insecticides

One of the possible solutions to the implications caused by the high doses of DEs is the combination of DEs with other substances thanks to the adsorptive nature of the DE particles. Indeed, the utilization of DEs as a carrier is a promising solution not only for the application of insecticides in reduced concentrations but also to combine at least two different modes of action, i.e., desiccation through the inert dusts and an additional action depending on the type of chemical (e.g., neurotoxic, etc.). Several studies have indicated a significant potential and even synergism of combinations of commercial DE formulations with residual insecticides. Wakil et al. [41] reported high mortality rates of *R. dominica* in wheat, rice, and maize treated with a combination of thiamethoxam and a commercial DE formulation, SilicoSec® (Biofa GmbH, Munsingen, Germany), in relatively low doses (0.25, 0.5, and 0.75 ppm for thiamethoxam and 100 ppm for SilicoSec®). The combination of 150 ppm of Protect-It® (Hedley Technologies Inc., Mississauga, ON, Canada) with 1.25, 2.5, or 5.0 ppm of imidacloprid resulted in higher mortality rates of different stored product insects than applications of these insecticides alone at almost all exposure intervals and commodities tested [42]. Ceruti and Lazzari [43] used 500 and 1000 ppm of Keepdry® (Irrigação Dias Cruz ME, Brazil) in combination with 0.5 or 1.0 g a.i./t of deltamethrin powder, which may represent an efficient control measure against the maize weevil *Sitophilus zeamais* Motschulsky (Coleoptera: Curculionidae) in stored corn, highlighting the potentials of having reduced residues of deltamethrin, as compared with using this active ingredient alone. Arthur [44] stated that an insecticide formulation (F2) containing 0.03% deltamethrin, 0.37% piperonyl butoxide, 0.95% chlorpyriphos-methyl, 10% mineral oil, and 88% Protect-It® was extremely effective in wheat, maize, and paddy rice at the rate of 100 ppm against *S. oryzae, S. zeamais, R. dominica* and *T. castaneum*. Awais et al. [45,46] tested three different doses of the DE formulation Concern (Wood Stream^{TM} Corporation, Lititz, PA, USA) combined with the Insect Growth Regulators (IGRs) lufenuron and tebufenozide against *T. castaneum* and the khapra beetle, *Trogoderma granarium* Everts (Coleoptera: Dermestidae) respectively, with the overall conclusions to specify that the combined use of DEs and IGRs is highly operative and beneficial for stored product insect control. A combination of the IGR S-methoprene and Protect-It® could also be a promising mixture as reported by Arthur [47]. In that study, the mixture had an additive effect and reduced the concentrations of both components required to suppress the progeny of *R. dominica* compared to the application of each insecticide alone [47]. In addition, SilicoSec® (25 ppm) and beta-cyfluthrin (0.125 or 0.25 ppm) acted synergistically for the control of *T. castaneum* and, especially, *S. oryzae* [48]. The long-term protection of a given insecticide is one of the key elements in stored-grain pest management, aiming to prevent new infestations and control the reproduction of the already existing individuals. Mixtures with DEs have the potential to enhance the residual efficacy of an insecticide. Wakil et al. [49] reported an increased mortality of adults of *R. dominica* over 9 months of wheat storage with applications of 200 ppm of SilicoSec® and 0.5 ppm thiamethoxam in comparison with the residual efficacy of thiamethoxam alone, which was decreased significantly 2 months after its application. Korunić et al. [50] applied a formulation containing a low quantity of DE and small amounts of deltamethrin and reported a high residual efficacy against *S. oryzae, R. dominica,* and *T. castaneum* even 12 months after the treatment. Wakil and Schmitt [51] also found that applications with 150 ppm of DEBBM (DE + bitterbarkomycin) plus 5.0 ppm imidacloprid were more effective than single insecticidal treatments for a period of five months, against all tested species on stored wheat.

5.3. Combination with Fungal Agents

Recently, extensive research focused on the adoption of entomopathogenic fungus species as an alternative approach to control insect pests of stored grain [52–57]. Fungal species such as *Beauveria bassiana* (Balsamo) Vuillemin (Ascomycota: Hypocreales), which is probably the most examined entomopathogenic fungus for stored product insects [58–64] has a complex interaction with cuticular lipids [65]. Results exalted the suitability of fungi as stored-product protectants but also pointed out their need for peculiar humid conditions to achieve satisfactory conidial adherence, germination, and penetration through the cuticle [66–68]. Increased humidity in stored commodities should be avoided [29], and hence, the fungal strains should be effective at drier conditions. Since DEs best perform under low humidity levels [2,25], the combination of fungi with DEs is very promising. The synergistic effect between DEs and entomopathogenic fungi expands the area for fungal spore penetration, increasing insect mycosis [40,62,69–73]. In addition, Batta [71] reported that the utilization of two different formulations of DE dusts, i.e., The Fossil Shield 90.0® (The Fossil Shield Co., Eiterfeld, Germany) and SilicoSec® (Agrinova GmbH, Obrigheim/Muhlheim, Germany), had a negligible effect on the viability of conidia of two fungal species. Dal Bello et al. [74] indicated the DE–fungal combinations to overcome some of the constraints in the use of fungi as biocontrol agents.

Applications of mixtures with these two ecologically compatible agents is a very appealing approach to IPM and can grant a more consistent management of multiple pest species under a wider range of environmental conditions.

The study of Athanassiou and Steenberg [70] demonstrated the potentials of these two agents together. The authors tested the insecticidal effect of *B. bassiana* combined with relatively low doses of Insecto® (Insecto Natural Products Inc., Costa Mesa, CA, USA), SilicoSec®, and PyriSec® (Biofa Gmbh, Germany), reporting a high level of control against the granary weevil *Sitophilus granarius* (L.) (Coleoptera: Curculionidae) under a broad range of temperatures and relative humidity levels [70]. In another published work by Wakil et al. [73], the application of 15 and 30 ppm of DEBBM combined with three doses of *B. bassiana* considerably increased adult mortality of *R. dominica*, especially at increasing temperatures and longer exposure intervals compared with DEBBM and *B. bassiana* alone. The synergistic interaction between Protect-It® and *B. bassiana* against several major stored-product insect species was also proved in laboratory bioassays [65,75]. Shafighi et al. [76] mentioned the high "speed of kill" of the combination of low doses of SilicoSec® when combined with entomopathogenic fungi against *T. castaneum*. Rizwan et al. [77] reported that the combination of the commercial DE formulation Diafil 610 (Celite Corporation, Lompoc, CA, USA) with *B. bassiana* had a suppressive effect on progeny (F1) production of the same beetle species. In field trials conducted on small farms, the treatment with mixtures of DE and *B. bassiana* outperformed the analogous combinations with imidacloprid after six months of storage [51].

Metarhizium anisopliae (Metschnikoff) Sorokin (Deuteromycotina: Hyphomycetes) and *Paecilomyces fumosoroseus* (=*Isaria fumosorosea*) (Wise) Brown & Smith (Ascomycota: Hypocreales) have also become a test subject for their insecticidal efficacy when combined with DEs, with reports to be in accordance with their potentials as control agents against several insect species, providing also long-term protection when applied in a variety of stored grains [40,61,76,78,79]. The virulence of *P. fumosorosea* integrated with DEBBM was shown to be an effective control measure for *R. dominica* in stored wheat [80]. *Nomuraea rileyi* (Farl.) Samson (Ascomycota: Hypocreales) and *Lecanicillium lecanii* (Zimm.) Zare & W.Gams (Ascomycota: Hypocreales) along with natural or modified DE formulations have been reported to show insecticidal, repellent, and ovicidal effects against *Bruchidius incarnatus* (Boheman) (Coleoptera: Chrysomelidae) and *R. dominica* under a variety of temperature and relative humidity conditions [81].

5.4. Combination with Botanicals

Plant extracts, essential oils, and other plant-based products are all ingredients with the potential to control stored-product insects [82,83]. However, their utilization is sometimes challenging due to their instability and high recommended doses. Combining them with DEs may enhance their properties, pursuing better insecticidal performances at lower doses and under a wide range of conditions. Several studies have been conducted toward this direction, using compounds from several sources. Bitterbarkomycin (BBM), a plant extract from the roots of *Celastrus angulatus* Max (Celastraceae), is known for its strong insecticidal and antifeedant activity against several insect species; low doses of DEBBM led to high mortality rates of *S. oryzae*, *S. zeamais*, *T. castaneum*, *R. dominica*, and *C. ferrugineus* in stored wheat [28,34]. Two DE formulations enhanced with abamectin, a macrocyclic lactone produced either directly by the actinomycete *Streptomyces avermitilis* or generated through semisynthetic modifications [84], were found to have high insecticidal properties against stored-product insects at rates as low as 75–125 ppm [28].

Constraints of the use of essential oils, such as their poor penetration, strong odor, lack of persistence, and high concentration requirements could be reduced if combined with DEs. Yang et al. [85] tested a combination of essential oil derived from *Allium sativum* L. (Amaryllidaceae) with 250 ppm of a DE formulation, reporting a strong synergistic effect and high initial efficacy against *S. oryzae*. Ziaee et al. [86] examined the synergistic/antagonistic interaction between *Carum copticum* (L.) (Apiaceae) essential oil with natural DE formulations of Iranian deposits against *T. confusum* and *S. granarius*, reporting the potentials of the combination for use in IPM programs. The same authors also stated that the essential oil increased the DE efficacy by increasing insect's locomotion activity through the particles and, at the same time, DEs reduced the oil concentration for the satisfactory protection of stored products. A new insecticide formulation using Celatom MN 23 (Celatom Diatomaceous Earth Functional Additives Technical Data Sheet, EP Minerals, Reno, NV, USA) enhanced with essential oil extracted from *Anethum graveolens* L. (Apiaceae) has been also examined by Korunić and Fields [87] and found to be effective in controlling four stored-product beetle species at lower doses and with far fewer negative effects on bulk density than using the DE alone. On the contrary, Campolo et al. [88] reported an antagonistic effect of *Citrus sinensis* (L.) Osbeck (Rutaceae) peel essential oil when admixed with the DE formulation Protector (Intrachem Bio, Grassobbio, Lombardy, Italy). Paponja et al. [89] developed an enhanced DE formulation admixing SilicoSec® with several botanicals (essential oil lavender, corn oil, and bay leaves dust) and silica gel, reporting higher mortality of all three insect species tested. Successful formulations of DEs and botanicals for the control of storage pests may be expected soon, but further testing is required to determine the duration of efficacy, cost of formulations, testing for their effect on non-target organisms, human safety, and effects on end-use quality.

6. DEs and Their Application in Urban, Agricultural, and Medical Environments

In the following paragraphs, we reviewed the studies that have investigated the efficacy of DEs against urban, medical, and agricultural pests [90–92]. Against these important pest groups, the insecticidal activity of DEs has been examined both when applied alone or in combination with entomopathogenic fungi or botanicals [42,93], following the same approach shown in the above-reported paragraphs dealing with stored product pest control. As a general trend, it has been noted that the biological activity of DEs increased when combined with entomopathogenic fungi [42,94].

6.1. DEs to Control Urban Pests

Insects and mites have successfully adapted over the years to the urban environment thanks to their ability to utilize food resources and harborages with humans [91]. These arthropod species can also transmit pathogenic organisms to food, as well as damages to house structures [91,95–97]. The control of arthropod vectors and pests, including urban ones, is challenging because of their strong reproductive ability, adaptability, and growing resistance to insecticides [98]. In addition, the adoption of insecticides in indoor areas is hazardous for human health [91]. Recently, several studies investigated the adoption of DE as an alternative to insecticides, highlighting their efficacy on different urban pests through different application scenarios [91,96,99–103] (Table 1).

The efficacy of DEs has been widely investigated on cockroaches, which are a worldwide public health pest that causes water and food contamination through transmitting pathogens mechanically, such as different forms of gastroenteritis [91,96]. A study compared the mortality of adult males and nymphs of the German cockroach, *Blattella germanica* L. (Blattodea: Ectiobidae), when DEs are applied as dry formulations or with the addition of water [96]. Mixing DE with water reduced the DE effectiveness, and the LC_{50} value was found to be 10 times lower if compared with dry DEs [96]. Similar results have been also found in stored product pests treated with dry DEs or with DEs formulated in water [30]. Overall, the bioactivity of DEs is inversely proportional to the water content and relative humidity [1]. To overcome the limitations related to high relative humidity conditions, mixing DEs with highly hydrophobic silanes may be a solution [99]. As reported by Faulde et al. [99], when DEs are mixed with hydrophobic silanes, a complete control of *B. germanica* could be achieved under humid conditions (R.H. > 80%) within 11 days [99]. In this work, it has been reported that the highest mortality rate of *B. germanica* (100% after 110 h) was achieved with the commercial DE Fossil-Shield® 90.0 S White, whose hydrophobicity increased by 3% Aerosil® with 1,1,1-trimethyl-*N*-trimethyl silane [99]. The same modified DEs led to the complete eradication of American cockroach, *Periplaneta americana* L. (Blattodea: Blattidae), and the silverfish, *Lepisma saccharina* L. (Thysanura: Lepismatidae), within 10 days, but the complete population suppression was not achieved in the case of the oriental cockroach, *Blatta orientalis* L. (Blattodea: Blattidae) [100]. These results highlight that cockroach susceptibility to DEs not only varies according to its formulations and their modifications, but it is also species-dependent [99,100]. Overall, hydrophobized DEs are more effective on certain cockroach species because of the higher absorption capacity of their cuticular waxes and the subsequent death by desiccation [99].

Thanks to their properties, DEs may act as physical barriers for arthropod pest intrusions and can be used to forecast the occurrence of subterranean termites that threaten housing construction and forest trees [101]. A study conducted by Gao et al. [101] showed that *Reticulitermes chinensis* Snyder (Rhinotermitidae: Blattodea) adult workers were not able to penetrate a 3 mm layer of dry DEs, suppressing their tunneling behavior, and died as a consequence of their movement. As reported by Ahmed et al. [103], mixing the soil with biofertilizers and DE increased the mortality and reduced the gallery length of another subterranean termite species, *Coptotermes heimi* (Wasmann) (Rhinotermitidae: Blattodea). On the other hand, DEs cannot be used as a barrier to prevent penetration of the soil surface by *Coptotermes formosanus* Shiraki (Rhinotermitidae: Blattodea), who was fully able to penetrate a DEs layer in laboratory bioassays [104]. Interestingly, although highly effective for the control of subterranean termites and cockroaches, DEs do not seem to be the most efficient inert dusts to control the pharaoh ant, *Monomorium pharaonis* (L.) (Hymenoptera: Formicidae)—a notorious domestic pest, for which the adoption of chemical-based insecticides is not recommended, particularly when ants infest crowded buildings such as hospitals [105]. Van Den Noortgate et al. [106] highlighted that the efficacy of DEs was lower if compared with various porous powders. For instance, zeolites ordered mesoporous silica material, and carbon black performed better than the DE benchmark material, especially the activated carbon powder (ACP) whose survival median time was almost four times shorter than that of the DEs (LT_{DE}: 95 min; LT_{ACP}: 25 min) [106].

Table 1. Local and commercial diatomaceous earths (DEs) evaluated against immature and adult stages of arthropods of urban interest. In addition to the mortality rates, the SiO_2 content (%) and the diameter of particles (μm) are reported. NA = not available data.

Pest Species	Family	Order	Developmental Stage	Tested DE	SiO_2 Content (%)	Ø Particles (μm)	Formulation	Mortality Rates	References	Notes
Blatta lateralis	Blattidae	Blattodea	Nymph	Turco 000	83–95	1–10	Dry	>90% after 12 h	[91]	Local commercialized DEs; 1 g/m^2 of DEs
Blatta lateralis	Blattidae	Blattodea	Nymph	Turco 004	83–95	10–30	Dry	>90% after 20 h	[91]	Local commercialized DEs; 1 g/m^2 of DEs
Blatta lateralis	Blattidae	Blattodea	Nymph	Turco 020	83–95	43–65	Dry	>90% after 24 h	[91]	Local commercialized DEs; 1 g/m^2 of DEs
Blatta orientalis	Blattidae	Blattodea	Adult + Nymph	Fossil Shield 90.0 S White®	0.35% (*w*/*v*)	5	Dry	70.6% on day 10	[99]	
Blattella germanica	Ectobiidae	Blattodea	Nymph	NA	NA	NA	Dry	LC_{50}: 4.2380 $g/m^{2(*)}$ LC_{50}: 5.2148 $g/m^{2(**)}$ LC_{50}: 12.9034 $g/m^{2\,(***)}$	[96]	
Blattella germanica	Ectobiidae	Blattodea	Adult (♂)	NA	NA	NA	Dry	LC_{50}: 8.0307 g/m^2 LC_{90}: 167.7116 g/m^2	[96]	No report if the $LC_{50/90}$ were at 24 h, 48 h, or 72 h
Blattella germanica	Ectobiidae	Blattodea	Nymph 2nd stage	NA	NA	NA	Dry + Water (50 mL)	LC_{50}: 20.0358 $g/m^{2(*)}$; LC_{50}: 7.9173 $g/m^{2\,(**)}$ LC_{50}: 6.3729 $g/m^{2\,(***)}$	[96]	
Blattella germanica	Ectobiidae	Blattodea	Adult (♂)	NA	NA	NA	Dry + Water (50 mL)	LC_{50}: 7.4093 g/m^2 LC_{90}: 91.2063 g/m^2	[96]	No report if the $LC_{50/90}$ were at 24 h, 48 h, or 72 h
Blattella germanica	Ectobiidae	Blattodea	Adult	BGN-1 (Local Turkish DEs)	NA	NA	Dry	100% mortality after 2 days (dose 5 g/m^2 and 10 g/m^2) on all type of floors	[102]	Ceramic tiles, Concrete floor, and parquet
Blattella germanica	Ectobiidae	Blattodea	Adult + Nymph	Fossil Shield 90.0 S W	0.35% (*w*/*v*)	5	Dry	100% mortality on day 6	[99]	
Blattella germanica	Ectobiidae	Blattodea	Adult + Nymph	Diamol KMT SilicoSec® Fossil Shield 90.0® Fossil Shield 90.0 W® Fossil Shield 90.0 S® Fossil Shield 90.0 S W® Fossil Shield 95.0®	0.35–0.40 (*w*/*v*)	5–7	Dry	Daily motility: control > Diamol KMT SilicoSec > FS 90.0 > FS 90.0 W > FS 90.0 S = FS 95.0 FS 90.0SW	[100]	
Coptotermes formosanus	Rhinotermitidae	Blattodea	Adult	Local DE	NA	NA	Dry	38.75% ± 6.60	[104]	No decrease in tunneling behavior
Coptotermes heimi	Rhinotermitidae	Blattodea	Adult	NA	NA	NA	Dry	At the highest dose of biofertilizer the mortality was lower than 40%	[103]	DEs were added to the soil + biofertilizers

Table 1. *Cont.*

Pest Species	Family	Order	Developmental Stage	Tested DE	SiO_2 Content (%)	Ø Particles (µm)	Formulation	Mortality Rates	References	Notes
Lepisma saccharina	Lepismatidae	Thysanura	Adult	Fossil Shield 90.0 S White®	0.35% (*w*/*v*)	5	Dry	100% mortality on day 9	[99]	
Lepisma saccharina	Lepismatidae	Thysanura	Adult	Fossil Shield 90.0 S White®	0.35% (*w*/*v*)	5	Dry	Low motility in both control and treated species	[99]	
Monomorium pharaonis	Formicidae	Hymenoptera	Adult	Lumino®	NA	NA	Dry	Lethal time: 95 minutes	[106]	No evidence if LT_{50} or LT_{90}
Periplaneta americana	Blattidae	Blattodea	Adult + Nymph	Fossil Shield 90.0 S White®	0.35% (*w*/*v*)	5	Dry	100% mortality on day 8	[99]	
Periplaneta americana	Blattidae	Blattodea	Adult	K14 (local turkish DEs)	NA	NA	Dry	100% mortality after 11 days (dose 40 g/m^2) on all type of floors	[107]	Ceramic tiles, Concrete floor, and laminate
Reticulitermes chinensis	Rhinotermitidae	Blattodea	Adult workers	NA	99	25–45	Moisture and dry DEs	100% after 6 hours when used dried DEs	[101]	10% and 25% of moisture led to low mortality rates Tunneling behavior is reduced in DEs moisture at 10%, 25%, and 50% Worker termites cannot penetrate a 3 mm layer of DEs

(*) LC_{50} calculated at 24 h; (**) LC_{50} calculated at 48 h; (***) LC_{50} calculated at 72 h.

6.2. DEs to Control Arthropod Pests and Vectors of Medical and Veterinary Importance

In recent years, several studies have investigated the use of DEs for managing pest and vector species of medical and veterinary importance [90,108–112] (Table 2). Many arthropods can play a pivotal role in the transmission of pathogens and eventually cause diseases in a wide range of vertebrates, including humans, livestock, pets, and wildlife [97]. Herein, we focus on some studies carried out to prove the efficacy of DEs alone or in combination with entomopathogenic fungi against arthropod pests and vectors of public health importance [90,108–112]. Many studies have investigated the adoption of DE to control bed bugs, *Cimex lectularius* L. (Hemiptera: Cimicidae), which are obligatory hematophagous insects that feed commonly on humans [95,97]. Apart from blood sucking, bed bugs are responsible for a range of emotional problems, anxiety, and sleeplessness [90]. As for other urban pests, there is no longer an absolute method to control/eradicate bed bugs, and the management relies either on the use of chemicals, such as pyrethroids, or on non-chemical tools, such as steam [95]. Given their low impact on mammals, DEs recently seem to be more of an option in bed bug control [90,113]. Several commercial DEs are known to be effective in the control of bed bugs, such as Bed Bug Kill [113], DE 51 [108], Mother Earth® D [90,108,114], Alpine® [90,114], Pro-Active®, DX13™-dust, and aerosol [90]. As reported by Akhtar and Isman [90], who evaluated the efficacy of several commercial DEs, their activity depends on the content of amorphous silicon dioxide and the dimension of the particles, resulting in the higher effectiveness of one DE to another. The efficacy of DEs may be increased by the addition of a dispersal agent, such as a bed bug alarm pheromone, which enhances bed bug crawling activity, increases bed bug locomotor activity, and thereby causes a higher contact with DE [115]. In addition, DE can be horizontally transferred from a treated bed bug to an untreated one [108]. This phenomenon is typical of gregarious insects and can facilitate the spread of DEs toward spaces that are hard to reach, contributing to the management of public health pests [108].

DEs have a recent use in the control of kissing bugs, *Triatoma infestans* (Klug) (Hemiptera: Reduviidae), which is a vector of *Trypanosoma cruzi*, causing Chagas' disease [97]. To date, several studies emphasize the efficacy of entomopathogenic fungi to control this species [116,117], but low humidity seems to be a limiting factor for fungal infection [117]. In addition, entomopathogenic fungi do not induce quick and high mortality as synthetic insecticides [117]. For these reasons, combining entomopathogenic fungi with oils and DEs may be a solution. The combination of DE + oil eventually enhances the adhesion and spread of particles (DE and conidia) on the lipophilic cuticle. In addition, the fungal development may be favored by the higher moisture provided by the abrasive action of DE and the subsequent trapping of moisture, and lastly, the oil serves as a nutrient source for the fungi [117]. The efficacy of the mixture is already well established, especially against stored-product pests [116–120], as detailed above. In laboratory bioassay, the mixture of *B. bassiana* and a commercial DE caused high mortality rates in all nymph instars and adults of *T. infestans*, ranging from 82% to 100% [120], but the same mixture elicited only 52.4% of *T. infestans* death in a field test in Northern Argentina [116]. Another study highlighted that the efficacy of the commercial DE KeepDry® toward *T. infestans* nymphs is highly increased when combined with vegetable oil and *M. anisopliae* (IP 46), even at a R.H. level of 75% [117]. The same combination was also effective in the control of the yellow fever mosquito, *Aedes aegypti* (L.) (Diptera: Culicidae), a vector or dengue, chikungunya, and Zika virus in the tropical and subtropical regions [109]. Overall, the combination of DE, entomopathogenic fungi, and a mineral/vegetable oil may represent a promising tool for the development of effective management strategies against *T. infestans* and *A. aegypti*. The combination of *M. anisopliae* (IP 119) with the commercial DE KeepDry® has been also successfully evaluated toward the cattle tick, *Rhipicephalus microplus* (Canestrini) (Ixodida: Ixodidae) [121]. The microsclerotia of *M. anisopliae* were incorporated in pellets containing inorganic materials, such as vermiculite, DE, and SiO_2 [121]. Overall, the pellets formulated with *M. anisopliae* microsclerotia effectively suppressed *R. microplus* in laboratory tests, demonstrating a promising pellet formulation for targeting the non-parasitic stage of this tick on the pasture [121]. Pellets can represent a possible upgrade of conventional granules thanks to their properties: a higher dose uniformity, higher mechanical resistance, narrower particle size distribution, and higher capacity of active incorporation [121]. The combination of fungal spores of *B. bassiana* and commercial DE resulted in significantly increased efficacy against blood-sucking poultry red mite *Dermanyssus gallinae* (De Geer) (Mesostigmata: Dermanyssidae) [94], which is a worldwide hematophagous ectoparasite in poultry farming [122] that is also responsible for the transmission of avian influenza viruses and *Salmonella enterica* ssp. *enterica* (S.) ser. Enteritidis and other important enterobacteria [123,124]. Although DEs were highly effective in the control of *D. gallinae*, Kilping and Steenberg [93] highlighted that four commercial DEs (SilicoSec®, Diamol®, Protect-It®, and Fossil Shield 90.0®) elicited avoidance behavior and repellence of the mites on the treated substrate. The authors noticed that the more effective the DE is, the greater the repellent activity. Consequently, the repellent activity had an impact on the efficacy of the inert dusts since mites will avoid treated surfaces. Furthermore, the dry conidia of *B. bassiana* also elicited a repellent response to poultry red mites both when applied on its own and when admixed with a low dose of the commercial DE Diamol [93].

In field bioassays, a liquid formulation of DEs elicited high mortality rates of the poultry red mite population [108]. A gradual reduction of the mite population (34% on day 7 to 53.5% on day 14; over 90% on days 21–28) was observed when the application of DE was combined with the mechanical cleaning [108]. The cleaning physically removes the mites and might also help the liquid formulation to stick better to surfaces than when covered with dust [108]. Overall, the adoption of liquid DEs is advisable mainly because product wastes are reduced, and an easier and safer application is provided [125]. Interestingly, DEs were found not to be as efficient as other products, such as kaolin and sulfur, to control the northern fowl mite, *Ornithonyssus sylvarium* (Canestrini & Fanzago) (Mesostigmata: Macronyssidae), which is another threat for hens [125,126]. Testing out the liquid formulations of DE, Martin and Mullens [126] noticed that a significant reduction of the northern fowl mite population occurred when DE was applied for two consecutive weeks, and the highest reduction of mite population was achieved with high concentrations of sulfur ($\geq$5.3%). Although DE effectiveness was found to be lower, in general, their use in dust boxes seems to enhance bird natural dustbathing behaviors, which translates into an increase in animal welfare and a reduction in the use of pesticides [127].

Table 2. Local and commercial diatomaceous earths (DEs) evaluated against immature and adult stages of arthropods of medical and veterinary relevance. In addition to the mortality rates, the SiO_2 content (%) and the diameter of particles (µm) are reported. NA = not available data; IP 46 = entomopathogenic fungus *Metarhizium anisopliae*.

Pest/Vector Species	Family	Order	Developmental Stage	Tested DE	SiO_2 Content (%)	Ø particles (µm)	Formulation	Mortality Rates	References	Notes
Aedes aegypti	Culicidae	Diptera	Adults	Keep Dry®	86	NA	DE DE + NDE + GIP46 + DE IP46 + DE + NIP46 + DE+ G	DE LT_{50}: 10.4 days DE + N LT_{50}: 8.9 days DE + G LT_{50}: 9.8 days IP46 + DE LT_{50}: 5.8 days IP46 + DE + N LT_{50}: 5.8 days IP46 + DE+ G LT_{50}: 5.9 days	[109]	G: Graxol® (vegetable oil) N: Naturol® (mineral oil)
Cimex lectularius	Cimicidae	Hemiptera	All stages	Alpine®Mother Earth®Pro-Active®DX13™ dust	NA	NA	Dry	LD_{50} (g m^{-2}) Alpine®: 4.48 after 24 h LD_{50} (g m^{-2}) Mother Earth®: 0.18 after 24 h LD_{50} (g m^{-2}) Pro-Active®: 2.26 after 24 h LD_{50} (g m^{-2}) DX13™ dust: 0.17 after 24 h	[90]	DX13™ was horizontally transferred from dead bed bugs to the untreated one
Cimex lectularius	Cimicidae	Hemiptera	All stages	DX13™ aerosol	NA	NA	Aerosol	Residual mortality (%) DX13™ aerosol: 81% after 21 days (72 h)	[90]	Mortality of bugs on the treated mattress after 32 weeks was 75%, 90%, and 100% after 24, 48, and 72 h
Cimex lectularius	Cimicidae	Hemiptera	Adult	DE 51	NA	NA	Dry	LC_{50} (mg) 24.4 and 5.1 at 48 h and 216 h	[108]	LC_{50} was calculated based on the transmission from a treated bug to an untreated one
Cimex lectularius	Cimicidae	Hemiptera	Nymph + Adult	DE 51	NA	NA	Dry	LC_{50} (mg) unexposed nymph 8.1 LC_{50} (mg) treated adults 6.4	[108]	Treated adults get in contact with untreated nymphs
Cimex lectularius	Cimicidae	Hemiptera	Nymph/Females	NA	NA	NA	Dry	Low mortality rates	[115]	The addition of alarm pheromone increased the movement of bed bugs throughout the Petri dish
Cimex lectularius	Cimicidae	Hemiptera	Adult	Bed Bug Killer®	NA	NA	Dry	(A) LT_{50} 7.42 days (B) LT_{50} 8.12 days	[113]	(A) Resistant strain (B) Susceptible strain
Cimex lectularius	Cimicidae	Hemiptera	All stages	Mother-Earth D®	NA	NA	Dry	94% of mortality after 10 days	[114]	Alpine® (0.25% Dinotefuran + 95% Diatomaceous Earth) has been also investigated, but its efficacy was lower than DE
Triatoma infestans	Reduviidae	Hemiptera	Nymphs	Keep Dry®	86	NA	DE DE + oil DE + IP46 DE + IP46 + oil	DE: 7.5% after 10 days DE + oil: 5.0% after 10 days DE + IP46: 100% after 10 days DE + IP46 + oil: 100% after 10 days	[117]	Cumulative mortality DE + IP46: LT_{50} = 5.7 days DE + IP46 + oil: LT_{50} = 4.5 days

Table 2. *Cont.*

Pest/Vector Species	Family	Order	Developmental Stage	Tested DE	SiO_2 Content (%)	Ø particles (μm)	Formulation	Mortality Rates	References	Notes
Triatoma infestans	Reduviidae	Hemiptera	Eggs	Keep Dry®	86	NA	DE DE + oil DE + IP46 DE + IP46 + oil	DE + oil eclosion: 92.5% H.R. 75% DE + oil eclosion: 83% H.R. >98% DE + IP46 + oil eclosion: 95% H.R. 75% DE + IP46 + oil no eclosion H.R. > 98%	[119]	Oil: Graxol®
Triatoma infestans	Reduviidae	Hemiptera	Nymphs	Keep Dry®	86	NA	*M. anisopliae* (IP 46) + DE + oil	(a) cumulative mortality: 100% (b) cumulative mortality: 5%	[119]	(a) H.R. > 98%, after 10 days and 24 h of exposition (b) H.R. = 75%, after 10 days and 24 h of exposition
Triatoma infestans	Reduviidae	Hemiptera	All stages	NA	NA	NA	DE + *B. bassiana*	Nymph 89.5–100%, MLT 5.1–8.3 days Adult 87.5%, MLT 10 days	[120]	MLT = mean lethal time
Ambloyomma americanum	Ixodidae	Ixodida	Larvae + Nymph	DeadZone	85	NA	Dry	Larval mortality: 100% after 6 h Nymphal mortality: 100% after 24 h	[128]	The DE was compared to a silica-gel based product
Ambloyomma americanum	Ixodidae	Ixodida	Larvae + Nymph	DeadZone	85	NA	Dry	Larval mortality: 84% after 24 h Nymphal mortality: 44.0% after 24 h	[128]	Highest dose: 10% of DE
Ceratophyllus idius	Ceratophyllidae	Siphonaptera	All stages	Drione Crawling Insect Killer	NA	NA	Dry	Lower number of fleas in nest treated with DE	[129]	38.12% diatomaceous earth as well as 0.2% pyrethrins and 1.0% piperonyl butoxide
Dermanyssus gallinae	Dermanyssidae	Acarina	All stages	Fisiocontrol™	86.2	<500	DE in water suspension	Topical mortality 95.4% Residual mortality 97.39%	[111]	Highest dose (10% of DE)
Dermanyssus gallinae	Dermanyssidae	Acarina	All stages	Fisiocontrol™	86.2	<500	DE + mechanical cleaning	Gradual reduction of mite population, over 90% at days 21–28	[111]	DE dose 10%
Dermanyssus gallinae	Dermanyssidae	Acarina	All stages	PosturaSec®	86.2	200	DE in water suspension	Immature stages: 98.9% (both doses) Adults: 98.8% (5% DE) 100% (10% DE)	[110]	
Dermanyssus gallinae	Dermanyssidae	Acarina	All stages	Silicosec®Ewazid® Silgur F46FS® Istant WFS 90.0 W	NA	NA	DE in water suspension	Silicosec® = 36.5% (24 h) Ewazid® Silgur F46 = 31% (24 h) Fossil Shied® Istant White = 100% (24 h) Fossil Shield® 90.0 White = 92.3% (24 h)	[112]	After 48 h, all the tested DE caused 100% of *D. gallinae* mortality

Table 2. *Cont.*

Pest/Vector Species	Family	Order	Developmental Stage	Tested DE	SiO_2 Content (%)	Ø particles (µm)	Formulation	Mortality Rates	References	Notes
Dermanyssus gallinae	Dermanyssidae	Acarina	All stages	Diamol KMTSilicoSec®FS 90.0®ProtectIt	NA	NA	Dry	Diamol KMT = 60% (LT50 = 3 days) SilicoSec® = 55% (LT50 = 3 days) Fossil Shield 90.0® = 30% (LT50: 3 days) ProtectIt® = 57% (LT50: 3 days)	[93]	The addition of the entomopathogenic fungi do not change the repellency of DEs
Dermanyssus gallinae	Dermanyssidae	Acarina	All stages	Diamol	NA	NA	Dry DE + *Beauveria bassiana*	Mortality 89.1% (H.R. 75%) Mortality 78.6% (H.R. 85%)	[94]	Synergistic interactions when applied simultaneously
Dermanyssus gallinae	Dermanyssidae	Acarina	Adult female	Diamol SilicoSec®	NA	NA	Dry	Low efficacy compared to a pure synthetic amorphous silica products	[130]	
Menacanthus stramineus	Menoponidae	Phthiraptera	All stages	Organic D/Earth®	NA	NA	DE mixed with sand (1:9)	60.4–95.2%	[126]	
Onythonyssus sylvarium	Macronyssidae	Acari	All stages	NA	NA	NA	DE suspended in deionized water	Low efficacy, it reduced the mite population only if applied for 2 consecutive weeks.	[125]	
Onythonyssus sylvarium	Macronyssidae	Acari	All stages	Organic D/Earth®	NA	NA	DE mixed with sand (1:9)	29.1–97.5%	[126]	Data refer to control over 4 weeks of dust box use
Onythonyssus sylvarium	Macronyssidae	Acari	All stages	Food-grade DE	NA	NA	Food-Grade DE mixed with sand (1:9)	When dust boxes were used, the northern fowl mite populations on flocks grew slowly (<100 mites)	[127]	
Protocalliphora spp.	Calliphoridae	Diptera	All stages	Drione Crawling Insect Killer	NA	NA	Drione Crawling Insect Killer	Lower number of fleas in nests treated with DE	[129]	38.12% DE + 0.2% pyrethrins + 1.0% piperonyl butoxide
Rhipicephalus microplus	Ixodidae	Ixodida	All stages	Keep Dry®	86	NA	IP 46 + DE (pellets)	The combination effectively suppressed the population of *R. microplus* and reduced the female oviposition period	[121]	Pellets: Vermiculite (AgroFloc) + DE (KeepDry®) + SiO_2

6.3. DEs to Control Crop Pests

Although DEs are not widely used by farmers to control arthropod pests, several studies reported their efficacy to control pests of agricultural interest such as soft-bodied insects, ants, and moths [42,131–134] (Table 3). It is generally advisable to use DE as an adjuvant rather than an active ingredient alone, considering the wide range of environmental conditions during the application [135]. Indeed, studies about DE effectiveness toward crop-damaging arthropods mainly focused on their use in combination with other products such as essential oils or entomopathogenic fungi [42,92,132,136,137]. For instance, a study conducted on the green peach aphid, *Myzus persicae* (Sulzer) (Hemiptera: Aphididae), reported that a DE + the essential oil of *Thymus capitatus* (L.) (Lamiaceae) caused mortality higher than 95% through in vitro bioassays [136]. The adoption of DE in solid form or suspended in water with neem oil extracted from *Azadirachta indica* A. Juss. (Meliaceae), also protects maize and tomato plants, causing a decrease in the number of larvae of the southern armyworm, *Spodoptera eridania* Stoll (Lepidoptera: Noctuidae) and fall armyworm, *Spodoptera frugiperda* Smith & Abbot (Lepidoptera: Noctuidae) [132]. In addition, Fossil Shield®, already adopted to control the red poultry mite [130], was proved to increase the efficacy of neem oil extract against cowpea aphid, *Aphis craccivora* Koch (Hemiptera: Aphididae) on the yardlong beans, *Vigna unguiculata* subsp. *sesquipedalis* L. (Fabaceae) [138]. Evaluating side effects on the aphid predator *Menochilus sexmaculatus* F. (Coleoptera: Coccinellidae), the toxicity of DE + neem oil was lower than that of the recommended chemical insecticide [138]. The same mixture, i.e., Fossil Shield® + neem oil, was evaluated against *M. persicae* (Sulzer) on globe artichoke *Cynara cardunculus* var. *scolymus* (L.) (Asteraceae) with promising results. The aphid population was reduced by 97% the day after the second spray [139]. Interestingly, the combination had a low impact on *M. persicae* common predators *Chrysoperla carnea* (Stephens) (Neuroptera: Chrysopidae), *Orius* spp. (Hemiptera: Antochoridae), *Coccinella* spp. (Coleoptera: Coccinellidae), and *Scymnus* spp. (Coleoptera: Coccinellidae), while *C. carnea* and *Orius* spp. were found to be more susceptible than the two coccinellids to that combination [139]. These results substantiated the findings by Ulrichs et al. [138], outlining the lower susceptibility of coccinellid predators to neem oil and DEs. In contact bioassays, DEs were also low risk toward predators of the spider mite, *Tetranychus urticae* Koch (Trombidiformes: Tetranychidae), such as *Phytoseiulus persimilis* Athias-Henriot (Mesostigmata: Phytoseiidae), *Neoseiulus fallacis* Garman (Mesostigmata: Phytoseiidae), and *Stethorus punctillum* (Weise) (Coleoptera: Coccinellidae), when DEs were tested in contact bioassay [140].

The synergistic interaction between DE and entomopathogenic fungi has been also evaluated toward an extremely wide range of agricultural pests [42,92]. For instance, a study on the western flower thrips, *Franklinella occidentalis* Pergande (Thysanoptera: Thripidae) highlighted that combining the entomopathogenic fungi, *Metarhizium flavoviride* (Gams and Rozsypal) (syn. *Metarhizium anisopliae* var. Acridum, pro parte) (Hyphomycetes: Deuteromycotina), with a commercial DE resulted in higher mortality of the thrips compared to the efficacies of each compound alone [92]. Synergistic interaction between DE and entomopathogenic fungi has been also reported for *T. infestans* [117], the silverleaf whitefly *Bemisia tabaci* (Gennadius) (Hemiptera: Aleyrodidae) [137], the cotton aphid, *Aphis gossypii* Glover (Hemiptera: Aphididae) [42], the indianmeal moth, *Plodia interpunctella* (Hübner), the almond moth, *Ephestia cautella* (Walker), and the Mediterranean flour moth, *Ephestia kuehniella* Zeller (Lepidoptera: Pyralidae) [141]. The insecticidal activity of DEs + fungi was evaluated toward the fire ant, *Solenopsis invicta* Buren (Hymenoptera: Formicidae) [131], which is responsible for a decline in production through direct predation on different plant parts (e.g. roots, fruits, flowers, stems), with reduction estimated to 15 to 33% in soybean, 20 to 35% in potato crops, or 50% in eggplant [141]. The effects of the combination of DE + *B. bassiana* toward healthy fire ants do not greatly increase the effect of *B. bassiana* alone, but testing DE alone in ants infected by *Thelohania solenopsae* (a common intracellular pathogen of fire ants) led to high insecticidal activity of DE, suggesting the synergistic interaction between *T. solenopsae* and DE [131].

7. DEs in Real-Scale Pest Management

The gradual withdrawal of active ingredients from the chemical-based pest management and the necessity to place insect control under the principles of IPM [142] led to the "re-evaluation" of inert dusts as a novel, effective, and sustainable management of arthropod pests and vectors all over the world. This is the main reason behind the recent popularity of DE formulations in pest management strategies [2], although commercial products ("Naaki" in Germany and "Neosyl" in England) have been available for stored-product protection since the 1930s [143]. In the following years, various enhanced or modified DE formulations have been created and evaluated, under extensive laboratory research, and some of them have reached the market as commercial formulations.

However, what are their potentials in praxis? Since several reports have questioned the compatibility of DEs with modern pest management programs, in this chapter, factors in terms of their potential impact in real scale applications (particularly as stored product protectants), including their utilization to reduce the standard application doses of residual insecticides and the role that DEs could have in resistance management will be discussed. Additional data of other relatively promising substances will be evaluated, aiming to expand the list of viable alternatives to hazardous chemicals that can be used as protectants in the food industry and beyond.

Virtually all the "classic" papers on the insecticidal efficacy of DEs examine formulations at a laboratory scale, with scarce data to be available in the literature regarding applications in large-scale scenarios. DEs have been approved for arthropod pest control, and commercial formulations are currently available as effective grain protectants. However, the grain industry is reluctant to use them for direct mixture with grains, as DE particles can adversely affect some physical and mechanical properties of the treated grain, obstructing their wider use as grain protectants. Indeed, for a satisfactory level of efficacy, the commercially available DE formulations should be applied at doses between 400 and 1000 ppm, but even in this case, adverse effects cannot be avoided [2,30]. However, several reports suggest that using DEs at concentrations lower than those indicated on the label could cause a sufficient reduction in the bulk density (test weight) of the grain [1,10,12,15]. Bulk density refers to a grading factor extensively used by the industry to determine the grain price, and its reduction through DE applications is of major importance. Korunić [10] examined 42 DE dusts from around the world and found significant correlations between DE insecticidal efficacy and adherence to kernels with bulk density reduction. In a later report, Korunić et al. [144] stated that the insecticidal efficacy and bulk density reduction could be linked by the capacity of a given DE to adhere to surfaces, which, eventually, is positively correlated with the insecticidal value of a given DE. Furthermore, when the DE particles are attached to the surface of the kernels, the spaces among the kernels increase, affecting their flowability, especially in mechanized handling systems. Jackson and Webley [145] found that when 0.5 g/kg of DE was applied on maize, the flow rate was reduced by about 39%. Apart from the grain industry, the milling industry has also expressed concerns about using DE formulations, as the presence of DE particles in the grain can damage the milling machinery through abrasive action. To overcome these limitations, new ways of DE applications have been proposed, intending to make the most of their advantages.

Table 3. Local and commercial diatomaceous earths (DEs) examined against immature and adult stages of arthropod pests of agricultural interest. In addition to the mortality rates, the SiO_2 content (%) and the diameter of particles (μm) are reported. NA = not available data.

Pest Species	Family	Order	Developmental Stage	Tested DE	SiO_2 Content (%)	Ø Particles (μm)	Formulation	Mortality Rates	References	Notes
Agrotis ipsilon	Noctuidae	Lepidoptera	IV instar larvae	Local Raw DE	NA	NA	DE suspended in water	Low concentration: 10% High concentration: 70%	[146]	Low concentration 1 g/L High concentration 4 g/L
Aphis craccivora	Aphididae	Hemiptera	All stages	Fossil Shield®	73	5–30	DE suspended in water + neem oil	Mean number of aphids/plant 2.2	[138]	Limited effects on the predator *Menochilus sexmaculatus*
Atta sexdens rubropilosa	Formicidae	Hymenoptera	Ant colonies	NA	NA	NA	Dry	Inactivity of the nests: 5.26–31.57%	[147]	
Bemisia argentifolii	Aleurodidae	Hemiptera	Nymphs	HYFLO®	NA	NA	DE + *Isaria fumosorosea* (612 strain)	Infected/dead nymphs ranged between 53% and 42.8%, day 4 and 6 respectively	[137]	
Epilachna vigintioctopunctata	Coccinellidae	Coleoptera	Larvae + Adults	Fossil Shield 90.0 S®	60–80	5–30	DE	FS 90.0 Adult: ≈75%; Larvae: ≈40%	[148]	After 48 h
Franklinella fusca	Thripidae	Thysanoptera	All stages	Celite®610 (Deadzone)	85	NA	DE DE + Orthene®	% of thrips/plant DE < DE + Orthene®	[149]	Average number of thrips per plant 2 days after insecticide application
Franklinella occidentalis	Thripidae	Thysanoptera	Adults	Puliantagai®	85	NA	(i) DE + *M. flavoridae* (ii) DE + *M. flavoridae* + Imidacloprid	(i) LT_{50} 3.77 days (ii) LT_{50} 4.23 days	[92]	
Franklinella occidentalis	Thripidae	Thysanoptera	Nymphs	Puliantagai®	85	NA	(i) DE + *M. flavoridae* (ii) DE + *M. flavoridae* + Imidacloprid	(i) LT_{50} 4.26 days (ii) LT_{50} 2.45 days	[92]	
Myzus persicae	Aphididae	Hemiptera	All stages	NA	NA	NA	DE + *Thymus capitatus* EO	Mortality 97.84%	[136]	After 24 h
Myzus persicae	Aphididae	Hemiptera	All stages	Fossil Shield®	60–80	5–30	DE alone DE + neem oil	Fossil-Shield 68.6% Fossil-Shield + neem oil: 96.4%	[139]	Limited effects on *Chrysoperla carnea*, *Orius* spp., *Coccinella* spp., and *Scymnus* spp.
Myzus persicae	Aphididae	Hemiptera	All stages	Pyrisec®	NA	NA	DE DE + *Paecilomyces lilacinus*	DE: 35–40% after 8 days DE + *P. lilacinus*: 54.83% after 8 days	[42]	25% pyrethrum, 3.1% pipronylbutaoxide, and 97.5% diatomaceous earth (SilicoSec®)
Myzus persicae	Aphididae	Hemiptera	All stages	Pyrisec®	NA	NA	DE DE + *Paecilomyces lilacinus*	DE: ≈40% after 10 days DE + *P. lilacinus*: ≈60% after 10 days	[42]	25% pyrethrum, 3.1% pipronylbutaoxide, and 97.5% diatomaceous earth (SilicoSec®)

Table 3. *Cont.*

Pest Species	Family	Order	Developmental Stage	Tested DE	SiO_2 Content (%)	Ø Particles (µm)	Formulation	Mortality Rates	References	Notes
Rhopalosiphum padi	Aphididae	Hemiptera	All stages	NA	NA	2.6	DE suspended in water	Wheat plant dusted with different dosages of DE did not show any visible injury	[133]	Reduction in chlorophyll content was observed in them.
Solenopsis invicta	Formicidae	Hymenoptera	Worker ants	NA	NA	NA	Dry + fungi	*Thelohania solenopsae* + DE: 89% *Beauveria bassiana* + DE: <50%	[131]	After 10 days
Spodoptera eridiana	Noctuidae	Lepidoptera	II instar larvae	KeepDry®	86	15	DE suspended (alone) DE suspended + neem oil	DE: 46.6% DE + neem: 93.7%	[132]	DE highest dose non-additive synergistic
Spodoptera exigua	Noctuidae	Lepidoptera	III instar larvae	Sayan®	92	50	DE suspended in water	Mortality: 59.25%, concentration of 20%	[150]	After 72 h
Spodoptera frugipreda	Noctuidae	Lepidoptera	II instar larvae	Dezone	85	NA	DE suspended in water	High maize grain yield 7387 kg/ha.	[134]	
Spodoptera frugiperda	Noctuidae	Lepidoptera	II instar larvae	KeepDry®	86	15	DE suspended (alone) DE suspended + neem oil	DE: 76.2% DE + neem: 66.6%	[132]	DE highest dose additive effect
Spodoptera littura	Noctuidae	Lepidoptera	III instar larvae	Fossil Shield 90.0®	60–80	5–30	DE suspended in water	≈70% after 48 h	[148]	
Tetranychus urticae	Acarina	Tetranychidae	Adult (♀)	DE_cide	67	NA	DE suspended in water	Contact mortality: 24.6%	[140]	Limited impact on TSSM predators (*Phytoseiulus persimilis*, *Neoseiulus fallacis* and *Stethorus punctillum*).

8. DE Applications for Structural Treatments

DEs leave no harmful residues in the surfaces applied and hence, applications could be carried out in food and processing facilities. Korunić et al. [12] reported that the treatment of hard wheat with either 50 or 300 ppm of Protect-It® had no significant effect on the milling, analytical, rheological, or baking quality, and these doses did not affect the properties for pasta production, while 100 to 900 ppm on barley showed no differences in malting quality characteristics. Desmarchelier and Dines [151] reported that treatments with Dryacide® did not affect flour quality, as determined by the volume of sponge cakes and the production of carbon dioxide by fermenting dough. Aldryhim [152] found no evidence of an adverse effect on wheat seed germination, wheat flour, and baking quality, using Dryacide®.

Another advantageous feature of DEs is their persistence and stability in a wide range of temperatures [153,154], as compared with contact insecticides [155,156]. Arthur [39] exposed adults of *T. castaneum* and *T. confusum* to filter papers containing 0.5 mg/cm^2 of Protect-It® and reported a positive effect of temperature and exposure interval on insect mortality, along with a negative effect of humidity.

The utilization of DEs has been addressed as a good way to strengthen the effects of heat treatments, as the exposed insects are expected to die earlier due to increased desiccation. According to Fields et al. [157], the complete control of *T. confusum* in an oat mill could be achieved after treatment at 41 °C for 13–22 h, when DEs are combined with heat. In contrast, heat alone caused the same results after 32–38 h exposure at a sufficiently higher temperature, 47 °C [157]. Additional data by Dowdy [158] and Dowdy and Fields [159] indicated that DEs appear to be of value in areas where lethal temperatures cannot be reached during heat treatment applications. Moreover, even after the treatments combining heat with DEs, delayed mortality may occur after a while for the remaining insects due to the residual toxicity of DEs [159].

Laboratory studies have been conducted to evaluate the insecticidal efficacy of different DE formulations when applied directly in different types of surfaces, such as concrete, ceramic, plywood, plastic, metal, etc. [39,160–164]. In general, lower doses are required on some surfaces, such as metal and glass, compared with surfaces with a rougher construction, such as wood and concrete [160,163–165]. Collins and Cook [162] reported that 5 g/m^2 of SilicoSec® was just as effective as 20 g/m^2 to achieve mortalities above 86% of different stored product insect and mite species after one week of exposure to glass and plastic surfaces. This observation is in accordance with the reports of Cook [166], Mewis and Ulrichs [37], and Athanassiou et al. [15], indicating that there is a limitation in the amount of DE particles the insects can pick up. In general, insects seem to pick up DE particles more easily if the formulation is equally applied onto the surface (e.g., on a Petri dish) and not adhered on the grain kernels [15]. In the latter case, DE particles are also likely to lose effectiveness by lipid absorption from the external part of the kernel [2].

Cleaning and sanitation before DE structural treatments is a key element in pest management practices since the presence of food may increase insect survival rates [159,167]. Arthur [39] using 0.5 mg/cm^2 of Protect-It® in plastic surface against *T. castaneum* and *T. confusum* emphasized the importance to eliminate the presence of food materials within the storage environment to maximize the effectiveness of the treatments. Dowdy [158] also addressed the impact of food in the effectiveness of treatments combining heat with some commercial DEs: Insecto®, Protect-It®, Concern®, and Natural Guard® (VPG Co-op Gardening Group, Inc., Bonham, TX, USA). Access to food significantly decreased insect mortality, providing an average between 21 and 88% of individuals fed and not fed respectively, 7 d after the treatment. Similar results have been published using other inert dust formulations, as food may provide water and nutrition that can lead to increased and prolonged insect survival, which may allow the continuance of the infestation for a certain period and the concomitant progeny production [168–170].

Although most laboratory studies tended to prefer dry DE applications over the use of slurry solutions [15,28,153,161,171], the reverse is probably more desirable in commercial practice. Slurries may be used easier in their application by the personnel, as there is a need to avoid exposure to the very dusty atmospheres created by dry-blown methods [153,165].

9. Other Relative Promising Substances

A plethora of other inert dusts has been also tested for their toxicity against arthropod species, with special reference to stored product pests. In general, inert dusts/materials can be categorized according to their chemical composition or level of activity in four wide groups: (*a*) clays, sand, kaolin, paddy husk ash, wood, and volcanic ash, (*b*) katelsous (rock phosphate and ground sulfur), lime (calcium hydroxide), limestone (calcium carbonate), and salt (sodium chloride), (*c*) synthetic silica aerogels produced by drying aqueous solutions of sodium silicate and (*d*) dusts containing natural silica, including DEs [1,2,172,173]. Zeolites (alkali metal aluminum silicates) have been also included in this group by Subramanyam and Roesli [2], since these substances have similar physical properties with DEs. In addition, Golob [173] divided the DE formulations into two groups, addressing the modified DE formulations that contain over 98% silicon dioxide (compared to the 90% silicon dioxide of the natural dusts) as the fifth group of inert dusts.

Zeolites are among the most promising alternatives to DEs, and their potentials in food and agriculture are well described by Eroglu et al. [5]. Nevertheless, regarding stored product protection, there are disproportionally few data as compared with DEs, although the interest for zeolites in stored product protection has been increased [174–178]. Zeolites' particle size effect, adherence to kernels, and influence on the test weight of grains have been examined by Rumbos et al. [177], showing similar trends with DEs. The results of these studies encourage further research to evaluate the use of zeolites as grain protectants but also to surface treatments or "crack and crevice" applications. Attempts have been also made to evaluate the insecticidal efficacy of other inert dusts, to use them in modern pest management at the post-harvest stages of durable agricultural commodities [2]. Even some of the currently existing DEs cannot be considered as pure DE formulations, as they contain additional inert materials that have a certain insecticidal action and can be drastically modified to obtain increased efficacy [11].

10. Conclusions and Future Challenges

The need to gradually withdraw from the chemical-based pesticide policies to more sustainable and ecological approaches is, at the present, one of the most challenging aspects of pest management. The current decrease in the registered pesticides will undoubtedly continue, increasing simultaneously the need to develop novel, effective but also ecologically compatible substances. On the other hand, the introduction of a new pesticide is a costly and long process, making the total overdrawn from the traditional protectants an unrealistic scenario. Thus, inert dusts such as DEs might have an important role to play in future pest management strategies, ensuring an abundant supply of safe and healthy food and feed. DEs hold great potential as carriers of common insecticides, minimizing the required application doses of the latter. In addition to synergistic effects, combined applications may also alleviate the negative effects of the substances and can be more compatible with the desired criteria for food safety and protection of human health and the environment. However, the introduction of other agents must be always appraised under the prism of the potential negative effects they may hold, such as for instance the adoption of entomopathogenic fungus agents [179]. Nevertheless, the application of insecticides and acaricides with a different mode of action may be a solution for the control of resistant arthropod populations, which is a hot topic in modern Integrated Pest Management [180–182]. Although there are plenty of data for their insecticidal and acaricidal properties, little progress has been made regarding the optimal processing of DE dusts used as insecticides and acaricides. Today, most commercial DE formulations are prepared through a basic process of quarrying, drying, and milling the mined heterogeneous rocks. This simplistic treatment leads eventually to formulations with great variability in their physicochemical characteristics, influencing simultaneously their insecticidal/acaricidal properties but also some properties of the commodity itself. Therefore, more specific methods of processing must be found to standardize the production of dusts bearing the most desirable features for increased insecticidal efficacy.

Application methods and systems of DEs are also an issue of major importance, requiring additional investigation. Even with the current DE formulations, different application techniques, such as using slurries or treating only partial layers of the food, should be explored. Thus, research should be conducted under a range of food-handling establishments to design effective protocols for pest management but also to determine the effects of sanitation on the performance of DE dusts. Such real-scale applications may highlight the potential of DEs and explore ways of integrating DE applications within the total pest/vector management program in food industry, agricultural and urban settings.

The data from laboratory studies underline the insecticidal and acaricidal value of DEs under a wide range of arthropods. Further analysis must be conducted toward this direction not only to identify all the target species but also to investigate the overall outcome of DEs in non-target species. Indeed, by examining the current literature, we observed that non-target effects of DE have been evaluated only on a limited number of natural enemies of crop pests, with special reference to aphidophagous coccinellids, lacewings, anthochorids, and Phytoseiidae mites, showing limited consequences for these important biocontrol agents. In this promising scenario, further research should be devoted in understanding the potential non-target effects of DE-based formulations.

Author Contributions: Conceptualization, G.B. and C.G.A.; Literature collection and analysis, V.Z., G.V.B., G.B., A.C. and C.G.A.; writing—original draft preparation, V.Z. and G.V.B.; writing—review and editing, V.Z., G.V.B., G.B., A.C. and C.G.A.; visualization, G.B., A.C., C.G.A.; supervision, G.B., A.C., C.G.A. All authors have read and agreed to the published version of the manuscript.

Funding: Valeria Zeni is partially funded by PRIMA iGUESSmed Project. This article has been co-financed by the European Regional Development Fund of the European Union and Greek national funds through the Operational Program Competitiveness, Entrepreneurship, and Innovation, under the call RESEARCH-CREATE-INNOVATE (project code: T2EΔK-03532).

Institutional Review Board Statement: Not applicable.

Informed Consent Statement: Not applicable.

Acknowledgments: We are grateful to Carey Yuan for her kind assistance during the preparation of this Invited Review for *Molecules*.

Conflicts of Interest: The authors declare no conflict of interest.

Sample Availability: Not applicable.

References

1. Korunić, Z. Diatomaceous earths, a group of natural insecticides. *J. Stored Prod. Res.* **1998**, *34*, 87–97. [CrossRef]
2. Subramanyam, B.; Roesli, R. Inert Dusts. In *Alternatives to Pesticides in Stored-Product IPM*; Subramanyam, B., Hagstrum, D.W., Eds.; Kluwer Academic Publishers: Dordreecht, Netherlands, 2000; pp. 321–380.
3. Fernandez, M.A.; Bellotti, N. Silica-based bioactive solids obtained from modified diatomaceous earths to be used as antimicrobial filler material. *Mater. Lett.* **2017**, *194*, 130–134. [CrossRef]
4. Bellotti, N.; Deyá, C. Chapter 14. Natural products applied to antimicrobial coatings. *Stud. Nat. Prod. Chem.* **2019**, *60*, 485–508.
5. Eroglu, N.; Emekci, M.; Athanassiou, C.G. Applications of natural zeolites on agriculture and food production. *J. Sci. Food Agric.* **2017**, *97*, 3487–3499. [CrossRef]
6. Arthur, F.H.; Puterka, G.J. Evaluation of kaolinite-based particle films to control *Tribolium* species (Coleoptera: Tenebrionidae). *J. Stored Prod. Res.* **2002**, *38*, 341–348. [CrossRef]
7. Athanassiou, C.G.; Kavallieratos, N.G.; Benelli, G.; Losić, D.; Usha Rani, P.; Desneux, N. Nanoparticles for pest control: Current status and future perspectives. *J. Pest Sci.* **2018**, *91*, 1–15. [CrossRef]
8. Korunić, Z. Overview of undesirable effects of using diatomaceous earths for direct mixing with grains. *Pestic. Fitomed.* **2016**, *31*, 9–18. [CrossRef]
9. Vayias, B.J.; Athanassiou, C.G.; Korunić, Z.; Rozman, V. Evaluation of natural diatomaceous earth deposits from south-eastern Europe for stored-grain protection: The effect of particle size. *Pest Manag. Sci.* **2009**, *65*, 1118–1123. [CrossRef]

10. Korunić, Z. Rapid assessment of the insecticidal value of diatomaceous earths without conducting bioassays. *J. Stored Prod. Res.* **1997**, *33*, 219–229. [CrossRef]
11. Baliota, G.V.; Athanassiou, C.G. Evaluation of a Greek diatomaceous earth for stored product insect control and techniques that maximize its insecticidal efficacy. *Appl. Sci.* **2020**, *10*, 6441. [CrossRef]
12. Korunić, Z.; Fields, P.G.; Kovacs, M.I.P.; Noll, J.S.; Lukow, O.M.; Demianyk, C.J.; Shibley, K.J. The effect of diatomaceous earth on grain quality. *Postharvest Biol. Technol.* **1996**, *9*, 373–387. [CrossRef]
13. Kavallieratos, N.G.; Athanassiou, C.G.; Paschalidou, F.G.; Andris, N.S.; Tomanovic, Z. Influence of grain type on the insecticidal efficacy of two diatomaceous earth formulations against *Rhyzopertha dominica* (F.) (Coleoptera: Bostrychidae). *Pest Manag. Sci.* **2005**, *61*, 660–666. [CrossRef]
14. Athanassiou, C.G.; Kavallieratos, N.G. Insecticidal effect, and adherence of PyriSec® in different grain commodities. *Crop Prot.* **2005**, *24*, 703–710. [CrossRef]
15. Athanassiou, C.G.; Kavallieratos, N.G.; Vayias, B.J.; Tomanović, Z.; Petrović, A.; Rozman, V.; Adler, C.; Korunić, Z.; Milovanović, D. Laboratory evaluation of diatomaceous earth deposits mined from several locations in central and southeastern Europe as potential protectants against coleopteran grain pests. *Crop Prot.* **2011**, *30*, 329–339. [CrossRef]
16. Rigaux, M.; Haubruge, E.; Fields, P.G. Mechanisms for tolerance to diatomaceous earth between strains of *Tribolium castaneum* (Coleoptera: Tenebrionidae). *Entomol. Exp. Appl.* **2001**, *101*, 33–39. [CrossRef]
17. Vayias, B.J.; Athanassiou, C.G.; Kavallieratos, N.G.; Buchelos, C.T. Susceptibility of different European populations of *Tribolium confusum* (Coleoptera: Tenebrionidae) to five diatomaceous earth formulations. *J. Econ. Entomol.* **2006**, *99*, 1899–1904. [CrossRef]
18. Vayias, B.J.; Athanassiou, C.G.; Buchelos, C.T. Evaluation of resistance development by *Tribolium confusum* du Val (Coleoptera: Tenebrionidae) to diatomaceous earth under laboratory selection. *J. Stored Prod. Res.* **2008**, *44*, 162–168. [CrossRef]
19. Athanassiou, C.G.; Kavallieratos, N.G.; Andris, N.S. Insecticidal effect of three diatomaceous earth formulations against adults of *Sitophilus oryzae* (Coleoptera: Curculionidae) and *Tribolium confusum* (Coleoptera: Tenebrionidae) on oat, rye, and triticale. *J. Econ. Entomol.* **2004**, *97*, 2160–2167. [CrossRef]
20. Athanassiou, C.G.; Vayias, B.J.; Dimizas, C.B.; Kavallieratos, N.G.; Papagregoriou, A.S.; Buchelos, C.T. Insecticidal efficacy of diatomaceous earth against *Sitophilus oryzae* (L.) (Coleoptera: Curculionidae) and *Tribolium confusum* Du Val (Coleoptera: Tenebrionidae) on stored wheat: Influence of dose rate, temperature and exposure interval. *J. Stored Prod. Res.* **2005**, *41*, 47–55. [CrossRef]
21. Nayak, M.K.; Daglish, G.J. Importance of Stored Product Insects. In *Recent Advances in Stored Product Protection*; Athanassiou, C.G., Arthur, F.H., Eds.; Springer: Berlin/Heidelberg, Germany, 2018; pp. 1–19.
22. Campbell, J.F.; Arthur, F.H.; Mullen, M.A. Insect management in food processing facilities. *Adv. Food Nutr. Res.* **2004**, *48*, 240–295.
23. Stejskal, V.; Hubert, J.; Aulocky, R.; Kucerova, Z. Overview of present and past and pest-associated risks in stored food and feed products: European perspective. *J. Stored Prod. Res.* **2015**, *64*, 122–132. [CrossRef]
24. Hagstrum, D.W.; Athanassiou, C.G. Improving stored product insect management: From theory to practice. *Insects* **2019**, *10*, 332. [CrossRef]
25. Vayias, B.J.; Athanassiou, C.G. Factors affecting effcacy of the diatomaceous earth formulation SilicoSec against adults and larvae of the confused beetle *Tribolium confusum* Du Val (Coleoptera: Tenebrionidae). *Crop Prot.* **2004**, *23*, 565–573. [CrossRef]
26. Pixton, S.W. Moisture content-its significance and measurement in stored products. *J. Stored Prod. Res.* **1967**, *3*, 35–47. [CrossRef]
27. Pixton, S.W.; Warburton, S. Moisture content/relative humidity equilibrium of some cereal grains at different temperatures. *J. Stored Prod. Res.* **1971**, *6*, 283–293. [CrossRef]
28. Athanassiou, C.G.; Korunić, Z. Evaluation of two new diatomaceous earth formulations, enhanced with abamectin and bitterbarkomycin, against four stored-grain beetle species. *J. Stored Prod. Res.* **2007**, *43*, 468–473. [CrossRef]
29. Athanassiou, C.G.; Arthur, F.H. Bacterial Insecticides and Inert Materials. In *Recent Advances in Stored Product Protection*; Athanassiou, C.G., Arthur, F.H., Eds.; Springer: Berlin/Heidelberg, Germany, 2018; pp. 83–98.
30. Fields, P.; Korunić, Z. The effect of grain moisture content and temperature on the efficacy of diatomaceous earths from different geographical locations against stored product beetles. *J. Stored Prod. Res.* **2000**, *36*, 1–13. [CrossRef]
31. Athanassiou, C.G.; Kavallieratos, N.G.; Tsakiri, J.B.; Xyrafidis, S.N.; Vayias, B.J. Effect of temperature and humidity on insecticidal effect of SilicoSec against *Ephestia kuehniella* (Lepidoptera: Pyralidae) larvae. *J. Econ. Entomol.* **2006**, *99*, 1520–1524. [CrossRef]
32. Aldryhim, Y.N. Combination of classes of wheat and environmental factors affecting the efficacy of amorphous silica dust, dryacide, against *Rhyzopertha dominica* (F.). *J. Stored Prod. Res.* **1993**, *29*, 271–275. [CrossRef]
33. Palyvos, N.E.; Athanassiou, C.G.; Kavallieratos, N.G. Acaricidal effect of a diatomaceous earth formulation against *Tyrophagus putrescentiae* (Astigmata: Acaridae) and its predator *Cheyletus malaccensis* (Prostigmata: Cheyletidae) in four grain commodities. *J. Econ. Entomol.* **2006**, *99*, 229–236. [CrossRef]
34. Athanassiou, C.G.; Korunić, Z.; Vayias, B.J. Diatomaceous earths enhance the insecticidal effect of bitterbarkomycin against stored-grain insects. *Crop Prot.* **2009**, *28*, 123–127. [CrossRef]
35. Rudolph, D. Occurrence, properties and biological implications of the active uptake of water vapour from the atmosphere in Psocoptera. *J. Insect Physiol.* **1982**, *28*, 111–121. [CrossRef]
36. Rudolph, D. Site, process and mechanism of active uptake of water vapour from the atmosphere in the Psocoptera. *J. Insect Physiol.* **1982**, *28*, 205–212. [CrossRef]

37. Mewis, I.; Ulrichs, C. Action of amorphous diatomaceous earth against different stages of the stored product pests *Tribolium confusum* (Coleoptera: Tenebrionidae), *Tenebrio molitor* (Coleoptera: Tenebrionidae), *Sitophilus granarius* (Coleoptera: Curculionidae) and *Plodia interpunctella* (Lepidoptera: Pyralidae). *J. Stored Prod. Res.* **2001**, *37*, 153–164. [PubMed]
38. Arthur, F.H. Toxicity of diatomaceous earth to red flour beetles and confused flour beetles (Coleoptera: Tenebrionidae): Effects of temperature and relative humidity. *J. Econ. Entomol.* **2000**, *93*, 526–532. [CrossRef]
39. Arthur, F.H. Impact of food source on survival of red flour beetles and confused flour beetles (Coleoptera: Tenebrionidae) exposed to diatomaceous earth. *J. Econ. Entomol.* **2000**, *93*, 1347–1356. [CrossRef]
40. Athanassiou, C.G.; Kavallieratos, N.G.; Vayias, J.B.; Tsakiri, J.B.; Mikeli, N.H.; Meletsis, C.M. Tomanović, Ž. Persistence and efficacy of *Metarhizium anisopliae* (Metschnikoff) Sorokin (Deuteromycotina: Hyphomycetes) and diatomaceous earth against *Sitophilus oryzae* (L.) (Coleoptera: Curculionidae) and *Rhyzopertha dominica* (F.) (Coleoptera: Bostrychidae) on wheat and maize. *Crop Prot.* **2008**, *27*, 1303–1311.
41. Wakil, W.; Riasat, T.; Lord, J.C. Effects of combined thiamethoxam and diatomaceous earth on mortality and progeny production of four Pakistani populations of *Rhyzopertha dominica* (Coleoptera: Bostrichidae) on wheat, rice and maize. *J. Stored Prod. Res.* **2013**, *52*, 28–35. [CrossRef]
42. Wakil, W.; Schmitt, T.; Kavallieratos, N.G. Performance of diatomaceous earth and imidacloprid as wheat, rice and maize protectants against four stored-grain insect pests. *J Stored Prod Res* **2021**, *91*, 101759. [CrossRef]
43. Ceruti, F.C.; Lazzari, S.M.N. Combination of diatomaceous earth and powder deltamethrin for insect control in stored corn. *Rev. Bras. Entomol.* **2005**, *49*, 580–583. [CrossRef]
44. Arthur, F.H. Evaluation of a new insecticide formulation (F2) as a protectant of stored wheat, maize, and rice. *J. Stored Prod. Res.* **2004**, *40*, 317–330. [CrossRef]
45. Awais, M.; Mansoor-ul-Hasan; Sagheer, M.; Asif, M.U.; Ali, Q.; Zaman, S. Efficacy of diatomaceous earth and insect growth regulators against *Tribolium castaneum* (Herbst) (Coleoptera: Tenebrionidae). *Sci. Lett.* **2019**, *7*, 59–67.
46. Awais, M.; Zeeshan, M.; Mansoor-ul-Hasan; Sagheer, M.; Asif, M.U.; Ali, Q.; Zaman, S. Combined effect of diatomaceous earth and two insect growth regulators against *Trogoderma granarium* (Coleoptera: Dermestidae). *Sci. Lett.* **2020**, *8*, 55–60.
47. Arthur, F.H. Evaluation of methoprene alone and in combination with diatomaceous earth to control *Rhyzopertha dominica* (Coleoptera: Bostrichidae) on stored wheat. *J. Stored Prod. Res.* **2004**, *40*, 485–498. [CrossRef]
48. Athanassiou, C.G. Toxicity of beta cyfluthrin applied alone or in combination with diatomaceous earth against adults of *Sitophilus oryzae* (L.) (Coleoptera: Curculionidae) and *Tribolium confusum* DuVal (Coleoptera: Tenebrionidae) on stored wheat. *Crop Prot.* **2006**, *25*, 788–794. [CrossRef]
49. Wakil, W.; Riasat, T.; Ashfaq, M. Residual efficacy of thiamethoxam, *Beauveria bassiana* (Balsamo) Vuillemin, and diatomaceous earth formulation against *Rhyzopertha dominica* F. (Coleoptera: Bostrychidae). *J. Pest Sci.* **2012**, *85*, 341–350. [CrossRef]
50. Korunić, Z.; Kalinovic, I.; Liska, A.; Hamel, D. Long Term Effectiveness of the Mixture of Diatomaceous Earth and Deltamethrin on Wheat. In Proceedings of the Tenth International Working Conference on Stored Product Protection, Estoril, Portugal, 27 June–2 July 2010; Carvalho, M.O., Fields, P.G., Adler, C.S., Arthur, F.H., Athanassiou, C.G., Campbell, J.F., Fleurat-Lessard, F., Flinn, P.W., Hodges, R.J., Isikber, A.A., et al., Eds.; Julius-Kühn-Archiv: Berlin, Germany, 2010; pp. 857–861.
51. Wakil, W.; Schmitt, T. Field trials on the efficacy of *Beauveria bassiana*, diatomaceous earth and Imidacloprid for the protection of wheat grains from four major stored grain insect pests. *J. Stored Prod. Res.* **2015**, *64*, 160–167. [CrossRef]
52. Dhuyo, A.R.; Ahmad, S. Evaluation of fungus *Beauveria bassiana* (Bals.) infectivity to the larger grain borer *Prostephanus truncatus* (Horn). *Pak. Entomol.* **2007**, *29*, 77–82.
53. Lord, J.C. Efficacy of *Beauveria bassiana* for control of *Tribolium castaneum* with reduced oxygen and increased carbon dioxide. *J. Appl. Entomol.* **2009**, *133*, 101–107. [CrossRef]
54. Phillips, T.W.; Throne, J.E. Biorational approaches to managing stored-product insects. *Annu. Rev. Entomol.* **2010**, *55*, 375–397. [CrossRef]
55. Ahmed, B.I. Potentials of entomopathogenic fungi in controlling the menace of maize weevil *Sitophilus zeamais* Motsch (Coleoptera: Curculionidae) on stored maize grain. *Arch. Phytopathol. Pflanzenschutz* **2010**, *43*, 107–115. [CrossRef]
56. Kavallieratos, N.G.; Athanassiou, C.G.; Aountala, M.M.; Kontodimas, D.C. Evaluation of the entomopathogenic fungi *Beauveria bassiana*, *Metarhizium anisopliae*, and *Isaria fumosorosea* for control of *Sitophilus oryzae*. *J. Food Prot.* **2014**, *1*, 4–17. [CrossRef]
57. Batta, Y.A. Recent advances in formulation and application of entomopathogenic fungi for biocontrol of stored-grain insects. *Biocontrol Sci. Technol.* **2016**, *26*, 1171–1183. [CrossRef]
58. Moore, D.; Lord, J.C.; Smith, S.M. Pathogens. In *Alternatives to Pesticides in Stored-Product IPM*; Subramanyam, B., Hagstrum, D.W., Eds.; Kluwer Academic Publishers: Dordreecht, Netherlands, 2000; pp. 193–227.
59. Lord, J.C. Low humidity, moderate temperature and desiccant duct favor the efficacy of *Beauveria bassiana* (Hyphomycetes: Moniliales) for the lesser grain borer *Rhyzopertha dominica* (Coleoptera: Bostrychidae). *Biol. Control* **2005**, *34*, 180–186. [CrossRef]
60. Akbar, W.; Lord, J.C.; Nechols, J.R.; Howard, R.W. Diatomaceous earth increases the efficacy of *Beauveria bassiana* against *Tribolium castaneum* larvae and increases conidia attachment. *J. Econ. Entomol.* **2004**, *97*, 273–280. [CrossRef]
61. Michalaki, M.P.; Athanassiou, C.G.; Kavallieratos, N.G.; Batta, Y.A.; Balotis, G.N. Effectiveness of *Metarhizium anisopliae* (MetschinkoV) Sorokin applied alone or in combination with diatomaceous earth against *Tribolium confusum* Du Val larvae: Influence of temperature, relative humidity and type of commodity. *Crop Prot.* **2006**, *25*, 418–425. [CrossRef]

62. Vassilakos, T.N.; Athanassiou, C.G.; Kavallieratos, N.G.; Vayias, B.J. Influence of temperature on the insecticidal effect of *Beauveria bassiana* in combination with diatomaceous earth against *Rhyzopertha dominica* and *Sitophilus oryzae* on stored wheat. *Biol. Control* **2006**, *38*, 270–281. [CrossRef]
63. Hansen, L.S.; Steenberg, T. Combining larval parasitoids and an entomopathogenic fungus for biological control of *Sitophilus granarius* (Coleoptera: Curculionidae) in stored grain. *Biol. Control* **2007**, *40*, 237–242. [CrossRef]
64. Wakil, W.; Ghazanfar, M.U. Entomopathogenic fungus as a biological control agent against *Rhyzopertha dominica* F. (Coleoptera: Bostrychidae) on stored wheat. *Arch. Phytopathol. Pflanzenschutz* **2010**, *43*, 1236–1242. [CrossRef]
65. Lord, J.C. Desiccant dusts synergize the effect of *Beauveria bassiana* (Hyphomycetes: Moniliales) on stored-grain beetles. *J. Econ. Entomol.* **2001**, *94*, 367–372. [CrossRef] [PubMed]
66. Moore, D.; Douro-Kpindou, O.K.; Jenkins, N.E.; Lomer, C.J. Effects of moisture content and temperature on storage of *Metarhizium flavoviride* conidia. *Biocontrol Sci. Technol.* **1996**, *6*, 51–61. [CrossRef]
67. Sheeba, G.; Seshardi, S.; Raja, N.; Janarthanan, S.; Ignacinutha, S. Efficacy of *Beauveria bassiana* for control of the rice weevil *Sitophilus oryzae* (L.) (Coleoptera: Curculionidae). *Appl. Entomol. Zool.* **2001**, *36*, 117–120. [CrossRef]
68. Batta, Y.A. Control of the lesser grain borer (*Rhyzopertha dominica* Fab., Coleoptera: Bostrichidae) by treatments with residual formulations of *Metarhizium anisopliae* (Metchnikoff) Sorokin (Deuteromycotina: Hyphomycetes). *J. Stored Prod. Res.* **2005**, *41*, 221–229. [CrossRef]
69. Kavallieratos, N.G.; Athanassiou, C.G.; Michalaki, M.P.; Batta, Y.A.; Rigatos, H.A.; Pashalidou, F.G.; Balotis, G.N.; Tomanović, Ž.; Vayias, B.J. Effect of the combined use of *Metarhizium anisopliae* (Metschinkoff) Sorokin and diatomaceous earth for the control of three stored-product beetle species. *Crop Prot.* **2006**, *25*, 1087–1094. [CrossRef]
70. Athanassiou, C.G.; Steenberg, T. Insecticidal effect of *Beauveria bassiana* (Balsamo) Vuillemin (Ascomycota: Hypocreaes) in combination with three diatomaceous earth formulations against *Sitophilus granarius* (L.) (Coleoptera: Curculionidae). *Biol. Control* **2007**, *40*, 411–416. [CrossRef]
71. Batta, Y.A. Control of main stored-grain insects with new formulations of entomopathogenic fungi in diatomaceous earth dusts. *Int. J. Food Eng.* **2008**, *4*, 1–16. [CrossRef]
72. Riasat, T.; Wakil, W.; Ashfaq, M.; Sahi, S.T. Effect of *Beauveria bassiana* mixed with diatomaceous earth on mortality, mycosis and sporulation of *Rhyzopertha dominica* on stored wheat. *Phytoparasitica* **2011**, *39*, 325–331. [CrossRef]
73. Wakil, W.; Riasat, T.; Ghazanfar, M.U.; Kwon, Y.J.; Shaheen, F.A. Aptness of *Beauveria bassiana* and enhanced diatomaceous earth (DEBBM) for control of *Rhyzopertha dominica* F. *Entomol. Res.* **2011**, *41*, 233–241. [CrossRef]
74. Dal Bello, G.; Fusé, C.; Juarez, P.; Pedrini, N.; Imaz, A.; Padin, S. Insecticidal effect of fenitrothion, diatomaceous earth and *Beauveria bassiana* against Coleopteran pests on stored grain. *Integr. Prot. Stored Prod. IOBC/WPRS Bull.* **2011**, *69*, 175–180.
75. Sedehi, A.; Sedaghatfar, E.; Modarres-Najafabadi, S.S. Studies on effect of the *Beauveria bassiana* on eggs and larvae of *Plodia interpunctella*. *Can. J. Basic Appl. Sci.* **2014**, *2*, 40–45.
76. Shafighi, Y.; Ziaee, M.; Ghosta, Y. Diatomaceous earth used against insect pests, applied alone or in combination with *Metarhizium anisopliae* and *Beauveria bassiana*. *J. Plant Prot. Res.* **2014**, *54*, 62–66. [CrossRef]
77. Rizwan, M.; Atta, B.; Rizwan, M.; Sabir, A.M.; Shah, Z.U.; Hussain, M. Effect of the entomopathogenic fungus, *Beauveria bassiana*, combined with diatomaceous earth on the red flour beetle, *Tribolium castaneum* (Herbst) (Tenebrionidae: Coleoptera). *Egypt J Biol Pest Control* **2019**, *29*, 27. [CrossRef]
78. Michalaki, M.P.; Athanassiou, C.G.; Steenberg, T.; Buchelos, C.T. Effect of *Paecilomyces fumosoroseus* (Wise) Brown and Smith (Ascomycota: Hypocerales) alone or in combination with diatomaceous earth against *Tribolium confusum* Jacquelin du Val (Coleoptera: Tenebrionidae) and *Ephestia kuehniella* Zeller (Lepidoptera: Pyralidae). *Biol. Control* **2007**, *40*, 280–286.
79. Sabbour, M.M.; Abd-El-Aziz, S.E.; Sherief, M.A. Efficacy of three entomopathogenic fungi alone or in combination with diatomaceous earth modifications for the control of three pyralid moths in stored grains. *J. Plant Prot. Res.* **2012**, *52*, 359–363. [CrossRef]
80. Riasat, T.; Wakil, W.; Yasin, M.; Kwon, Y.J. Mixing of *Isaria fumosorosea* with enhanced diatomaceous earth and bitterbarkomycin for control of *Rhyzopertha dominica*. *Entomol. Res.* **2013**, *43*, 215–223. [CrossRef]
81. Sabbour, M.M.; Abd-El-Aziz, S.E. Control of *Bruchidius incarnatus* and *Rhyzopertha dominica* using two entomopathogenic fungi alone or in combination with modified diatomaceous earth. *Elixir Int. J. Ent.* **2014**, *68*, 22239–22242.
82. Athanassiou, C.G.; Rani, P.U.; Kavallieratos, N.G. The Use of Plant Extracts for Stored Product Protection. In *Advances in Plant Biopesticides*; Singh, D., Ed.; Springer: New Delhi, India, 2014; pp. 131–147.
83. Campolo, O.; Giunti, G.; Russo, A.; Palmeri, V.; Zappalà, L. Essential Oils in Stored Product Insect Pest Control. *J. Food Qual.* **2018**, *2018*, 6906105. [CrossRef]
84. Fisher, M.H.; Mrozik, H. Chemistry. In *Ivermectin and Abamectin*; Campbell, W.C., Ed.; Springer: New York, NY, USA, 1989; pp. 1–23.
85. Yang, F.; Liang, G.; Xu, Y.; Lu, Y.; Zeng, L. Diatomaceous earth enhances the toxicity of garlic, *Allium sativum*, essential oil against stored-product pests. *J. Stored Prod. Res.* **2010**, *46*, 118–123. [CrossRef]
86. Ziaee, M.; Moharramipour, S.; Francikowski, J. The synergistic effects of *Carum copticum* essential oil on diatomaceous earth against *Sitophilus granarius* and *Tribolium confusum*. *J. Asia Pac. Entomol.* **2014**, *17*, 817–822. [CrossRef]
87. Korunić, Z.; Fields, P.G. Evaluation of three new insecticide formulations based on inert dusts and botanicals against four stored-grain beetles. *J. Stored Prod. Res.* **2020**, *88*, 101633. [CrossRef]

88. Campolo, O.; Romeo, F.V.; Malacrinò, A.; Laudani, F.; Carpinteri, G.; Fabroni, S.; Rapisarda, P.; Palmeri, V. Effects of inert dusts applied alone and in combination with sweet orange essential oil against *Rhyzopertha dominica* (Coleoptera: Bostrichidae) and wheat microbial population. *Ind. Crops Prod.* **2014**, *61*, 361–369. [CrossRef]
89. Paponja, I.; Rozman, V.; Liška, A. Natural formulation based on diatomaceous earth and botanicals against stored product insects. *Insects* **2020**, *11*, 613. [CrossRef]
90. Akhtar, Y.; Isman, M.B. Efficacy of diatomaceous earth and a DE-aerosol formulation against the common bed bug, *Cimex lectularius* Linnaeus in the laboratory. *J. Pest Sci.* **2016**, *89*, 1013–1021. [CrossRef]
91. Alkan, M.; Atay, T.; Ertürk, S.; Kepenekçi, I. Comparison of bioactivities of native diatomaceous earth against Turkestan cockroach [*Blatta Lateralis* Walker (Blattodea: Blattidae)] Nymphs. *Appl. Ecol. Environ. Res.* **2019**, *17*, 5897–5994. [CrossRef]
92. Ge, W.; Du, G.; Zhang, L.; Li, Z.; Xiao, G.; Chen, B. The Time–Concentration–Mortality Responses of Western Flower Thrips, *Frankliniella occidentalis*, to the Synergistic Interaction of Entomopathogenic Fungus *Metarhizium flavoviride*, Insecticides, and Diatomaceous Earth. *Insects* **2020**, *11*, 93. [CrossRef]
93. Kilpinen, O.; Steenberg, T. Repellent activity of desiccant dusts and conidia of the entomopathogenic fungus *Beauveria bassiana* when tested against poultry red mites (*Dermanyssus gallinae*) in laboratory experiments. *Exp. Appl. Acarol.* **2016**, *70*, 329–341. [CrossRef]
94. Steenberg, T.; Kilpinen, O. Synergistic interaction between the fungus *Beauveria bassiana* and desiccant dusts applied against poultry red mites (*Dermanyssus gallinae*). *Exp. Appl. Acarol.* **2014**, *62*, 511–524. [CrossRef]
95. Akhoundi, M.; Bruel, C.; Izri, A. A Harmful Effects of Bed Bug-Killing Method of Diatomaceous Earth on Human Health. *J. Insect Sci.* **2019**, *19*, 5. [CrossRef]
96. Hosseini, S.A.; Bazrafkan, S.; Vatandoost, H.; Abaei, M.R.; Ahmadi, M.S.; Tavassoli, M.; Shayeghi, M. The insecticidal effect of diatomaceous earth against adults and nymphs of *Blattella germanica*. *Asian Pac. J. Trop. Biomed.* **2014**, *4*, S228–S232. [CrossRef]
97. Di Giovanni, F.; Wilke, A.B.B.; Beier, J.C.; Pombi, M.; Mendoza-Roldan, J.A.; Desneux, N.; Canale, A.; Lucchi, A.; Dantas-Torres, F.; Otranto, D.; et al. Parasitic strategies of arthropods of medical and veterinary importance. *Entomol. Gen.* **2021**, *41*, 70176.
98. Pan, X.; Wang, X.; Zhang, F. New Insights into Cockroach Control: Using Functional Diversity of *Blattella germanica* Symbionts. *Insects* **2020**, *11*, 696. [CrossRef]
99. Faulde, M.K.; Scharninghausen, J.J.; Cavaljuga, S. Toxic and behavioural effects of different modified diatomaceous earths on the German cockroach, *Blattella germanica* (L.) (Orthoptera: Blattellidae) under simulated field conditions. *J. Stored Prod. Res.* **2006**, *42*, 253–263. [CrossRef]
100. Faulde, M.K.; Tisch, M.; Scharninghausen, J.J. Efficacy of modified diatomaceous earth on different cockroach species (Orthoptera, Blattellidae) and silverfish (Thysanura, Lepismatidae). *J. Pest Sci.* **2006**, *79*, 155–161. [CrossRef]
101. Gao, Y.; Yu, S.; Li, J.; Sun, P.; Xiong, M.; Lei, C.; Zhang, Z.; Huang, Q. Bioactivity of diatomaceous earth against the subterranean termite *Reticulitermes chinensis* Snyder (Isoptera: Rhinotermitidae). *Environ. Sci. Pollut. Res.* **2018**, *25*, 28102–28108. [CrossRef]
102. Özcan, K.; Tunaz, H.; Işikber, A.A.; Kubilay, E.M. Lethal Effect of Turkısh Dıatomaceous Earth (Bgn-1) agaınst Adults of German Cockroaches (*Blatella Germanıca* L.). Contact Pesticides, Residual Products, and Plant Extracts. In In Proceedings of the 12th International Working Conference on Stored Product Protection (IWCSPP), Berlin, Germany, 7–11 October 2018.
103. Ahmed, S.; Hassan, B.; Farooq, M.U. Effect of biofertilizers and diatomaceous earth on life and movement of subterranean termites under laboratory conditions. *Int. J. Trop. Insect Sci.* **2018**, *38*, 348–352. [CrossRef]
104. Grace, J.K.; Yamamoto, R.T. Diatomaceous earth is not a barrier to Formosan subterranean termites (Isoptera: Rhinotermitidae). *Sociobiology* **1993**, *23*, 25–30.
105. Lucky, A. Urban Ants of North America and Europe: Identification, Biology, and Management. *Syst. Entomol.* **2009**, *34*, 406–407. [CrossRef]
106. Van Den Noortgate, H.; Sree, S.P.; Ostyn, N.; Lagrain, B.; Roefaers, M.; Wenseleers, T.; Martens, J.A. Material properties determining insecticidal activity of activated carbon on the pharaoh ant (*Monomorium pharaonis*). *J. Pest Sci.* **2018**, *922*, 643–652. [CrossRef]
107. Al, N.; Tunaz, H.; Işikber, A.A.; Er, M.K. Lethal Effect Of Turkish Diatomaceous Earth (K14) Against Adults Of American Cockroaches (Periplaneta americana L.). In Proceedings of the International Biological, Agricultural And Life Science Congress, Lviv, Ukraine, 7–8 November 2019.
108. Akhtar, Y.; Isman, M.B. Horizontal Transfer of Diatomaceous Earth and Botanical Insecticides in the Common Bed Bug, *Cimex lectularius* L.; Hemiptera: Cimicidae. *PLoS ONE* **2013**, *8*, 75626.
109. Rodrigues, J.; Borges, P.R.; Fernandes, É.K.K.; Luz, C. Activity of additives and their effect in formulations of *Metarhizium anisopliae* s.l. IP 46 against *Aedes aegypti* adults and on post mortem conidiogenesis. *Acta Trop* **2019**, *193*, 192–198. [CrossRef]
110. Alves, L.F.A.; Oliveira, D.G.P.; Kasburg, C.R.; Nardelli, M.S. Acaricidal Activity Of Inert Powders Against The Poultry Red Mite *Dermanyssus gallinae* (De Geer, 1778) (Mesostigmata: Dermanyssidae). *Arch. Vet. Sci.* **2019**, *24*, 81–92.
111. Alves, L.F.A.; de Oliveira, D.G.P.; Pares, R.B.; Sparagano, O.A.E.; Godinho, R.P. Association of mechanical cleaning and a liquid preparation of diatomaceous earth in the management of poultry red mite, *Dermanyssus gallinae* (Mesostigmata: Dermanyssidae). *Exp. Appl. Acarol.* **2020**, *81*, 215–222. [CrossRef] [PubMed]
112. Ulrichs, C.; Han, Y.J.; Abdelhamid, M.T.; Mewis, I. Management of the poultry red mite, *Dermanyssus gallinae*, using silica-based acaricides. *Exp. Appl. Acarol.* **2020**, *822*, 243–254. [CrossRef] [PubMed]

113. Lilly, D.G.; Webb, C.E.; Doggett, S.L. Evidence of Tolerance to Silica-Based Desiccant Dusts in a Pyrethroid-Resistant Strain of *Cimex lectularius* (Hemiptera: Cimicidae). *Insects* **2016**, *7*, 74. [CrossRef] [PubMed]
114. Singh, N.; Wang, C.; Wang, D.; Cooper, R.; Zha, C. Comparative Efficacy of Selected Dust Insecticides for Controlling *Cimex lectularius* (Hemiptera: Cimicidae). *J. Econ. Entomol.* **2016**, *109*, 1819–1826. [CrossRef]
115. Benoit, J.B.; Phillips, S.A.; Croxall, T.J.; Christensen, B.S.; Yoder, J.A.; Denlinger, D.L. Addition of alarm pheromone components improves the effectiveness of desiccant dusts against *Cimex lectularius*. *J. Med. Entomol.* **2009**, *46*, 572–579. [CrossRef]
116. Pedrini, N.; Mijailovsky, S.J.; Girotti, J.R.; Stariolo, R.; Cardozo, R.M.; Gentile, A.; Juárez, M.P. Control of pyrethroid-resistant chagas disease vectors with entomopathogenic fungi. *PLoS Negl. Trop. Dis.* **2009**, *3*, 434. [CrossRef]
117. Luz, C.; Rodrigues, J.; Rocha, L.F.N. Diatomaceous earth and oil enhance effectiveness of *Metarhizium anisopliae* against *Triatoma infestans*. *Acta Trop.* **2012**, *122*, 29–35. [CrossRef]
118. Athanassiou, C.G.; Palyvos, N.E. Laboratory evaluation of two diatomaceous earth formulations against *Blattisocius keegani* fox (Mesostigmata, Ascidae) and *Cheyletus malaccensis* Oudemans (Prostigmata, Cheyletidae). *Biol. Control* **2006**, *38*, 350–355. [CrossRef]
119. Rodrigues, J.; Lobo, L.S.; Fernandes, E.K.K.; Luz, C. Effect of formulated *Metarhizium anisopliae* on eggs and eclosing nymphs of *Triatoma infestans*. *J. Appl. Entomol.* **2015**, *139*, 146–153. [CrossRef]
120. Forlani, L.; Pedrini, N.; Juarez, M.P. Contribution of the horizontal transmission of the entomopathogenic fungus *Beauveria bassiana* to the overall performance of a fungal powder formulation against *Triatoma infestans*. *Res. Rep. Trop. Med.* **2011**, *2*, 135.
121. Santos, T.R.; da Paixão, F.R.S.; Catão, A.M.L.; Muniz, E.R.; Ribeiro-Silva, C.S.; Taveira, S.F.; Luz, C.; Mascarin, G.M.; Fernandes, E.K.K.; Marreto, R.N. Inorganic pellets containing microsclerotia of *Metarhizium anisopliae*: A new technological platform for the biological control of the cattle tick *Rhipicephalus microplus*. *Appl. Microbiol. Biotechnol.* **2021**, *105*, 5001–5012. [CrossRef]
122. Kilpinen, O.; Roepstorff, A.; Permin, A.; Nørgaard-Nielsen, G.; Lawson, L.G.; Simonsen, H.B. Influence of *Dermanyssus gallinae* and *Ascaridia galli* infections on behaviour and health of laying hens (*Gallus gallus domesticus*). *Br. Poult. Sci.* **2005**, *46*, 26–34. [CrossRef]
123. Chu, T.T.H.; Murano, T.; Uno, Y.; Usui, T.; Yamaguchi, T. Molecular Detection of Avian Pathogens in Poultry Red Mite (Dermanyssus gallinae) Collected in Chicken Farms. *J. Vet. Med. Sci.* **2014**, *76*, 14–0253.
124. Sparagano, O.A.E.; Ho, J. Parasitic Mite Fauna in Asian Poultry Farming Systems. *Front. Vet. Sci.* **2020**, *7*, 400. [CrossRef]
125. Mullens, B.A.; Soto, D.; Martin, C.D.; Callaham, B.L.; Gerry, A.C. Northern fowl mite (*Ornithonyssus sylviarum*) control evaluations using liquid formulations of diatomaceous earth, kaolin, sulfur, azadirachtin, and *Beauveria bassiana* on caged laying hens. *J. Appl. Poult. Res.* **2012**, *21*, 111–116. [CrossRef]
126. Martin, C.D.; Mullens, B.A. Housing and dustbathing effects on northern fowl mites (*Ornithonyssus sylviarum*) and chicken body lice (*Menacanthus stramineus*) on hens. *Med. Vet. Entomol.* **2012**, *26*, 323–333. [CrossRef]
127. Murillo, A.C.; Mullens, B.A. Timing diatomaceous earth-filled dustbox use for management of northern fowl mites (Acari: Macronyssidae) in cage-free poultry systems. *J. Econ. Entomol.* **2016**, *109*, 2572–2579. [CrossRef]
128. Showler, T.A.; Flores, N.; Caesar, R.M.; Mitchel, R.D.; Perez de Leon, A.A.P. Lethal Effects of a Commercial Diatomaceous Earth Dust Product on *Amblyomma americanum* (Ixodida: Ixodidae) Larvae and Nymphs. *J. Med. Entomol.* **2020**, *57*, 1575–1581. [CrossRef]
129. Dawson, R.D. Efficacy of diatomaceous earth at reducing populations of nest-dwelling ectoparasites in tree swallows. *J. Field Ornithol.* **2004**, *75*, 232–238. [CrossRef]
130. Kilpinen, O.; Steenberg, T. Inert dusts and their effects on the poultry red mite (*Dermanyssus gallinae*). *Exp. Appl. Acarol.* **2009**, *481*, 51–62. [CrossRef]
131. Brinkman, M.A.; Gardner, W.A. Use of diatomaceous earth and entomopathogen combinations against the red imported fire ant (Hymenoptera: Formicidae). *Florida Entomol.* **2001**, *84*, 740–741. [CrossRef]
132. Constanski, K.C.; Zorzetti, J.; Santoro, P.H.; Hoshino, A.T.; Oliveira Janiero Neves, P.M. Inert powders alone or in combination with neem oil for controlling *Spodoptera eridania* and *Spodoptera frugiperda* (Lepidoptera: Noctuidae) larvae. *Semin. Agrar.* **2016**, *37*, 1801–1810. [CrossRef]
133. Singh, B.; Singh, V. Laboratory and field studies demonstrating the insecticidal potential of diatomaceous earth against wheat aphids in rice-wheat cropping system of Punjab (India). *Cereal Res. Commun.* **2016**, *44*, 435–443. [CrossRef]
134. Aniwanou, C.T.S.; Sinzogan, A.A.C.; Deguenon, J.M.; Sikirou, R.; Stewart, D.A.; Ahanchede, A. Bio-efficacy of diatomaceous earth, household soaps, and neem oil against *Spodoptera frugiperda* (Lepidoptera: Noctuidae) larvae in Benin. *Insects* **2021**, *12*, 18. [CrossRef]
135. Constantinescu-Aruxandei, D.; Lupu, C.; Oancea, F. Siliceous Natural Nanomaterials as Biorationals—Plant Protectants and Plant Health Strengtheners. *Agronomy* **2020**, *10*, 1791. [CrossRef]
136. Khaled, W.; Ben Fekih, I.; Harbaoui, I.; Boukhris-Bouhachem, S. Insecticidal activity assessment of Thymus capitatus essential oils in combination with natural abrasives against Myzus persicae. In Proceedings of the Second Africa-International Allelopathy Congress, Sousse, Tunisia, 16–19 November 2016.
137. Lozano-Contreras, M.G.; Maldonado-Blanco, M.G.; Elías-Santos, M.; González-Hernández, A.; Nava-Camberos, U. Application of *Isaria fumosorosea* blastospores produced in liquid culture for control of *Bemisia argentifolii* on cotton plants. *Southwest Entomol.* **2013**, *38*, 57–66. [CrossRef]

138. Ulrichs, C.H.; Mewis, I.; Schnitzler, W.H. Efficacy of neem and diatomaceous earth against cowpea aphids and their deleteriou effect on predating Coccinelidae. *J. Appl. Entomol.* **2001**, *125*, 571–575. [CrossRef]
139. El-Wakeil, N.E.; Saleh, S.A. Effects of neem and diatomaceous earth against *Myzus persicae* and associated predators in addition t indirect effects on artichoke growth and yield parameters. *Arch. Phytopathol. Plant Prot.* **2009**, *42*, 1132–1143. [CrossRef]
140. Shah, R.; Appleby, M. Testing the contact and residual toxicity of selected low-risk pesticides to *Tetranychus Urticae* Koch and its Predators. *Sarhad J. Agric.* **2019**, *35*, 1113–1121. [CrossRef]
141. Adams, C.T.; Banks, W.A.; Lofgren, C.S. Red Imported Fire Ant (Hymenoptera: Formicidae): Correlation of Ant Density with Damage to Two Cultivars of Potatoes (*Solanum tuberosum* L.). *J. Econ. Entomol.* **1988**, *81*, 905–909. [CrossRef]
142. Hagstrum, D.W.; Phillips, T.W. Evolution of stored-product entomology: Protecting the world food supply. *Annu. Rev. Entomol.* **2017**, *62*, 379–397. [CrossRef] [PubMed]
143. Parkin, E.A. Control of the granary weevil with finely ground mineral dusts. *Ann. Appl. Biol.* **1944**, *31*, 84–88. [CrossRef]
144. Korunić, Z.; Cenkowski, S.; Fields, P.G. Grain bulk density as affected by diatomaceous earths and application method. *Postharvest Biol. Technol.* **1998**, *13*, 81–89. [CrossRef]
145. Jackson, K.; Webley, D. Effects of Dryacide on the Physical Properties of Grains, Pulses and Oilseeds. In *Stored Product Protection, Proceedings of the Sixth International Conference on Stored Product Protection, Canberra, Australia, 1994*; Highley, E., Wright, E.J., Banks, H.J., Champ, B.R., Eds.; University Press: Cambridge, UK, 1994; pp. 635–637.
146. Gesraha, M.A.; Ebeid, A.R.; Salem, N.Y.M.; Abdou, W.L. Comparative Study on Some Biological Indices of *Agrotis ipsilon* (Lepidoptera: Noctuidae) Larvae Treated with Three Control Agents under Laboratory Conditions. *Annu. Res. Rev. Biol.* **2017**, *21*, 1–8. [CrossRef]
147. Ferreira-Filho, P.J.; Wilcken, C.F.; Neves, D.A.; Pogotto, M.H.F.A.D.; Carmo, J.B.; Guerreiro, J.C.; Serrao, J.E.; Zanuncio, J.C. Does Diatomaceous Earth Control Leaf-Cutter Ants (Hymenoptera: Formicidae) in the *Eucalyptus* Plantations? *J. Ec. Entomol.* **2015**, *108*, 1124–1128. [CrossRef]
148. Mucha-Pelzer, T.; Debnath, N.; Goswami, A.; Mewis, I.; Ulrichs, C. Bekämpfung von *Epilachna vigintioctopunctata* (F.) und Spodoptera litura (F.) mit Silikaten. *Gesunde Pflanzen* **2008**, *60*, 23–28. [CrossRef]
149. Mitchell, R.D.; Mott, D.W.; Dhammi, A.; Reisig, D.D.; Roe, R.M. Field Evaluation of a New Thrips Control Agent for Cotton: A Mechanical Insecticide. In Proceedings of the Beltwide Cotton Conferences, San Antonio, TX, USA, 3–5 January 2018.
150. Ebadollahi, A.; Sadeghi, R. Diatomaceous Earth and Kaolin as Promising Alternatives to the Detrimental Chemicals in the Management of *Spodoptera exigua*. *J. Entomol.* **2018**, *15*, 101–105. [CrossRef]
151. Desmarchelier, J.M.; Dines, J.C. Dryacide treatment of stored wheat: Its efficacy against insects, and after processing. *Aust. J. Exp. Agric.* **1987**, *27*, 309–312. [CrossRef]
152. Aldryhim, Y.N. Efficacy of the amorphous silica dust, Dryacide, against *Tribolium confusum* Duv. and *Sitophilus granarius* (L.) (Coloptera: Tenebrionidae and Curculionidae). *J. Stored Prod. Res.* **1990**, *26*, 207–210. [CrossRef]
153. McLaughlin, A. Laboratory Trials on Desiccant Dust Insecticides. In *Stored Product Protection, Proceedings of the Sixth International Conference on Stored Product Protection*; Highley, E., Wright, E.J., Banks, H.J., Champ, B.R., Eds.; University Press: Cambridge, UK, 1994; pp. 638–645.
154. Subramanyam, B.H.; Hagstrum, D.W. Resistance Measurement and Management. In *Integrated Management of Insects in Stored Products*; Subramanyam, B.H., Hagstrum, D.W., Eds.; Marcel Dekker Inc.: New York, NY, USA, 1995; pp. 231–398.
155. Musser, F.R.; Shelton, A.M. The influence of post-exposure temperature on the toxicity of insecticides to *Ostrinia nubilalis* (Lepidoptera: Crambidae). *Pest Manag. Sci.* **2005**, *61*, 508–510. [CrossRef]
156. Khan, H.A.A.; Akram, W. The effect of temperature on the toxicity of insecticides against *Musca domestica* L.: Implications for the effective management of diarrhea. *PLoS ONE* **2014**, *9*, 95636. [CrossRef]
157. Fields, P.; Dowdy, A.; Marcotte, M. Structural Pest Control: The Use of an Enhanced Diatomaceous Earth Product Combined with Heat Treatment for the Control of Insect Pests in Food Processing Facilities. Canada–United States Working Group on Methyl Bromide Alternatives. 1997. Available online: http://res.agr.ca/winn/Heat-DE.htm (accessed on 16 October 2021).
158. Dowdy, A.K. Mortality of red four beetle, *Tribolium castaneum* (Coleoptera: Tenebrionidae) exposed to high temperature and diatomaceous earth combinations. *J. Stored Prod. Res.* **1999**, *35*, 175–182. [CrossRef]
159. Dowdy, A.K.; Fields, P.G. Heat combined with diatomaceous earth to control the confused flour beetle (Coleoptera: Tenebrionidae) in a flour mill. *J. Stored Prod. Res.* **2002**, *38*, 11–22. [CrossRef]
160. Cook, D.A.; Collins, D.A.; Collins, L.E. Efficacy of diatomaceous earths, applied as structural treatments, against stored product insects and mites. *HGCA Proj. Re.* **2004**, *344*, 50.
161. Collins, D.A.; Cook, D.A. Laboratory evaluation of diatomaceous earths, when applied as dry dust and slurries to wooden surfaces, against stored-product insect and mite pests. *J. Stored Prod. Res.* **2006**, *42*, 197–206. [CrossRef]
162. Collins, D.A.; Cook, D.A. Laboratory studies evaluating the efficacy of diatomaceous earths, on treated surfaces, against stored-product insect and mite pests. *J. Stored Prod. Res.* **2006**, *42*, 51–60. [CrossRef]
163. Schöller, M.; Reichmuth, C. Field trials with the diatomaceous earth SilicoSec® for treatment of empty rooms and bulk grain. *Julius-Kühn-Archiv* **2010**, *425*, 899–905.
164. Ertürk, S.; Atay, T.; Toprak, U.; Alkan, M. The efficacy of different surface applications of wettable powder formulation of Detech®diatomaceous earth against the rice weevil, *Sitophilus oryzae* (L.) (Coleoptera: Curculionidae). *J. Stored Prod. Res.* **2020**, *89*, 101725. [CrossRef]

165. Gowers, S.L.; Le Patourel, G.N.J. Toxicity of deposits of an amorphous silica dust on different surfaces and their pick-up by *Sitophilus granarius* (Coleoptera: Curculionidae). *J. Stored Prod. Res.* **1984**, *70*, 25–29. [CrossRef]
166. Cook, D.A. The Efficacy of High Temperature and Diatomaceous Earth Combinations Against Adults of the Red Flour Beetle Tribolium Castaneum (Coleoptera: Tenebrionidae) and the Grain Weevil Sitophilus Granarius (Coleoptera: Curculionidae). In *Proceedings of the BCPC International Congress—Crop Science and Technology*; The British Crop Protection Council: Alton, UK, 2003; pp. 445–450.
167. Cox, P.D.; Parish, W.E. Effects of refuge content and food availability on refuge-seeking behavior in *Cryptolestes ferrugineus* (Stephens) (Coleoptera: Cucujdiae). *J. Stored Prod. Res.* **1991**, *27*, 135–139. [CrossRef]
168. Vrba, C.H.; Arai, H.P.; Nosal, M. The effect of silica aerogel on the mortality of *Tribolium confusum* (Duval) as a function of exposure time and food deprivation. *Can. J. Zool.* **1983**, *61*, 1481–1486. [CrossRef]
169. Loschiavo, S.R. Availability of food as a factor in the effectiveness of a silica aerogel against the merchant grain beetle (Coleoptera: Cucujidae). *J. Econ. Entom.* **1988**, *81*, 1237–1240. [CrossRef]
170. White, N.D.G.; Loschiavo, S.R. Factors affecting survival of the merchant grain beetle (Coleoptera: Cucujidae) and the confused flour beetle (Coleoptera: Tenebrionidae) exposed to silica aerogel. *J. Econ. Entom.* **1989**, *82*, 960–969. [CrossRef]
171. Bridgeman, B. Application technology and usage patterns of diatomaceous earth in stored product protection. Stored-product Protection. In Proceedings of the Seventh International Working Conference on Stored-product Protection, Beijing, China, 14–19 October 1998; Zuxun, J., Quan, L., Yongsheng, L., Xianchang, T., Lianghua, G., Eds.; pp. 785–789.
172. Banks, J.H.; Fields, P.G. Physical Methods for Insect Control in Stored-Grain Ecosystems. In *Stored-Grain Ecosystems*; Jayas, D.S., White, N.D.G., Muir, W.E., Eds.; CRC Press: New York, NY, USA, 1995; pp. 353–409.
173. Golob, P. Current status and future perspective for inert dusts for control of stored product insects. *J. Stored Prod. Res.* **1997**, *33*, 69–79. [CrossRef]
174. Kljajić, P.; Andrić, G.; Adamović, M.; Bodroža-Solarov, M.; Marković, M.; Perić, I. Laboratory assessment of insecticidal effectiveness of natural zeolite and diatomaceous earth formulations against three stored-product beetle pests. *J. Stored Prod. Res.* **2010**, *46*, 1–6. [CrossRef]
175. Kljajić, P.; Andrić, G.; Milan, A.; Golić, M.P. Possibilities of application of natural zeolites in stored wheat grain protection against pest insects. *J. Process. Energy Agric.* **2011**, *15*, 12–16.
176. Bodroža-Solarov, M.; Kljajić, P.; Andrić, G.; Filipčev, B.; Šimurina, O.; Golić, P.M.; Adamovic, M. Application of principal component analysis in assessment of relation between the parameters of technological quality of wheat grains treated with inert dusts against rice weevil (*Sitophilus oryzae* L.). *Pestic. Fitomed.* **2011**, *26*, 385–390. [CrossRef]
177. Rumbos, C.I.; Sakka, M.; Berillis, P.; Athanassiou, C.G. Insecticidal potential of zeolite formulations against three stored grain insects, particle size effect, adherence to kernels and influence on test weight of grains. *J. Stored Prod. Res.* **2016**, *68*, 93–101. [CrossRef]
178. Floros, G.; Kokkari, A.; Kouloussis, N.; Kantiranis, N.; Damos, P.; Filippidis, A.; Koveos, D. Evaluation of the natural zeolite lethal effects on adults of the bean weevil under different temperatures and relative humidity regimes *J. Econ. Entom.* **2017**, *111*, 482–490. [CrossRef]
179. Westwood, G.S.; Huang, S.W.; Keyhani, N.O. Molecular and immunological characterization of allergens from the entomopathogenic fungus *Beauveria bassiana*. *Clin. Mol. Allergy.* **2006**, *4*, 12. [CrossRef]
180. Chen, C.Y.; Chen, M.F. Susceptibility of field populations of the lesser grain borer, *Rhyzopertha dominica* (F.), to deltamethrin and spinosad on paddy rice in Taiwan. *J. Stored Prod Res.* **2013**, *55*, 124–127. [CrossRef]
181. Daglish, G.J.; Nayak, M.K. Prevalence of resistance to deltamethrin in *Rhyzopertha dominica* (F.) in eastern Australia. *J. Stored Prod Res.* **2018**, *78*, 45–49. [CrossRef]
182. Nayak, M.K.; Daglish, G.J.; Phillips, T.W.; Ebert, P.R. Resistance to the fumigant phosphine and its management in insect pests of stored products: A global perspective. *Ann. Rev. Entomol.* **2020**, *65*, 333–350. [CrossRef] [PubMed]

MDPI
St. Alban-Anlage 66
4052 Basel
Switzerland
Tel. +41 61 683 77 34
Fax +41 61 302 89 18
www.mdpi.com

Molecules Editorial Office
E-mail: molecules@mdpi.com
www.mdpi.com/journal/molecules

www.ingramcontent.com/pod-product-compliance
Lightning Source LLC
Chambersburg PA
CBHW041417210926
51850CB00013B/3295
* 9 7 8 3 0 3 6 5 3 2 3 9 4 *